BRASSINOSTEROIDS
A New Class of Plant Hormones

BRASSINOSTEROIDS
A New Class of Plant Hormones

V. A. Khripach
Institute of Bio-organic Chemistry
Academy of Sciences of Belarus
Minsk, Belarus

V. N. Zhabinskii
Institute of Bio-organic Chemistry
Academy of Sciences of Belarus
Minsk, Belarus

A. E. de Groot
Department of Organic Chemistry
Wageningen Agricultural University
Wageningen, The Netherlands

ACADEMIC PRESS

San Diego   London   Boston   New York   Sydney   Tokyo   Toronto

This book is printed on acid-free paper. ∞

Copyright © 1999 by ACADEMIC PRESS

All Rights Reserved.
No part of this publication may be reproduced or transmitted in any form or by any means, electronic or mechanical, including photocopy, recording, or any information storage and retrieval system, without permission in writing from the publisher.

Academic Press
a division of Harcourt Brace & Company
525 B Street, Suite 1900, San Diego, California 92101-4495, USA
http://www.apnet.com

Academic Press
24-28 Oval Road, London NW1 7DX, UK
http://www.hbuk.co.uk/ap/

Library of Congress Catalog Card Number: 98-87079

International Standard Book Number: 0-12-406360-8

PRINTED IN THE UNITED STATES OF AMERICA
98 99 00 01 02 03 ML 9 8 7 6 5 4 3 2 1

Contents

	Preface	ix
	Abbreviations	xiii
I.	INTRODUCTION	1
II.	BRASSINOSTEROIDS (BS) IN NATURE	7
	A. Historical Aspects	7
	B. Nomenclature, Structures, and Classification	8
	C. Related Steroids	21
III.	ISOLATION AND IDENTIFICATION	25
IV.	SPECTRAL PROPERTIES	33
	A. NMR Spectroscopy	33
	B. Mass Spectrometry	43
	C. X-Ray Analysis	49
	D. Other Methods	53
V.	BIOSYNTHESIS AND METABOLISM	55
	A. Introduction	55
	B. The Early C6-Oxidation Pathway	57
	C. The Late C6-Oxidation Pathway	60
	D. Experiments with BS Mutants	62
	E. Biotransformations of BS	65

VI. BASIC SYNTHETIC METHODS AND FORMAL SYNTHESES   73

    A. Introduction   73
    B. Synthesis of the Cyclic Part   76
        1. Starting with Δ^5-Sterols   77
        2. Starting with Ergosterol   85
        3. Starting with Hyodeoxycholic Acid   87
    C. Synthesis of the Side Chains   88
        1. With Preservation of a Native Carbon Skeleton   89
        2. With Reconstruction of the Native Carbon Skeleton of the Side Chain   97

VII. SYNTHESES OF NATURAL BS   137

    A. Syntheses of Brassinolide   138
        1. Starting with Stigmasterol   139
        2. Starting with Crinosterol   151
        3. Starting with Pregnenolone   152
        4. Starting with Hyodeoxycholic Acid Derivatives   153
    B. Syntheses of Epibrassinolide   154
    C. Syntheses of Homobrassinolide   160
    D. Syntheses of 28-Norbrassinolide   165
    E. Syntheses of BS Containing a $\Delta^{24(28)}$-Double Bond   170
    F. Syntheses of 6-Deoxo-BS   174
    G. Syntheses of BS with Modifications in Ring A   177
    H. Syntheses of Isotopically Labeled BS   181

VII. SYNTHESES OF BS ANALOGS   185

    A. Analogs with the Side Chain of a Starting Material   186

	B. Analogs with Different Functional Groups in the Side Chain	**189**
	1. 22,23-Dihydroxy Derivatives	189
	2. 22,23-Epoxides	191
	3. Analogs with Additional or without Hydroxy Group(s)	192
	C. Analogs with a Modified Carbon Skeleton in the Side Chain	**195**
	1. With Side Chains Epimeric at C-20	195
	2. With Side Chains Epimeric at C-24	196
	3. Analogs with a Shortened Side Chain	197
	4. Analogs with an Ester, Ether, or Amide Function in the Side Chain	199
	5. Analogs with a Heterocyclic Moiety in the Side Chain	201
	6. Other Analogs	203
	D. Analogs with a Modified Cyclic Part	**204**
	1. Analogs with an Additional Double Bond in the Cyclic Part	204
	2. Ring B Hetero Analogs	206
	3. Analogs with Modifications in Ring A	214
	E. Esters and Ethers	**216**
IX.	**PHYSIOLOGICAL MODE OF ACTION OF BS**	**219**
	A. Growth and Development	**220**
	B. Interaction with Other Phytohormones	**236**
	C. Effect on Cell Membranes	**240**
	1. H^+-Pump Activation and Electrical Properties	240
	2. Effect on Chemical Composition. Membrane-Protective Action	248
	3. Membrane Permeability and Transport	255
	D. Effect on Protein and Nucleic Acid Metabolism. Stress Response	**263**

E.	Resistance to Diseases	277
F.	Effect on the Photosynthetic Apparatus	285
G.	Mechanism of Action, Reception, and Transport	290
H.	Other Effects of BS	295

X. **BIOASSAYS AND STRUCTURE-ACTIVITY RELATIONSHIPS OF BS** — 301

    A. Bioassays — 301
    B. Structure-Activity Relationships — 309

XI. **PRACTICAL APPLICATIONS AND TOXICOLOGY** — 325

    A. Applications of BS — 325
        1. Cereals — 328
        2. Legumes — 338
        3. Potato and Vegetables — 339
        4. Miscellaneous — 340
    B. Toxicology of BS — 345

Appendix — 347

References — 373

Index — 445

Preface

Plants possess the ability to biosynthesize a large variety of steroids, but it was not until 1979 that steroids with plant hormonal activity were discovered. American scientists were able to isolate a new steroidal lactone called brassinolide from bee-collected pollen of *Brassica napus* L. Two years later, castasterone was found in insect galls of *Castanea crenata* spp. To date, more than 40 structurally and functionally related steroids in plants are known, and this group of compounds is now characterized as brassinosteroids (BS). BS are present in nearly every part of the plant, with the highest concentration in the reproductive organs (pollen and immature seeds). They demonstrate various kinds of regulatory actions on growth and development of plants, such as stimulation of cell enlargement and cell division, accretion of biomass, yield and quality of seeds, and plant adaptability. At the molecular level, BS change the gene expression and the metabolism of nucleic acids and proteins. For these reasons, BS are now considered a new group of plant hormones.

The extremely high activity of BS has attracted the attention of many specialists in the fields of analytical and synthetic chemistry, biochemistry, plant physiology, and agriculture. Several problems connected with BS had to be solved after their discovery. First, it was necessary to elaborate methods for isolation and identification as well as for biological testing to guide the isolation procedure. Also very important was progress in the chemical synthesis of BS and their analogs. This was necessary to elucidate the structures, to study the biosynthesis, and to clarify the structure-activity relationships in the BS series. Later, synthesis became important for obtaining sufficient quantities for biological testing and for searching for stable, economically attractive compounds with high activity for practical application in agriculture.

It is very important to realize that BS are naturally occurring compounds, ubiquitous in plants that are used for human and animal nutrition. BS are metabolized in the usual ways, and they have played roles in the long combined evolution of plants and animals. This fact provides relative assurance of the safety of BS in the very low doses found naturally in plants.

Chapter II reviews all the BS known today, and a new system of abbreviated names related to their structures is proposed. The isolation procedures and spectroscopic structure determination of BS are documented in Chapters III and IV. The biosynthesis of BS and their metabolic transformations are treated in Chapter V. Synthetic approaches to BS and the syntheses of their most important natural representatives and analogs are discussed in Chapters VI, VII, and VIII. Much knowledge has been gathered about the physiological mode of action of BS, and the present state of the art is described in Chapter IX. Special attention has been devoted to investigations carried out in the former Soviet Union and described in Russian journals and patents. This literature is practically inaccessible to an international readership, and this monograph will be of help in filling this void. To obtain a more precise understanding of the structural requirements for high activity in BS, much attention has been given to structure activity relationships in Chapter X. This will allow researchers to predict the activities of new analogs and to design BS with the best synthetic cost/activity ratio.

Chapter XI covers the substantial amount of data indicating that the application of BS allows a significant increase in the yields of most crops. Parallel with increasing crop yields, a substantial improvement in their quality takes place in many cases. Based on these properties, the natural phytohormone epibrassinolide (EBl) has been approved for agricultural use in countries of the former Soviet Union since 1992.

Preface

The world literature on BS now includes more than 1000 publications. These are spread among one monograph, conference proceedings, and general and specialized reviews on the distribution of BS in plants; their isolation, structure, and synthesis; plant physiology; and practical application of BS. To the best of our knowledge, the bulk of this literature has been considered in compiling this monograph.

The authors thank the many people who have contributed to this work. First, we sincerely thank Dr. Natalia Khripach, not only for the way she took care of our mood and health but also for her substantial contribution to the biological part of the book and for many stimulating discussions. We also acknowledge the valuable contributions of Prof. Dr. Linus van der Plas and the staff of the Laboratory of Plant Physiology of the Wageningen Agricultural University and Dr. Harro Bouwmeester of the AB-DLO Institute in Wageningen for reading the chapters on plant physiology. We extend our appreciation to Dr. Margarita I. Zavadskaya for help in the preparation of the manuscript and for her kind and fruitful discussions throughout this work. Thanks are also due to Drs. Natalia N. Malevannaya, Ivan K. Volod'ko, Alexander I. Zabolotny, Alexander N. Vedeneev, Natalia N. Vlasova, Ludmila N. Kalituho, and Zhanna E. Mazets for generously supplying data and participating in discussions on the particular problems of BS action in plants. We also thank Dr. Vladimir B. Petukhov for fruitful discussion on the prospects of studying BS in vertebrates, Dr. Alexander S. Lyakhov for discussion and for providing of X-ray analysis data, and Dr. Han Zuilhof, Bep van Veldhuizen, and Dr. Maarten Posthumus for their contributions in the areas of NMR and mass spectroscopy.

Abbreviations

Substances, Reagents, and Solvents

ABA	Abscisic acid
AD-mix-β	A mixture of (DHQD)$_2$-PHAL, K$_2$OsO$_2$(OH)$_4$, K$_3$Fe(CN)$_6$, and K$_2$CO$_3$
AIBN	Azobis(isobutyronitrile)
9-BBN	9-Borabicyclo[3.3.1]nonane
BS	Brassinosteroids
CHP	Cumene hydroperoxide
CSA	Camphor-10-sulfonic acid
DBU	1,5-Diazabicyclo[4.3.0]non-5-ene
DCC	1,3-Dicyclohexylcarbodiimide
DEAD	Diethyl azodicarboxylate
DHP	Dihydropyran
DIBAH	Diisobutylaluminum hydride
DMAP	4-(Dimethylamino)pyridine
L-DET	(L)-Diethyl tartrate
DME	1,2-Dimethoxyethane
DMF	*N,N*-Dimethylformamide
DMSO	Dimethyl sulfoxide
DHQD CLB	Dihydroquinidine *p*-chlorobenzoate
DHQD NAP	Dihydroquinidine 9-(1'-naphthyl) ether
DHQD PHN	Dihydroquinidine 9-phenanthryl ether
DHQ CLB	Dihydroquinine *p*-chlorobenzoate
(DHQD)$_2$-PHAL	Dihydroquinidine 1,4-phthalazinediyl diether
DPTBSCl	Dipropyl-*tert*-butylsilyl chloride
GA	Gibberellic acid
HMPA	Hexamethylphosphoramide
IAA	Indole-3-acetic acid

LDA	Lithium diisopropylamide
MCPBA	*m*-Chloroperbenzoic acid
NCS	*N*-Chlorosuccinimide
NMO	*N*-Methylmorpholine *N*-oxide
PCC	Pyridinium chlorochromate
PPTA	4-Phenyl-1,2,4-triazoline-3,5-dione
PPTS	Pyridinium toluene-*p*-sulfonate
Py	Pyridine
TBAF	Tetrabutylammonium fluoride
TDTAP	Tetradecyltrimethylammonium permanganate
TFD	Methyl(trifluoromethyl)dioxirane
THF	Tetrahydrofuran

Protecting Groups and Ligands

Ac	Acetyl
acac	Acetylacetonate
Bn	Benzyl
Bu	*n*-Butyl
Bz	Benzoyl
dmpe	1,2-Bis(dimethylphosphino)ethane
dppp	1,3-Bis(diphenylphosphino)propane
EE	1-Ethoxyethyl
MEM	(2-Methoxyethoxy)methyl
MOM	Methoxymethyl
Ms	Methanesulfonyl (mesyl)
Ph	Phenyl
Pr	*n*-Propyl
TBDMS	*tert*-Butyldimethylsilyl
Th	Thienyl
THP	Tetrahydropyranyl
TMS	Trimethylsilyl
Ts	*p*-Toluenesulfonyl (tosyl)
Tol	Tolyl (4-MeC$_6$H$_4$)

CHAPTER I

INTRODUCTION

Until recently it was assumed that the main plant growth processes were controlled by 5 types of phytohormones, namely the auxins, the cytokinins, the gibberellins, abscisic acid, and ethylene (Fig. 1).

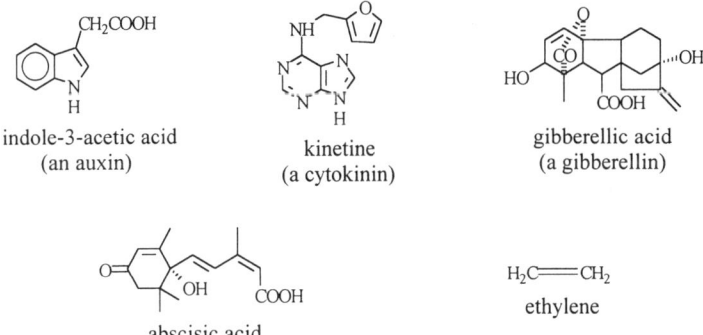

Fig. 1. The structures of some phytohormones.

The role of steroids as hormones of mammals has been known since 1930 and steroidal hormones have been found also in insects and fungi. Plants possess the ability to biosynthesize a large variety of steroids and a hormonal function was repeatedly postulated for plants also. However, it was not until 1979 that steroids with hormonal functions were discovered in plants. In that year American scientists reported a new steroidal lactone called brassinolide from bee-collected pollen of *Brassica napus* L. Two years later castasterone was isolated from insect galls of *Castanea crenata* spp. (Fig. 2). To date, more than 40 structurally and functionally related steroids have been isolated from natural sources, and this group of compounds is indicated now as brassinosteroids (BS) and considered as a new group of plant hormones.

BS demonstrate various kinds of regulatory activities on the growth and development of plants, such as stimulation of cell enlargement and cell division, lamina inclination, bending of leafs at the joints, and changes in membrane potentials. At the molecular level BS change the gene expression and the metabolism of nucleic acids and proteins.

BS are widespread in nature and are found in gymnosperms, monocotyledons, dicotyledons, and algae. BS are present in nearly every part of the plant, with the highest concentrations in the reproductive organs (pollen and immature seeds).

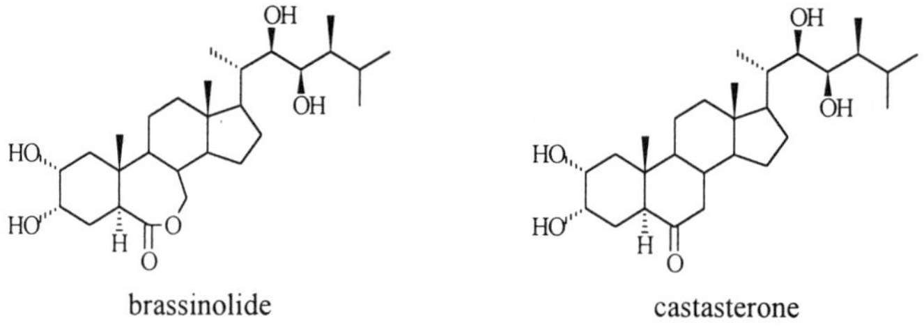

Fig. 2. The first brassinosteroids.

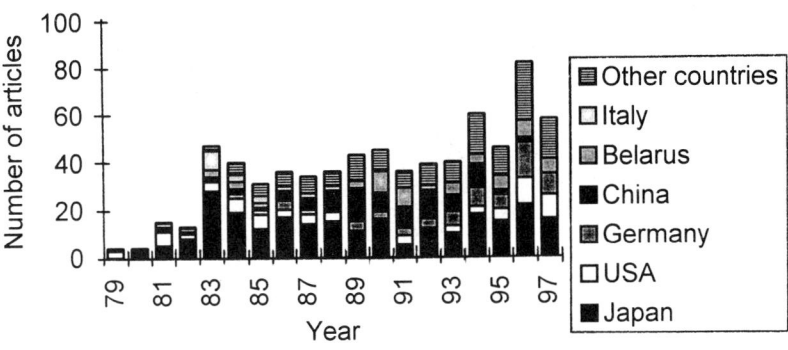

Fig. 3. The number of articles on BS.

Treatment of plants with BS at the appropriate stage of their development results in an increase of crop yield and, in some cases, in an increase of its quality. These effects can be achieved by applying BS at doses of 20-50 mg/ha and that is much less than those for the usual plant growth stimulators. For practical application the ability of BS to increase the resistance of plants to unfavorable factors of the environment, such as extreme temperatures, salinity, drought, or pesticides, is important also.

The extremely high activity of BS has attracted the attention of many specialists in the fields of chemistry, biology, and agriculture (Fig. 3). As result of large-scale scientific programs, started initially in the U.S. and Japan and later in the USSR, Germany, and China (Fig. 4), many problems connected with BS

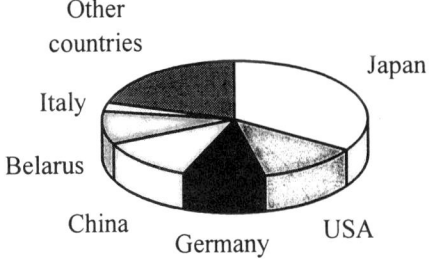

Fig. 4. The contribution of scientists from different countries to the total number of articles.

chemistry and biology have been solved since 1980.

Plants contain BS at such low concentrations (10^{-5}-10^{-11}%) that elaboration of specific methods for their analysis became necessary. Although the first BS were identified after isolation as pure compounds, later on the presence of many BS was proven by their synthesis as reference compounds, in combination with GC-MS analysis. Progress in BS chemistry and biology would have been impossible without sensitive biological tests. In the beginning tests originally developed for other phytohormones were used. When research on BS was well established, a number of specific tests were developed for this group of phytohormones.

Because of the low BS contents in plants, the only source of these compounds for biological studies and practical purposes is chemical synthesis. The progress in the chemical synthesis of BS and their analogs has been important and has led to economically feasible approaches that have brought their practical application in agriculture within reach. To date, only one attempt has been made to realize the total synthesis of BS but also the partial synthesis of these complex compounds, with 10-13 chiral centers, is a rather difficult task. Investigations in this direction have led to the elaboration of many new synthetic methods, especially for the construction of the side chain. Much attention has been devoted to syntheses of BS starting from abundantly available natural steroids. A number of sterols can be used for this purpose, and the best choices are of course those that contain the same carbon skeleton as the final compound. In this way carbon-carbon bond-forming reactions can be avoided and the synthesis can be performed in a more simple and less expensive way. This is especially important for BS that have chances for practical application in agriculture.

Since their discovery in 1979, a great deal of information on BS has been accumulated, and the necessity for a monograph in this field has become evident. The data presented here form a reliable basis for further development of BS research in a theoretical and practical way. Further investigations will contribute to theories about the role of steroids in the exchange of biological information in all living organisms. BS research is based on a vast experience in the field of steroids and phytohormones. Many problems proved to be common for steroids and phytohormones and the results obtained with BS are of interest for many scientists working in related disciplines.

The present book is the first monograph in English that covers all aspects of BS research, including isolation and identification of BS, distribution in the plant kingdom, biosynthesis and metabolism, synthesis, biological activity, and practical application. In this book also much information on this subject from countries of the former Soviet Union has been made available for international readership. The authors hope that this book will stimulate scientists to do research on brassinosteroids, being one of the new and promising subjects in the area of natural product chemistry and plant physiology.

CHAPTER II

BRASSINOSTEROIDS (BS) IN NATURE

A. HISTORICAL ASPECTS

The first report about the possible existence of a new type of plant hormone was published in 1968 in Japan (Marumo et al., 1968; Abe and Marumo, 1991). From 430 kg of fresh leaves of *Distylium racemosum* Sieb. et Zucc., an evergreen tree known as "Isunoki" in Japan, three active fractions indicated as *Distylium* factor A_1 (751 μg), A_2 (50 μg), and B (236 μg) were isolated. All three fractions proved to be much more active in the rice lamina inclination test than IAA. However, the limited amount of material did not allow chemical identification of the individual compounds at that time.

Later the isolation of an oil fraction from rape pollen *Brassica napus* L. with strong plant growth activity was reported (Mitchell *et al.*, 1970; Mandava *et al.*, 1978; Steffens, 1991). This fraction was called "brassins" and the authors suggested that brassins could be a new plant hormone. The bean second-internode bioassay was used for detection and monitoring of the biologically active compounds. Brassins induced elongation of the second internode at an average of 155 mm at a dose of 10 µg per plant, whereas untreated plants grew only 12 mm. Histological studies revealed a response different from that for gibberellin A_3.

Isolation of the new hormone and elucidation of its structure turned out to be a difficult task. At first it was supposed that brassins were β-glycosides of fatty acids. A special program was initiated by the U.S. Department of Agriculture in 1974 (Steffens, 1991) and, taking into account the low content of the unknown compound in the natural source, about 227 kg of bee-collected rape pollen (*Brassica napus* L.) was processed. The chemical part of the program resulted in the isolation of a few crystals (4 mg) of a new compound from 40 kg of rape pollen. X-ray crystallographic analysis showed that the new compound was a steroidal lactone with the structure of brassinolide (Grove *et al.*, 1979). The discovery of brassinolide prompted intensive studies on the isolation and identification of new members of this group. To date, more than 40 BS have been found and identified in plants.

B. NOMENCLATURE, STRUCTURES, AND CLASSIFICATION

The term "brassinosteroids" originates from the Latin name for rape - *Brassica napus* L. Most of the compounds of this class have received their names in a similar manner, derived from the name of the plant from which they were isolated or identified for the first time. In many cases these names are the basis for a mixed type of nomenclature, in which different prefixes, suffixes,

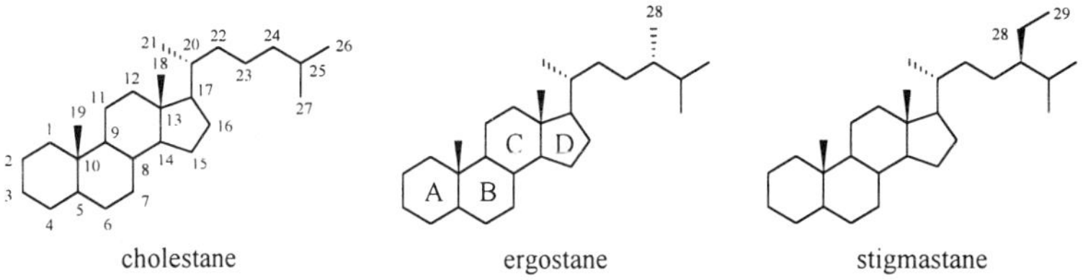

BRASSINOSTEROIDS IN NATURE

Fig. 5. The basic chemical structures used for the full chemical names of BS.

signs, and ciphers are added. The suffix "-olide" means that the molecule of the compound contains a lactone moiety (e.g., brassinolide); the suffix "-one" is characteristic for 6-ketobrassinosteroids (e.g., brassinone).

The full chemical names of BS are rather long ("Nomenclature of Steroids", 1989). Structures of steroids having the same carbon skeleton (cholestane, ergostane, stigmastane) are used as the basis for BS nomenclature (Fig. 5). In principle, cholestane may be used as a basis for any name of BS provided that the substituent at C-24 and its configuration are indicated (e.g., (24S)-methylcholestane = ergostane).

The base structure defines the stereochemistry of all chiral centers except that of the chiral center at C-5, which determines the fusion of rings A and B, and those of the carbon atoms bearing additional substituents. It is common for steroids that the configuration of substituents in rings A, B, and C is indicated as β for those above the plane of the tetracyclic skeleton and as α when the substituents are below that plane.

(24R)-24-ethyl...
24α-ethyl...

(24S)-24-ethyl...
24α-ethyl...

Fig. 6. One has to be aware of the fact that sometimes the priority of the substituents, and consequently the indication of the configuration (Brewster, 1986), may change without an actual change of the configuration of the chiral center.

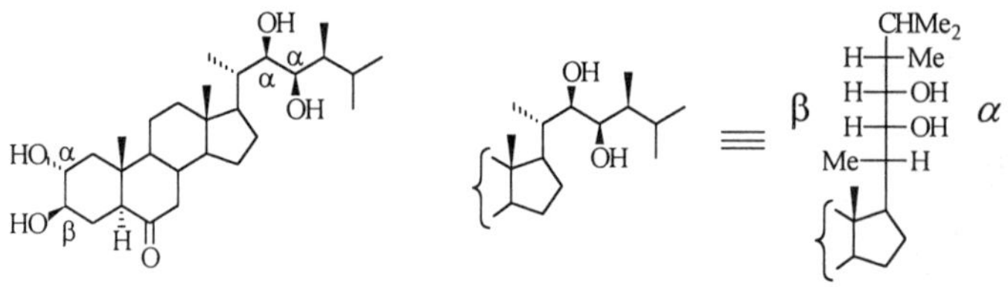

Fig. 7. Principles of the Fieser-Plattner nomenclature.

Two conventions are used to indicate the configuration of functional groups in the side chain. The first is the Cahn-Ingold-Prelog convention (Cahn et al., 1966), which will be followed as a rule in this book (Fig. 6).

The Fieser-Plattner nomenclature (Fieser and Fieser, 1959) can be used for such cases to facilitate understanding. According to these rules, the longest C-17 alkyl substituent is placed upward from ring D and projected in the plane of the drawing (Fig. 7). Functional groups pointing upward appear on the right side in the projection and are denoted as α, and functional groups pointing downward appear on the left side and are indicated as β. Note that for the cyclic part of the molecule the indications α and β have a different meaning. Functional groups pointing upward are indicated as β, and functional groups pointing downward are indicated as α.

Trivial names of BS that differ from the base structure only in the configuration of one chiral center are obtained by the prefix "epi-", e.g., 3-epibrassinolide and 24-epicastasterone. An increase in the number of carbon atoms by one methylene group is denoted by the prefix "homo-", and a similar decrease by the prefix "nor-". For names of compounds containing alkyl substituents at C-24, the numbers before the prefixes may be omitted, e.g., 24-epibrassinolide = epibrassinolide.

As already mentioned, more than 40 BS have been identified to date and it can be expected that many more members of this class will be discovered in the near future (Yokota, 1984, 1985, 1987a,b, 1995, 1997; Adam and Marquardt,

1986; Lakhvich et al., 1990; Kim, S.-K., 1991; Yokota et al., 1991; Takatsuto, 1994a; Fujioka and Sakurai, 1997b). Mostly the nomenclature of a new class of compounds is a mix of established nomenclature rules combined with names derived from the source from which the compound has been isolated. In addition, abbreviations, short names, and acronyms are introduced, become popular, and add to the confusion. Sooner or later it becomes clear that some order has to be brought into the collection, and proposals for systematization of the nomenclature for the new class of compounds begin to appear in the literature. In this respect we do not want to disappoint our readers and we propose a systematic classification of BS that, in our opinion, will be suitable for present and future use.

This classification is first of all based on established nomenclature but it takes into account the habit of chemists and plant physiologists to use abbreviations and trivial names. Also elements of the metabolic pathways and biological activities will not be neglected and will be logically incorporated. This has led to an approach in which the most active and first identified representative of this class of compounds - i.e., brassinolide (Bl) - is taken as the basic structure of the system (Fig. 8).

- As a general indication for the class of brassinosteroids the capitals BS will be used for the natural products. All nonnatural synthetic brassinosteroid-like compounds should be indicated as BS analogs.

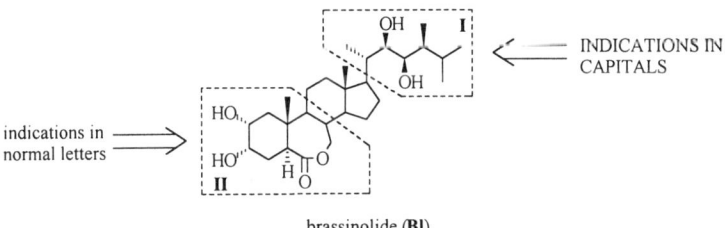

Fig. 8. The great diversity in structures is found within the areas **I**, the side chain, and **II**, the cyclic part. These parts of the molecule correspond with the places where the important metabolic transformations take place.

- In the structures of BS, as represented by brassinolide (Bl), several common structural elements can be recognized. All known BS have a trans-fused AB ring system and a normal steroid CD ring system with a substituted side chain at C-17.

- A first classification of the structures of BS is based on the number of carbon atoms that are present in the molecule:

the 27C compounds are indicated with the letters NB, derived from norbrassinolide,

the 28C compounds are indicated with the letter B, derived from brassinolide,

the 29C compounds are indicated with the letters HB, derived from homobrassinolide.

- A second classification can be made by the obvious division of the structure of BS in the side chain and the cyclic part. All further indications with respect to the side chain are specified with capitals, and all indications with respect to the cyclic part of the molecule are specified with normal letters.

- With respect to the side chain, further characterization can be indicated in an implicit or explicit way. Many structural derivatives of brassinolide concentrate around C-24 so when no carbon atom in the side chain is mentioned explicitly, the structural variation is connected with C-24. All other structural variations at carbon atoms in the side chain are indicated with the number of that atom followed by the characterization of the modification in capital letters. For example, 24-epibrassinolide will be indicated as EB and not as 24-EB; the structure with an extra methyl group at C-25 is indicated as 25-HB. The letter D is used to denote the most common $\Delta^{24(28)}$-dehydro derivatives. A deoxy compound will be indicated by using the full expression, so cathasterone can be characterized with respect to the side chain with the indication 23-deoxy-B. Configurations other than those in brassinolide will be indicated with the normal *R,S* nomenclature. For instance, 22*S*-Bl is the indication for that BS that is epimeric at C-22 with respect to brassinolide (Bl).

- With respect to the cyclic part, again the situation as in brassinolide is used as the reference point with respect to ring A; the lactone moiety in ring B is indicated with the small letter l. So brassinolide will be fully characterized with the indication Bl. This indication will be used throughout this book and is close to other short names that are in use for this compound in the literature.

- A ketone in ring B will be indicated with the letter k, so another well-known BS, castasterone, will be characterized by the indication Bk. Absence of this ketone group will be indicated with the letter d, the short name for functional groups that are not present, as in de(oxy) or de(oxa).

- Further deviations in ring A with respect to the situation in brassinolide (Bl) are connected with the configurations of the two hydroxyl groups, indicated with $2\beta,3\beta$ for the corresponding β-hydroxy compounds. The absence of the C-2 or C-3 hydroxyl group will be indicated with the letter 2d (for 2-deoxy) or 3d (for 3-deoxy), and the presence of a ketone at C-3 will be indicated as 3k.

- All indications for the cyclic part will be placed behind the indications for the side chain.

- The presence or absence of other functional groups will be indicated using the normal chemical rules for nomenclature.

It should be noticed that indications such as diepi- or triepi- are not in accordance with the normal rules for nomenclature and the use of these indications should be avoided.

This set of nomenclature rules is summarized in Table I.

The BS identified in nature to date are collected in Table II. All BS are characterized as well by their shorthand description and in this way the relation between the characterization and the structural formula becomes evident. It is relatively easy to translate the shorthand description into the chemical formula and vice versa.

Apart from free BS, several types of conjugates have been found as natural hormones. Glycosides bearing the sugar moiety at the 2-, 3-, 23-, 25-, and 26-hydroxyl group have been identified in the course of metabolic studies of

TABLE I

Codes for a Shorthand Description of BS

Code	Meaning
In capitals	*Side chain*
NB	27C norbrassinolide-like side chain
B	28C-brassinolide-like side chain
HB	29C 28-homobrassinolide-like side chain
E	epi at C-24
D	dehydro, $\Delta^{24(28)}$
25HB	29C 25-homobrassinolide
In normal letters	*Cyclic part*
l	lactone, brassinolide-like cyclic part
k	ketone, castasterone-like cyclic part
3k	ketone at C-3
d	6-deoxocastasterone-like cyclic part
2d, 3d	2-deoxy, 3-deoxy, absence of the indicated hydroxyl group
2β,3β	β-position of the indicated hydroxyl group

brassinolide and epibrassinolide. Teasterone 3-myristate represents the first example of a fatty acid type conjugate. Similar 3-OH derivatives (laurate, myristate, and palmitate) have been characterized as metabolites of epibrassinolide. A list of BS found in natural sources as native compounds is outlined in Table III.

Chemical classification is not the only possibility for arranging BS. A division based on biosynthesis may provide insight into their mutual relationship. Analysis of the structures of all natural BS currently known, combined with the level of their biological activity and the data on their biosynthesis and metabolism, allows the following groups to be distinguished:

1) Compounds in which one or more functions characteristic for Bl are absent. These compounds are not as active as those of the second group and are considered to be their biosynthetic precursors.

2) BS with a full set of functional groups in the molecule (2α,3α- and 22R,23R-diol functions, B-homo-7-oxa-6-keto- or 6-ketone). These compounds reveal the highest level of biological activity. Brassinolide and castasterone are the most important among the natural BS, taking into account the wide distribution of these compounds and their high biological activity. The term "real" brassinosteroids may be applied to them, which means that these compounds act in plants as hormones. The existence of similar compounds with a full set of functions but with a different carbon skeleton in the side chain (epibrassinolide, epicastasterone, homobrassinolide, (24S)-ethylbrassinone, norbrassinolide, brassinone) may be explained by a beginning of the biosynthesis with brassicasterol, β-sitosterol, or 22-dehydrocholesterol instead of campesterol. These other sterols are present in plant membranes, and the enzyme systems responsible for the biosynthesis of the real BS apparently are not really selective and can accept more than one sterol as substrate.

3) Metabolites of BS. Compounds with a trans-diol group in ring A and BS conjugates belong to this group.

It is obvious that the list of natural BS is far from complete. Just in the immature seeds of *Phaseolus vulgaris*, more than 60 BS were found (Kim, S.-K., 1991), and the structures of most of these still remain to be determined. Another study of this natural source (Kim, S.-K. *et al.*, 1994b) revealed the presence of three new 2-deoxy-BS. It was assumed that these could be (3ξ,22ξ,23ξ)-2-deoxy-25-methyldolicholide, (3ξ,22ξ,23ξ)-2-deoxy-dolichosterone, and (3ξ,22ξ,23)-2-deoxy-24-ethylbrassinone. Also *Lilium longiflorum* was mentioned to contain a new conjugate - teasterone 3-laurate - along with the known teasterone 3-myristate (Yasuta *et al.*, 1995). Studies on fully grown seeds of *Pisum sativum* (Yokota *et al.*, 1996a) demonstrated the occurrence of a new BS with an extra hydroxyl group whose structure has not yet been established.

TABLE II
Major Structural Types of Natural BS

AB cycles	Side chain	27C-BS NB	28C-BS B	28C-BS EB	28C-BS BD	29C-BS HB	29C-BS 25HB	29C-BS HBD	29C-BS 25HBD
l		norbrassinolide **NBl**	brassinolide **Bl**	epibrassinolide **EBl**	dolicholide **BDl**	homobrassinolide **HBl**		homodolicholide **HBDl**	
l2d			2-deoxybrassinolide **Bl2d**						
d2d3k			3-dehydro-6-deoxoteasterone **Bd2d3k**						
d2d			6-deoxotyphasterol **Bd2d**						
d		6-deoxo-28-norcastasterone **NBd**	6-deoxocastasterone **Bd**	6-deoxo-24-epicastasterone **Ebd**	6-deoxodolichosterone **BDd**			6-deoxohomodolichosterone **HBDd**	6-deoxo-25-methyldolichosterone **25HBDd**

(*continues*)

TABLE II (continued)

AB cycles	Side chain	27C-BS NB	28C-BS B	28C-BS EB	28C-BS BD	29C-BS HB	29C-BS 25HB	29C-BS HBD	29C-BS 25HBD
d2β			3-epi-6-deoxo-castasterone **Bd2β**						
k2d3β			teasterone **Bk2d3β**			homoteasterone **HBk2d3β**			3-epi-2-deoxy-25-methyldolichosterone **25HBDk2d3β**
k2d3k			3-oxoteasterone **Bk2d3k**						
k2d			typhasterol **Bk2d**			homotyphasterol **HBk2d**			2-deoxy-25-methyldolichosterone **25HBDk2d**
k		brassinone **NBk**	castasterone **Bk**	epicastasterone **Ebk**	dolichosterone **BDk**	homocastasterone **HBk**	25-methyl-castasterone **25HBk**	homodolichosterone **HBDk**	25-methyldolichosterone **25HBDk**

(continues)

TABLE II (continued)

Side chain / AB cycles	27C-BS NB	28C-BS B	28C-BS EB	29C-BS BD	29C-BS HB	29C-BS 25HB	29C-BS HBD	29C-BS 25HBD
k3β		3-epicastasterone **Bk3β**	3,24-diepicastasterone **EBk3β**					
k2β		2-epicastasterone **Bk2β**						2-epi-25-methyldolichosterone **25HBDk2β**
k2β3β		2,3-diepicastasterone **Bk2β3β**						2,3-diepi-25-methyldolichosterone **25HBDk2β3β**
k2,3epoxy		secasterone **Bk2,3epoxy**						

TABLE III
A List of Natural BS

N	Trivial name	Proposed abbreviation	First isolation/ identification
1	brassinolide	Bl	Grove et al., 1979
2	castasterone	Bk	Yokota et al., 1982a
3	dolicholide	BDl	Yokota et al., 1982b
4	brassinone	NBk	Abe et al., 1983
5	(24S)-ethylbrassinone (homocastasterone)	HBk	Abe et al., 1983
6	norbrassinolide	NBl	Abe et al., 1983
7	dolichosterone	BDk	Baba et al., 1983
8	homodolichosterone	HBDk	Baba et al., 1983
9	homodolicholide	HBDl	Yokota et al., 1983b
10	6-deoxocastasterone	Bd	Yokota et al., 1983d
11	6-deoxodolichosterone	BDd	Yokota et al., 1983d
12	typhasterol	Bk2d	Schneider et al., 1983; Yokota et al., 1983a
13	homobrassinolide	HBl	Ikekawa et al., 1984
14	teasterone	Bk2d3β	Abe et al., 1984a
15	23-O-β-D-glycopyranosyl-25-methyldolichosterone	23-gly-25HBDk	Yokota et al., 1986
16	2,3-diepicastasterone	Bk2β3β	Takahashi et al., 1987
17	3,24-diepicastasterone	EBk3β	Takahashi et al., 1987
18	2,3-diepi-25-methyldolichosterone	25HBDk2β3β	Takahashi et al., 1987
19	epibrassinolide	EBl	Yokota et al., 1987a
20	2-epicastasterone	Bk2β	Takahashi et al., 1987
21	3-epicastasterone	Bk3β	Takahashi et al., 1987

(*continues*)

TABLE III (continued)

N	Trivial name	Proposed abbreviation	First isolation/ identification
22	24-epicastasterone	Ebk	Yokota et al., 1987a
23	3-epi-2-deoxy-25-methyldolichosterone	25HBDk2d3β	Yokota and Takahashi, 1987
24	2-deoxy-25-methyldolichosterone	25HBDk2d	Yokota and Takahashi, 1987
25	6-deoxohomodolichosterone	HBDd	Yokota et al., 1987b
26	2-epi-25-methyldolichosterone	25HBDk2β	Takahashi et al., 1987
27	25-methyldolichosterone	25HBDk	Kim, S.-K. et al., 1987
28	6-deoxo-25-methyldolichosterone	25HBDd	Kim, S.-K., 1991
29	3-epi-6-deoxocastasterone	Bd2β	Kim, S.-K., 1991
30	3-epi-1α-hydroxycastasterone	1α-hydroxy-Bk3β	Kim, S.-K., 1991
31	23-O-β-D-glycopyranosyl-2-epi-25-methyldolichosterone	23-gly-25HBDk2β	Kim, S.-K., 1991
32	1β-hydroxycastasterone	1β-hydroxyBk	Kim, S.-K., 1991
33	homoteasterone	HBk2d3β	Schmidt et al., 1993b
34	25-methylcastasterone	25HBk	Taylor et al., 1993
35	3-oxoteasterone	Bk2d3k	Abe et al., 1994; Yokota et al., 1994
36	teasterone 3-myristate	Bk2d3β myristate	Asakawa et al., 1994
37	cathasterone	B23deoxyk2d3β	Fujioka et al., 1995a
38	3-dehydro-6-deoxoteasterone	Bd2d3k	Griffiths et al., 1995a
39	6-deoxotyphasterol	Bd2d	Griffiths et al., 1995a
40	6-deoxo-24-epicastasterone	Ebd	Spengler et al., 1995
41	6-deoxo-28-norcastasterone	NBd	Spengler et al., 1995
42	2-deoxybrassinolide	HBl2d	Schmidt et al., 1995c
43	homotyphasterol	Bk2d	Abe et al., 1995a
44	secasterone	Bk2,3epoxy	Voigt et al., 1995

C. RELATED STEROIDS

There are many natural steroids containing one, two, or more structural units that are characteristic for BS. The presence of a 2,3-diols function is rather common in naturally occurring steroids. Thus, some 2β,3β-diols are found in sapogenins like samogenin (Djerassi and Fishman, 1955), the structure of which is depicted in Fig. 9. A defensive steroid lucibufagenin isolated from the firefly *Photinus pyralis* (Goetz et al., 1981) possesses the same structural element.

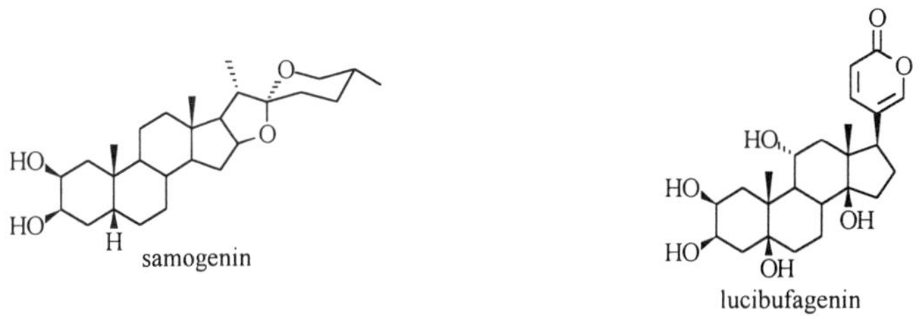

Fig. 9. Examples of natural steroids containing the 2β,3β-diol function.

Figure 10 demonstrates the occurrence in nature of steroid compounds containing the 2α,3α-diol function, which is more characteristic for BS. The antifeedant azedarachol was found in *Melia azedarach* (Nakatani et al., 1985) and cholest-5-ene-2α,3α,7β,15β,18-pentol 2,7,15,18-tetraacetate was isolated in a small amount from the marine hydroid *Eudendrium glomeratum* (Fattorusso et al., 1985).

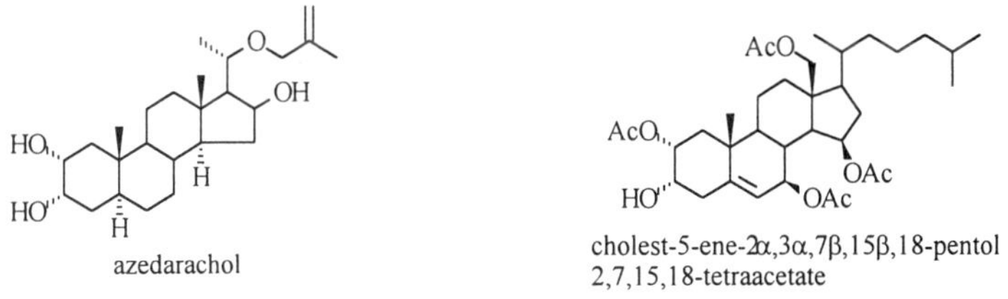

Fig. 10. Examples of natural steroids containing the 2α,3α-diol function.

Substances containing a lactone moiety in ring B are not characteristic for natural steroids other than BS. Asterasterol A, isolated from an Antarctic starfish of the family Asteriidae (De Marino et al., 1997), possesses a 6-oxa-7-keto lactone ring B and with respect to that structural moiety, it is a regioisomer of normal BS.

Numerous 6-ketones are widely distributed in both the animal and plant kingdoms and some of them are shown in Fig. 11. Pinnasterol was isolated from the red alga *Laurencia pinnata* (Fukuzawa et al., 1981), diaulusterol B was found in skin extracts of the dorid nudibranch *Diaulula sandiegensis* (Williams et al., 1986), and chiograsterone (Sauer and Takeda, 1970) and chiograsterol A (Takeda et al., 1965) were identified as steroidal constituents of *Chionographus japonica*.

BS have many structural elements in common with ecdysteroids. The latter are insect moulting hormones, but at the same time they are widespread in plants, probably as plant-protecting agents (Akhrem et al., 1973; Koolman, 1989; Akhrem and Kovganko, 1989). The characteristic features of ecdysteroids are a diol function in ring A, a Δ^7-6-ketone in ring B, a 14α-hydroxyl group, and

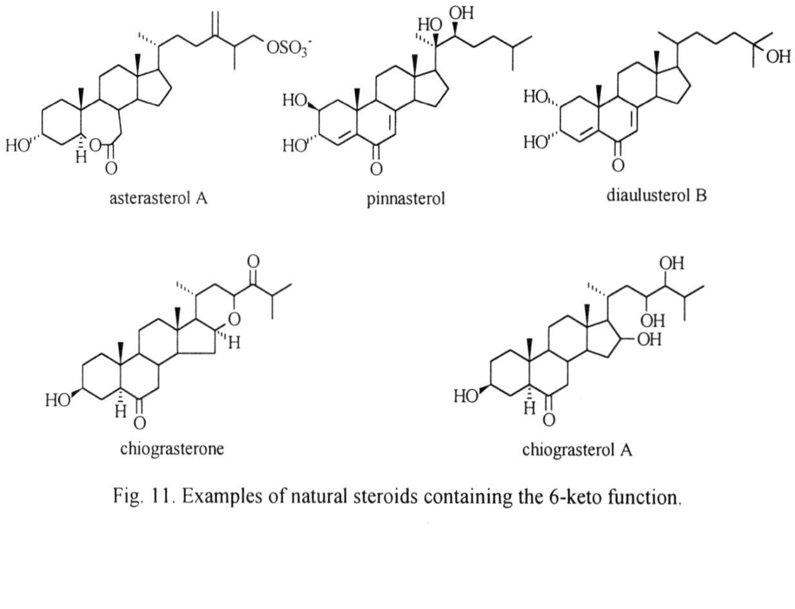

Fig. 11. Examples of natural steroids containing the 6-keto function.

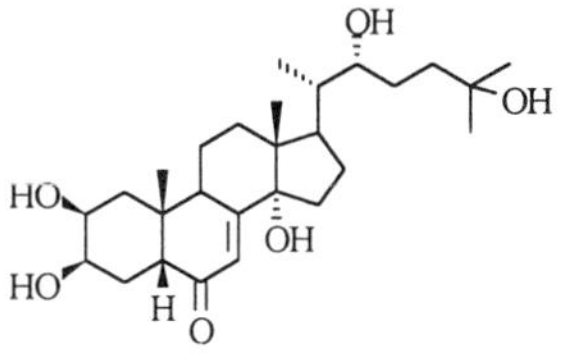

α-ecdysone

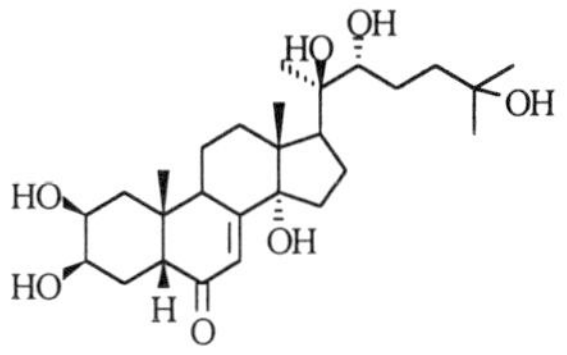

ecdysterone

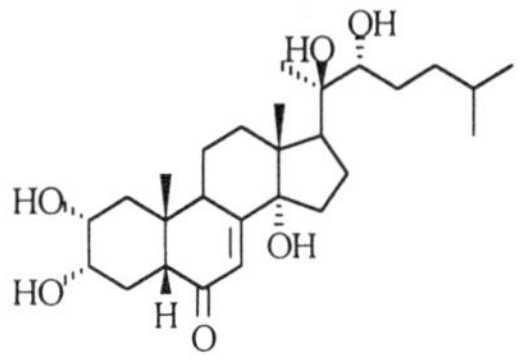

ponasterone B

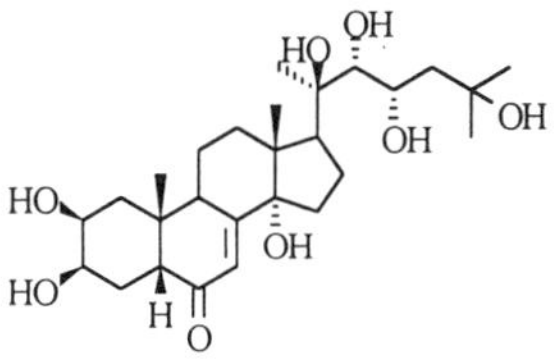

gerardiasterone

Fig. 12. Examples of ecdysteroids.

a polyoxygenated side chain (Fig. 12). Cis fusion of rings A and B and a 2β,3β-diol function are more characteristic for ecdysteroids. The chemical similarity of BS and ecdysteroids has caused a natural interest in the effects of BS in insects. It was found that BS did influence the development of insects; moreover, the type of activity could be both agonistic and antagonistic (Adam *et al.*, 1986; Hetru *et al.*, 1986; Richter *et al.*, 1987; Lehmann *et al.*, 1988; Richter and Adam, 1991; Richter and Koolman, 1991; Machackova *et al.*, 1995; Charrois *et al.*, 1996; Luu and Werner, 1996). The biological activity of ecdysteroids in typical plant bioassays was investigated as well (Dreier and Towers, 1988).

CHAPTER III

ISOLATION AND IDENTIFICATION

The identification of BS in plant raw material is a complex and multistep process of extraction, chromatographic purification, monitoring of the biological activity of the fractions, and other operations. Sensitive and highly specific biological tests play an important role in guiding the correct choice of active fractions in the course of the isolation process. The bean second-internode bioassay was applied mostly in the early period of BS investigations (Thompson *et al.*, 1981, 1982). Later, the rice lamina inclination test was used in most procedures (Wada *et al.*, 1984); the wheat unrolling test has been applied as well (Wada *et al.*, 1985).

Isolation of BS in a pure form is tedious and time consuming. An example of a relatively simple isolation process is that shown for castasterone (Yokota *et al.*, 1982a) in Scheme 1. The insect galls of the chestnut tree *Castanea* spp. (40 kg)

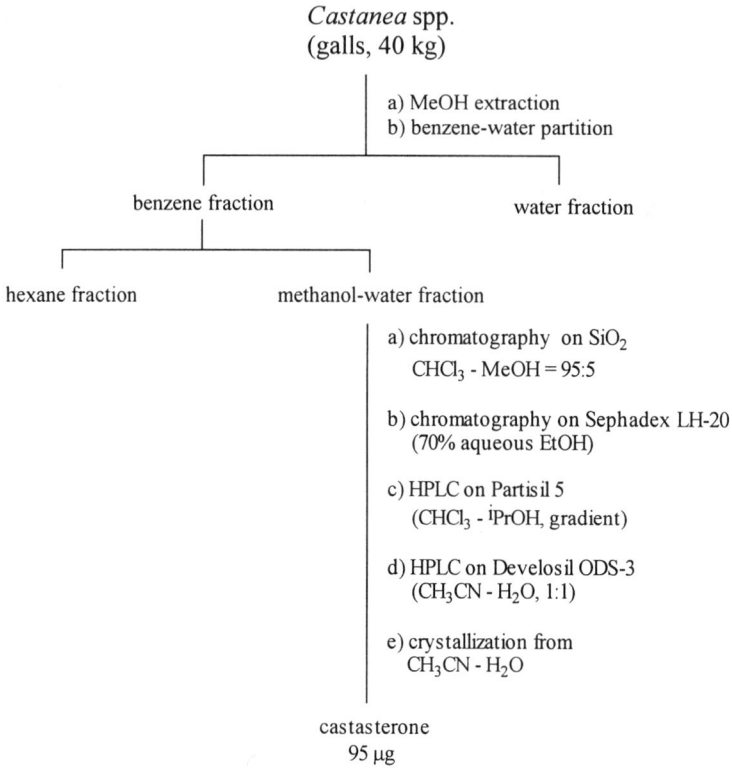

SCHEME 1

were extracted with methanol followed by partition of the extract between benzene and water. The organic layer was again partitioned between hexane and 90% aqueous methanol. The methanol fraction was then submitted to chromatography on SiO_2 and on Sephadex. Additional purification by HPLC on Partisil 5 and on Develosil ODS-3 and finally crystallization from aqueous acetonitrile afforded 95 µg of castasterone.

In Scheme 2 the more complicated process that was used for the isolation of the glycoside of 25-methyldolichosterone is depicted (Yokota et al., 1986). A sequence of extraction procedures gave three fractions, two of which showed a strong BS activity. The first fraction was found to contain free BS. TLC analysis of the second fraction showed the presence of active compounds with a smaller

ISOLATION AND IDENTIFICATION

mobility than that of free BS. A final purification by various types of chromatography led to the isolation of the first BS conjugate, 23-*O*-β-D-glycopyranosyl-25-methyldolichosterone.

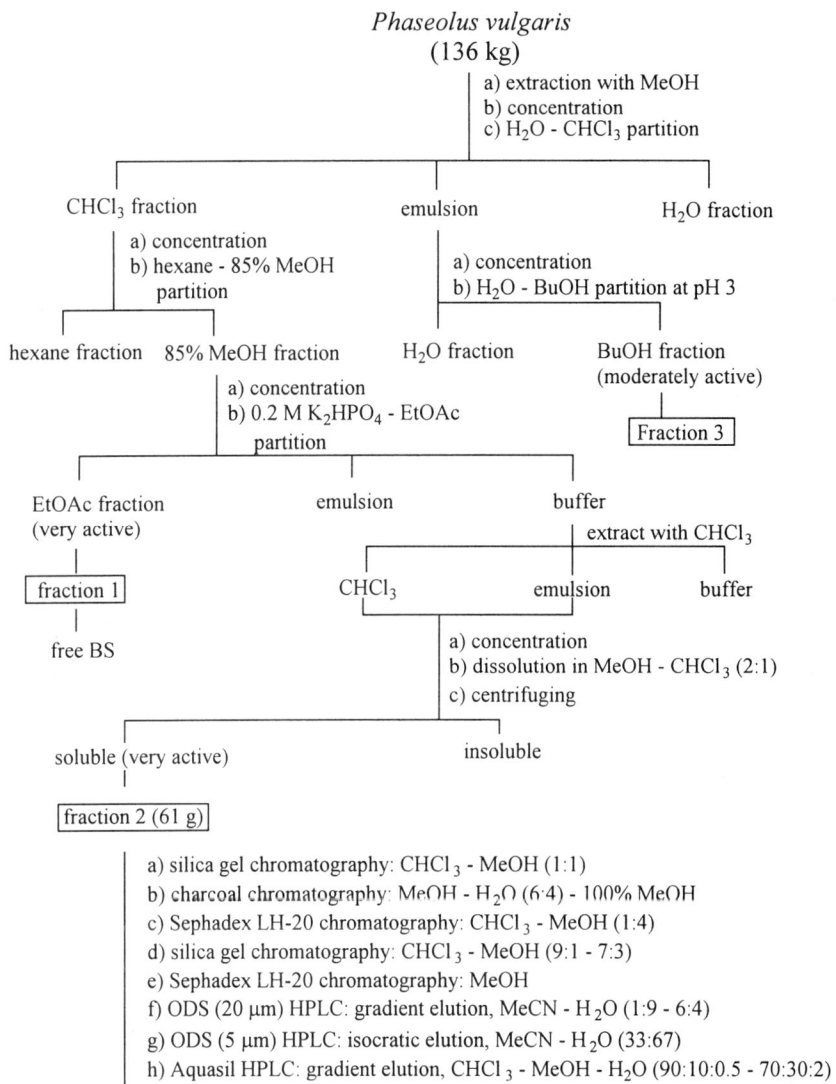

SCHEME 2

The preparation of samples from sources of biological origin could be improved by using immobilized phenylboronic acid gel (Gamoh *et al.*, 1994). BS with the 22,23-diol group react to the corresponding cyclic boronates on the immobilized gel. That allows the removal of all other impurities and the parent BS can be set free by treatment with H_2O_2 and acetonitrile.

Because of the very low BS content in plants, it is not always possible to isolate them in quantities that are sufficient for the recording of all spectral characteristics. In such cases the identification of BS has to be done by making derivatives and comparing these with separately synthesized authentic samples. Gas chromatography in combination with mass spectrometry (GC-MS) was employed in most cases for the identification of natural BS (Ikekawa *et al.*, 1983, 1984; Ikekawa and Takatsuto, 1984). A variant of GC-MS is selected-ion monitoring (GC-MS-SIM), where only one or more ions with a certain molecular mass of a known compound are used for registration (Ikekawa *et al.*, 1983; Allevi *et al.*, 1988).

Usually methaneboronic acid derivatives of BS are used for spectroscopic analysis. Methaneboronic acid is a reagent that is specific for a vicinal diol group and the methaneboronates (Scheme 3) can be obtained by simple treatment of BS with methaneboronic acid in pyridine (~30 min) at 70 °C (Takatsuto *et al.*, 1982b; Ikekawa *et al.*, 1984; Takatsuto, 1994a). In BS with separate hydroxyl group(s) application of a combined variant of protection, for example, as a methaneboronate-trimethylsilyl derivative, is necessary (Takatsuto and Ikekawa, 1986a). The formed derivatives exhibit moderate volatility, are thermally stable, and are suitable for GC analysis.

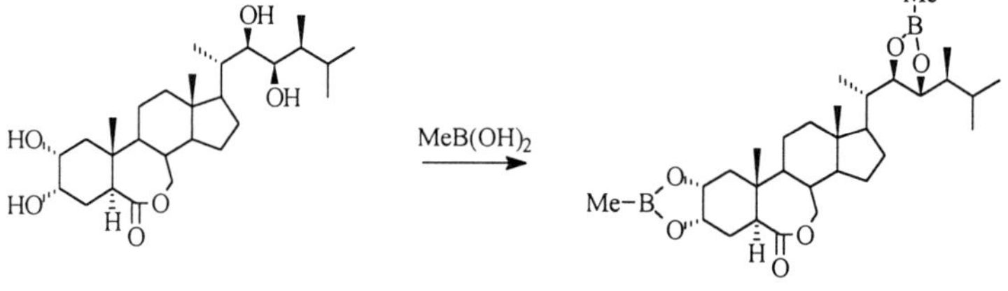

SCHEME 3

Electron impact ionization is a standard method for ionization of BS derivatives. However, chemical ionization is about 100 times more sensitive (Ikekawa *et al.*, 1983), which allows the detection of less than 1 ng of BS.

A good alternative for GC-MS is a method based on high-performance liquid chromatography (HPLC) of BS derivatives (Takatsuto and Gamoh, 1990; Gamoh *et al.*, 1992, 1996a,b; Gamoh and Takatsuto, 1994). UV detection was used first with naphthaleneboronic acid (Gamoh *et al.*, 1988) as a reagent for derivatization (Fig. 13). The detection limit was about 100 pg per injection. Experiments with derivatives suitable for fluorimetric detection, such as cyclic ethers of 9-phenanthreneboronic acid (Gamoh *et al.*, 1989a), 1-cyanoisoindole-2-*m*-phenylboronic acid (Gamoh and Takatsuto, 1989; Shim *et al.*, 1996), *m*-aminophenylboronic acids (Gamoh and Takatsudo, 1990), phenylboronic acid (Liu, G.-Q. *et al.*, 1995), and (dansylamino)phenylboronic acid (Gamoh *et al.*, 1989b, 1990a; Motegi *et al.*, 1994), show even better results. The BS derivatives can be detected down to 50-20 pg per injection. Another useful variant of HPLC analysis of BS relies on derivatization with ferroceneboronic acid, which enables electrochemical detection (Gamoh *et al.*, 1990b).

Immunological methods for BS analysis were developed as well (Horgen *et al.*, 1984; Yokota *et al.*, 1990b; Taylor *et al.*, 1993). Monoclonal antibodies to

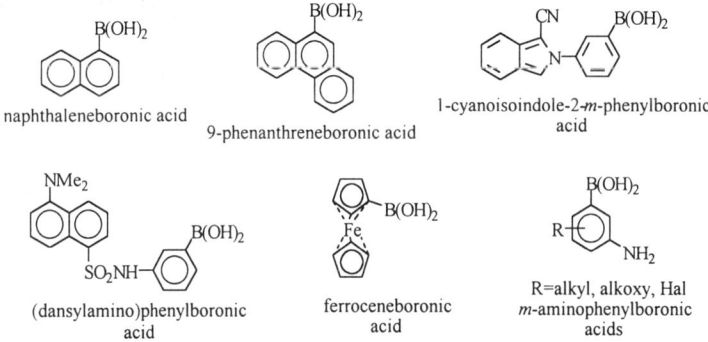

Fig. 13. Reagents used for derivatization of BS for HPLC analysis.

epibrassinolide were obtained from CAF_1 mice and used for investigation of *Brassica napus* tissues (Horgen *et al.*, 1984). However, this radioimmunoassay proved to be not specific enough. Better results were obtained by Yokota *et al.* (1990b) when an antiserum against castasterone was prepared by conjugating its carboxymethoxylamine oxime with bovine serum albumin and immunizing a rabbit. In a radioimmunoassay, this antiserum could recognize brassinolide and castasterone with a detection limit of about 0.3 pmol. There was also a 20-50% cross-reactivity with other natural BS. Radioimmunoassays and enzyme-linked immunosorbent assays were applied for the analysis of endogenous BS in different parts of *Phaseolus vulgaris* (Yokota *et al.*, 1990b) as well as for the detection of BS in the pollen of *Lolium perenne* L. (Taylor *et al.*, 1993) in the course of a study of the pollen developmental sequence.

Studies on the occurrence of BS in different plants showed that they can be found in monocots, dicots, and gymnosperms. It is now generally accepted (Takatsuto, 1991; Sakurai and Fujioka, 1993; Takatsuto, 1994a) that BS are present in all higher plants. In addition, BS were found in the green alga *Hydrodictyon reticulatum* (Yokota *et al.*, 1987a) and in the fern *Equisetum arvense* L. (Takatsuto *et al.*, 1990a), which indicates that they can be found in lower plants as well. However, in these species the wide occurrence of BS still has to be proven.

Plant tissues differ greatly with respect to BS content; the highest levels are observed in plant reproductive tissues, especially in pollen, which contains 10-100 μg/kg (10^{-6}-10^{-5}%) of BS (Takatsuto, 1994a). Immature seeds were found to be another rich source of BS, with amounts of 1-100 μg/kg (10^{-7}-10^{-5}%). The endogenous level of BS in shoots and leaves is much lower, these parts contain only 10-100 ng/kg (10^{-9}-10^{-8}%) of BS. It was noted also that young growing tissues contain more BS than old ones. A remarkably high BS level was observed in crown gall cells of *Catharanthus roseus* transformed by different strains of *Agrobacterium tumefaciens*. The total amount of various BS in these cells was found to be 30-40 µg/kg ($(3\text{-}4)\cdot 10^{-6}$%) (Park, K.-H. *et al.*, 1989; Sakurai *et al.*, 1991). This can offer a good possibility for the production of BS (Saimoto *et al.*, 1989).

Most plants contain castasterone as the main BS together with comparable amounts of brassinolide. The ratio of these two hormones may be different in the course of plant development. Thus the ratio brassinolide/castasterone for resting seeds of *Raphanus sativus* L. was 0.4, whereas this value for germinated seeds reached 1.1 (Schmidt *et al.*, 1991). In some plants the biosynthetic precursors of brassinolide are the major BS, as in the case of pollen of orange, tulip, and lily (Abe, 1991), which contain up to 100 µg/kg of typhasterol.

CHAPTER IV

SPECTRAL PROPERTIES

A. NMR Spectroscopy

NMR spectroscopy plays a key role in the structure determination of BS. The method is unsurpassed for detailed structure determination in solution. The fast progress in this area during the past two decades is connected to the introduction of high-field cryomagnets and computer data processing. Traditional proton and carbon spectra, along with special techniques such as nuclear Overhauser effect (NOE), DEPT, and two-dimensional spectra (^{1}H-^{1}H and ^{1}H-^{13}C COSY), have resulted in a situation that unequivocal assignment of all centers of a molecule is a matter of time and size of the sample. A number of articles on the subject have been published about the NMR spectroscopy of natural BS and their isomers

(Porzel et al., 1992; Ando et al., 1993; Porzel, 1996) and about synthetic intermediates (Gonzalez et al., 1986; Hazra et al., 1993b). An impressive example of the effective utilization of modern methods was demonstrated by Adam et al. (1996b). Specially developed detectors have made it possible to record ^{1}H spectra with samples of less than 0.2 mmol, whereas ^{13}C spectra could be recorded with samples down to about 2 mmol.

As a rule, BS with more than three hydroxyl groups are hardly soluble in $CDCl_3$. Therefore it is necessary to record spectra of these compounds in other solvents, mostly in C_5D_5N. This can cause an inconvenience, since direct comparison of spectral data of the final compounds with spectra obtained in $CDCl_3$ for intermediate products becomes impossible. The problem can be solved by taking spectra at increased temperature (ca. 50 °C) (Mori et al., 1984) or in strongly diluted solutions.

The analysis of steroid spectra begins, as a rule, with the identification of the angular methyl groups (Table IV). The position of the signal of the 18-Me is rather constant (δ 0.61-0.72)[*], as the nearest neighborhood of this group for most natural BS is practically the same. In the case of the 22S,23S-diols the signal is a bit shifted to lower field (δ ca. 0.74). The chemical shift of the 19-Me is highly influenced by the character of the substituents in the cyclic part of the molecule. For the 2α,3α-dihydroxy-6-ketones, it appears at δ 0.75-0.76 and it practically does not shift going to the 2-deoxy-3α-hydroxy derivatives. The position of the signal of this 19-Me can be used for the determination of the configuration of the diol group in ring A. Thus the spectrum of (2S,3R)-25-methyldolichosterone exhibits this signal at δ 0.98, which is characteristic for 2β,3β-diols.

Introduction of a lactone group in ring B results in a low-field shift of the 19-Me of 0.15-0.18 ppm, to δ 0.91-0.93. Signals of angular methyl groups in

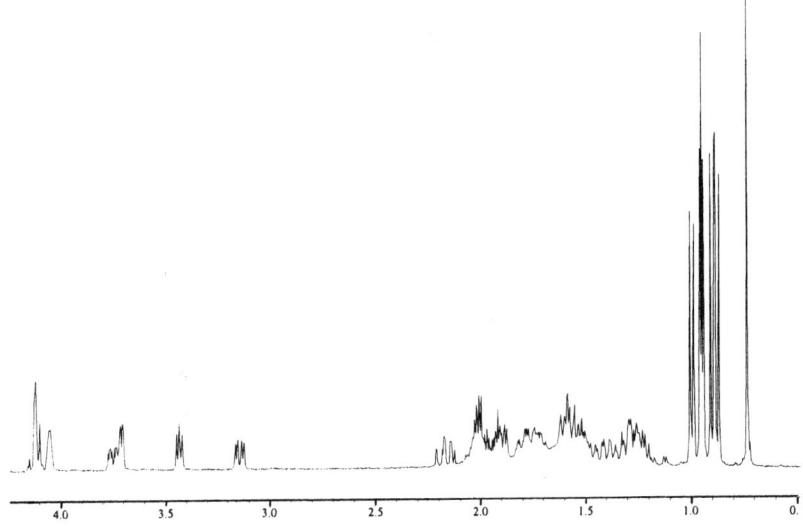

Fig. 14. ¹H NMR spectrum of epibrassinolide (CDCl₃, 400 MHz).

spectra recorded in C₅D₅N show a shift to lower field of 0.10-0.15 ppm compared to those taken in CDCl₃. Methyl groups of side chains appear as three-proton multiplets at δ 0.84-1.14 (doublets with J 6-8 Hz for 21-, 26-, 27-, and 28-Me and a triplet for 29-Me). These signals lie very close together and often they overlap (Fig. 14).

$\Delta^{24(28)}$-BS exhibit a doublet of the 21-Me at δ 0.91-0.96 and show the signals of the 26/27-Me at δ 1.06-1.14. Assignment of the methyl terminal groups can be confirmed by double-resonance experiments. Irradiation of the allylic C-25 proton at δ 2.27 results in collapse of doublets of the 26/27-Me into singlets (Baba *et al.*, 1983). The difference in chemical shifts of the 26- and 27-methyl groups demonstrates the rigidity of the steroidal side chain - even the rotation around the C-24 - C-25 bond is restricted. The signal of the 29-Me can be

---

♦ NMR spectra in CDCl₃ are discussed, if not indicated otherwise.

identified relatively easily, as it appears at δ 0.95 as a triplet with J 7.5 Hz. For $\Delta^{24(28)}$-BS this group exhibits a doublet at δ 1.71-1.75 with J 7.0-7.1 Hz.

The region of δ 1.1-2.0, with the signals of most of the methine and methylene protons, is excessively overcrowded. ^{1}H-^{1}H COSY 2D NMR offers a method for a detailed analysis of this region (Porzel *et al.*, 1992; Ando *et al.*, 1993; Adam *et al.*, 1996b). An example is shown in Fig. 15.

Signals of protons adjacent to carbonyl groups are situated in the region δ 2.00-3.18. Thus the 7β-H in the NMR spectrum of typhasterol appears at δ 2.31 as a double doublet (J 13, 4.5 Hz) (Yokota *et al.*, 1983a). The signal of the 5-H shows a low-field shift as a result of a 1,3-diaxial interaction between the 5α-H and the substituent at C-3. Spectra of the tetraols show the signal of the 5α-H as

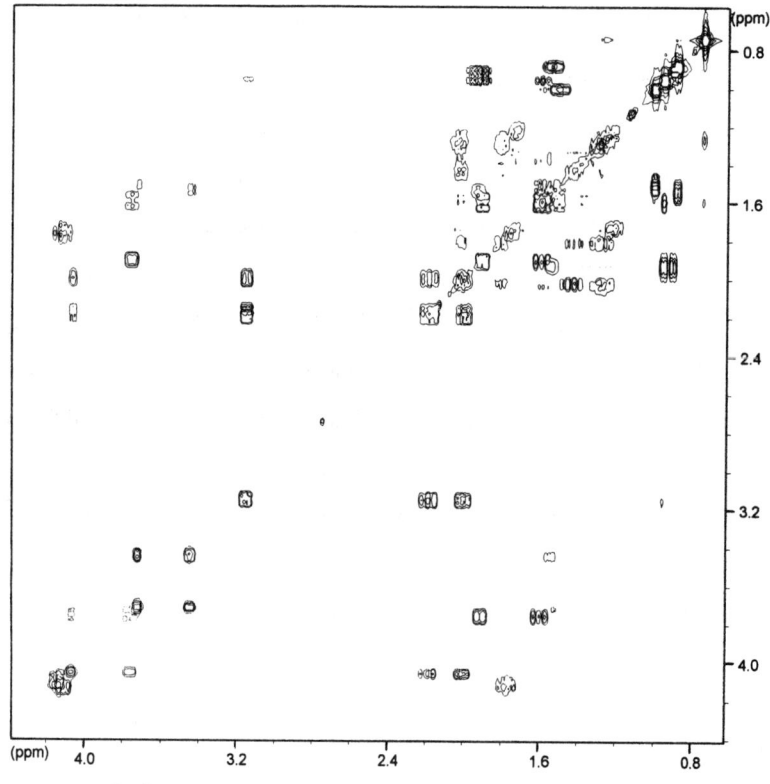

Fig. 15. ^{1}H-^{1}H COSY 2D NMR spectrum of epibrassinolide (CDCl$_3$, 400 MHz).

a double doublet in the region δ 2.54-2.69. The largest coupling constant (*J* 10-14 Hz) is due to a trans-diaxial interaction with the β-proton at C-4; the lowest one (*J* 4-8 Hz) is caused by the axial-equatorial interaction with the 4α-H. For typhasterol this proton exhibits a quintet (Schneider *et al.*, 1983; Yokota *et al.*, 1983a) or a triplet (Takatsuto *et al.*, 1984b) at δ 2.73-2.76.

Introduction of a lactone moiety in ring B results in a low-field shift of the 5-H and 7-H signals. Multiplicity and coupling constants are not changed significantly compared to those of the corresponding ketones. The position of the 5α-H signal depends greatly on the configuration of the substituent at C-3. For 3β-hydroxy lactones (natural series) it resonates at δ 2.84-2.86. Compounds with the opposite configuration of the hydroxyl group (3α-alcohols and 2α,3α-diols) exhibit the signal of the 5α-proton at δ 3.09-3.18. An additional low-field shift is due to 1,3-diaxial interaction between the 3α-H and the 5α-H. The proton at C-5 in the isomeric lactones (B-homo-6-oxa-7-ketones) appears as a double doublet (*J* 11-12, 4-5 Hz) at δ 4.20-4.62.

Signals of the 7-H in polyhydroxy lactones are usually unresolved and appear as two-proton multiplets at δ 4.00-4.13 for natural BS and at δ 2.40-2.60 for their regioisomers.

The region of δ 3.32-4.17 contains signals of the methine protons from carbon atoms connected with hydroxyl groups. In ring A the 2,3-cis-diol group is a common element in structures of brassino- and ecdysteroids. The configuration of this 2,3-cis-diol group in BS is mostly opposite to that in ecdysteroids, but because of the difference in AB ring fusions (trans for brassinosteroids and cis for ecdysteroids), the hydroxyl groups in both cases are equatorial at C-2 and axial at C-3. Hence, common rules that were established for spectral data of ecdysteroids (Akhrem *et al.*, 1973) hold also for BS. First of all this concerns the multiplicity of the protons at C-2 and C-3. These protons couple with each other and with two protons at C-1 and C-4. It is known that the coupling constant of vicinal protons in the cyclohexane ring depends on their mutual orientation. For equatorial-equatorial and equatorial-axial interaction, this constant is 2-6 Hz; for diaxial interaction, it is 11-12 Hz.

The signals of the methine protons, belonging to the secondary carbon atoms of the $2\alpha,3\alpha$-diol group, are seldomly well resolved; usually they appear as two multiplets. Examination of the form of the signal (half-width of a line) allows its unequivocal assignment. A wide multiplet with a W/2 of 20-24 Hz is characteristic for an axial proton at C-2, whereas the equatorial proton at C-3 exhibits a narrow signal with a half-width of 8-10 Hz. A good resolution of these signals could be achieved for some compounds. The axial proton at C-2 in the spectrum of castasterone (Mori et al., 1984) exhibits a double doublet at δ 3.77. The coupling constants J 6.1 and 3.2 Hz are due to axial-equatorial interactions with the β-protons at C-1 and C-3. The large constant (J 11.2 Hz) is caused by a trans-diaxial interaction with the α-proton at C-1.

The presence of a signal of the equatorial 3-H as a double triplet can be explained by the fact that its interactions are identical with two of the three protons. Small coupling constants specify the absence of trans-diaxial interactions with other protons.

Similar rules are observed for BS with $2\beta,3\beta$-hydroxyl groups. Thus the proton at C-2 in the ^{1}H NMR spectrum of (2S,3R)-25-methyldolichosterone exhibits a double doublet with coupling constants J 5.0 and 2.2 Hz, which certify its equatorial orientation (Mori and Takeuchi, 1988). The double doublet at δ 3.63 (formed as the result of coupling with two equatorial protons and one axial proton) belongs to the axial 3α-H.

The chemical shifts of the protons in the A and B rings of BS practically do not depend on the character of the side chain. The chemical shifts of these protons are given in Table IV.

Additional information about the structure of BS can be obtained from analysis of the spectra of their derivatives; in some cases the signals of methine protons become better resolved. Acetylation of the 2α,3α-diol group results in a low-field shift of the corresponding signals (1.0-1.3 ppm). The signal of the 2-H is moved to the region δ 4.78-4.93, whereas the multiplet of the 3-H is shifted to δ 5.30-5.34.

Assignment of the methine protons of the side chain (Table V) is more difficult but it can be solved for $\Delta^{24(28)}$-BS relatively easily. In the ^{1}H NMR spectra of dolicholide and dolichosterone, the 22-H and 23-H appear as doublets with J 8 Hz (Baba et al., 1983). Comparison of the chemical shifts of these protons with the appropriate data for brassinolide and homobrassinolide (Yokota et al., 1982a; Baba et al., 1983) shows that the signal with δ 4.03 should be assigned to the proton at C-23. For protons at C-22 the difference of the chemical shifts for brassinolide and dolicholide is much less. The multiplicity of the signal is caused by interaction of the protons with each other. The proton at C-22, moreover, couples with the proton at C-20, although the coupling constant J_{20}-J_{22} usually is small (0-2 Hz), which results only in widening of the signal. Occasionally, for example for homodolicholide (Yokota et al., 1983b), the value of this constant reaches 4 Hz. An additional splitting of this signal, as result of a long-range correlation (J 0.5-2 Hz), has been described for BS also (Mori et al., 1984).

For the 24-alkyl-BS the multiplicity of the 22-H is practically identical to that of dolichosterone. The proton at C-23 exhibits a double doublet; the smaller coupling constant is caused by correlation with the proton at C-24. Sometimes (e.g., for homobrassinolide) this interaction leads just to widening of the doublet (Yokota et al., 1982a).

The olefinic protons of the $\Delta^{24(28)}$-double bond in BS resonate in the region δ 5.03-5.52. Compounds with a 24-methylene group exhibit two singlets at δ 5.03-5.08 (δ 5.08-5.15 for the 25-methyl derivatives). Unsaturated C-29 brassinosteroids contain a one-proton quartet at δ 5.46-5.52 (J 7 Hz).

TABLE IV

Typical Values of Chemical Shifts of BS (Cyclic Part)

Cyclic part	18-H	19-H	2-H	3-H	5-H	7α-H/ 7β-H
	0.65-0.72	0.91-0.93	3.70-3.73	4.01-4.03	3.09-3.13	4.05-4.13
	0.69-0.71	0.89		4.17-4.18	3.18	4.10
	0.69-0.71	0.92		3.48-3.56	2.84-2.86	4.01/4.06
	0.61-0.69	0.75-0.76	3.76-3.78	4.04-4.06	2.67-2.69	ca. 2.00
	0.61-0.69	0.73-0.76		4.12-4.17	2.72-2.76	2.01-2.03/ 2.30-2.31
	0.67-0.68	0.75-0.76		3.45-3.58	2.22	-/2.33
	0.68	0.81	3.60	3.39	2.32	
	0.70	1.04	4.25		2.64	2.39 2.00
	0.61-0.68	0.79-0.81	3.76-3.77	3.95-3.96		
	0.69-0.72	0.96-0.99				

(*continues*)

TABLE IV (*continued*)

Cyclic part	18-H	19-H	2-H	3-H	5-H	7α-H/ 7β-H
	0.66-0.68	0.80-0.81	3.16	3.24		
	0.62-0.68	0.94-0.98	4.02-4.04	3.63	2.21	-/2.31
	0.70	0.89		4.21	4.62	2.49
	0.70	0.92		3.45-3.58	4.21-4.26	2.40-2.43/2.54

A characteristic feature of the NMR spectra of the $\Delta^{24(28)}$-BS is the presence of a C-25 proton signal at δ 2.26-2.27 (for the C-28 series) or δ 2.74-2.76 (for the C-29 series). The chemical shift of this proton was used to prove the Z geometry of the $\Delta^{24(28)}$-double bond in homodolichosterone (Sakakibara and Mori, 1983c). The authors compared the data for this C-25 proton in homodolichosterone (δ 2.27) with those in fucosterol (δ 2.2, for the *E* isomer) and isofucosterol (δ 2.88, for the *Z* isomer). NOE difference measurements served as an additional argument in favor of the *Z* configuration. Irradiation of the C-28 proton at δ 5.52 led to a 5% increase for the signals of the C-22 and C-23 protons.

NMR spectroscopy is a valuable tool for studying the preferential side chain conformations of BS in solution. This is especially important for investigations of hormone-receptor interactions. It can provide additional information and serve as a basis for molecular modeling methods (McMorris *et al.*, 1994; Brosa

TABLE V

Typical Values of Chemical Shifts of BS (Side Chain)

Side chain	21-H	22-H	23-H	25-H	26/27-H	28-H	29-H
(OH, OH)	0.90-0.92	3.52-3.58	3.69-3.73	1.63-1.65	0.93-0.95 0.98-1.03	0.85	
(OH, OH)	0.90-0.99	3.67-3.71	3.35-3.42	1.89-1.90	0.86-0.88 0.92-0.93	0.84-0.85	
(OH, OH)	0.95-0.96	3.62-3.63	4.03	2.26	1.08-1.09 1.10-1.11	5.04-5.05 5.07-5.08	
(OH, OH)	0.95-1.01	3.58-3.60	3.68-3.72		0.83-0.84 0.90-0.91		
(OH, OH)	0.92-0.93	3.67-3.69	3.92-3.96	2.74-2.76	1.06 1.14	5.46-5.52	1.71-1.72
(OH, OH)	0.85-0.92	3.77-3.82	3.47-3.48		0.94	0.91 1.00	
(OH, OH)	0.96	3.71-3.82	4.05-4.06	1.11	1.11	5.08-5.09 5.15	
(OH, OH)	1.01	3.56	3.70	1.69	0.87 0.96	0.89	
(OH, OH)	1.02	3.60	3.58	2.09	0.88 0.95	1.36/1.09	0.96

et al., 1996a). Analysis of the vicinal coupling constants using the Karplus relationship allows one to draw some conclusions. Cerny and Budesinsky (1990) suggested a gauche arrangement of the hydrogen atoms around the C-22 - C-23 bond for 24R-BS based on the vicinal coupling constant J_{22-23} = 5-9 Hz. More precise and extensive studies of brassinolide and epibrassinolide have been performed (Stoldt *et al.*, 1997; Porzel *et al.*, 1997). All the vicinal coupling constants of the side chain were determined and the corresponding dihedral angles were calculated. In addition, quantitative 2D NOE measurements and molecular dynamics simulations were carried out. The obtained results showed that brassinolide was able to maintain a preferential conformation. The conformation of epibrassinolide around the C-22 - C-23 bond was suggested to be more flexible, which could explain its weaker biological activity.

B. MASS SPECTROMETRY

Mass spectrometry of organic compounds is one of the important methods for the establishment of their structure. It has two advantages over other analytical techniques - high sensitivity and speed of analysis. The special importance of this method for natural products such as BS is evident since they are accessible for researchers in very minute amounts. The mass spectra of BS give information about molecular weight and molecular formula and make it possible to investigate peculiarities of the structure of the cyclic part and of the side chain of a molecule. Mass spectra are most commonly obtained using electron impact ionization at 70 eV (70 eV EI).

A characteristic feature of the mass spectra of BS is the absence or the very low intensity of the molecular ion peak. Along with ions corresponding to cleavage of carbon - carbon of bond $[M-X]^+$, appropriate ions caused by one-, two-, and three-step dehydrations $[M-X-nH_2O]^+$ are formed. Usually in mass spectra of steroids there are also $[M-X-CH_3]^+$ peaks, formed as result of the loss of one of the angular methyl groups.

An example of an EI mass spectrum of a BS is shown in Fig. 16. The spectrum of epibrassinolide shows a hardly distinguishable molecular ion peak and also the appropriate dehydration peaks. The peaks appropriate to cleavage of the C-24 - C-25 bond are also of a low intensity. The most intense peaks in the spectrum of tetrahydroxy lactones are the peaks corresponding to ions formed by

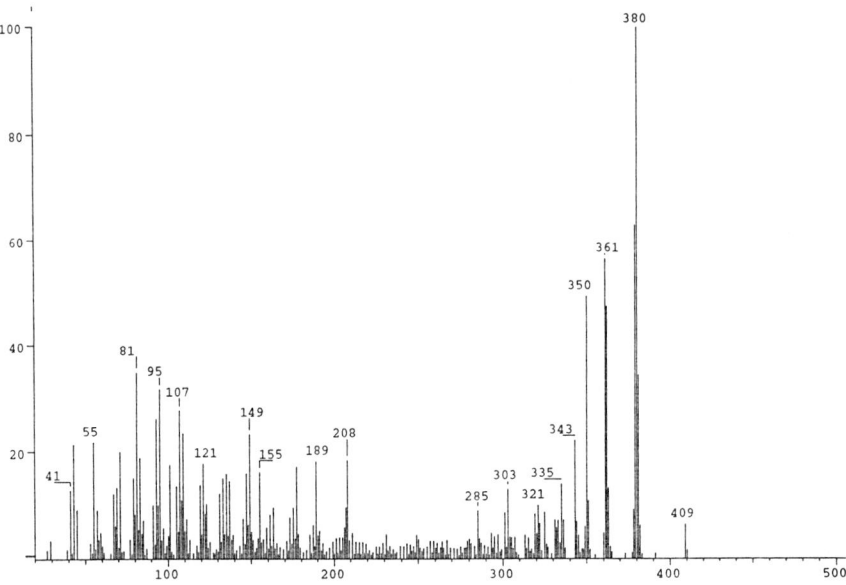

Fig. 16. Mass spectrum of epibrassinolide (EI, 70 eV).

C-22 - C-23 bond cleavage with or without H-transfer and to ions formed by subsequent loss of one or more water molecules.

In Table VI data of fragmentation patterns of the major natural BS are given.

The disadvantage of 70 eV EI mass spectra is the absence or the extremely low intensity of the molecular ion peak. Other ionization methods such as chemical ionization (CI), field desorption (FD), or fast atom bombardment (FAB) can be applied to overcome this problem.

Most characteristic in CI mass spectra are the molecular ion peaks $[M+H]^+$. Dehydration peaks corresponding to the loss of one or two molecules of water are present also. Thus brassinolide exhibits peaks at m/z 481 ($[M+H]^+$, 100%) as well as dehydration peaks at m/z 463 and 445 (Takatsuto et al., 1984b).

In field desorption mass spectra molecular ion peaks $[M+1]^+$, dehydration peaks $[M+1-18]^+$, and peaks corresponding to cleavage of carbon-carbon bonds of the side chain can be distinguished (Table VII). Although the method cannot compete with CI-MS in sensitivity (Caballero et al., 1996, 1997), it allows one to determine the molecular weight of BS without prior derivatization.

TABLE VI
EI-MS Fragment Ions of Natural BS

BS	Characteristic peaks	Reference
brassinolide	480 [M]$^{+\cdot}$, 465, 462, 447, 409, 380 (C-22 - C-23 fission accompanied by H-transfer, 100%), 379, 361, 350, 343, 331, 325, 322, 313, 307, 303, 285, 177, 173, 155, 131, 101, 71, 43	Takatsuto et al., 1984b
homo-brassinolide	476 [M-18]$^{+\cdot}$, 461, 409, 380, 379, 361, 350, 177, 145, 115, 85	Takatsuto and Ikekawa, 1982
homo-castasterone	460 [M-18]$^{+\cdot}$, 393, 364, 345, 327, 287, 263, 245, 175, 173, 145, 115, 85	Takatsuto and Ikekawa, 1982
dolicholide	379 ([M-99]$^+$, C-22 - C-23 fission), 361, 343, 331, 325, 321, 303, 285, 100 ([99+H]$^{+\cdot}$, 100%), 85, 43	Takatsuto and Ikekawa, 1983b
dolichosterone	363 ([M-99]$^+$, C-22 - C-23 fission), 345, 333, 315, 305, 287, 269, 100 (base, [99+H]$^{+\cdot}$), 85	Takatsuto and Ikekawa, 1983b
castasterone	446 [M-18]$^{+\cdot}$, 394, 393 ([M-71]$^+$, C-22 - C-23 fission), 364 ([M-101+H]$^{+\cdot}$, C-22 - C-23 fission accompanied by H-transfer, 100 %), 363, 345, 327, 287, 263, 245, 175, 173, 155, 147, 107, 101, 95, 43	Takatsuto et al., 1984b
norbrassinolide	443 [M-18-15]$^+$, 430 [M-2x18]$^{+\cdot}$, 415, 394, 380 ([M-87+H]$^{+\cdot}$, C-22 - C-23 fission accompanied by H-transfer), 362 (100%), 349, 344, 321 (C-17 - C-20 fission), 303, 285, 117, 87	Takatsuto et al., 1984b
epicastasterone	446 [M-18]$^{+\cdot}$, 394, 393, 364, 345, 327, 287, 263, 245, 175, 173, 155, 147, 107, 101, 95, 43	Takatsuto and Ikekawa, 1984a

TABLE VII

FD-MS Fragment Ions of Natural BS

BS	Base peaks	Reference
brassinolide	483, 482, 481 [M+1]$^+$, 463, 380, 379, 349, 131, 101	Aburatani et al., 1985b
homo-castasterone	479 [M+1]$^+$, 461 [M+1-18]$^+$, 393 ([M-85]$^+$, C-23 - C-24 fission), 363 ([M-115]$^+$, C-22 - C-23 fission), 333 ([M-145]$^+$, C-20 - C-22 fission), 145, 115	Takatsuto and Ikekawa, 1982
homo-brassinolide	495 [M+1]$^+$, 477 [M+1-18]$^+$, 379 ([M-115]$^+$, C-22 - C-23 fission), 349 ([M-145]$^+$, C-20 - C-22 fission], 145, 115	Takatsuto and Ikekawa, 1982

The use of negative ion mass spectroscopy, which gives characteristic fragments of ions [M-H]$^-$ and [M-4H]$^-$, has also been described for BS (Schmidt et al., 1986a,b).

Because of the very low BS content in plants, a special microanalytical technique had to be developed for effective identification. Derivatization of BS with methaneboronic acid (Takatsuto et al., 1982b; Ikekawa et al., 1984; Takatsuto, 1994a; Takatsuto and Ikekawa, 1986a) to give volatile and thermally stable methaneboronates can be effective in this case. Methaneboronic acid is a specific reagent for BS containing one or two diol groups, and formation of methaneboronates (MB) or bismethaneboronates (BMB) facilitates considerably the isolation of BS from complex natural mixtures. In the case of 2-deoxy-BS an additional derivatization to trimethylsilanes is necessary (Takatsuto and Ikekawa, 1986a). Methaneboronate derivatives give more intense molecular ion peaks for 6-ketones than for the corresponding lactones. An example of EI mass spectrum of a BMB of epibrassinolide is shown in Fig. 17.

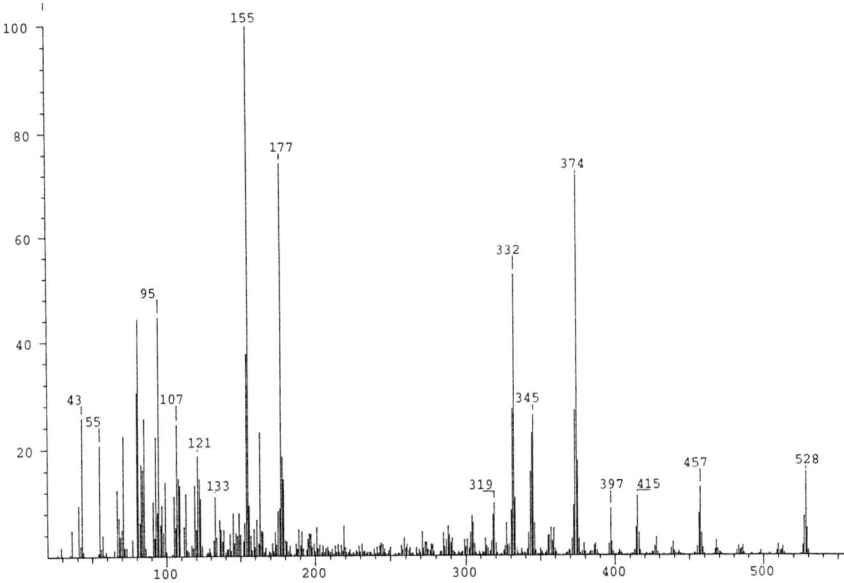

Fig. 17. Mass spectrum of epibrassinolide bismethaneboronate (EI, 70 eV).

A substituent at C-24 (alkyl or alkenyl) greatly influences the character of the fragmentation. For saturated BS the most characteristic are peaks resulting from cleavage of the C-23 - C-24, C-20 - C-22, and C-17 - C-20 bonds (Ikekawa *et al.*, 1984; Takatsuto, 1994a). The fragment ions for tetrahydroxy lactones at m/z 457 and for tetrahydroxy ketone derivatives at m/z 441 are the result of C-23 - C-24-fission (Scheme 4). These peaks are of a low intensity and in the description of a spectrum they are often omitted. The fragment ions at m/z 374 and 358 caused by cleavage of the C-20 - C-22 bond accompanied by hydrogen transfer are common for tetrahydroxy lactone and tetrahydroxy ketone bismethaneboronates.

For 6-deoxo-BS, peaks at m/z 343 correspond to this type of fragmentation. The low-molecular-mass part of the BS spectra contains intense peaks corresponding to C-20 - C-22 fission, and the molecular mass of the resulting fragments depends on the substituent at C-24.

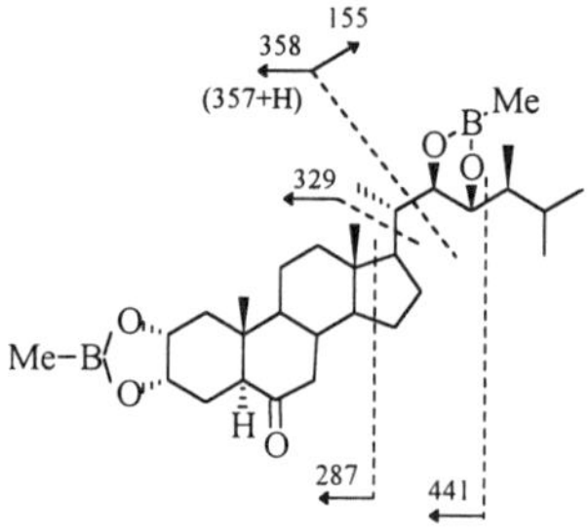

castasterone bismethaneboronate 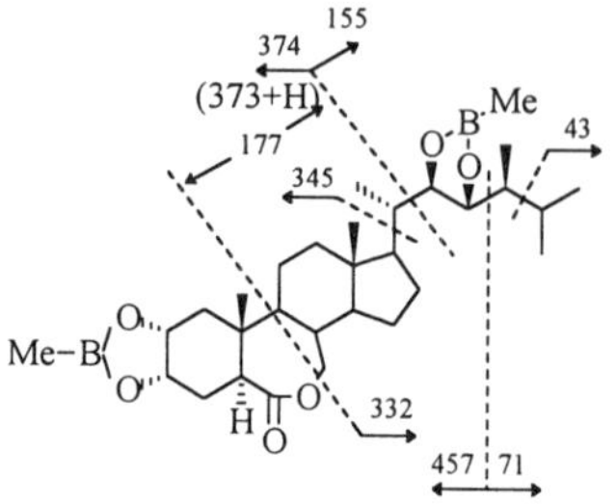   brassinolide bismethaneboronate

SCHEME 4

Fragmentation of bismethaneboronate derivatives of $\Delta^{24(28)}$-unsaturated BS is different (Ikekawa et al., 1984; Takatsuto, 1994a). Along with peaks resulting from the cleavage of the C-24 - C-25, C-20 - C-22, and C-17 - C-20 bonds, an additional set of fragment ions, caused by fission of the cyclic moiety of the side chain, is observed (Scheme 5). In the case of unsaturated BS, cleavage of the C-17 - C-20 bond is accompanied by a two-hydrogen transfer. For saturated BS this is not characteristic; instead one-hydrogen transfer takes place after C-20 - C-22 fission.

Mass spectra of boronate derivatives of BS also contain fragment ions [M-15]$^+$ and [M-29]$^+$ resulting from the elimination of one or two methyl groups, accompanied by one-hydrogen transfer. For the 3α-TMS derivatives a peak [M-18]$^{+\cdot}$ is observed (Takatsuto and Ikekawa, 1986a), which is due to the loss of

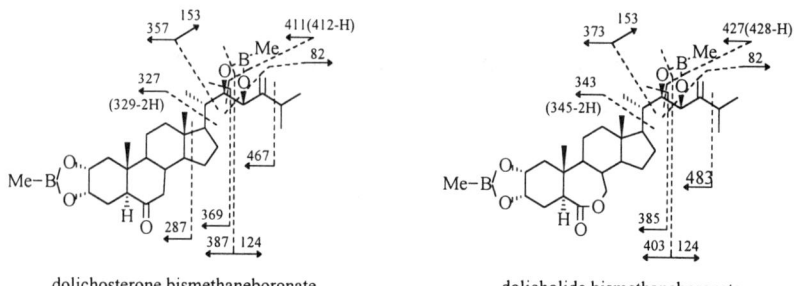

dolichosterone bismethaneboronate   dolicholide bismethaneboronate

SCHEME 5

water from the enol form of the 6-keto group together with a hydrogen of the 3α-substituent. In addition, mass spectra of TMS derivatives contain characteristic fragment ions [M-90]⁺ caused by elimination of trimethylsilanol.

Mass spectra with chemical ionization of BS boronate derivatives reveal as base peak fragment ions [M+H]⁺. Besides, peaks [M+H-60]⁺ arising from the elimination of methaneboronic acid as well as weak peaks caused by cleavage of the C-17 - C-20 and C-20 - C-22 bonds are present. A double bond in the side chain only causes minor changes in the fragmentation pattern.

C. X-RAY ANALYSIS

The role X-ray analysis had in the beginning of studies on BS is different from its role now. Only X-ray analysis could provide reliable and convincing information about the structure of brassinolide and congeners. After that, the data obtained by NMR, mass, and IR spectroscopy formed the basis for the establishment of structures of related compounds. Now the use of X-ray analysis for BS is important just for the elucidation of subtle details of their structure and for structural elucidation of synthetic intermediates.

So it was evident that the isolation of the first member of this new group of phytohormones, i.e., brassinolide, was accompanied by an X-ray analysis (Grove *et al.*, 1979). The article contained crystallographic data (see Table VIII) in addition to a computer-generated perspective drawing of brassinolide. Soon afterward, the synthesis of epibrassinolide was accomplished (Thompson *et al.*, 1979), and some of its X-ray data were reported. In the course of the (22*S*,23*S*)-homobrassinolide synthesis (Mori *et al.*, 1982), an X-ray analysis became necessary to clarify the stereochemistry of the side chain. This was carried out for the bisacetonide of (22*S*,23*S*)-homocastasterone (Fig. 18).

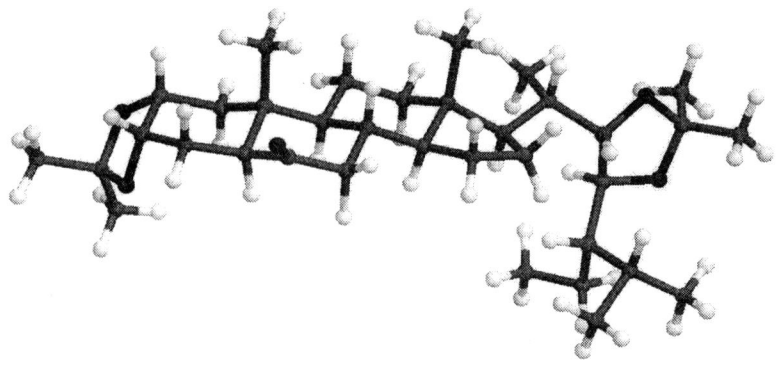

Fig. 18. A computer-generated drawing of the bisacetonide of (22S,23S)-homocastasterone.

An X-ray analysis of (22S,23S)-homobrassinolide (Kutschabsky *et al.*, 1990) allowed a better insight into the conformation of this molecule. Especially the influence of the lactone ring B that is characteristic for many BS on the conformation became clear. This ring has a chair conformation (Fig. 19), which

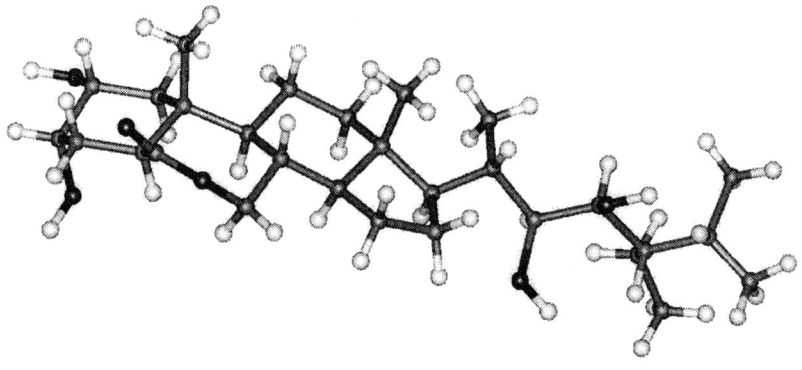

Fig. 19. A computer-generated drawing of (22S,23S)-homobrassinolide.

is partly stabilized by the C(6)-O(7) double bond. The latter is 0.133 nm shorter than the normal C=O bond. The torsion angle C(7)O(7)C(6)C(5) is only 4.5°. Both six-membered rings (A and C) exist in a chair conformation, the five-membered ring D has an envelope, or "a half-chair", conformation, and the side chain is maximally stretched.

X-ray analysis was used also for structural elucidation of BS synthetic intermediates (Table VIII). The most convincing proof of the unusual stereochemistry of hydroxylation of Δ^2-6-ketones was obtained by an X-ray analysis of 2α,3α-diacetoxycholest-4-en-6-one (Akhrem *et al.*, 1990a). It proved to be useful for assignment of the configuration at C-22 in the synthesis of fluorinated BS (Jin *et al.*, 1993) and for confirmation of the side chain structure in the synthesis of brassinolide (McMorris *et al.*, 1996). However, in the two last cases only computer-generated drawings of the molecules were given without any crystallographic data. Extensive X-ray studies have been carried out in the course of the synthesis of the side chain *via* the nitrile oxide approach (Litvinovskaya *et al.*, 1996b, 1997; Khripach *et al.*, 1998).

TABLE VIII

The Crystallographic Data of Natural BS and Related Compounds

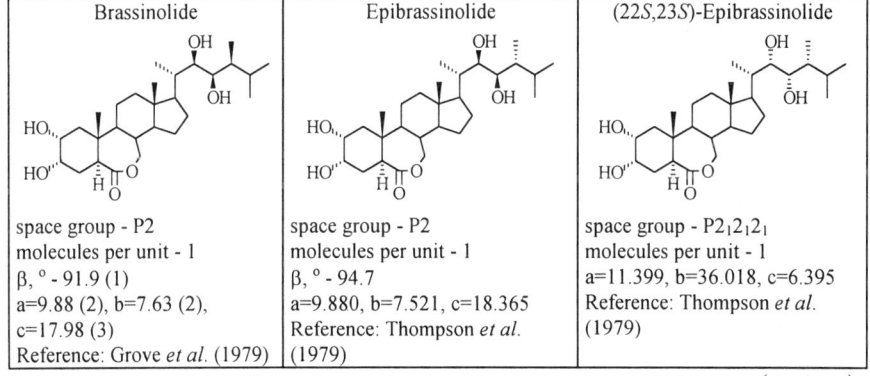

Brassinolide	Epibrassinolide	(22S,23S)-Epibrassinolide
space group - P2	space group - P2	space group - P2$_1$2$_1$2$_1$
molecules per unit - 1	molecules per unit - 1	molecules per unit - 1
β, ° - 91.9 (1)	β, ° - 94.7	a=11.399, b=36.018, c=6.395
a=9.88 (2), b=7.63 (2),	a=9.880, b=7.521, c=18.365	Reference: Thompson *et al.* (1979)
c=17.98 (3)	Reference: Thompson *et al.* (1979)	
Reference: Grove *et al.* (1979)		

(*continues*)

TABLE VIII (continued)

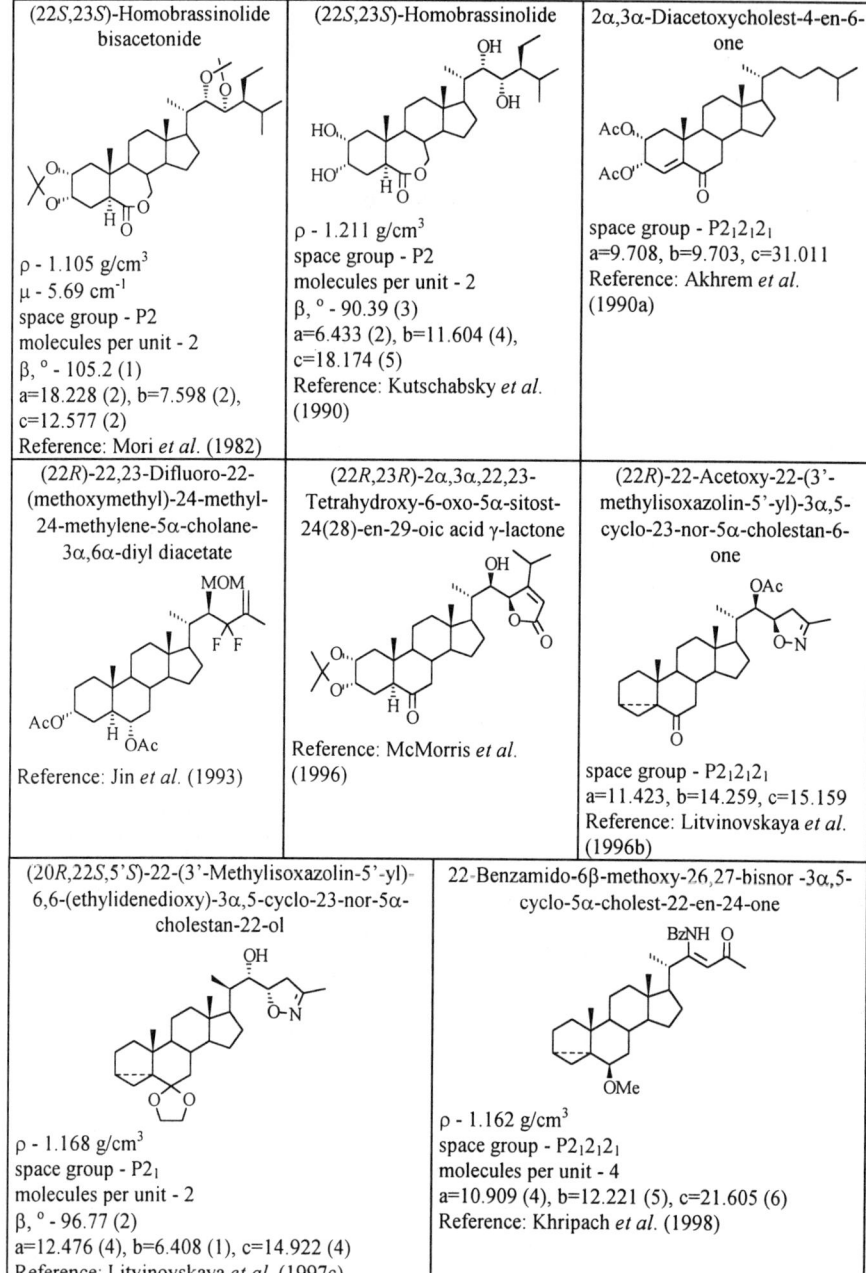

(22S,23S)-Homobrassinolide bisacetonide	(22S,23S)-Homobrassinolide	2α,3α-Diacetoxycholest-4-en-6-one
ρ - 1.105 g/cm^3 μ - 5.69 cm^{-1} space group - P2 molecules per unit - 2 β, ° - 105.2 (1) a=18.228 (2), b=7.598 (2), c=12.577 (2) Reference: Mori et al. (1982)	ρ - 1.211 g/cm^3 space group - P2 molecules per unit - 2 β, ° - 90.39 (3) a=6.433 (2), b=11.604 (4), c=18.174 (5) Reference: Kutschabsky et al. (1990)	space group - P2$_1$2$_1$2$_1$ a=9.708, b=9.703, c=31.011 Reference: Akhrem et al. (1990a)
(22R)-22,23-Difluoro-22-(methoxymethyl)-24-methyl-24-methylene-5α-cholane-3α,6α-diyl diacetate	(22R,23R)-2α,3α,22,23-Tetrahydroxy-6-oxo-5α-sitost-24(28)-en-29-oic acid γ-lactone	(22R)-22-Acetoxy-22-(3'-methylisoxazolin-5'-yl)-3α,5-cyclo-23-nor-5α-cholestan-6-one
Reference: Jin et al. (1993)	Reference: McMorris et al. (1996)	space group - P2$_1$2$_1$2$_1$ a=11.423, b=14.259, c=15.159 Reference: Litvinovskaya et al. (1996b)
(20R,22S,5'S)-22-(3'-Methylisoxazolin-5'-yl)-6,6-(ethylidenedioxy)-3α,5-cyclo-23-nor-5α-cholestan-22-ol	22-Benzamido-6β-methoxy-26,27-bisnor -3α,5-cyclo-5α-cholest-22-en-24-one	
ρ - 1.168 g/cm^3 space group - P2$_1$ molecules per unit - 2 β, ° - 96.77 (2) a=12.476 (4), b=6.408 (1), c=14.922 (4) Reference: Litvinovskaya et al. (1997c)	ρ - 1.162 g/cm^3 space group - P2$_1$2$_1$2$_1$ molecules per unit - 4 a=10.909 (4), b=12.221 (5), c=21.605 (6) Reference: Khripach et al. (1998)	

D. OTHER METHODS

CD and IR spectroscopy are not so important nowadays for structural studies of BS compared with the methods described above. At the first stage of BS investigations CD spectroscopy was used to confirm the presence of trans-fused AB rings in the molecule. Castasterone revealed a negative maximum (Θ=-5475) at 292 nm (Yokota et al., 1982a), and a similar spectrum ($[\Theta]_{294nm}$=-4660) was obtained for typhasterol (Yokota et al., 1983a).

Nonambiguous configurational assignment of the 22,23-diol side chain could be done by examination of CD spectra of the corresponding bisnaphthoates in acetonitrile (Kawamura et al., 1997a,b). The bisnaphthoates of the 22R,23R series revealed negative couplets with large amplitudes (A values) exceeding -400. The derivatives of the 22S,23S series demonstrated positive CD curves with A values around +80. The 22R,23R and 22S,23S derivatives of natural BS (epicastasterone and epibrassinolide) were also successfully differentiated. It was interesting that the stereochemistry of a substituent at C-24 did not affect the CD signs. A 50 to 100-fold enhancement of the sensitivity was achieved by employing fluorescence-detected dichroic (FDCD) spectroscopy.

The *in situ* obtained complexes of $[Mo_2(OAc)_4]$ with vicinal diols were shown to be suitable for determination of their absolute configuration by CD spectroscopy (Frelek et al., 1997). The Cotton effect at around 300 nm was dependent on the torsion angle in the O-C-C-O moiety of the complexes.

Application of CD spectroscopy for elucidation of the stereochemistry of the asymmetric centers at C-20 and C-22 of hydroxy isoxazolines, useful intermediates for preparation of BS and analogs, was investigated by Garbuz et al. (1994).

Infrared analysis is suitable to reveal the presence of functional groups such as hydroxyl (3100-3600 cm^{-1}), ketone (1710-1715 cm^{-1}), or lactone (1730-1740 cm^{-1}). IR spectra of some natural BS are given in the Appendix.

CHAPTER V

BIOSYNTHESIS AND METABOLISM

A. INTRODUCTION

Proposals for the biosynthesis of BS have been reported since 1981, when Wada and Marumo mentioned possible pathways to $2\alpha,3\alpha$-diols. A second biosynthetic route was suggested by Mandava (1988), who proposed a route to brassinolide starting from plant sterols. The first brassinolide biosynthesis, partly supported by experimental evidence, was reported by Yokota *et al.* (1991). Since then, considerable progress has been achieved in this field, first and foremost by the efforts of Japanese scientists (Abe *et al.*, 1994; Choi, Y.-H. *et al.*, 1993, 1996, 1997; Fujioka and Sakurai, 1995, 1997a,b; Sakurai and Fujioka, 1993, 1997a,b; Sakurai *et al.*, 1991, 1996; Suzuki *et al.*, 1993a, 1994a,c, 1995a,b;

Yokota, 1997; Yokota *et al.*, 1990a). As a result, many aspects of the biosynthesis of brassinolide have been clarified.

Intensive investigations were carried out on crown gall cells generated from *Catharanthus roseus* by transformation with various strains of *Agrobacterium tumefaciens* carrying different Ti-plasmids (Sakurai *et al.*, 1991). It should be pointed out that the production of BS in these cells was not a direct result of these transformations. BS biosynthesis takes places in normal cells as well, but less intensively. Cultured cells of *Catharanthus roseus* proved to be especially useful for biosynthetic studies of BS, because brassinolide and castasterone are produced in amounts comparable with those observed for pollen or immature seeds. This study led to the isolation of a number of new BS and ultimately resulted in the elucidation of the biosynthesis of brassinolide in this plant.

The major sterol components in *Catharanthus roseus* are campesterol and/or 22-dihydrobrassicasterol; together they make up about 50% of the total sterol fraction (Suzuki *et al.*, 1995b). It is natural to assume that campesterol will be the starting material for the biosynthesis of brassinolide in view of the similarity of their carbon skeletons.

However, experiments with *Phaseolus vulgaris* showed (Kim, S.-K. *et al.*, 1988) that the major sterols of this plant (up to 88%) were those carrying a 24-ethyl or a 24-ethylidene group (sitosterol, stigmasterol, isofucosterol). The corresponding 24-ethyl-BS were also found in *Phaseolus vulgaris*, but the content of 24-methyl-BS was much higher. That means that the oxidation reactions in *Phaseolus vulgaris* are much more selective for 24-methyl and 24-methylenesterols compared with those for the 24-ethyl and 24-ethylidene derivatives.

The existence of similar BS with a full set of functional groups but with a different carbon skeleton in the side chain (epibrassinolide, epicastasterone, homobrassinolide, (24*S*)-ethylbrassinone, norbrassinolide, brassinone) may be explained by the co-occurrence of campesterol as starting material for the biosynthesis of brassinolide together with other sterols such as brassicasterol, β-sitosterol, and 22-dehydrocholesterol. The formation of BS from these sterols apparently is possible and may be explained by the low specificity of the enzyme systems that are involved in these biotransformations.

B. THE EARLY C6-OXIDATION PATHWAY

Feeding experiments on cultured cells of *Catharanthus roseus* with [^{13}C]- and/or [^{14}C]methylcholesterol followed by GC-mass spectrometric analysis of the biosynthetic products have led to the identification of 24-methylcholestanol, 6α-hydroxy-24-methylcholestanol, and 6-oxo-24-methylcholestanol (Suzuki *et al.*, 1995b). Obviously these compounds are intermediates in the first stages of the biosynthetic route (Scheme 6) from campesterol to brassinolide, which is referred to as the "early C6-oxidation pathway" (Sakurai and Fujioka, 1997b; Yokota, 1997).

The identification of cathasterone (Fujioka *et al.*, 1995a) showed that the pathway to the 22R,23R-glycol system needed at least two steps. The occurrence of two other possible intermediates, e.g., (23S)-hydroxy-6-oxocampestanol and Δ^{22}-6-oxocampestanol, was not detected. It was demonstrated that [26,28-^{2}H$_6$]cathasterone was transformed into [26,28-^{2}H$_6$]teasterone and [26,28-^{2}H$_6$]typhasterol (Fujioka *et al.*, 1995a). 3-Dehydroteasterone was found to be an intermediate in the reversible conversion between teasterone and typhasterol (Suzuki *et al.*, 1994c); the equilibrium of this process is shifted to typhasterol (Suzuki *et al.*, 1994a). This is rather unusual because a transformation of the more stable equatorial 3β-alcohol in teasterone has to take place into the less stable axial 3α-alcohol in typhasterol.

Feeding experiments with labeled teasterone and typhasterol confirmed that both compounds were involved in the formation of castasterone and brassinolide (Suzuki *et al.*, 1994a). The conversion of castasterone into brassinolide has been

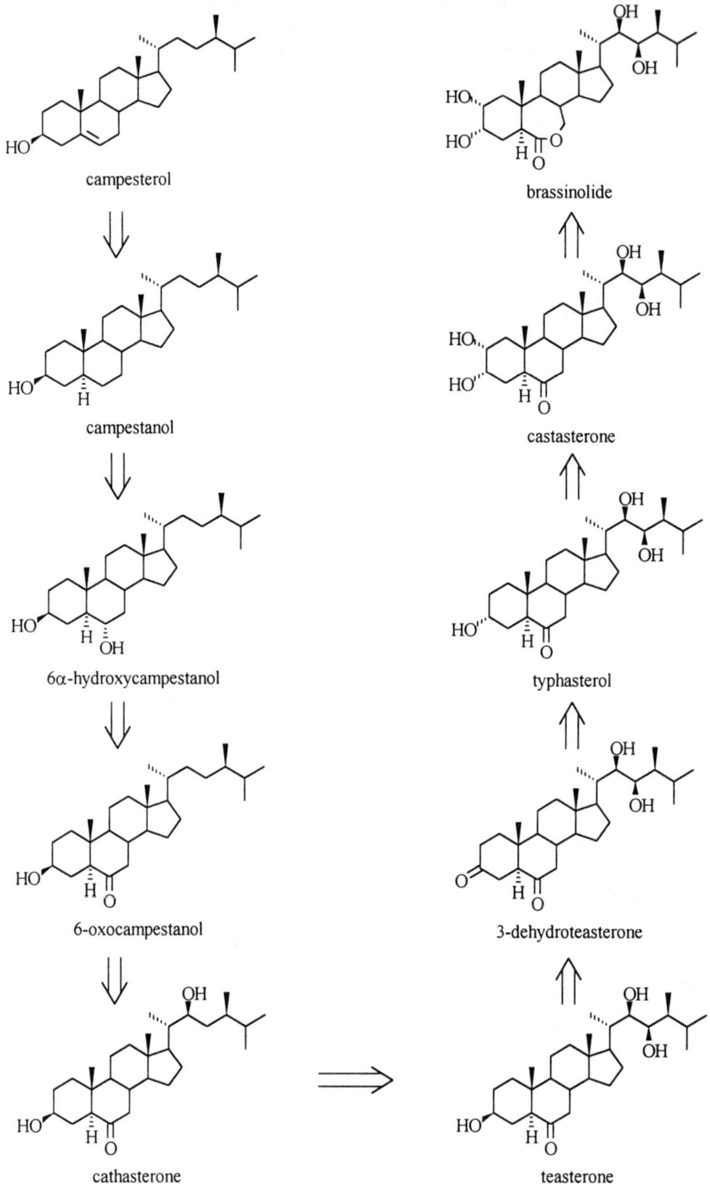

SCHEME 6

proven for cultured cells of *Catharanthus roseus* (Yokota *et al.*, 1990a, 1991). However, it does not take place in tissues of mung bean or rice (Yokota *et al.*, 1991). In experiments with rice (Abe, 1991), the applied 6-ketones affected the lamina joints just as fast as the corresponding lactones, without an observable lag phase. Perhaps the 6-ketones can act as hormones themselves in these cases. Studies on lily mature pollen (Abe *et al.*, 1996) revealed that both conjugated (laurate and myristate) and unconjugated forms of teasterone were present in that material; no other BS conjugates were detected. The experiments showed that teasterone esterified with myristic acid was metabolized to a number of free unconjugated BS along with about 60% of the unmetabolized myristate. That means that these conjugates of teasterone act as storage compounds in the BS biosynthesis and that this conjugation is a reversible reaction. Evidently only free teasterone is able to take part in the next stage of the brassinolide biosynthesis. A possible role of BS conjugates in controlling the level of free hormones was postulated (Yasuta *et al.*, 1995).

The proposed pathway of brassinolide biosynthesis is in good agreement with the relative activity of its biogenetically related congeners (Fujioka *et al.*, 1995b). Each metabolic step leads to an increase of the biological activity in the rice lamina inclination test (Fig. 20). An exception is the more or less equal activity of teasterone and typhasterol, which reflects perhaps the reversibility of this biosynthetic step. A dramatic difference (500 times) is observed between the activity of 6-oxocampestanol and cathasterone, emphasizing the importance of the hydroxylation of the side chain for the biological activity of these compounds.

Formation of 3-epicastasterone from castasterone was observed in seedlings of *Catharanthus roseus*, this in contrast to observations in cultured cells (Suzuki *et al.*, 1995a). This process may play a role in the inactivation of castasterone in plants. Similar studies on the biosynthesis of brassinolide have been performed in seedlings of *Nicotiana tabacum* and *Oryza sativa* (Suzuki *et al.*, 1995a).

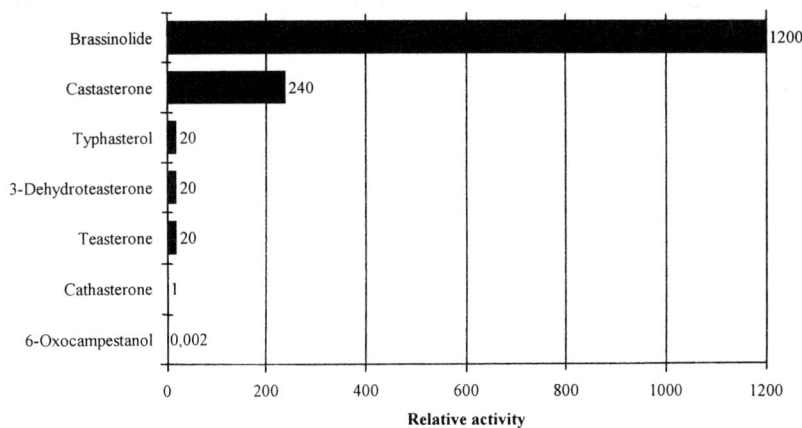

Fig. 20. Relative activity of brassinolide and its biogenetic precursors.

Despite the fact that the most detailed and large-scale investigations were carried out on cultured cells of *Catharanthus roseus* only, a similar pathway to brassinolide may be expected for other plants as well. Co-occurrence of brassinolide, castasterone, typhasterol, and teasterone has been established also for *Phaseolus vulgaris* (Kim, S.-K., 1991), *Citrus unshiu* and *Thea sinensis* (Abe, 1991), *Lilium elegans* (Suzuki et al., 1994a,b), and *Lilium longiflorum* (Abe, 1991; Abe et al., 1994).

C. THE LATE C6-OXIDATION PATHWAY

The isolation of BS such as 6-deoxocastasterone or 3-dehydro-6-deoxoteasterone (Griffiths et al., 1995a,b) confirms that different pathways for the biosynthesis of brassinolide may operate independently (Choi, Y.-H. et al., 1997; Fujioka et al., 1998) and an alternative biosynthetic route called the "late C6-oxidation pathway" (Scheme 7) was proposed also. The transformation of 6-deoxocastasterone into castasterone, which confirmed the existence of such a pathway, was demonstrated in seedlings of *Catharanthus roseus* and in cultured cells and seedlings of *Nicotiana tabacum* and *Oryza sativa* (Choi, Y.-H. et al.,

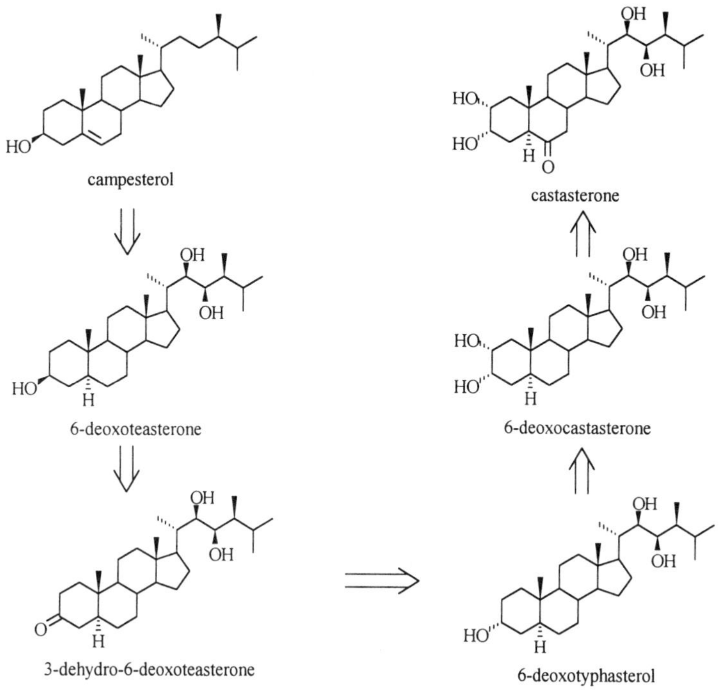

SCHEME 7

1996). It has been shown also that the conversion of 6-deoxoteasterone and 6-deoxotyphasterol into teasterone and typhasterol does not take place in cultured cells of *Catharanthus roseus* (Choi, Y.-H. *et al.*, 1997). This indicate that only 6-deoxocastasterone is subjected to the late C6 oxidation in plants.

The identification of an "early" and a "late" C6-oxidation pathway for the biosynthesis of brassinolide reflects our present knowledge of BS biosynthesis as it occurs mainly in cultured cells of *Catharanthus roseus*. The existence of other biosynthetic routes cannot be excluded (Schneider *et al.*, 1997). The transformation of castasterone into brassinolide is for instance not found in tissues of the mung bean (Yokota *et al.*, 1991). An intriguing question poses the co-occurence of 24*R*- and 24*S*-BS in several plants (Schmidt *et al.*, 1993a; Schneider *et al.*, 1997; Kauschmann *et al.*, 1997). This may be explained by an inversion of the configuration at C-24 and/or by formation of both isomers from a common $\Delta^{24(28)}$ precursor.

D. EXPERIMENTS WITH BS MUTANTS

Tremendous progress has been achieved in the understanding of many subtle details of the biosynthesis of BS and their possible mode of action in the course of investigations with *Arabidopsis thaliana* L. mutants (Chory *et al.*, 1996; Clouse, 1996a,b, 1997; Clouse and Zurek, 1991; Clouse *et al.*, 1993, 1996a,b, 1997; Fujioka *et al.*, 1997; Kauschmann *et al.*, 1996a,b, 1997; Szekeres *et al.*, 1996; von Arnim *et al.*, 1997; Xu W. *et al.*, 1995, 1996). In fact, these studies are the most convincing evidence for the hormonal role of BS in plants.

Two types of *Arabidopsis thaliana* L. mutants can be distinguished. The lesions in genes that code for enzymes that are involved in BS biosynthesis result in reduction of the level or even the elimination of the hormone. Such mutants can be rescued to the wild-type phenotype by exogenous application of BS (BS-deficient mutants). Hormone-insensitive mutants constitute another type of BS mutants. Their phenotype cannot be restored by exogenous treatment with hormones. They result from lesions in genes encoding for the receptor(s) of BS and/or the signal transduction system. Both types of mutants have similar phenotypes.

Studies on BS-deficient mutants have led to the identification of several genes encoding for the biosynthesis of brassinolide. The DET2 gene of a recessive mutant *det2* was investigated by Li *et al.* (1996b, 1997). It was shown that the DET2 gene encoded for a reductase involved in the brassinolide biosynthesis. Sequence analysis revealed a 38 to 42% sequence identity of the DET2 gene to that of mammalian 5α-reductase involved in the transformation of testosterone into dihydrotestosterone. The similarity was even higher (from 54 to 60%) taking into account the conservative substitutions. In addition, the human steroid 5α-reductase, when expressed in the *det2* mutant, was able to substitute the DET2 in the brassinolide biosynthesis (Li *et al.*, 1997). The mutant could be rescued by exogenous application of brassinolide but not by gibberellins or auxins (Li *et al.*, 1996). These experiments unambiguously showed that BS deficiency in these *det2* mutants was caused by blocking the conversion of campesterol into campestanol caused by the defect in the DET2 gene. Its orthology with mammalian 5α-reductase (Li *et al.*, 1997) demonstrates the similarity in plant and mammalian signaling systems.

Additional studies (Fujioka *et al.*, 1997) showed that the DET2 gene acted at the stage of reduction of 24*R*-methylcholest-4-en-3-one. Its concentration was three times higher, whereas the concentrations of the other measured BS were less than 10% of that for wild-type plants. The investigations showed also that at least one other enzyme must exist in *Arabidopsis*, responsible for the transformation of $\Delta^{5,6}$-steroids into the 3-oxo-$\Delta^{4,5}$ precursors of BS.

The *cpd* mutant of *Arabidopsis* was isolated by T-DNA tagging (Koncz *et al.*, 1996; Szekeres *et al.*, 1996). The CPD gene was demonstrated to encode for a cytochrome of a novel P450 family, CYP90. The studies revealed its homology with other steroid hydroxylases (24% for rat testosterone-16α-hydroxylase and 19% for human progesterone-21-hydroxylase). The *cpd* mutant could be restored to a wild type by a number of BS, including brassinolide biosynthetic precursors such as teasterone, 3-dehydroteasterone, typhasterol, and castasterone. However, BS lacking the C-23 hydroxyl group (cathasterone, 6α-hydroxycampestanol, 6 oxocampestanol, Δ^{22}-6-oxocampestanol, and (22*R*,23*R*)-epoxy-6-oxocampestanol) and other plant growth factors (auxins, gibberellins, cytokinins, abscisic acid, methyljasmonate, and salicylic acid) were not able to alter the *cpd* mutant phenotype to the wild type. The obtained results evidence that the CPD gene is involved in the biosynthesis of brassinolide at the stage of C-23 hydroxylation. It may be either the transformation of cathasterone into teasterone (early C6-oxidation pathway) and/or the conversion of 6-deoxocathasterone into 6-deoxoteasterone (late C6-oxidation pathway).

An *Arabidopsis* mutant *dwf4* was found to be defective in the 22-hydroxylation step (Choe *et al.*, 1997). The mutant plants could be rescued by application of 22-hydroxylated BS including cathasterone and 6-deoxocathasterone. The gene DWF4 responsible for the mutation was cloned. The corresponding enzyme (designated CYP90B1) was shown to be 43% identical to the enzyme CYP90A involved in the 23-hydroxylation of steroids.

Further studies have led to the identification of *cbb1*, *cbb3* (Kauschmann *et al.*, 1996a,b, 1997), and *dim* (Klahre and Chua, 1996; Klahre *et al.*, 1997) BS deficient mutants. The brassinolide deficiency in the *dim* mutant may be caused by blocking of the teasterone $\Rightarrow$ typhasterol transformation or by reduction of the biosynthesis of campesterol (Klahre *et al.*, 1997). The same biosynthetic steps may be blocked in the *cbb1* mutant.

Apart from brassinolide-deficient mutants, a number of brassinolide-insensitive mutants have been identified for *Arabidopsis thaliana* L. (Clouse *et al.*, 1996a,b, 1997; Kauschmann *et al.*, 1996a,b, 1997; Li and Chory, 1997; Li *et al.*, 1996a,b). These mutants are extremely useful in BS perception and signal transduction investigations.

BS-sensitive mutants have been found also for the garden pea *Pisum sativum* (Nomura *et al.*, 1997; Yokota *et al.*, 1996b, 1997). Among these the *lkb* mutant is considered to be the BS-deficient mutant in which the biosynthesis at the step before the formation of teasterone formation is blocked. An elevated level of isofucosterol is characteristic for the mutant. This together with a drastically reduced content of sitosterol and campesterol can be considered as an argument in favor of restriction of the biosynthesis of BS in the transformation of 24-methylenecholesterol into campesterol. Another BS-sensitive *lk* mutant is likely to suffer from a blockade of biosynthesis of campestanol in *Pisum sativum* from campesterol.

E. BIOTRANSFORMATIONS OF BS

The metabolism of castasterone and brassinolide was studied on explants of mung bean *Vigna radiata* seedlings (Yokota et al., 1991; Suzuki et al., 1993b). It was demonstrated that castasterone was converted into unknown (mainly nonglucosidic) metabolites in this plant rather than into brassinolide. Incubation of mung bean with brassinolide **1** led to formation of a new glycoside, which was characterized as 23-*O*-β-D-glycopyranosyloxybrassinolide **45** (Scheme 8). The

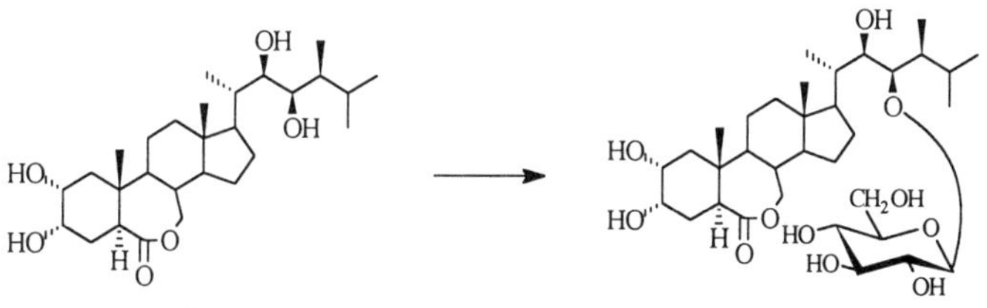

SCHEME 8

latter turned out to be rather active in the rice lamina inclination test, probably because of the liberation of free brassinolide in the test system. The isolation from seeds of *Phaseolus vulgaris* of two similar glycosides (23-*O*-β-D-glycopyranosyloxy-25-methyldolichosterone and its 2-epimer) may indicate that the described glucosidation reaction can be an important process in BS storage and/or BS deactivation in plants (Kim, S.-K., 1991).

Metabolic transformations of ^{14}C-labeled epibrassinolide were investigated in cucumber and wheat (Nishikawa et al., 1994, 1995a,b; Nishikawa and Abe, 1996). [^{14}C]Epibrassinolide **19** was found to be metabolized in seedlings of cucumber *Cucumis sativus* L. (Nishikawa et al., 1995a) into tetrahydroxy lactone **46** (Scheme 9). A number of less and more polar metabolites were detected, which were supposed to be conjugates of **46** with fatty acid(s) and/or sugar(s).

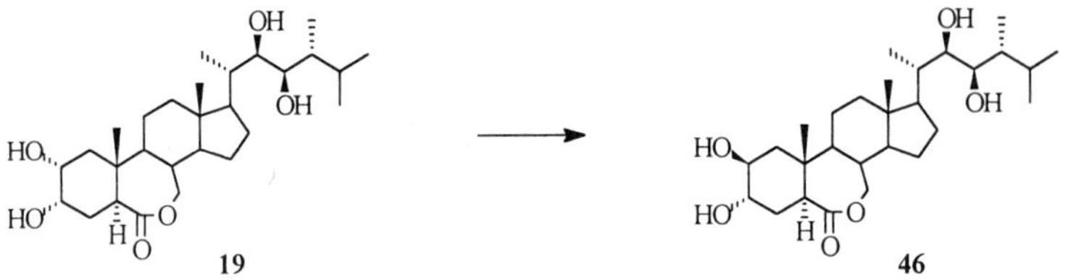

SCHEME 9

Interesting results have been obtained in the course of metabolic studies of epibrassinolide and epicastasterone in cell suspension cultures of *Licopersicon esculentum* (Schneider *et al.*, 1994, 1996; Hai *et al.*, 1995, 1996; Adam *et al.*, 1996a,b) and *Ornithopus sativus* (Kolbe *et al.*, 1994, 1995, 1996, 1997; Adam *et al.*, 1996b). Epibrassinolide, exogenously applied to cultures of *Licopersicon esculentum* (Schneider *et al.*, 1994; Hai *et al.*, 1995), was converted into two polar metabolites, 25-*O*-β-D-glycopyranosyloxy-24-epibrassinolide **49** and 26-*O*-β-D-glycopyranosyloxy-24-epibrassinolide **50** (Scheme 10). The corresponding aglycons **47** and **48** could not be detected in the cell cultures.

At the same time, incubation of both aglycons in the cell cultures led to complete glycosidation. Pentaol **48** showed moderate activity in the rice lamina

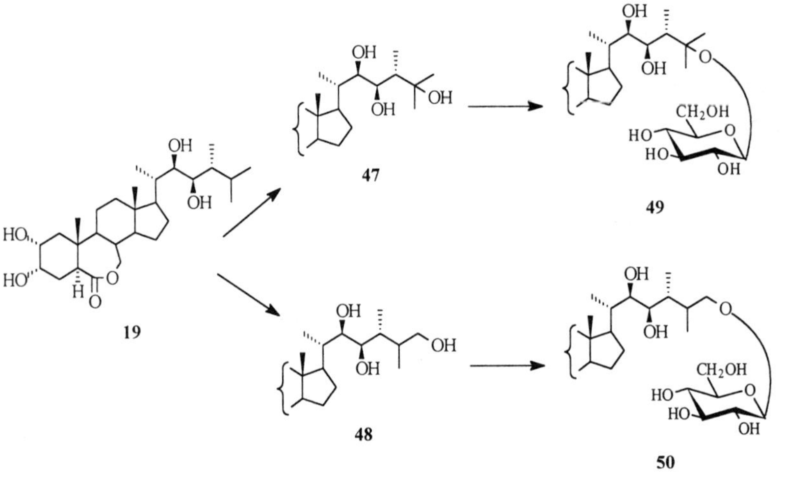

SCHEME 10

inclination test, but 25-hydroxy-24-epibrassinolide **47** revealed an extraordinarily high activity in this test (about 10 times more active than epibrassinolide). This finding was considered at first as an indication that hydroxylation at C-25 may be not a detoxification but an activation step, similar to the vitamin D group (Zhu, G.-D., and Okamura, W.H., 1995).

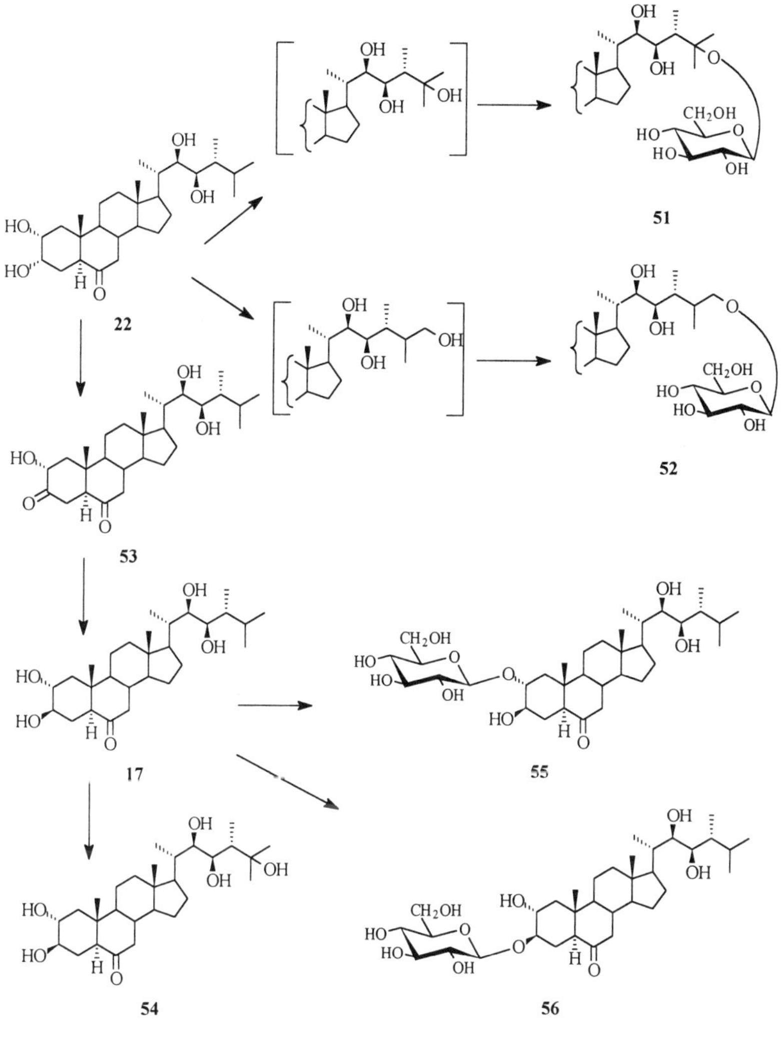

SCHEME 11

However, additional and more detailed investigations (Kauschmann et al., 1997) revealed that in fact the 25-hydroxylation was a BS-deactivation step. The activity of most 25-hydroxy-BS studied (25-hydroxybrassinolide, 25-hydroxy-24-epibrassinolide, and 25-hydroxy-24-epicastasterone) was about 100-fold lower than that of the corresponding 25-deoxy compounds.

Hydroxylation at C-25 and C-26 was shown to be catalyzed by two different enzymes (Winter et al., 1996, 1997); one of these is assumed to be cytochrome P450 dependent. Inhibitors of cytochrome the P450 such as clotrimazole and ketoconazole affected only the C-25 hydroxylation step, whereas the hydroxylation at C-26 was not influenced. It should be pointed out that both the C-25 and C-26 hydroxylases show a high specificity. Only brassinolide and epibrassinolide could be used as the substrates and no hydroxylation was observed in BS analogs such as (22S,23S)-homobrassinolide, ecdysone, 20-hydroxyecdysone, and even epicastasterone.

Similar results with respect to the hydroxylation and subsequent glycosidation of the side chain were obtained for the metabolism of epicastasterone in *Lycopersicon esculentum* (Hai et al., 1996). Both the 25- and 26-glycosides **51** and **52** were found in the cell suspension cultures (Scheme 11). However, the whole process was in this case accompanied by additional transformations in ring A. Three nonglucosidic products, **53**, **17**, and **54**, were isolated from the hydrolyzed part of the extract. One of them, tetraol **17**, is known to be a natural BS and is found in *Phaseolus vulgaris* (Kim, S.-K., 1991). Isolation of the diketone **53** can be considered as an indication for the mechanism of epimerization at C-3. Apart from hydroxylation at C-25, the tetraol **17** was found to be glycosidated by *Lycopersicon esculentum* on the 2α- and 3β-hydroxyl groups to give the glycosides **55** and **56**.

Application of epibrassinolide to cell suspension cultures of *Ornithopus sativus* followed by analysis of the extracts led to the isolation of another set of metabolites (Kolbe et al., 1994, 1995, 1996; Adam et al., 1996b). In the first

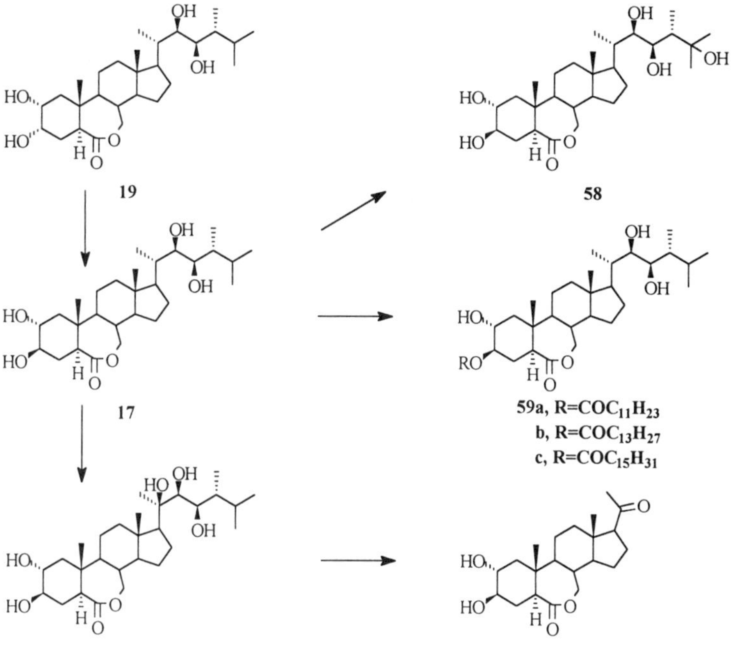

SCHEME 12

stage epibrassinolide **19** was transformed into the tetrahydroxy lactone **17** (Scheme 12), which represents a branching point in the metabolic process. Hydroxylation at C-25 gave the pentaol **58** (minor reaction). Oxidation at C-20 furnished the pentaol **57**, which rapidly underwent side chain cleavage between C-20 and C-22 to give the pregnane-type metabolite **60**. Analysis of the lipophilic fraction of the cell extract showed the presence of three fatty acid conjugates **59a-c** (laurate, myristate, and palmitate). Their role was assumed to be one of compartmentation of BS in membrane structures (Adam et al., 1996b). Fatty acid conjugates are relatively new members of the BS family. To date, only teasterone (Asakawa et al., 1994, 1996) and typhasterol (Yasuta et al., 1995) fatty acid esters have been detected in plants as native conjugates.

Epicastasterone **22** was shown to be metabolized in a similar way in cell suspension cultures of *Ornithopus sativus* (Kolbe et al., 1994, 1995, 1996).

Epimerization at C-3 led to the tetraol **17** (Scheme 13) and esterification provided the same fatty acid derivatives **62a-c** as for epibrassinolide. The corresponding degradation of the side chain *via* the (20*R*)-alcohol **61** furnished the pregnane **63**. Unlike the metabolism of epibrassinolide, compound **63** underwent reduction at C-6 to give the alcohol **64**.

BS metabolites with a disaccharide moiety were found in *Licopersicon esculentum* L. (Kolbe *et al.*, 1997). Exogenously applied epiteasterone **65a** was shown to be converted into the corresponding diglucosides 3-*O*-β-D-glycopyranosyl-(1→6)-β-D-glycopyranoside **65b** and 3-*O*-β-D-glycopyranosyl-(1→4)-β-D-galactopyranoside **65c** (Scheme 14). The process was relatively fast, because about 45% of the exogenously applied epiteasterone was conjugated within 5 h.

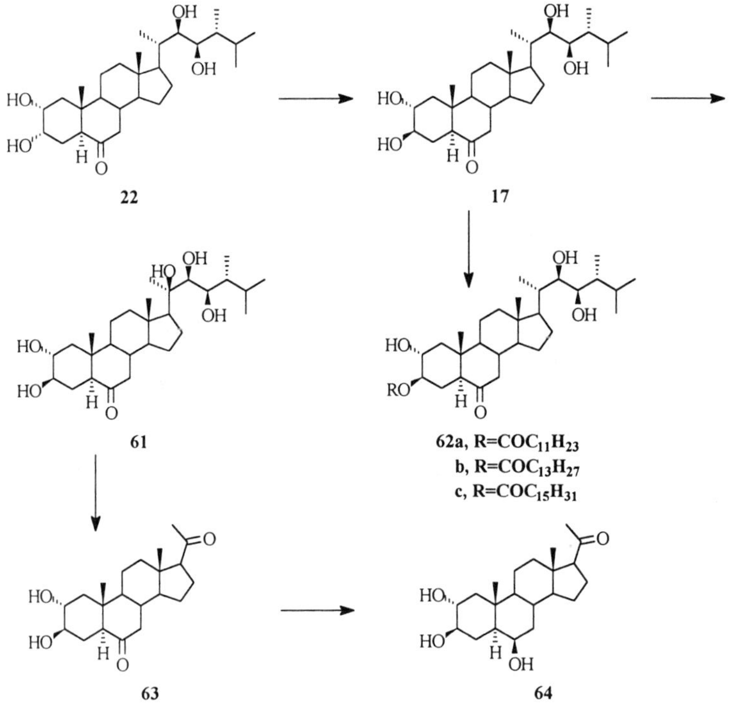

62a, R=COC$_{11}$H$_{23}$
b, R=COC$_{13}$H$_{27}$
c, R=COC$_{15}$H$_{31}$

SCHEME 13

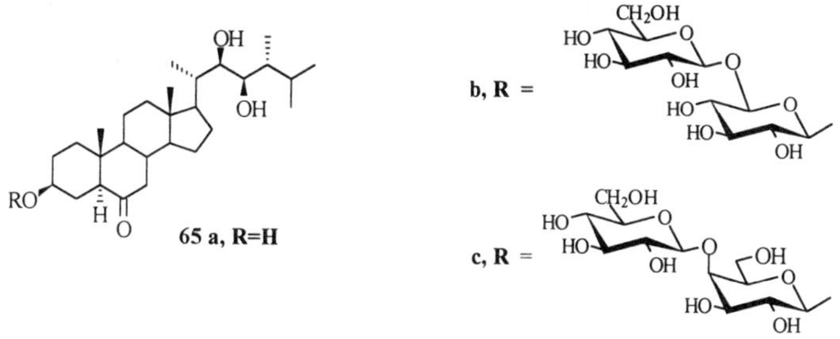

SCHEME 14

Adam *et al.* (1996b) have postulated two principal pathways of BS metabolism in plants, both of which are likely to be inactivation processes:

i) oxidation of the terminal part of the side chain followed by glucosidation;

ii) degradation of the side chain *via* hydroxylation at C-20 and subsequent oxidative cleavage of the C-20 - C-22 bond to give pregnane-like steroids. This process is accompanied by epimerization at C-3 and sometimes combined with fatty acid or glucose conjugation.

Biotransformations of BS by living systems other than plants were studied also. Although the production of BS by microorganisms has not been described until now, microbial transformations are well-known and a useful tool for the preparation of various other steroids (Akhrem and Titov, 1965). Transformation of BS analog **66** with a saturated side chain by the microorganism *Mycobacterium vaccae* was studied by Vorbrodt *et al.* (1991). Two compounds of the androstane series **67** and **68** were isolated from the extract after fermentation (Scheme 15).

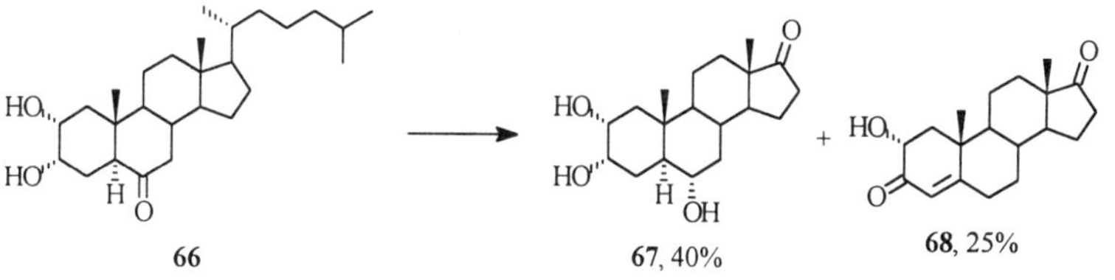

SCHEME 15

Incubation of epibrassinolide and epicastasterone with the fungus *Cunninghamella echinulata* led to oxidation of ring C to give the corresponding 12β-hydroxylated BS (Voigt *et al.*, 1993a; Adam *et al.*, 1989). Microbial hydroxylation of 24-epicastasterone by the fungus *Cochliobolus lunatus* furnished the corresponding 15β-hydroxylated derivative (Voigt *et al.*, 1993b).

CHAPTER **VI**

BASIC SYNTHETIC METHODS AND FORMAL SYNTHESES

A. INTRODUCTION

There are two conceptually different approaches to the synthesis of steroids, total synthesis and partial synthesis. The first approach implies the synthesis of the whole steroid molecule from simple precursors. The second approach relies on the transformation of starting materials that already possess the tetracyclic fragment characteristic for steroids.

The construction of the tetracyclic steroid skeleton is one of the more intriguing problems in synthetic organic chemistry (Akhrem and Titov, 1967). Despite the great number of methods developed, this approach has proven to be feasible only for the preparation of relatively simple steroids. However, it is essential for the search for new steroidal drugs with modified skeletons and also it is a good opportunity for the elaboration of new synthetic methods. Only one attempt to the total synthesis of BS has been reported to date (Berthon *et al.*, 1994; Diziere *et al.*, 1994; Zoller *et al.*, 1997).

Practically all syntheses of BS are partial syntheses (Adam, 1987; Lakhvich *et al.*, 1991; Khripach *et al.*, 1993a; Back, 1995), which imply the introduction of the necessary functional groups into the tetracyclic fragment and the construction of the side chain. The implementation of the first task is based mainly on experience accumulated during the syntheses of other steroids, first of all the ecdysteroids. For the construction of the characteristic side chain of BS, new methods had to be developed.

The first problem that has to be solved in partial syntheses of BS is the choice of an appropriate starting material, preferably one with the same carbon skeleton. It is important to have functional groups in the starting compound which can be transformed into those that are characteristic for BS. For large-scale preparations the availability of the starting material is a determining factor.

The most suitable starting compounds for partial syntheses are sterols containing a Δ^{22}-double bond (Fig. 21), such as stigmasterol **69** and ergosterol **72**. They can be used for the preparation of the corresponding 29C- and 28C-brassinosteroids in procedures that do not need a reconstruction of the native carbon skeleton of the side chain. The presence of the Δ^{22}-double bond also

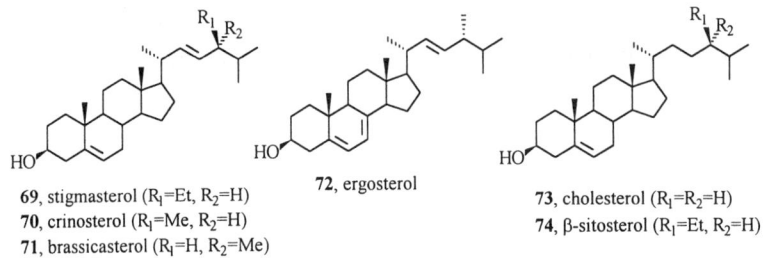

69, stigmasterol (R_1=Et, R_2=H)
70, crinosterol (R_1=Me, R_2=H)
71, brassicasterol (R_1=H, R_2=Me)

72, ergosterol

73, cholesterol (R_1=R_2=H)
74, β-sitosterol (R_1=Et, R_2=H)

Fig. 21. The most widely used sterols for preparation of natural BS and analogs.

gives a possibility to replace part of the side chain *via* the 22-aldehydes. Attractive starting materials would be 22-dehydrocampesterol (crinosterol) **70** and brassicasterol **71**; unfortunately, these compounds are not so easily available as stigmasterol or ergosterol (Matsumoto *et al.*, 1983; Konai *et al.*, 1984a,b, 1985; Khripach *et al.*, 1988; Kuriyama *et al.*, 1989). Cholesterol **73** and β-sitosterol **74** are abundantly available, but the saturated side chain in these molecules is difficult to functionalize. Employment of a mixture of sterols from soybean oil (containing sitosterol and stigmasterol) or *Brassica campestris* oil (containing sitosterol, campesterol, and brassicasterol) may be interesting from a practical point of view (Mitra *et al.*, 1984b,c; Zhang, H. *et al.*, 1988, 1989; Neeland and Towers, 1989).

Some other possibilities are offered by derivatives of cholic acids, e.g., hyodeoxycholic acid **75**, and by steroids of the androstane and pregnane series such as androstenolone **76** and pregnenolone **77** (Fig. 22), which are abundantly

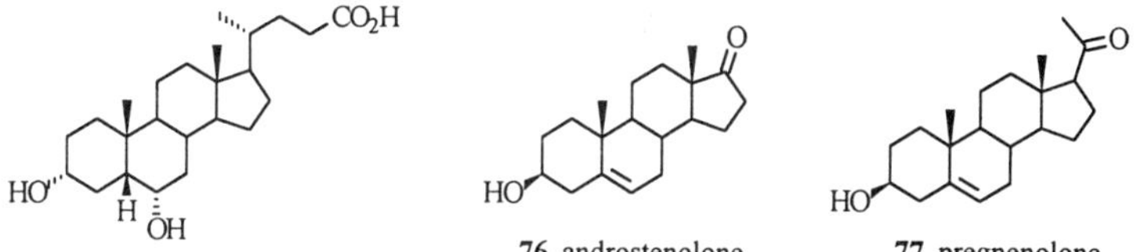

75, hyodeoxycholic acid **76**, androstenolone **77**, pregnenolone

Fig. 22. Alternative starting materials for preparation of natural BS and analogs.

available because of their use for the production of pharmaceutical steroid hormones.

Steroidal alkaloids and sapogenins, e.g., solasodine **78** (Fig. 23) and diosgenin **79**, in principle also can be used as starting compounds, but also in

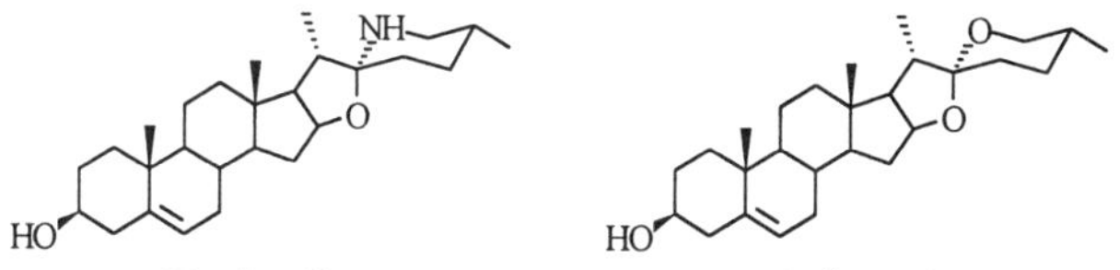

78, solasodine **79**, diosgenin

Fig. 23. Starting materials for preparation of BS analogs.

these molecules the side chain is not suitably functionalized and their use is sensible only for the preparation of the corresponding analogs.

B. SYNTHESIS OF THE CYCLIC PART

For the construction of the cyclic part of BS, steroids have been used that contain functionalized AB rings as depicted in Scheme 16. The most commonly used starting steroid is of type **80**, such as sterols, pregnenolone, androstenolone, steroidal alkoloids, and sapogenins.

Type **81** precursor is represented by ergosterol, which has been used mainly for the synthesis of epibrassinolide and some related BS. In some procedures reduction of ergosterol is performed as one of the first steps and in this way it is converted in a precursor of type **80**. Type **82** starting material is represented by derivatives of hyodeoxycholic acid.

Most syntheses of BS are directed toward the preparation of compounds with the 2α,3α-diol function in ring A and a lactone function in ring B because these BS are the most active.

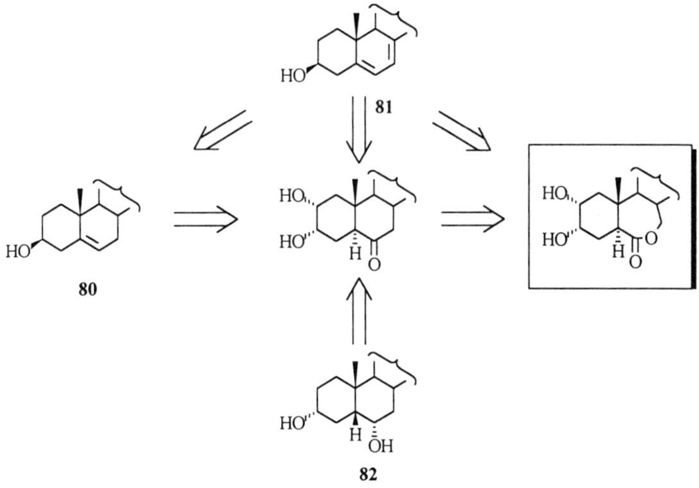

SCHEME 16

BS with a carbonyl group at C-6 are in most cases the direct precursors for the corresponding lactones, because the Baeyer-Villiger oxidation can be easily accomplished as a last step in the synthesis. Syntheses of 6-deoxobrassinosteroids often also involve the 6-keto derivatives; the carbonyl function is then removed in one of the last stages of the sequence.

1. Starting with Δ^5-Sterols

The most widely used methods for formation of the 2,3-dihydroxy-6-keto functionality in BS are known from classical steroid chemistry and have been improved for ecdysteroid syntheses. In these methods the C-6 carbonyl group is introduced first, followed by the construction of the Δ^2-double bond and its hydroxylation.

The most straightforward method for the synthesis of C-6 carbonyl compounds from Δ^5-sterols is the hydroboration of the Δ^5-double bond followed by oxidation of the intermediate alcohol **84** as shown in Scheme 17 (Fung and Siddall, 1980; Ishiguro *et al.*, 1980; Takatsuto *et al.*, 1981, 1984b; Kametani *et al.*, 1985, 1986, 1988a; Suntry, Ltd., 1981). This method is not compatible with compounds containing additional double bonds in the molecule. A more effective and universal method is based upon the specific rearrangement of 3β-tosyl or 3β-mesyl derivatives of Δ^5-steroids into alcohols **87** (Scheme 18), the so-called isosteroidal rearrangement (Steele and Mosettig, 1963; Aburatani *et al.*, 1984a,b, 1985a,b, 1986; Akhrem *et al.*, 1985c, 1989c; Anastasia *et al.*, 1983c,d; Khripach *et al.*, 1990j; Kondo and Mori, 1983; Lakhvich *et al.*, 1985a-

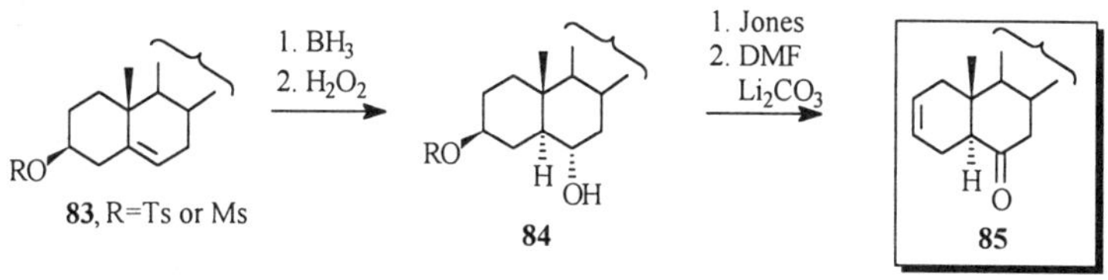

SCHEME 17

c; Mitra et al., 1984a; Mori et al., 1981, 1982, 1984; Mori, 1980a,b; Okada and Mori, 1983a,b; Sakakibara and Mori, 1982, 1983b,c, 1984; Sakakibara et al., 1982; Sumitomo Chem. Co., Ltd., 1982, 1983; Takatsuto and Ikekawa, 1982, 1984a, 1986e; Thompson et al., 1979, 1980a).

Usually this rearrangement is carried out in aqueous acetone in the presence of potassium acetate or bicarbonate and leads to a mixture of the alcohols **83** (R=H) and **87** (~15:85) as a result of the reaction of water with the intermediate carbocation **86** at C-3 or at C-6. In most cases the mixture is submitted directly to Jones oxidation to ketone **88**.

A one-step transformation of the 3β-tosylates (mesylates) into ketone **88** (Takatsuto and Ikekawa, 1984a) is possible also by heating **83** in DMSO in the presence of sodium acetate. The yield of the ketone **88** is rather low (~50%), and the Δ^4-6-ketone and the $\Delta^{3,5}$-diene were isolated as byproducts.

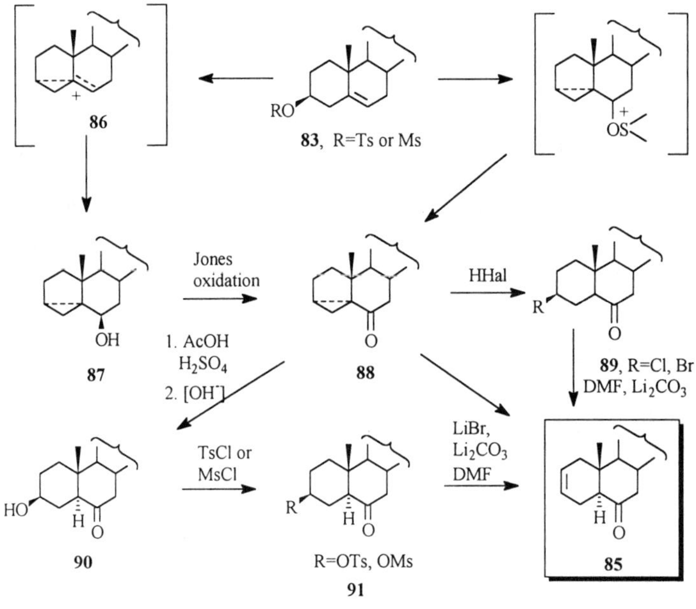

SCHEME 18

The cyclopropane in **88** gives the opportunity for an easy transformation into the corresponding 3-substituted 6-ketones **89** by opening of the ring with hydrochloric acid (Anastasia *et al.*, 1983c,d), hydrobromic acid (Akhrem *et al.*, 1984a, 1985b, 1987b; Anastasia *et al.*, 1983b; Yuya *et al.*, 1985b), or a mixture of acetic and sulfuric acids (Thompson *et al.*, 1981; Aburatani *et al.*, 1987b; Takatsuto and Ikekawa, 1986b,e). Further dehydrohalogenation of **89** provides the Δ^2-6-ketone **85**. Similar routes based on the elimination of sulfonic acids in **91** can be used as well (Thompson *et al.*, 1981; Fung and Siddall, 1980; Ishiguro *et al.*, 1980; Takatsuto *et al.*, 1981, 1984b; Kametani *et al.*, 1986, 1988a; Kametani and Honda, 1987d; Takatsuto and Shimazaki, 1992; Nowak *et al.*, 1994). Dehydration of 3β-alcohols **90** with $CuSO_4$ on SiO_2 is another possibility for synthesizing Δ^2-derivatives (Huang, L.-F. and Zhou, W.-S., 1994; Brosa *et al.*, 1996b).

A number of one-step transformations of ketone **88** to the Δ^2-6-ketone **85** have been reported. The method was investigated first for the synthesis of ecdysteroids (Barton *et al.*, 1970, 1984) and involved the heating of 3α,5-cyclo-6-oxosteroids **88** in sulfolane in the presence of TsOH (Mori, 1980b; Mori *et al.*, 1982; Takatsuto and Ikekawa, 1984a). Improvements of the method were found in treatment with pyridinium bromide (Anastasia *et al.*, 1985a), sodium (Aburatani *et al.*, 1986, 1987a), or lithium bromide (Yuya *et al.*, 1984b; Kametani and Honda, 1987c) in the presence of TsOH in DMF under reflux and with pyridinium tosylate in toluene (Hayashi *et al.*, 1986) or in DMF (Watanabe *et al.*, 1988). It has been shown that these transformations proceed *via* the Δ^3-6-ketones, which rearrange under the reaction conditions to the desired Δ^2-6-ketones (Takatsuto *et al.*, 1988a), and Δ^4-6-ketones are formed as byproducts (Mitra and Kapoor, 1985).

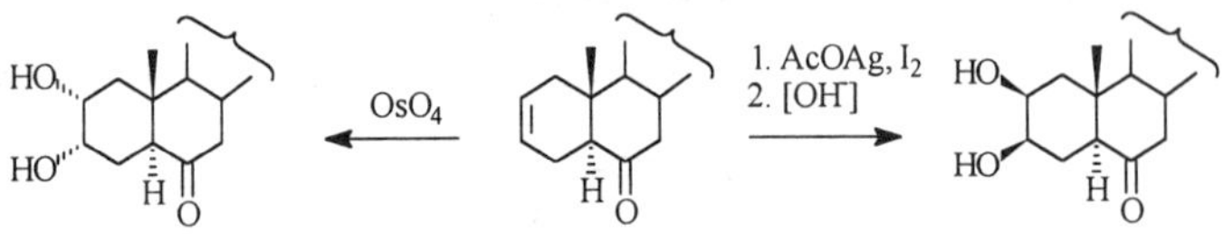

SCHEME 19

One of the most common procedures for the conversion of the Δ^2-double bond into the 2α,3α-diol functionality is oxidation with OsO$_4$ (Shoppee et al., 1957; Van Rheenen et al., 1976; Schröder, 1980). The reaction can be carried out in equimolecular and in catalytic variants; in the last case N-methylmorpholine N-oxide is used as the co-oxidant. More detailed investigations of the reaction revealed the formation of 3-5%, and in some cases up to 14% (Aburatani et al., 1987b; McMorris et al., 1996), of the isomeric 2β,3β-diols. Some BS contain a 2β,3β-diol function and for an effective introduction of these β-hydroxyl groups the Woodward hydroxylation of Δ^2-steroids can be used (Scheme 19).

There have been several attempts to elaborate alternatives for the preparation of the 2α,3α-diols that avoid the use of the expensive and toxic OsO$_4$. It has been found that the introduction of an additional Δ^4-double bond changes the stereochemistry of the Woodward hydroxylation (Akhrem et al., 1981, 1983b,c, 1988a, 1989d, 1990a), and the 2α,3α-diols are obtained as the main product (Scheme 20). It is interesting to note that the hydroxylation with OsO$_4$ of $\Delta^{2,4}$-6-ketones also gives products with the opposite configuration, i.e., the 2β,3β-diols.

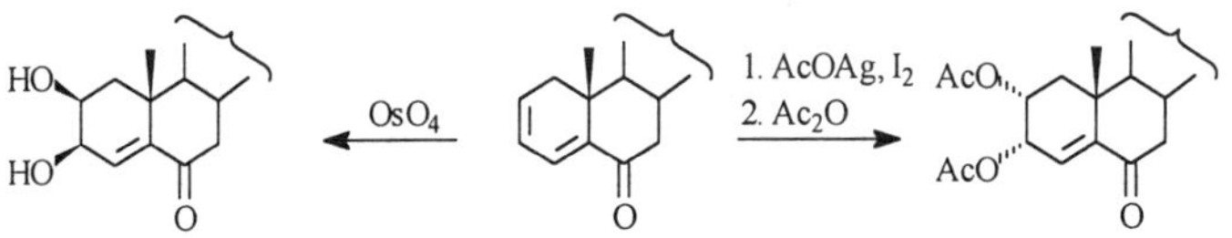

SCHEME 20

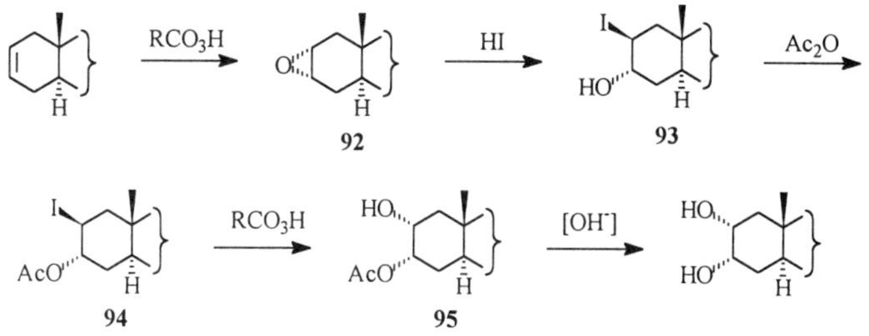

SCHEME 21

Another alternative for the synthesis of 2α,3α-diols is *via* epoxidation of Δ²-steroids, which gives only the 2α,3α-epoxide **92** (Scheme 21) (Cerny, 1989). Treatment with HI leads to the trans-diaxial iodohydrin **93**. Protection of the 3α-OH is necessary before the substitution of the adjacent iodine by a hydroxyl group can be carried out. Treatment of the iodo acetate **94** with peracid gives the monoacetate **95**, which was saponified to afford the 2α,3α-diol in a total yield up to 69%.

The formation of the 3β-hydroxy function can be easily effected by opening of the cyclopropane in compounds like **88** using acetic and sulfuric acids followed by deacetylation as shown at Scheme 22 (Thompson *et al.*, 1979; Wada

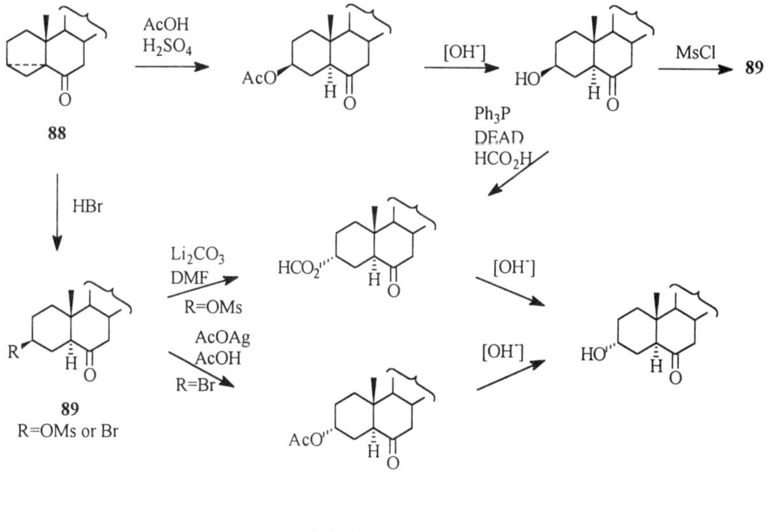

SCHEME 22

and Marumo, 1981; Takatsuto and Ikekawa, 1982, 1986b; Aburatani et al., 1987b). 2-Deoxybrassinosteroids are also available through nucleophilic substitution of the 3β-mesyl (tosyl) or 3β-bromo derivatives with inversion at C-3 (Takatsuto and Ikekawa, 1986b,e, 1987b; Takatsuto et al., 1984b, 1988b). More effective proved to be the reaction of bromide **89** (R=Br) with silver acetate in acetic acid (Aburatani et al., 1987b) followed by hydrolysis. The configuration of the hydroxyl group at C-3 can also be inverted using the Mitsunobu reaction (Voigt et al., 1996c). Another approach to the preparation of 2-deoxy-BS is the hydride reduction of 2α,3α-epoxides (Kovganko and Ananich, 1995).

The Baeyer-Villiger oxidation of 6-oxosteroids **96** with CF_3CO_3H is the standard method for the introduction of the lactone moiety into ring B (Adam and Marquardt, 1986; Kohout, 1984; Akhrem et al., 1983a; Tachimori, 1986a; Yuya and Takeuchi, 1986, 1987; Sato et al., 1987a,b; Abe and Yuya, 1991; Lakhvich et al., 1991). The reaction proceeds in general with formation of both regioisomers. The ratio of the 7-oxa and 6-oxa derivatives **97** and **98** depends strongly on the substituents in ring A (Ahmad et al., 1970; Takatsuto and Ikekawa, 1983a). It has been shown that the oxidation of 6-ketosteroids with a methyl group or a hydrogen atom at C-3 gives mainly the 6-oxa derivatives **98**. This is in accordance with the migratory aptitude for alkyl groups. For compounds with electron-withdrawing groups at C-1, C-2, or C-3, the 7-oxa derivatives **97** are the main products. Compounds with two electron-withdrawing groups show a cumulative effect. The ratio of the isomers is in favor of the

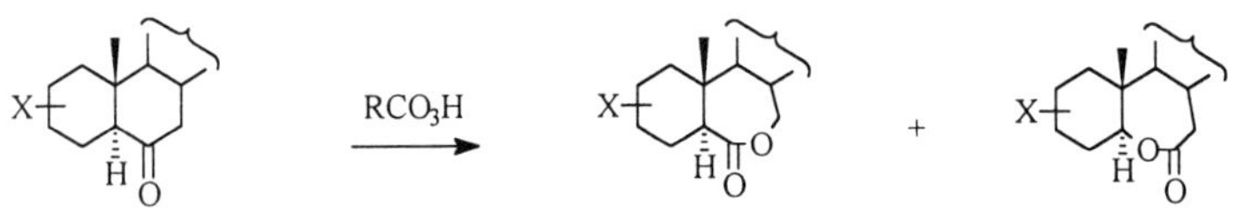

SCHEME 23

desired 7-oxa lactone **97**; for the 2α,3α-diacetates this ratio is 86:14, and for the 2α,3α-diols it is 92:8 (Takatsuto and Ikekawa, 1983a). For this reason the Baeyer-Villiger oxidation usually is carried out as the last step in the synthesis. *m*-Chloroperbenzoic acid can be used also for the Baeyer-Villiger oxidation of 6-oxosteroids (Mori, 1980a; Sakakibara and Mori, 1983b; Takatsudo and Hayashi, 1987).

It has been shown that the oxidation of 6-oxosteroids with a 3α,5-cyclopropane system by peracids in the presence of water can afford the Δ^7-6-oxo derivatives instead of the expected lactones (Akhrem *et al.*, 1984b, 1990b).

The main disadvantage of the use of peracids for the lactonization of 6-oxosteroids is the possible formation of regioisomers. Several attempts have been made to circumvent this problem by using B-seco derivatives **102** (Brooks and Ekhato, 1982; Mori *et al.*, 1984; Zhou, W.-S. *et al.*, 1988a; Zhou, W.-S., 1989). The appropriate 6-ketones **96** as well as the dienone **99** can serve as starting material (Scheme 24). Bromination of ketone **96** followed by

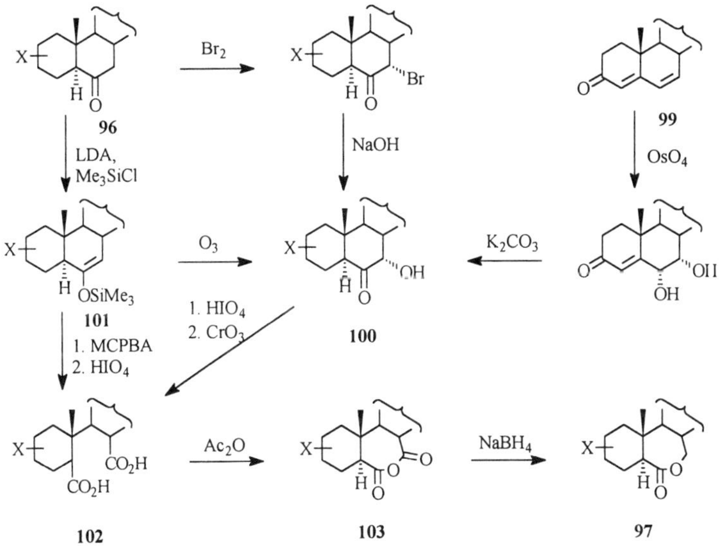

SCHEME 24

nucleophilic substitution of the bromine gives the 7-hydroxy-6-ketone **100** (Brooks and Ekhato, 1982). The latter can be prepared also by silylation of the kinetic enolate of ketone **96** followed by ozonolysis of the intermediate silyl enol ether **101** (Mori *et al.*, 1984; Zhou, W.-S. *et al.*, 1988a; Zhou, W.-S., 1989) or by osmylation of the dienone **99** (see Scheme 27) followed by treatment with base (Khripach *et al.*, 1995d). Periodate oxidation of ketol **100**, ring closure, and regioselective reduction of the anhydride **103** finally give the lactones **97** (Brooks and Ekhato, 1982).

Ozonolysis of the 5-cyclo silyl enol ether **104** led to a mixture of 7-hydroxy ketones **105** and B-seco derivative **106** (Scheme 25). Cleavage of the hydroxy ketone **105** with periodic acid afforded quantitatively the aldehydo acid **106**, which was further transformed into lactone **107** by hydride reduction and cyclization (Zhou, W.-S. *et al.*, 1988a). All attempts to transform the cyclo lactone **107** into the Δ^2-steroid **109** failed, so the detour *via* **108** was necessary to achieve the synthesis of **109**.

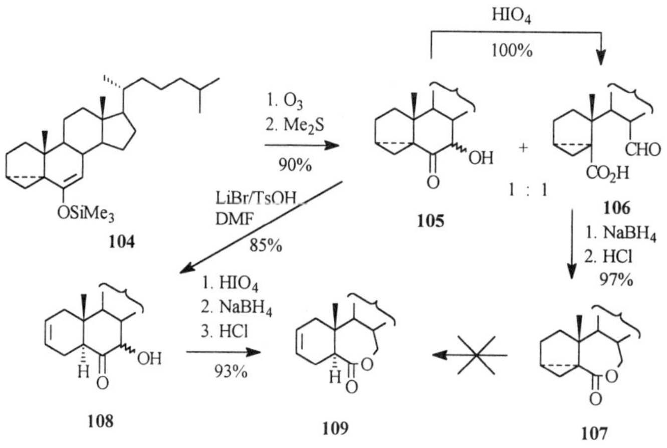

SCHEME 25

2. Starting with Ergosterol

The presence of an additional Δ^7-double bond in ergosterol **72** makes a reduction step necessary to reach the correct oxidation state for ring B. A straightforward method is the direct reduction of the 5,7-diene system with dissolving metals (Li, Na, K). Unfortunately, the conditions for selective formation of the desired Δ^5-isomer **71** have not yet been found (Anastasia *et al.*, 1984b; Akhrem *et al.*, 1989c; Khripach *et al.*, 1991e); the best ratio of Δ^5-and Δ^7-isomers **71** and **110** that has been obtained was 76:20 (Barton *et al.*, 1984).

Reduction of the Diels-Alder adduct **111** with lithium in ethylamine (Anastasia *et al.*, 1983c) also gives a mixture of the Δ^5- and Δ^5-alkenes **71** and **110** (3:2), which can be tosylated to give the corresponding tosylates. Only the tosylate **112** is capable of giving the isosteroidal rearrangement, and the desired **113** can be easily separated from the unchanged tosylate of the alcohol **110**.

The formation of the Δ^7-isomer **110** can be avoided by reduction of the $\Delta^{4,6}$-3-ketone **115** that is available from ergosterol *via* Oppenauer oxidation and isomerization of the intermediate $\Delta^{4,7}$-3-ketone **114** (Scheme 27) (Khripach *et al.*, 1992b). Treatment of **115** with lithium in liquid ammonia then gives brassicasterol **71**.

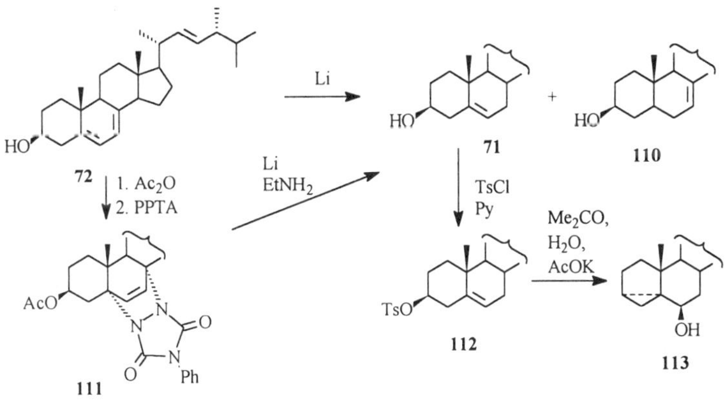

SCHEME 26

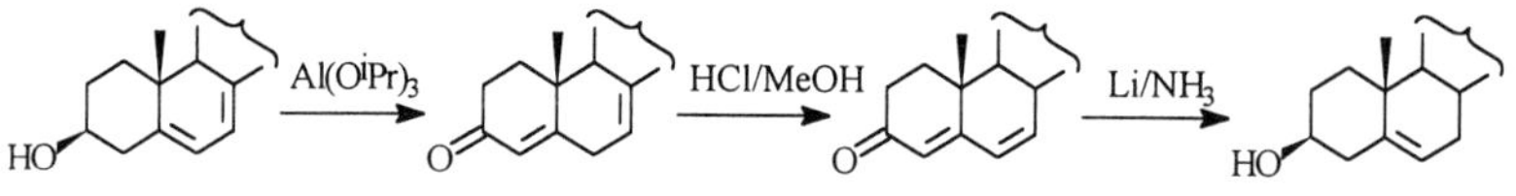

SCHEME 27

The best method for the transformation of ergosterol into BS uses the isosteroidal rearrangement (Thompson et al., 1979, 1980a,b; Khripach et al., 1991g; McMorris and Patil, 1993), which proceeds approximately 900 times faster in tosylate **116** than in the corresponding tosylates of 3β-hydroxy-Δ^5-steroids (Fieser and Fieser, 1959). Rearrangement of the mesylate **117** produces fewer byproducts and in this way alcohol **118** can be obtained in 89% yield. This

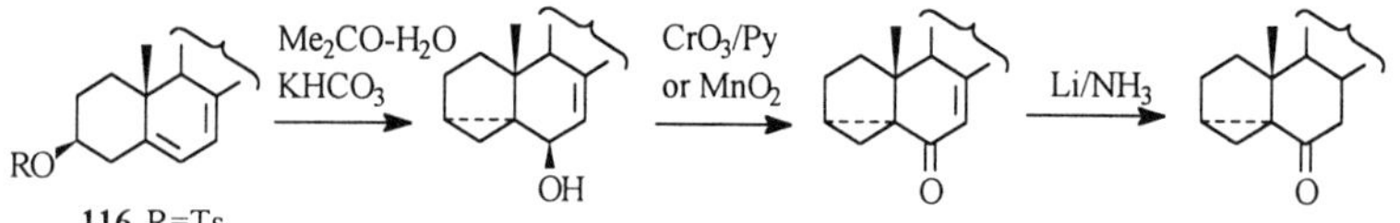

SCHEME 28

alcohol is a very unstable compound and sensitive to acidic conditions; therefore CrO_3 (Anastasia et al., 1976) in pyridine or MnO_2 (Barton et al., 1970) has to be used for its oxidation to **119**. In this approach the Δ^7-double bond is removed in a later stage by treatment of the enone **119** with lithium in liquid ammonia at low temperature (-60 °C) (Dauben and Deviny, 1966; Traven et al., 1991b).

Another solution for the reduction of Δ^7-6-ketones starts with the initial oxidation of ergosterol to the hydroxy enone **121** (Ferrer et al., 1990). Treatment with lithium in liquid ammonia results in saturation of the Δ^7-double bond with simultaneous deoxygenation of the 5α-hydroxyl group and deprotection of the 3β-hydroxyl group (Scheme 29). The alcohol **122** could be dehydrated with copper sulfate absorbed on silica gel. An improved variant of this synthesis consists of the reduction of hydroxy enone **121** with an excess of lithium to diol

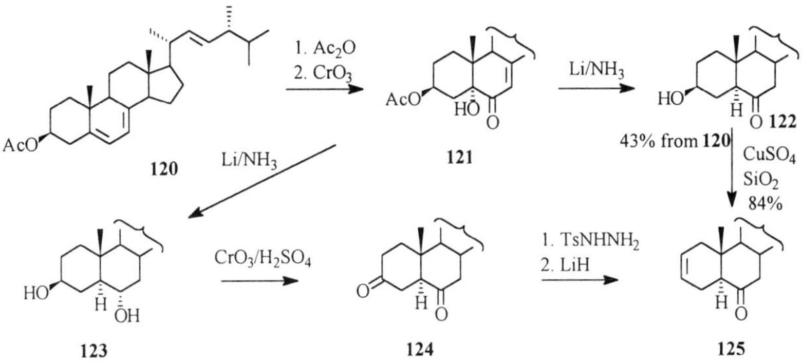

SCHEME 29

123 (Brosa et al., 1996b). Oxidation with Jones reagent followed by a selective reduction of diketone **124** via its tosylhydrazone afforded the dienone **125** in 40% overall yield from ergosterol.

3. Starting with Hyodeoxycholic Acid

The presence of two hydroxyl groups at C-3 and C-6 in the cyclic part of hyodeoxycholic acid makes it an attractive starting material for BS synthesis (Zhou, W.-S., 1989, 1994; Zhou, W.-S. et al., 1988a,b, 1989a-c, 1990a,b, 1991a,b, 1992a,b; Zhou, W.-S. and Wang, Z.-Q., 1991; Zhou, W.-S., and

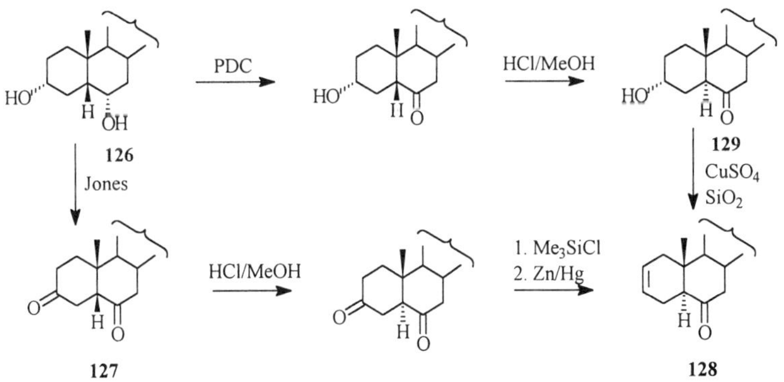

SCHEME 30

Huang, L.-F., 1992; Zhou, W.-S., and Tian, W.-S., 1984, 1985, 1987; Tian, 1984; Shen, Z.-W., and Zhou, W.-S., 1990; Huang, L.-F. *et al.*, 1993). Jones oxidation of diol **126** gives the corresponding diketone **127**, which upon epimerization at C-5 can be subjected to reductive elimination to give the enone **128**. Selective oxidation of the diol **126** can be effected also with pyridinium dichromate and epimerization at C-5 again leads to the enone **128** *via* the hydroxy ketone **129** (Zhou, W.-S., and Tian, W.-S., 1987; Zhou, W.-S., 1989; Zhou, W.-S. *et al.*, 1990a).

C. SYNTHESIS OF THE SIDE CHAINS

The construction of the side chain with its array of chiral centers is probably the most difficult part in the synthesis of BS. BS share this problem with vitamin D and its analogs, and it has long attracted the attention of chemists working in the field of steroid chemistry. The synthetic approaches toward BS side chains can be divided into two principal groups, depending on the starting material:

i) syntheses with preservation of a native carbon skeleton;

ii) syntheses which involve the formation of new carbon-carbon bonds in the side chain.

The shortest way to synthesize the brassinolide side chain (**131→130**) is the direct hydroxylation of the Δ^{22}-double bond in a suitable sterol (Scheme 31). Unfortunately, this approach is of limited use because of the scarcity of the corresponding sterol (crinosterol), but it is obvious that it can be used for the preparation of homobrassinolide from stigmasterol and of epibrassinolide from ergosterol or brassicasterol. Brassinolide can be synthesized starting from ergosterol or brassicasterol as well (**135→130**), but then inversion of the configuration at C-24 will be necessary.

Most BS syntheses that use the second approach have to partially rebuild the side chain, and many of these syntheses are based on the reaction of various

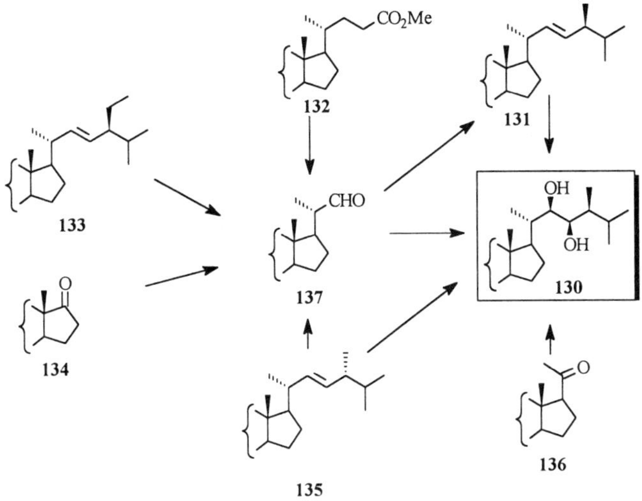

SCHEME 31

metal-organic compounds with 22-aldehydes (**137→130**). The common procedure for the preparation of C-22 aldehydes is ozonolysis of Δ^{22}-sterols, usually stigmasterol (**133→137**), but 17-oxosteroids (**134→137**) and derivatives of cholic acids (**132→137**) can be used also. Alternative routes for the synthesis of BS are available by transformation of cholic acid (**132→130**) or pregnane derivatives (**136→130**).

1. With Preservation of a Native Carbon Skeleton

Construction of the side chain of BS is relatively easy when Δ^{22}-steroids with the same carbon skeleton are used as starting material. In this case hydroxylation of the double bond with OsO_4 (Criegee *et al.*, 1942) or with a mixture of silver acetate and iodine (Woodward and Brutcher, 1958) can be used. These methods are normally complementary with respect to stereochemistry, but the peculiarity of Δ^{22}-steroids is that both methods give the same result, and in accordance with Murphy's law, the main products are the nonnatural 22*S*,23*S*-diols.

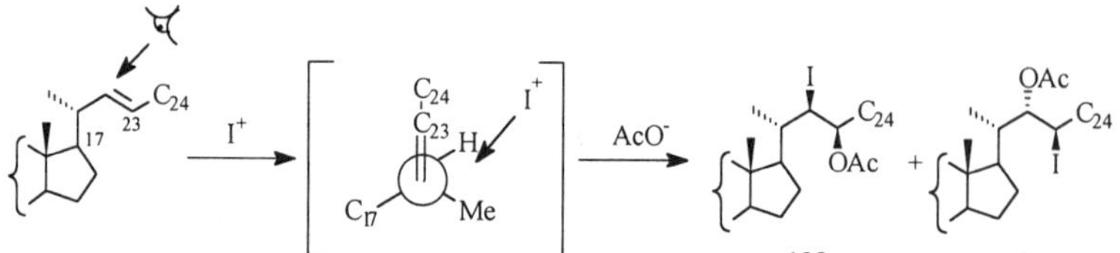

SCHEME 32

The stereochemistry of these hydroxylations can be explained by the preferential conformation of the steroid side chain, which is known to be relatively rigid (Barton *et al.*, 1972; Nakane *et al.*, 1975; Ishiguro and Ikekawa, 1975; Sakakibara and Mori, 1982). Iodine attacks the double bond from the less hindered side to form two iodohydrins **138** and **139** (Scheme 32), neither of which can be used for the synthesis of 22*R*,23*R*-diols (Sakakibara and Mori, 1982).

The reaction of the Δ^{22}-olefin **140** with OsO$_4$ can give two possible intermediates **141** and **142** (Sakakibara and Mori, 1982) (Scheme 33). In the case of intermediate **141** the bulky isopropyl group will be oriented toward the steroid skeleton and will have strong steric interaction with the 16-methylene

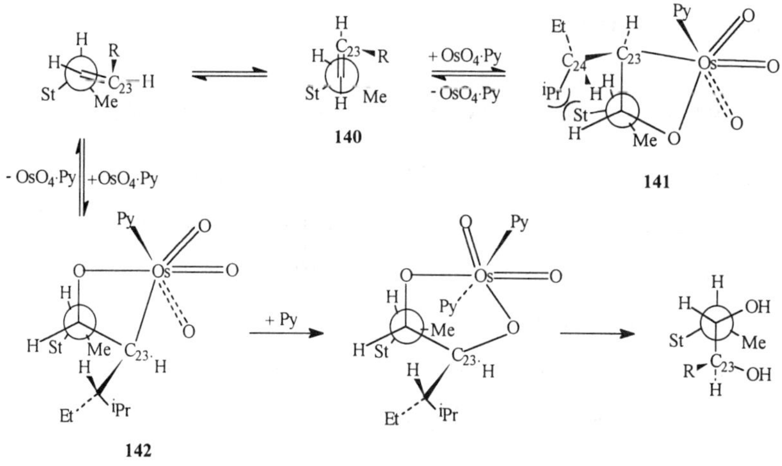

SCHEME 33

group. Moreover, the complex **141** will be destabilized by the interaction of the C-24 hydrogen with the bulky osmium atom. The formation of intermediate **142** is much more favorable, because the ethyl group will be located at the less hindered side, away from the steroid skeleton, and the isopropyl group will be remote from it.

In practice, the hydroxylation of Δ^{22}-steroids with OsO_4 leads to the formation of both isomers (Scheme 34) with a ratio of diols **143** and **144** that

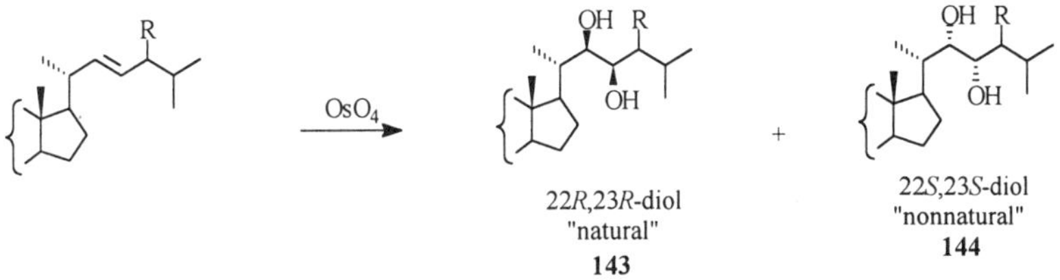

SCHEME 34

depends on the reaction conditions and on the size and configuration of the substituent at C-24 (see Table IX).

In the absence of chiral catalysts the ratio of isomers is unfavorable for the synthesis of the natural configuration of the hydroxyl groups in BS, but this oxidation was repeatedly used for the preparation of the nonnatural 22S,23S isomers of BS (Mori et al., 1982; Thompson et al., 1981, 1982; Anastasia et al., 1983d; Takatsuto et al., 1983a; Akhrem et al., 1984a, 1985a,b, 1987b; Zhou, W.-S., and Tian, W.-S., 1985; Wada and Marumo, 1981; Anastasia et al., 1985b, 1986a; Takatsuto and Muramatsu, 1988).

The synthesis of the natural 22R,23R-diols can be considerably improved (Ikekawa and Zhao, 1991) when Sharpless hydroxylation is applied for the oxidation of Δ^{22}-steroids (Jacobsen et al., 1988). Thus derivatives of quinidine (DHQD CLB, DHQD PHN, and (DHQD)$_2$-PHAL) (Fig. 24) promote the formation of the 22R,23R-diols of the natural series, whereas derivatives of quinine (DHQ CLB) are practically selective for the hydroxylation to the

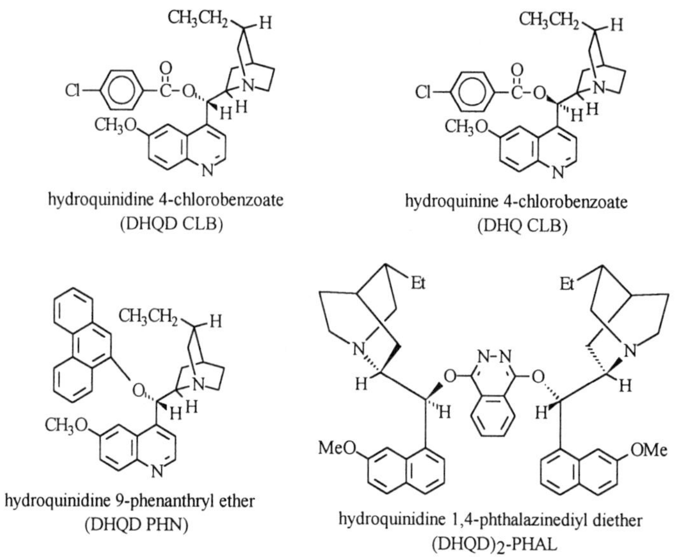
hydroquinidine 4-chlorobenzoate
(DHQD CLB)
hydroquinine 4-chlorobenzoate
(DHQ CLB)
hydroquinidine 9-phenanthryl ether
(DHQD PHN)
hydroquinidine 1,4-phthalazinediyl diether
(DHQD)2-PHAL

Fig. 24. Chiral catalysts for asymmetric hydroxylation of Δ^{22}-olefins.

nonnatural 22*S*,23*S*-diols. The reaction time may be decreased and a better yield of the diols achieved by addition of methanesulfonamide (Jacobsen *et al.*, 1988; Huang, L.-F. *et al.*, 1995).

The effect of the chiral catalysts is dependent on the substituent at C-24. With a methyl group at C-24 about 10% of the unwanted 22*S*,23*S*-diol was formed in most cases (Table IX). Hydroxylation of stigmastane derivatives (Scheme 35), with an ethyl group at C-24, gave a considerable amount of the 22*S*,23*S* isomer and also overoxidation was observed (Hellrung *et al.*, 1997).

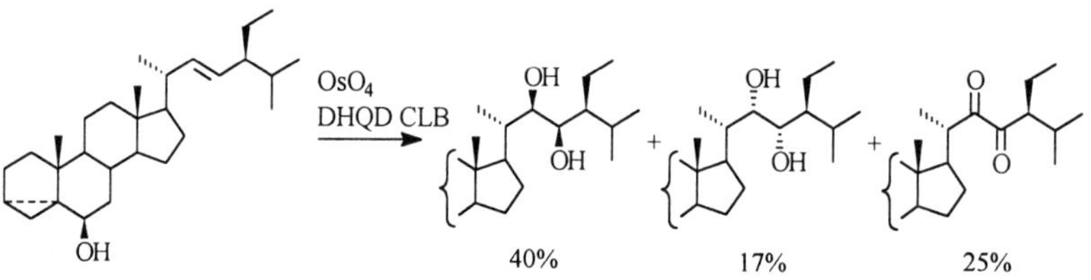
OsO4
DHQD CLB
40%
17%
25%

SCHEME 35

TABLE IX
Hydroxylation of Δ^{22}-Steroids with OsO_4

Side chain	Catalyst	% 143	% 144	References
	-	13	87	Thompson et al., 1982
	-	6	94	Wada and Marumo, 1981
	-	4	96	Takatsuto and Ikekawa, 1982
	-	13	87	Barton et al., 1970
	-	4	96	Brosa et al., 1992
	DHQD CLB	28	72	Brosa et al., 1992
	DHQD CLB	57-60	40-43	Sun et al., 1991
	DHQD PHN	60-69	31-40	Brosa et al., 1992
	(DHQD)$_2$-PHAL	80	20	Marino et al., 1996
	(DHQD)$_2$-PHAL	50	50	Centurion et al., 1996
	DHQ CLB	0	100	Sun et al., 1991
	-	34	66	Takatsuto et al., 1983a
	-	18	82	Thompson et al., 1982
	-	13	87	Anastasia et al., 1983d
	DHQD CLB	89	11	Zhou, W.-S. et al., 1991a; Sun et al., 1991
	DHQ CLB	0	100	Sun et al., 1991
	(DHQD)$_2$-PHAL	0	100	Marino et al., 1996
	-	50	50	Thompson et al., 1982; Anastasia et al., 1983c
	-	37	63	Takatsuto and Ikekawa, 1984a
	-	25	75	Akhrem et al., 1989c
	DHQD CLB	89	11	Sun et al., 1991
	DHQD CLB	90	10	McMorris and Patil, 1993
	DHQ CLB	10	90	Sun et al., 1991
	(DHQD)$_2$-PHAL	90-91	9-10	McMorris and Patil, 1993
	-	100	0	Marino et al., 1996
	DHQD PHN	93	7	Huang, L.-F. et al., 1993
	-	30	70	Thompson et al., 1982
	-	29	71	Hirano et al., 1984
	-	33	67	Takatsuto and Ikekawa, 1987a
	-	20	80	Khripach et al., 1990j
	-	30	70	Thompson et al., 1982
	-	33	64	Huang, L.-F. and Zhou, W.S., 1994
	DHQD CLB, DHQD NAP, DHQD PHN	90	10	Huang, L.-F. and Zhou, W.S., 1994

(*continues*)

TABLE IX (continued)

Side chain	Catalyst	% 143	% 144	References
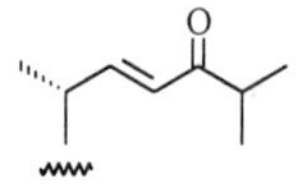 [structure: α,β-unsaturated ketone]	-		79% of 144 was isolated	Khripach et al., 1993c
[structure: CO₂Me alkene]	-	11	89	Zhou, W.-S. et al., 1991a
	DHQD CLB	80	20	Zhou, W.-S. et al., 1991a
[structure: OH, OTHP alkene]	-	24	76	Honda et al., 1993
	DHQD CLB	9	91	Honda et al., 1993
	DHQ CLB	87	13	Honda et al., 1993

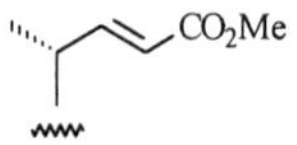

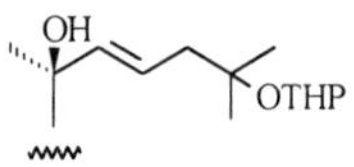

An alternative route to the 22R,23R-diols involves formation of the 22R,23R-epoxides as intermediates. It proved to be more effective for the synthesis of the natural BS than simple oxidation with OsO_4 but less effective than the hydroxylation with this reagent in the presence of chiral amines (Mori et al., 1981, 1982; Sakakibara et al., 1982; Takatsuto and Ikekawa, 1982; Sakakibara and Mori, 1982; Lakhvich et al., 1985a,b).

Epoxidation of Δ^{22}-steroids gives a mixture of both epoxides with the 22R,23R-epoxides **145** (Scheme 36) as the main products (Nakane et al., 1975; Ishiguro and Ikekawa, 1975; Piatak and Wicha, 1978; Hirano et al., 1984; Zeelen, 1984). The ratio of the isomers **145** and **146** depends on the substituent at C-24 and its configuration, ranging from 2:1 for R=H (Ishiguro and Ikekawa, 1975) to 1.5:1 for R=(24S)-Me (Mori et al., 1982) to (2.2-1.1):1 for R=(24S)-Et (Takatsuto and Ikekawa, 1982; Brosa and Miro, 1997).

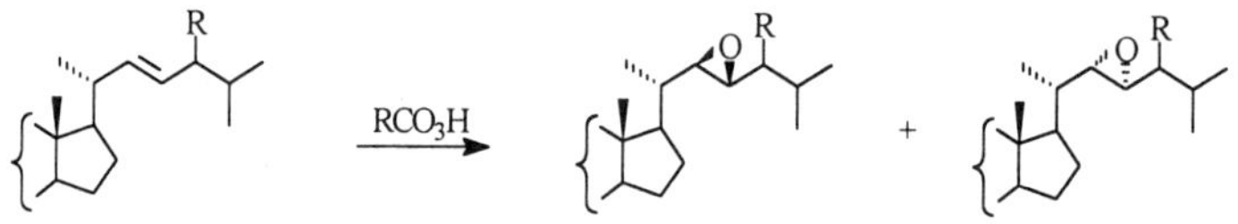

SCHEME 36

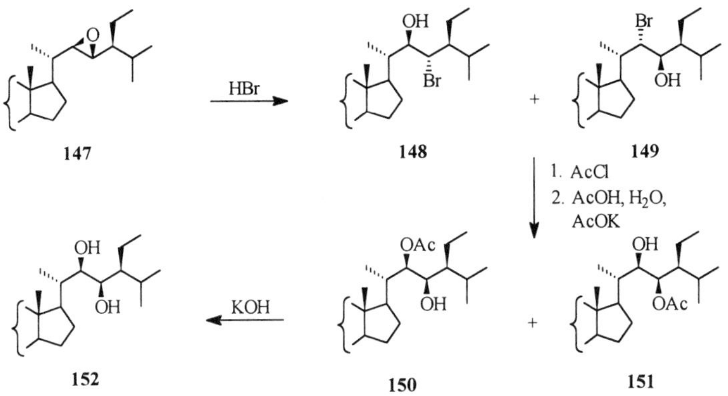

SCHEME 37

The transformation of epoxide **147** to diol **152** (Scheme 37) illustrates the details of the method. Treatment of **147** with HBr gives a mixture of bromohydrins **148** and **149**; it is not necessary to separate these compounds because both ultimately lead to the same product. Acetylation of this mixture followed by nucleophilic substitution of the bromines with inversion of configuration gives a mixture of hydroxy acetates **150** and **151**; saponification of this mixture affords the 22R,23R-diol **152**.

Opening of the epoxy ring can also be achieved with the phenyl selenide anion (Sakakibara and Mori, 1983b,c; Sumimoto Chem. Co., Ltd., 1983b; Takatsuto and Ikekawa, 1986d; Takatsuto, 1988), which allows the construction of an additional double bond as illustrated in Scheme 38 for alcohol **153** and its regioisomer **154**. Treatment of the mixture with hydrogen peroxide afforded an inseparable mixture of the allylic alcohols **155** and **156**. Epoxidation of this mixture followed by treatment with aluminum isopropoxide led to a mixture of the diols **157** and **158**, which could be separated.

The synthesis of a brassinolide-like side chain directly from ergosterol has to include an inversion of the configuration at C-24, and this can be achieved *via* the allylic alcohol **162** (Anastasia *et al.*, 1984b; Khripach *et al.*, 1994a). One method to prepare this alcohol involves the diene **159** (Anastasia *et al.*, 1984b),

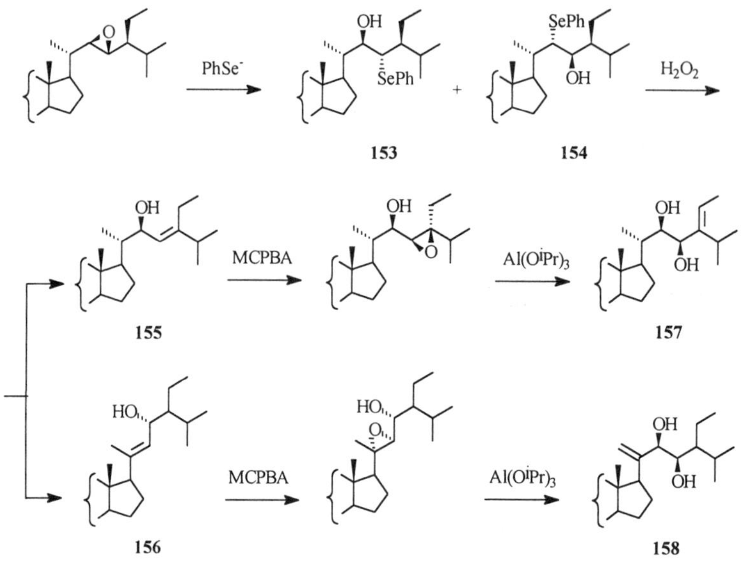

SCHEME 38

where a stereoselective addition with molecular oxygen gave a mixture of epimeric C-22 epidioxides (Scheme 39). The main isomer **160** was reduced over a Lindlar catalyst to afford the diol **161**. Deoxygenation of the primary alcohol group in diol **161** *via* reduction of the corresponding mesylate furnished the allylic alcohol **162**.

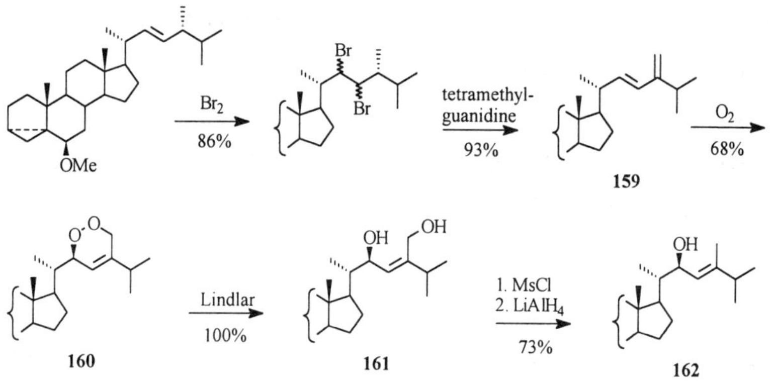

SCHEME 39

An alternative route to **162** (Scheme 40) involves epoxidation of the olefin **163** followed by epoxide ring opening (Khripach *et al.*, 1994a; Voigt *et al.*, 1997). In both steps also the corresponding isomers are obtained, and this is a disadvantage of this approach. The Jones oxidation of bromohydrin **164** led smoothly to the bromo ketone **165**, which was dehydrobrominated to give the enone **166**. Transformation of enones such as **166** into the allylic alcohol **162** is known (see Scheme 46).

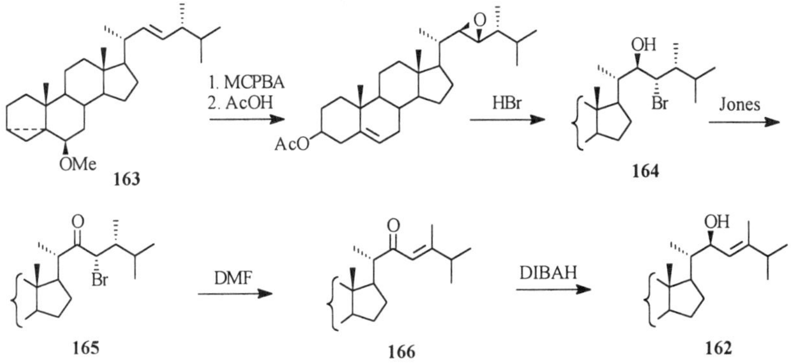

SCHEME 40

2. With Reconstruction of the Native Carbon Skeleton of the Side Chain

The effective construction of a BS side chain with its 4 consecutive chiral centers is a rather difficult task. To date, this problem cannot be considered to be solved satisfactorily for the synthesis of brassinolide. Synthesis of a BS side chain can be achieved starting from 17-ketones **134**, 20-ketones **136**, 22-aldehydes **137**, and 24-esters hyodeoxycholic acid **132** (Scheme 41). All these approaches will be described in the next paragraphs.

Steroids possessing fragments **134**, **136**, and **132** are available from natural sources or may be obtained by microbiological breakdown of natural steroids, whereas 22-aldehydes **137** are prepared chemically.

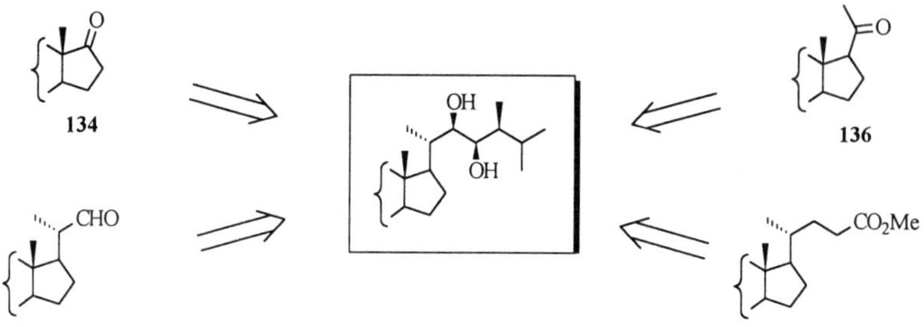

134 136 137 132

SCHEME 41

a. Starting with 17-ketones. Utilization of steroids of the androstane series implies the creation of additional chiral centers at C-17 and C-20. That makes the whole synthetic route longer, but at the same time it gives an opportunity for the preparation of BS analogs that are functionalized in ring D and at C-21. Androstane derivatives **134** can be easily transformed into the *Z*-olefins **167** (Trost and Verhoeven, 1978; Batcho *et al.*, 1981a,b; Riediker and Schwartz, 1981), which are good substrates for a stereoselective ene reaction with various aldehydes and unsaturated compounds (Trost and Verhoeven, 1978; Dauben and Brookhart, 1981; Nakai and Mikami, 1989b; Houston *et al.*, 1993). The ene reaction of **167** with formaldehyde leads to the homoallylic alcohol **168** with the natural stereochemistry at C-20 (Batcho *et al.*, 1981a; Johnson *et al.*, 1984; Hazra *et al.*, 1990, 1992, 1993a, 1994; Kobayashi *et al.*, 1994). Its reduction proceeds from the less hindered side of the steroid skeleton to give the corresponding saturated alcohol as the only product (Scheme 42), and Jones oxidation then affords the aldehyde **169**.

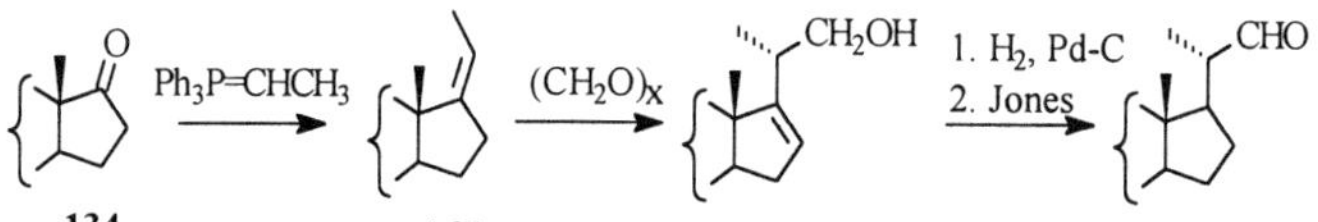

134 167 168 169

SCHEME 42

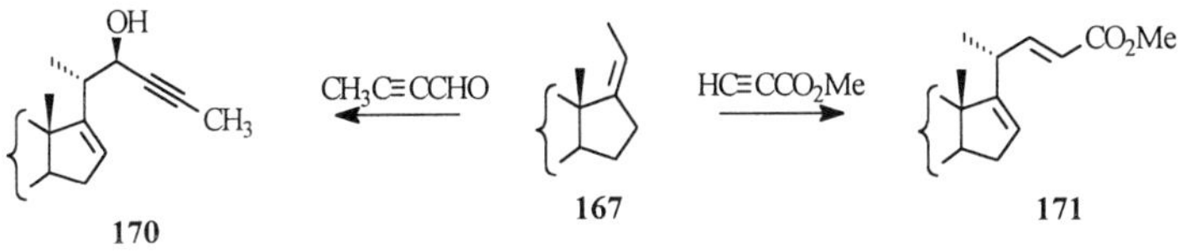

SCHEME 43

A stereoselective ene reaction of the Z-olefin **167** with other aldehydes may be used also for simultaneous construction of the chiral centers at C-20 and C-22 (Scheme 43) to give intermediates such as alcohol **170** (Mikami *et al.*, 1988a,b). The ene reaction of olefin **167** with acetylenes affords Δ^{22}-steroids such as **171** (Batcho *et al.*, 1981a,b; Dauben and Brookhart, 1981; Baggiolini *et al.*, 1982).

The corresponding 16α-hydroxy derivatives **172** may be subjected to various sigmatropic rearrangements to give similar compounds (Tanabe and Hayashi, 1980; Koreeda *et al.*, 1980; Dauben and Brookhart, 1981; Mikami *et al.*, 1985a). Thus, the [2,3]Wittig rearrangement of the ether derived from **172** leads to the homoallylic alcohol **173** (Castedo *et al.*, 1985). The hydroxyl function at C-22 can be introduced by this method as well by rearrangement of the ethers **174** (Koreeda and Ricca, 1986; Mikami *et al.*, 1985b, 1986; Granja, 1991). Rearrangement of the ether **174** when R=SiMe₃ gives exclusively the alcohol **175** (Mikami *et al.*, 1985b), whereas the reaction of **174** with R=H led to the

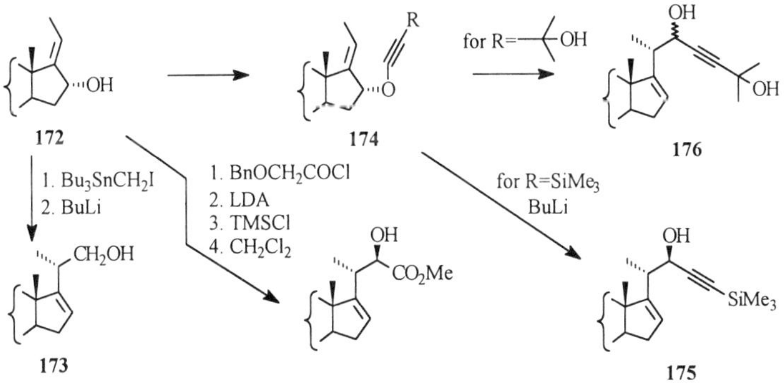

SCHEME 44

compound with opposite configuration at C-22. Rearrangement of the ether **174** when R=C(CH$_3$)OH affords a mixture of both epimers **176** at C-22 (Granja, 1991).

b. Starting with 20-ketosteroids. Compounds in this group cannot be considered as very convenient starting materials for BS synthesis. Formation of the asymmetric center with the correct stereochemistry at C-20 is the main problem that has to be solved. Nevertheless several attempts to utilize 20-ketosteroids have been reported.

The key stage in the polyoxygenated steroid side chain synthesis developed by Kametani *et al.* (1985, 1986, 1988a) was the addition of the anion derived from 3-isopropyltetronic acid **178** to the 20-ketone **177** (Scheme 45). The reaction proceeded with formation of a mixture (91:9) of two adducts obtained as their methoxymethyl ethers in 84% overall yield. The best method for the elimination of the tertiary alcohol **179** proved to be *via* its 20-trifluoroacetate derivative. The elimination with DBU afforded a mixture (82:18) of 5-ylidenetetronates which could be separated. The catalytic hydrogenation of **180**

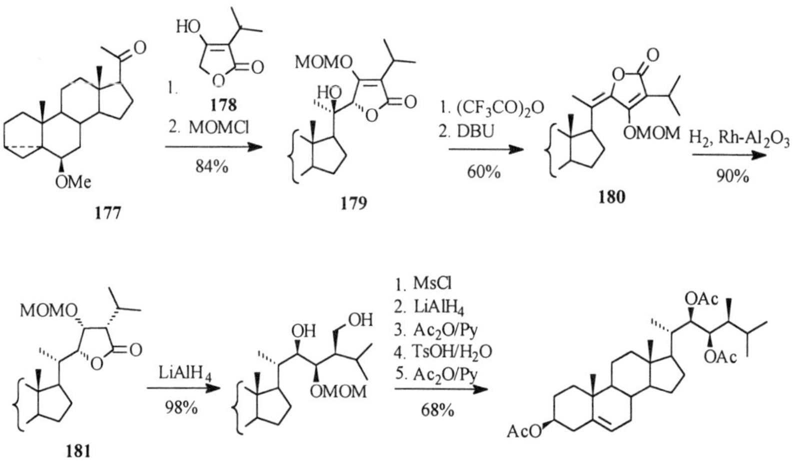

SCHEME 45

BASIC SYNTHETIC METHODS AND FORMAL SYNTHESES 101

over rhodium-alumina is a crucial point in the whole sequence, because in this step all four asymmetric centers of the brassinolide side chain are formed. The reduction proceeded in high yield and gave the desired compound **181**. The remaining steps included reductive opening of the lactone ring followed by removal of the primary hydroxyl group at C-28.

20-Ketosteroids have been used for the preparation of allylic alcohols as depicted in Scheme 46 (Takahashi *et al.*, 1985). Reduction of the 3-tetrahydropyranyl ether of pregnenolone **182** gave a mixture of two alcohols (6.6:1). The major isomer was converted into the corresponding tosylate **183**, which was substituted by the protected cyanohydrin **184**. Hydrolysis of the alkylated product furnished the enone **185**, which was reduced with diisobutylaluminum hydride to a mixture of the allylic alcohols **186** (anti-Cram product) and **187** in a ratio of 97:3 (85% yield). Treatment of enone **185** with L-Selectride gave the Cram product **185** as the main isomer in a ratio of 7:93 (72% yield).

Starting from 20-ketone **188** a synthesis of enone **193**, a useful intermediate for the synthesis of brassinolide, has been elaborated (Akhrem *et al.*, 1989a; Khripach *et al.*, 1993d). The 1,3-dipolar cycloaddition reaction of the acetylenic

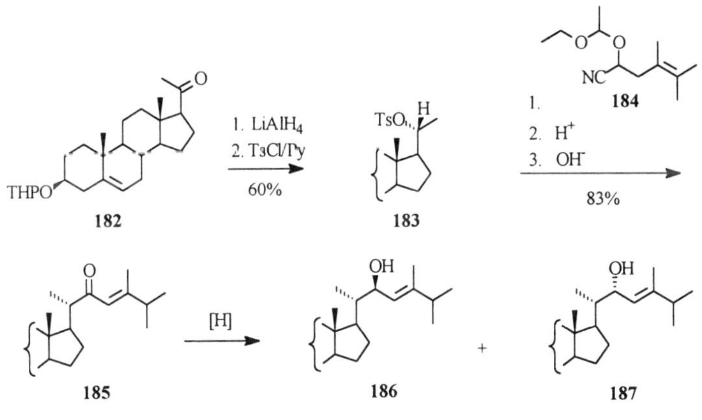

SCHEME 46

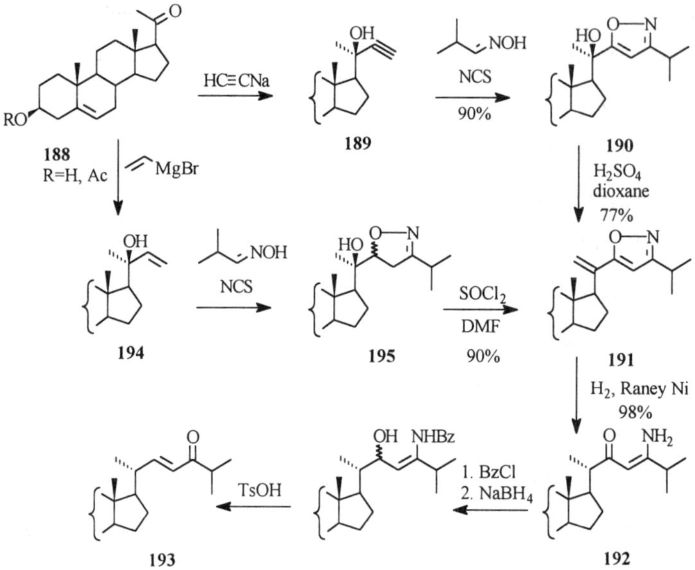

SCHEME 47

alcohol **189** with isobutyronitrile oxide to hydroxy isoxazole **190** is the key step in this approach (Akhrem et al., 1987a, 1989a). Dehydration of **190** led to the $\Delta^{20(22)}$-isoxazole **191** and the reductive cleavage of this isoxazole proceeded with simultaneous reduction of the exo-methylene double bond to afford the enamino ketone **192** (Scheme 47). The hydrogenation reaction proved to be very selective also in this case and enamino ketone **192**, with the natural configuration at C-20, was isolated as the sole product (Khripach et al., 1993d). Its transformation into the enone **193** was achieved via benzoylation, reduction of the 22-keto group, and dehydration.

The reaction of the allylic alcohol **194** with isobutyronitrile oxide gave a mixture of the isoxazolines **195**, epimeric at C-22 (Akhrem et al., 1989b; Khripach et al., 1990c). This mixture could be transformed also in good yield into the methylene isoxazole **191** by treatment with thionyl chloride in dimethylformamide (Khripach et al., 1990a,b, 1992a).

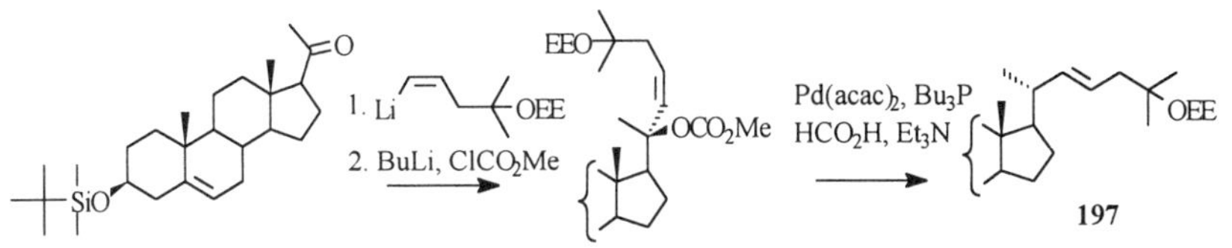

SCHEME 48

Palladium-catalyzed hydrogenolysis of allylic carbamates derived from 20-ketones led to a method of side chain construction that allowed the preparation of both 20R and 20S isomers, depending on the geometry of the starting olefin (Mandai *et al.*, 1992a,b, 1994). Thus, the reaction of the carbamate **196** containing a Z-double bond gave the Δ^{22}-olefin **197** as the major product along with 6-9% of the corresponding isomer at C-20.

c. Starting with 22-aldehydes. Very convenient intermediates for the construction of new BS side chains are 22-aldehydes **137**, which have been used extensively for that purpose. Ozonolysis of Δ^{22}-steroids or other oxidative cleavage reactions (Adam *et al.*, 1983; Schönecker *et al.*, 1984a,b, 1989; McMorris *et al.*, 1996) are the obvious methods for the preparation of these 22-

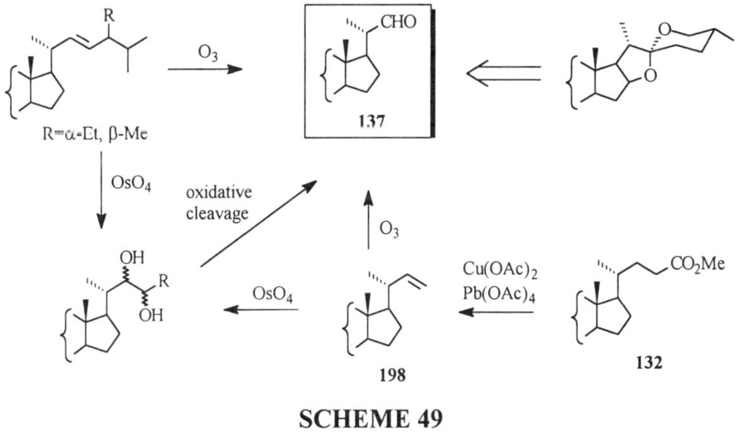

SCHEME 49

aldehydes. Preparation of **137** starting from hyodeoxycholic acid derivatives **132** has been achieved *via* oxidative decarboxylation followed by ozonolysis of the resulting olefin **198** (Zhou, W.-S., and Tian, W.-S., 1985; Tian *et al.*, 1989). Diosgenin has been used as starting material for **137** also (Liu, X. *et al.*, 1989).

A large number of approaches to BS side chain construction have been elaborated starting from 22-aldehydes **137**, and they can be divided roughly into 9 groups, according to the type of key intermediate:

- homopropargylic alcohols **199**
- propargylic alcohols **200**
- allylic alcohols **201**
- Δ^{22}-olefins **202**
- hydroxy furans **203**
- hydroxy lactones **204**

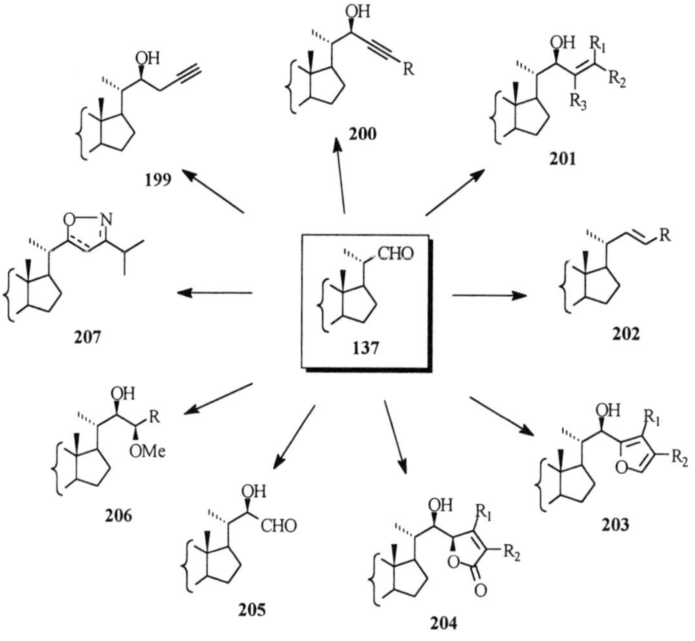

SCHEME 50

- hydroxy aldehydes **205**
- hydroxy ethers **206**
- isoxazole/isoxazolines **207**

The type of intermediate defines the general synthetic strategy, which as a rule is quite unique for each group.

A common feature in the intermediates of the type **199-201** and **203-206** is the presence of the 22*R*-hydroxyl group. This is a required element in the BS side chain and it serves as a chiral handle for the introduction of other functional groups at C-23 and C-24. Therefore the problem of stereoselective synthesis of 22*R*-alcohols is a very crucial one for preparation of BS. In general, these 22-alcohols are prepared *via* addition of various nucleophiles to the 22-aldehydes, but alternative methods have been elaborated also.

1) Stereochemical aspects of the synthesis of 22-alcohols. Addition of metal-organic compounds to 22-aldehydes is the method of choice for preparation of 22-alcohols. The addition generally proceeds with formation of a mixture of the 22α- and 22β-alcohols **208** and **209**; the ratio depends on the type of nucleophile and on the reaction conditions (Scheme 51).

Normally Cram products **208** are the main isomers, but in general the ratio of alcohols **208** and **209** covers a wide range and depends also on the substituent (Fig. 25) in the α-position of the anionic center (Table X).

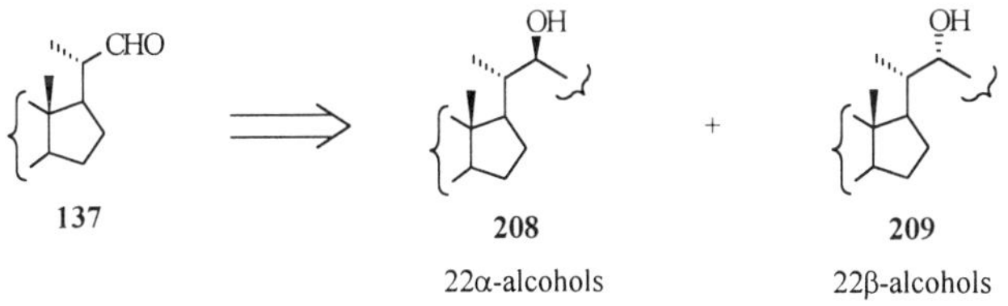

137 ⇒ 208 + 209
        22α-alcohols   22β-alcohols

SCHEME 51

TABLE X
Addition of Metal-Organic Reagents to 22-Aldehydes

Reagent	% 208	% 209	Total yield, %	References
⟩≡—Li	60	40	100	Takatsuto et al., 1982a
	60	40	73.5	Takatsuto et al., 1984b
	50	50	76	Ishiguro et al., 1980
			-	Aburatani et al., 1985b, 1987b; Sakakibara and Mori, 1983a
⟩≡—MgBr	50	50	-	Hirano and Djerassi, 1982; Anastasia et al., 1983a
	60	40	65	Anastasia et al., 1983d
	52	48	-	Preus and McMorris, 1979
MOMO⟩≡—Li	50	50	85	Sardina et al., 1983
	50	50	90	Sardina et al., 1986
THPO⟩≡—Li	55	45	38	Eguchi et al., 1989
Li structure	70	30	77	Wilson et al., 1992
Li structure	89	11	82	Shu and Djerassi, 1981
H Me / Li SiMe$_2$Ph	75	25	64	Hayami et al., 1983
H / Li	64	36	74	Wiersig et al., 1979
H / Li	82	18	61	Wiersig et al., 1979
Li OSiEt$_3$	87	13	61	Torneiro et al., 1997
Li OMOM	88	12	88	Torneiro et al., 1997
Li SiMe$_3$	91	9	97	Khripach et al., 1993g
Li pTol—S	92	8	84	Marino et al., 1996

(continues)

TABLE X (continued)

Reagent	% 208	% 209	Total yield, %	References
(Li, S(=O)pTol allyl, isomer A)	80	20	75	Marino et al., 1996
(Li, S(=O)pTol allyl, isomer B)	60	40	86	Marino et al., 1996
(dithiane-Li)	88	12	89	Hazra et al., 1994; Takatsuto et al., 1981; Takatsuto and Ikekawa, 1983a,b, 1987a
(2-furyl-Li)	70 / 74 / 72	30 / 26 / 28	96 / / 90	Kametani et al., 1989; Honda et al., 1990; Zhou, W.-S., and Ge, C.-S., 1989
(3-isopropyl-2-furyl-Li)	82	18	99	Kametani et al., 1988b
(3-methyl-2-furyl-Li)	75	25	-	Honda et al., 1990; Tsubuki et al., 1992a
(3-methyl-2-furyl-OLi)	88	12	74	Donaubauer et al., 1984
(4-isopropyl-furan-OLi)	89	11	90	Donaubauer et al., 1984
[Bu–Al–$\cdot$]Li⁺	85	15	54 / 36 / -	Fung and Siddall, 1980; Mori et al., 1984; Okada and Mori, 1983b

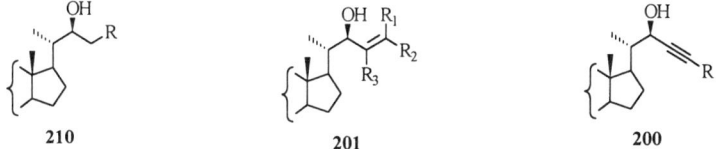

Fig. 25. Three types of C-22 alcohols can be produced in the addition reaction. The alcohols **210** are of little practical interest unless activating groups are present at C-24. Nevertheless such intermediates have been used for the preparation of 23-deoxy-BS (Ihara Chem. Ind. Co., Ltd.,

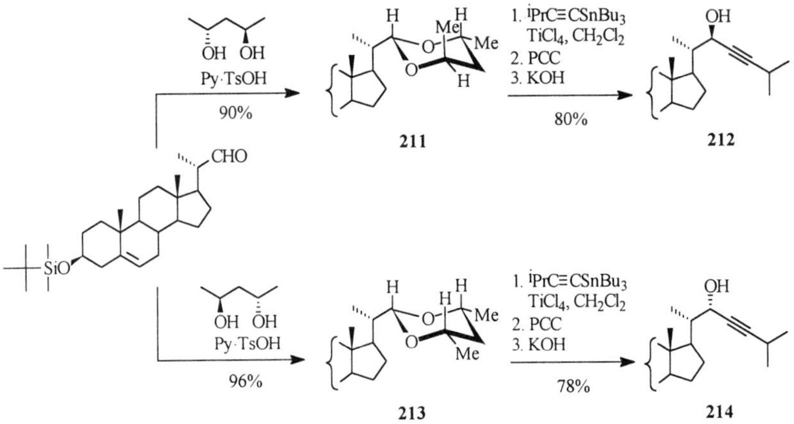

SCHEME 52

1982a). Propargylic alcohols **200** can be easily transformed into the corresponding allylic alcohols **201** and it is natural to discuss both types of intermediates together.

An effective method to control the stereochemistry of addition is the utilization of chiral acetals as carbonyl component (Yamamoto, Y., 1985a-c; Yamamoto, Y., and Yamada, J., 1986; Yamamoto, Y. et al., 1986a,b, 1991). In the case of the compound **211** (Scheme 52) the acetal group has the same directing effect as the rest of the steroid molecule. As a result the alcohol **212** is practically the only product of the addition. Less obvious results have been obtained for acetal **213**. The acetal fragment defines the stereochemistry of the addition in this case, and alcohol **214** is the main product of the reaction (the ratio **214:212** = 9:1). In contrast, boron and silicon organic compounds gave predominantly the Cram alcohols (Yamamoto, Y. et al., 1991).

A large number of acetals and dioxanone derivatives were studied in reaction with stannane **215** (Granja et al., 1993; Castedo et al., 1987). Depending on the substituent R, both 22R- and 23S-alcohols could be isolated as practically the only products of the addition reaction (Scheme 53).

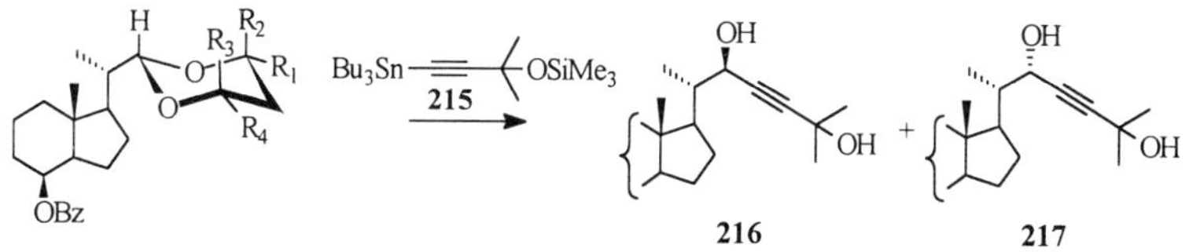

Ratio **216:217**: $R_1=R_3=Me$, $R_2=R_4=H$ (87:1); $R_1=R_2=R_3=R_4=H$ (3:2); $R_1=R_3=H$, $R_2=R_4=Me$ (1:12); $R_1=R_2=R_3=H$, $R_4=Me$ (1:2.6); $R_1=Me$, $R_2=R_3=R_4=Me$ (0:100); $R_1=Me$, $R_2=H$, $R_3R_4=O$ (45:1).

SCHEME 53

Apart from the possibility of controlling the stereochemistry in the addition stage, it is of course also possible to transform the epimers into each other. This may be done by mesylation of the undesired alcohol followed by nucleophilic substitution with potassium superoxide (Takatsuto *et al.*, 1984b; Mori *et al.*, 1984; Amann *et al.*, 1987; Nakamura and Kuwajima, 1985) or by the Mitsunobu reaction (Aburatani *et al.*, 1986). Better results have been achieved *via* oxidation of the mixture of 22-alcohols followed by hydride reduction of the resulting 22-ketones **218** (Table XI).

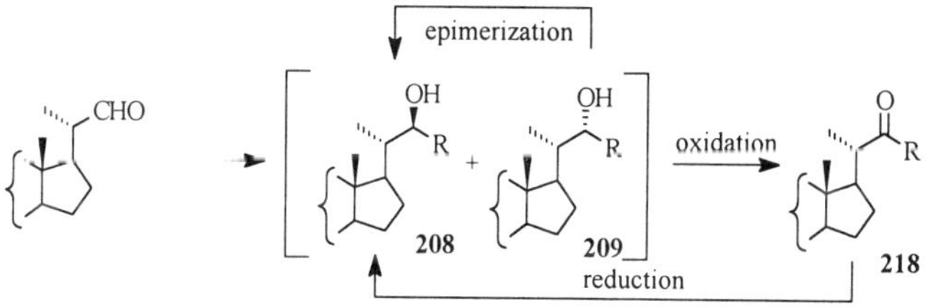

SCHEME 54

TABLE XI
Stereoselectivities in the Hydride Reduction of 22-Ketones 218

Side chain	Reagent(s)	% 208	% 209	Reference
(structure: ketone with alkyne-isopropyl)	NaBH$_4$	60	40	Mori et al., 1984
(structure: ketone-alkyne-CH$_2$OMOM)	LiAlH$_4$, (L)-(−)-N-methylephedrine	94	6	Sardina et al., 1986
	LiAlH$_4$, (L)-(−)-N-methylephedrine	93	7	Sardina et al., 1983
	LiAlH$_4$, (D)-(+)-N-methylephedrine	29	71	Sardina et al., 1983
(structure: ketone-alkyne-CH$_2$OTBDMS)	R-Alpine-Borane	99	1	Midland and Kwon, 1984
	S-Alpine-Borane	13	87	Midland and Kwon, 1984
	L-Selectride	8	92	Midland and Kwon, 1984
(structure: ketone-alkyne-CH$_3$)	R-Alpine-Borane	99	1	Midland and Kwon, 1984
	S-Alpine-Borane	27	73	Midland and Kwon, 1984
(structure: ketone-furyl)	NaBH$_4$	21	79	Tsubuki et al., 1992b
	NaBH$_4$	20	80	Zhou, W.-S., and Ge, C.-S., 1989
	DIBAH	36	64	Tsubuki et al., 1992b
	DIBAH	42	58	Zhou, W.-S., and Ge, C.-S., 1989
	L-Selectride	24	76	Tsubuki et al., 1992b
	K-Selectride	25	75	Zhou, W.-S., and Ge, C.-S., 1989
(structure: ketone-methylfuryl)	LiAlH$_4$	6	94	Tsubuki et al., 1992b

BASIC SYNTHETIC METHODS AND FORMAL SYNTHESES

2) Via homopropargylic alcohols. Addition of propargylic anions to aldehyde **219** afforded homopropargylic alcohols that together with some interesting seleno-organic reactions led to allylic alcohol **162** as depicted in Scheme 55 (Back *et al.*, 1989; Back and Krishna, 1991). Treatment of aldehyde **219** with 3-lithio-1-(trimethylsilyl)propyne followed by desilylation with tetrabutylammonium fluoride gave a mixture (1.32:1) of the 22*R*- and 22*S*-alcohols **220**. The mixture of compounds **220** was then subjected to free-radical selenosulfonation with *Se*-phenyl *p*-tolueneselenosulfonate to give the alcohols **221**, which were epimeric at C-22. Base-catalyzed isomerization of the mixture **221** afforded the allylic sulfones **222**. Their oxidation followed by selenoxide elimination produced a mixture of allenic sulfones, and at this point the 22*R* isomer **223** could be purified by flash chromatography. Introduction of the terminal isopropyl group was achieved *via* addition of cuprate reagent to allene **223** to give the sulfone **224**. Reductive desulfonylation furnished the desired allylic alcohol **162** along with alcohol **225** and diene **226**.

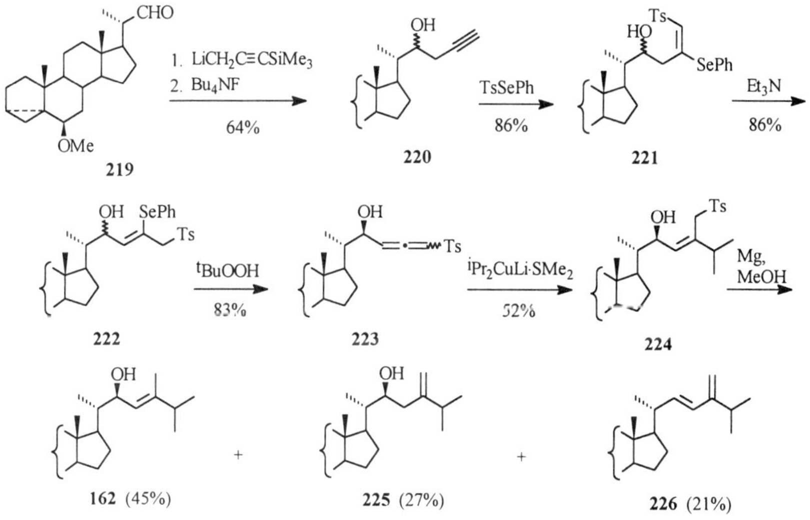

SCHEME 55

3) Via propargylic and allylic alcohols

a) Preparation. 22-Allylic and -propargylic steroid alcohols are flexible intermediates that can be transformed easily into steroid side chains with a 22R,23R-diol function and an alkyl substituent at C-24 with the correct stereochemistry. It is not surprising that this approach has attracted the attention of many chemists for the synthesis of BS side chains.

Propargylic alcohols **200** can be easily transformed into both Z- and E-allylic alcohols **227** or **228**, so the approaches easily merge. Partial hydrogenation of the acetylenic alcohols **200** over a Lindlar catalyst gives in nearly quantitative

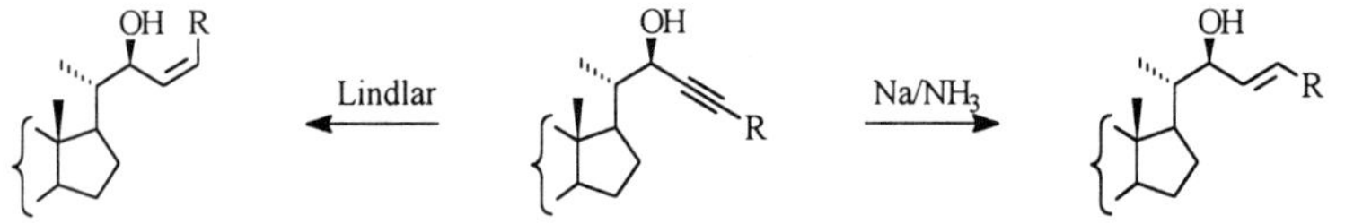

SCHEME 56

yield the Z-allylic alcohol (Anastasia et al., 1983d; Takatsuto et al., 1984b; Takatsuto and Ikekawa, 1986b,c,e; Aburatani et al., 1987b), whereas reduction with sodium in liquid ammonia proceeds with formation of the E-olefins **228**.

The low stereoselectivity in the addition of lithium isopropylacetylide to 22-aldehyde **229** can be compensated by the possibility of transforming both isomeric propargylic alcohols **230** and **231** into the desired Δ^{22}-olefin **234** (Zhou, W.-S., and Tian, W.-S., 1985; Zhou, W.-S., 1989). Depending on the method of reduction of the triple bond, Z- or E-allylic alcohols **232** or **233** were obtained (Scheme 57). Both alcohols led after treatment of the corresponding carbamates with methyllithium to the same olefin **234**.

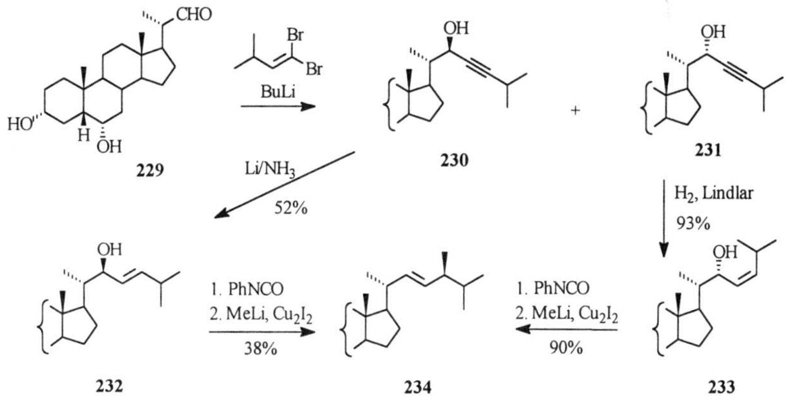

SCHEME 57

An approach based on addition of aluminum organic reagents to 22-aldehydes has been proposed at the first stage of BS investigations (Scheme 58).

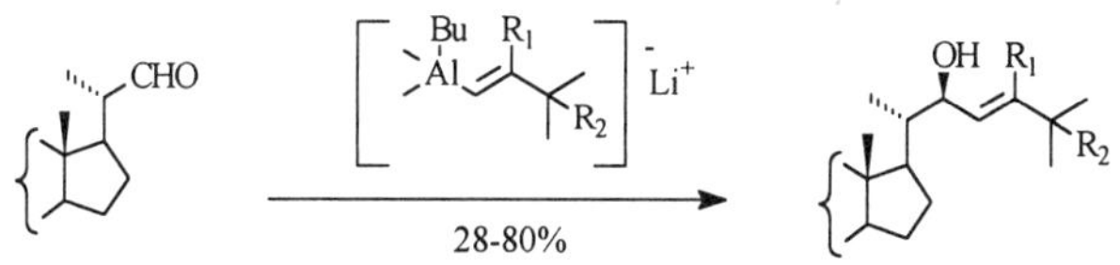

R_1=Me, Et; R_2=H, Me

SCHEME 58

Its disadvantage is the low yield of the desired 22-alcohol, especially for highly functionalized molecules (Fung and Siddall, 1980; Mori *et al.*, 1984; Mori and Takeuchi, 1988).

Syntheses of allylic alcohols *via* acetylenic alcohols suffer however from a low stereoselectivity in the addition step. An attempt to overcome this drawback involves a sulfenate-sulfoxide rearrangement in combination with an equilibrium as the key step (Zhou, W.-S., and Shen, Z.-W., 1991). The sulfoxides **237** and **238** gave the same anion **239** upon treatment with LDA, and alkylation with MeI afforded the methyl derivative **240** (Scheme 59). Allylic alcohol **241** was obtained after rearrangement of sulfoxide **240** in 47% overall yield from **235** or

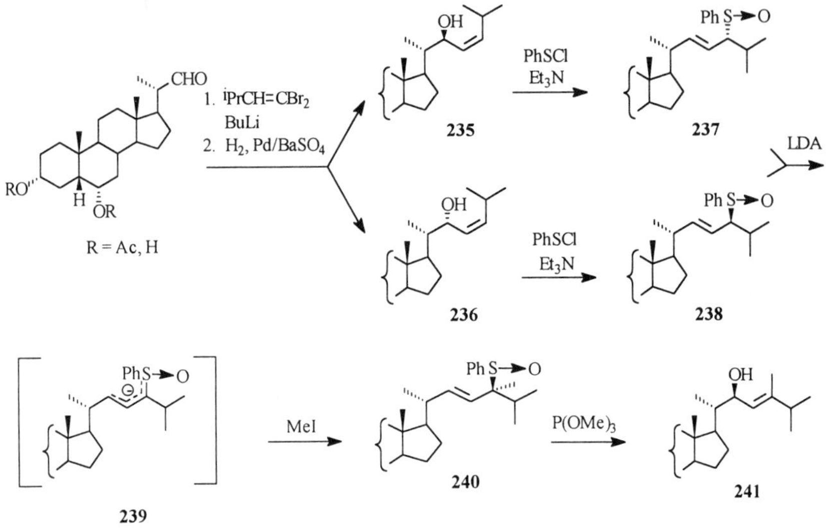

SCHEME 59

in 40% overall yield from **236** along with a small amount of the corresponding epimer at C-22.

A two-step route to allylic alcohol **243** has been investigated (Khripach *et al.*, 1995a, 1996d) that involved the addition of the anion derived from sulfone **242** to aldehyde **219** to construct the carbon skeleton of the side chain (Scheme 60). Desulfonization with lithium in liquid ammonia was accompanied by migration of the $\Delta^{24(28)}$-double bond to give the desired allylic alcohol **243**.

The presence of a bulky substituent in the α-position of the anionic center of **245** enabled a considerable increase of the stereoselectivity of the addition

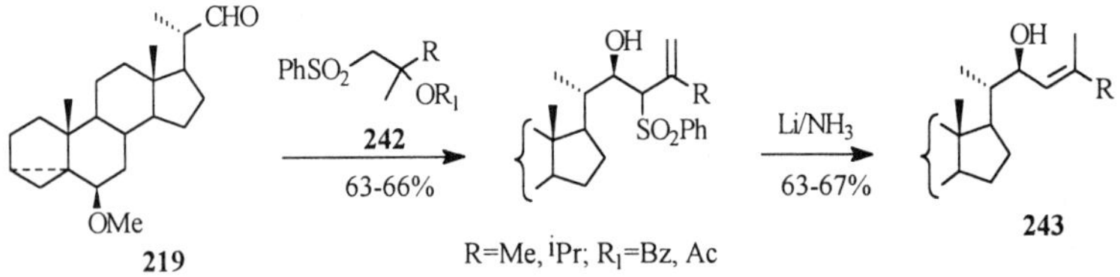

SCHEME 60

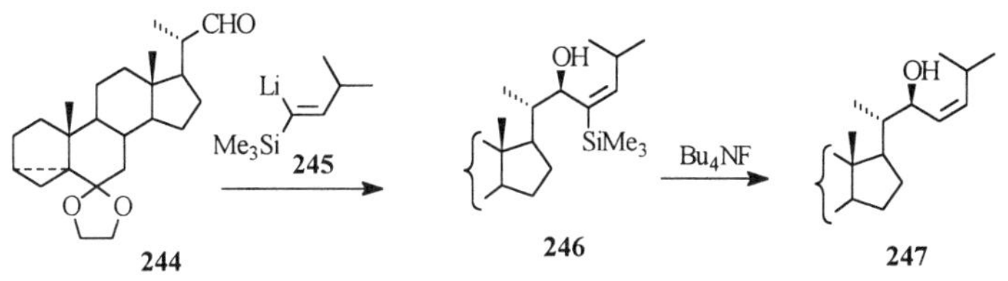

SCHEME 61

(Khripach et al., 1990i, 1993g). Thus, the reaction of aldehyde **244** with lithium vinyl silane **245** gave the alcohol **246** along with a small amount of its 22-epimer (10:1) in 97% total yield (Scheme 61). Desilylation of **246** afforded the desired allylic alcohol **247**.

The introduction of a three-carbon fragment *via* addition to aldehyde **219** (Scheme 62) can be achieved by addition of one of the most simple vinyl anions in a reasonable stereoselectivity (Back et al., 1991, 1993, 1997a). Arsenic ylides

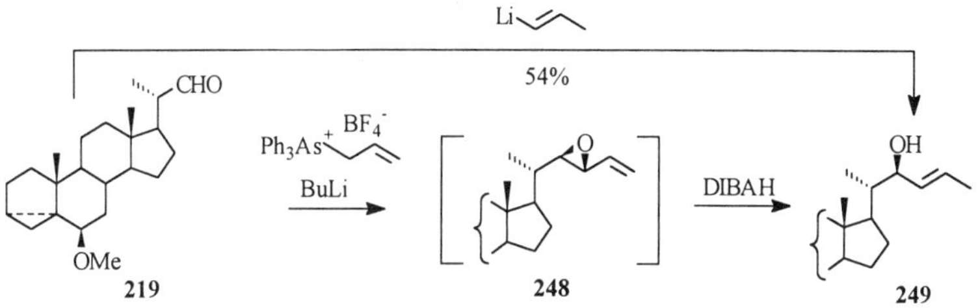

SCHEME 62

provide another alternative for preparation of the alcohol **249** (Werner et al., 1996). Treatment of the aldehyde **219** with propenyltriphenylarsonium tetrafluoroborate led to formation of epoxide **248**, which after hydride reduction gave the allylic alcohol **249** in 75% overall yield and over 99% ee.

b) Transformation into the BS side chain. At this point some methods will be presented on how to convert these C-22 allylic alcohols into the side chain as it is present in many BS. Most approaches use epoxidation of the Δ^{23}-double bond as the first step because it proceeds stereoselectively, guided by the directing effect of the adjacent 22R-hydroxy group, to the desired 23R,24R-

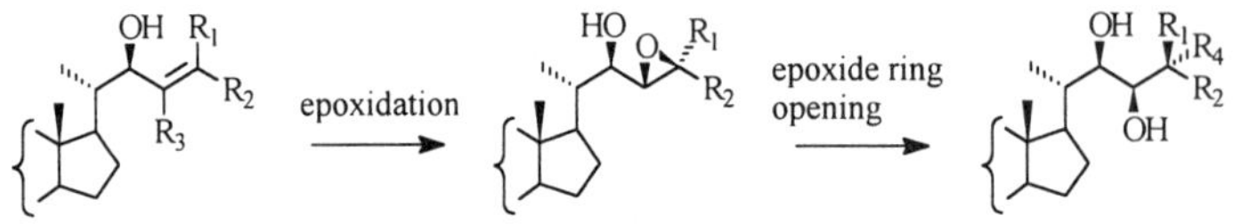

SCHEME 63

R_1 and R_2=alkyl, H, SiMe$_2$Ph; R_3=H, SiMe$_3$, R_4=alkyl

epoxides as major products (Scheme 63). The reaction has been studied in detail (Back and Baron, 1996) with the allylic alcohols **201** with various substituents R_1-R_3 and an explanation for the observed stereoselectivity has been presented.

When R_1 and R_2 are both alkyl groups (usually R_1=Me or Et and R_2=^{i}Pr or ^{t}Bu), no further C-C bond formation stages will be necessary to complete the side chain, and it can be easily finished by an anti-Markovnikov reduction of the hydroxy epoxides **250** (Scheme 64). Intermediates such as **250** are also important for the preparation of $\Delta^{24(28)}$-BS **251**.

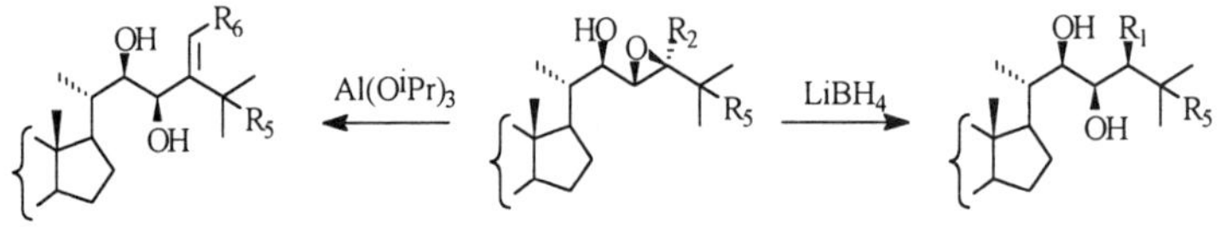

R_2=Me or Et, R_5=H or Me, R_6=H or Me

SCHEME 64

Some synthetic schemes rely upon more available allylic alcohols **201** with only one alkyl substituent at C-24 (R_2=R_3=H). In this case the whole sequence should include another C-C bond formation that goes together with opening of the epoxide (Scheme 65). The nucleophilic fragment that has to be brought into the side chain may be either the 28-methyl (Fung and Siddall, 1980; Okada and Mori, 1983b; Mori *et al.*, 1984; Mori and Takeuchi, 1988; Hazra *et al.*, 1994) or the terminal isopropyl group (Hayami *et al.*, 1983; Teijin, Ltd., 1983; Back *et al.*, 1991, 1993, 1997a).

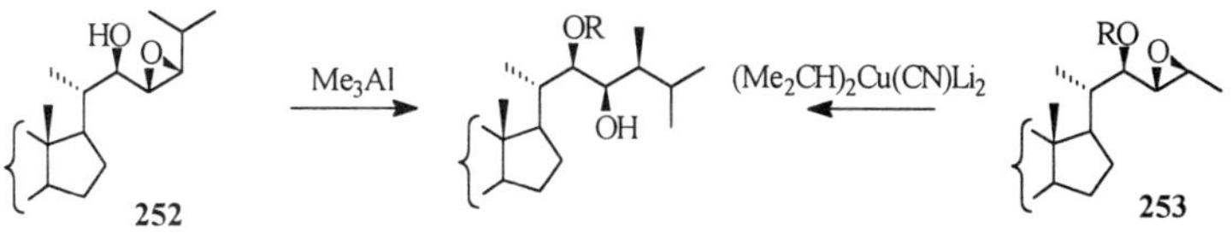

SCHEME 65

R=H, Bn

A roundabout way for opening of the epoxide ring *via* the nitrile **254** has been elaborated in some approaches to brassinolide as shown in Scheme 66 (Ishiguro *et al.*, 1980; Takatsuto *et al.*, 1984b).

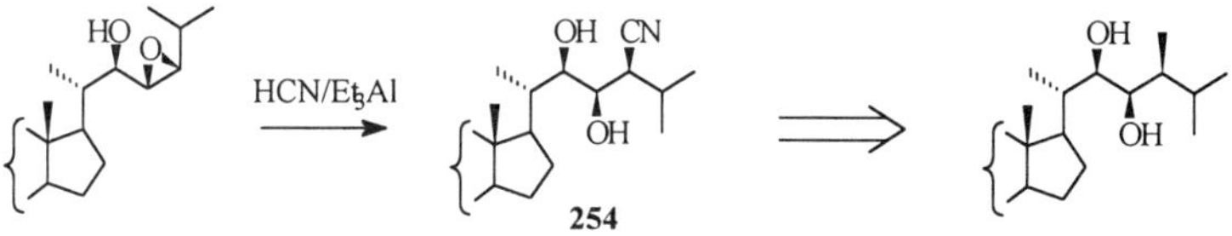

SCHEME 66

4) **Via Δ^{22}-olefins.** Progress in the asymmetric hydroxylation of olefins makes these Δ^{22}-steroids convenient intermediates for the synthesis of BS side chains. There is no necessity to prepare steroids of the stigmastane or ergostane series starting from 22-aldehydes, because these compounds are much more easily available from the corresponding sterols from nature. In contrast, chemical synthesis is practically the only source of Δ^{22}-steroids with a 24α-methyl group. Many synthetic approaches to BS side chains involved formation of Δ^{22}-steroids as intermediates; some of them have already been mentioned in the preceding paragraph. The C-22 allylic alcohols, described above, are often transformed into Δ^{22}-olefins (Scheme 67), sometimes together with the stereoselective

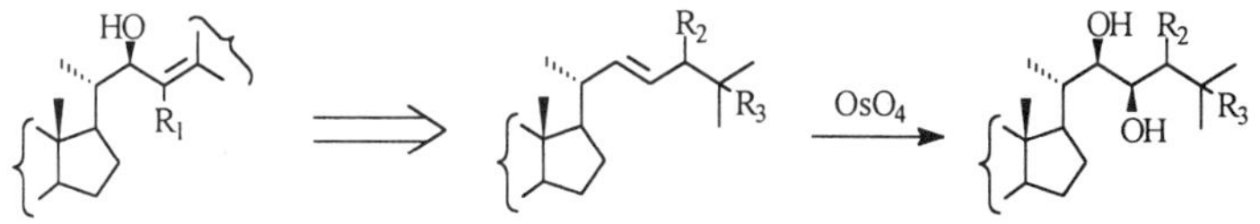

R_1=H, pTol; R_2=Me, Et; R_3=H, OH

SCHEME 67

introduction of an additional substituent at C-24.

An effective method which is widely used also for other groups of steroids for the construction of Δ^{22}-unsaturated side chains starting from the allylic alcohols is the Claisen rearrangement of appropriate precursors (Anastasia and Fiecchi, 1981; Fujimoto et al., 1984, 1985; Gilhooly et al., 1982; Hirano and Djerassi, 1982; Meadows and Williams, 1980; Morris et al., 1981; Preus and McMorris, 1979; Riccio et al., 1990; Shu and Djerassi, 1981; Takatsuto and Ikekawa, 1986b,c,e; Wiersig et al., 1979). Also in this way a substituent can be introduced at C-24, along with the construction of the Δ^{22}-double bond. Two examples of such an approach are shown in Scheme 68 for the synthesis of crinosterol and its derivatives (Anastasia et al., 1983a,d). The Claisen rearrangement of allylic alcohol **255** with triethyl orthopropionate afforded a mixture of two olefinic esters **256**, epimeric at C-25 (Anastasia et al., 1983a). The esters were reduced into a methyl group in the usual way. An alternative route consisted of the Claisen rearrangement of allylic alcohol **257**. Transformation of the resulting ester **258** into the olefin **260** was achieved by reduction to the aldehyde **259** and decarbonylation over a Wilkinson's catalyst (Anastasia et al., 1983d).

The stereospecific S_N2' substitution of allylic carboxylates with cuprate reagents was shown to allow such a controlled introduction of the alkyl

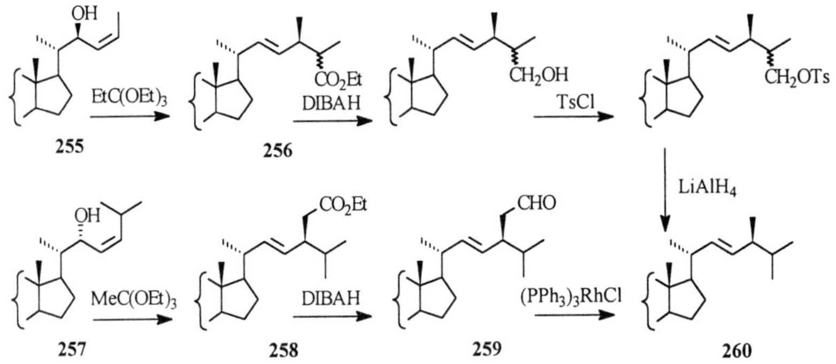

SCHEME 68

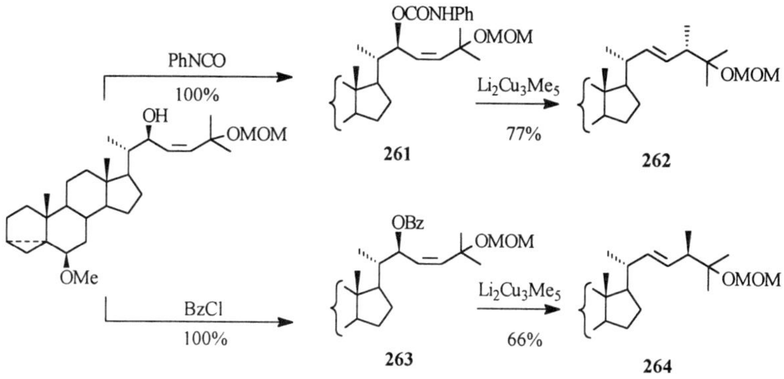

BASIC SYNTHETIC METHODS AND FORMAL SYNTHESES 119

SCHEME 69

substituent at C-24 (Sardina et al., 1983, 1986). Thus, the reaction of the carbamate **261** with cuprate reagent gave only the product of syn-S_N2' substitution **262** (Scheme 69), whereas the benzoate **263** afforded the olefin **264** as a result of an anti-S_N2' substitution.

Difficulties with the introduction at the C-24 position of substituents other than methyl groups prompted the authors (Torneiro et al., 1997) to develop an improvement of this approach (Scheme 70). The S_N2' substitution of the derivative **265** with various Grignard reagents in the presence of CuCN and LiCl proceeded stereoselectively to give the olefins **266**.

Another variant of the S_N2' substitution for stereoselective construction of the C-24 chiral center was proposed by Marino et al. (1996). Condensation of the aldehyde **219** with the α-lithiovinyl sulfide **267** gave the allylic alcohol **268** as the major isomer, together with a small amount of the anti-Cram product

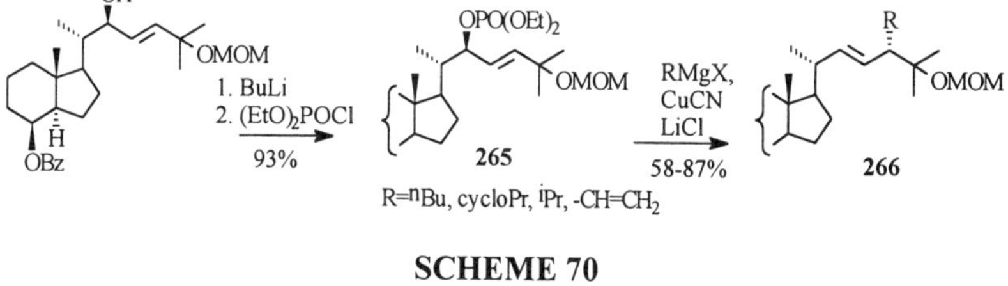

SCHEME 70

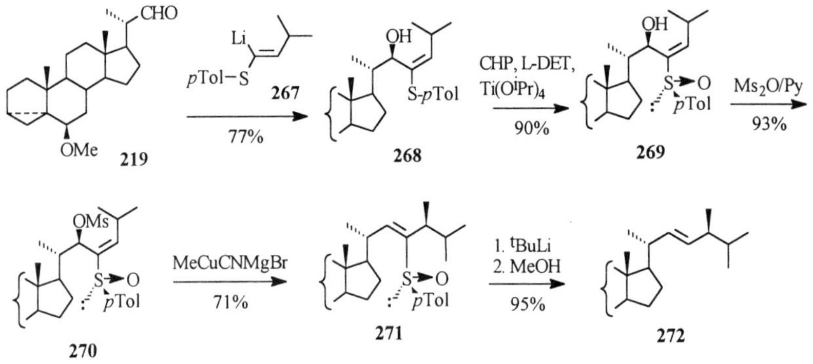

SCHEME 71

(92:8) in 84% overall yield (Scheme 71). Asymmetric oxidation of **268** with cumene hydroperoxide in the presence of titanium isopropoxide and L-(+)-diethyl tartrate afforded (S_s)-sulfoxide **269** as the sole product. The key stage of the whole methodology was the S_N2' substitution of the allylic mesyloxy sulfinyl group in **270** by methyl cyanocuprate to give a mixture (85:15) of the desired (22E,24R)-olefin **271** along with its 22Z,24S isomer. Reductive desulfurization of the sulfoxide **271** led to the olefin **272** in nearly quantitative yield.

The Wittig olefination reaction is the most obvious route to Δ^{22}-steroids starting from 22-aldehydes. It has been extensively explored for synthesis of steroids, e.g., vitamins D (Zhu, G.-D., and Okamura, W.H., 1995), and there are a few examples of its use for the preparation of the side chain of BS. A main drawback is the low stereoselectivity of the reaction, resulting in formation of both E- and Z-olefins (Piatak and Wicha, 1978; Bogoslovskii et al., 1978; Salmond and Barta, 1978; Zhabinskii et al., 1996). The situation is much better for resonance-stabilized ylides, where E-olefins may be obtained as the only isomers (Sucrow and Littmann, 1976; Vanderah and Djerassi, 1978; Kihira and Hoshita, 1985; Calverley, 1987).

Good results were obtained for the reaction of 22-aldehydes with arsonium organic compounds (Huang, Y.-Z. et al., 1988; Shen, Z.-W. et al., 1990; Zhou,

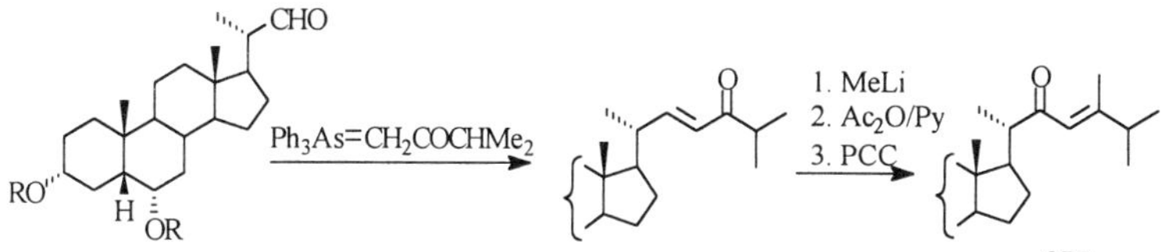

SCHEME 72

W.-S. et al., 1989a, 1990b). Treatment of the aldehydes **229** or **273** with isobutylcarbonylarsonium ylide gave the α,β-unsaturated ketone **274** (Scheme 72). Its reaction with methyllithium furnished the corresponding tertiary allylic alcohol. Oxidation gave the transposed unsaturated ketone **275** and this could be easily transformed into brassinolide precursors (see Scheme 46).

A similar approach to enones such as **275** has been explored by Hazra et al. (1996). The Horner-Wadsworth-Emmons reaction of the 22-aldehyde with diethyl (3-methyl-2-oxobutyl)phosphonate gave the desired product in 92% yield, whereas the Wittig olefination with (3-methyl-2-oxobutyl)triphenylarsonium bromide and (3-methyl-2-oxobutyl)triphenylphosphonium bromide furnished the enone in 71 and 32% yield, respectively.

A good alternative to the Wittig reaction is the Julia olefination (Scheme 73). It was first used in steroid chemistry for the synthesis of vitamin D (Kocienski et al., 1978; Yamada et al., 1984; Kutner et al., 1987, 1993; Choudhry et al., 1993) and proved to be useful also for the side chain construction in BS (Sakakibara et al., 1982; Mori et al., 1982; Khripach et al., 1990j,k, 1994b). Addition of the anion derived from phenyl sulfone to the 22-aldehyde **137** followed by acetylation gave four acetoxy sulfones **276**, isomeric at C-22 and C-23. Treatment of the mixture with sodium amalgam resulted in formation of practically only the *E*-olefin. Synthesis of Δ^{22}-olefins could also be achieved via the Ramberg-Bäcklund rearrangement (Schmittberger and Uguen, 1996, 1997). The procedure included alkylation of the thiol **277**, chlorination of the resulting sulfide **278**, oxidation, and base treatment of the intermediate chlorsulfone.

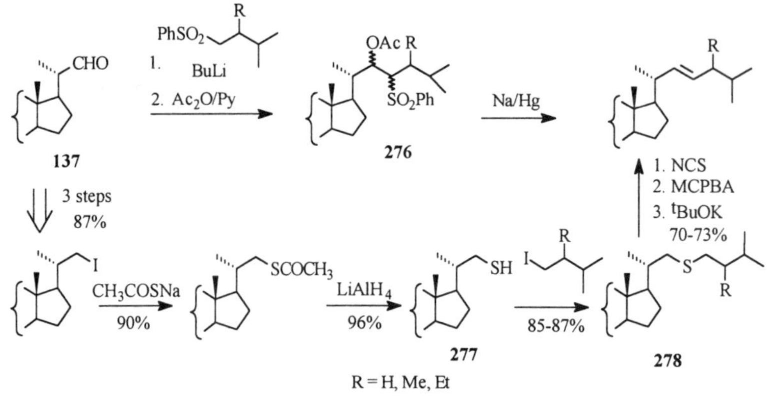

SCHEME 73

Together the Julia olefination and the Ramberg-Bäcklund rearrangement constitute a convergent synthesis of the desired molecules from two less complicated subunits. The sulfone **282** and the iodide **281** can both play the role of the second partner in coupling to the steroid part. Though both compounds have a rather simple structure compared to BS, their preparation, especially in optically pure form, is somewhat troublesome (Tachibana *et al.*, 1992; Schmittberger and Uguen, 1997). In the BS series it was solved first by Mori *et al.* (1982) starting from (*R*)-(+)-citronellic acid **279** (Scheme 74). The acid **280** could be prepared also from a lower homolog by one-carbon elongation (Levene and Marker, 1935) or by asymmetric synthesis (Mukaiyama *et al.*, 1978).

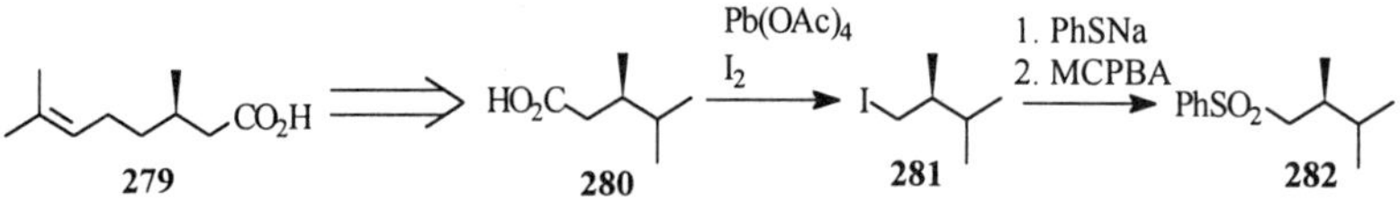

SCHEME 74

5) Via hydroxy furans. Syntheses based on this approach rely on the reaction of various 2-lithiofurans **283** with 22-aldehydes **137** followed by rearrangement of the major isomer **284** into an anomeric mixture of cyclic

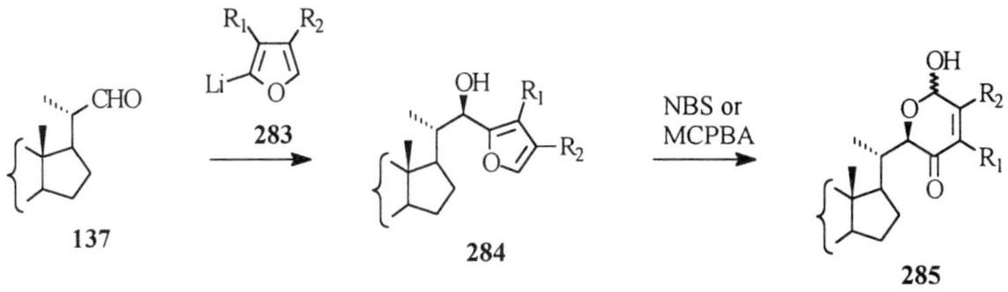

SCHEME 75

R_1=H, ^{i}Pr; R_2=Me, H

hemiacetals **285** (Scheme 75). The further strategy depends on the substituents R_1 and R_2 and a synthetic target.

Treatment of the acetals **286** with pyridinium chlorochromate gave the lactones **287** or **289** (Scheme 76). Starting from lactone **287**, no further C-C bond formation is necessary for preparation of the side chain of brassinolide from intermediate **288** (Kametani *et al.*, 1988b). Starting from lactone **289**, the introduction of one methyl group is necessary by conjugate addition. After reduction and deprotection, intermediate **290** is obtained, which is suitable for the construction of the side chain of brassinolide (Honda *et al.*, 1990; Honda and Tsubuki, 1990; Tsubuki *et al.*, 1992a).

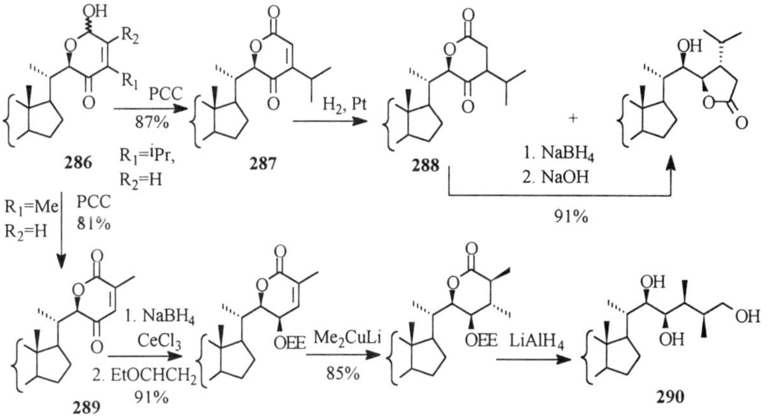

SCHEME 76

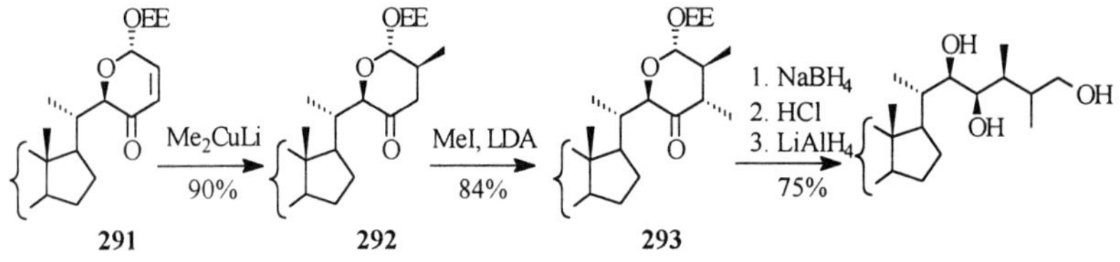

SCHEME 77

The stereoselective introduction of two methyl groups will be necessary for syntheses starting from acetals **286** when R_1 and R_2 are hydrogen atoms (Kametani et al., 1989; Zhou, W.-S., and Ge, C.-S., 1989; Zhou, W.-S. et al., 1993). This may cause problems because the stereochemistry of the alkylation is greatly influenced by substituents in the pyranone ring (Zhou, W.-S. et al., 1993). An example of a successful synthesis of the brassinolide side chain is demonstrated by Kametani et al. (1989). Conjugate addition to enone **291** proceeded stereoselectively and so did the alkylation of **292**. The stereoselective reduction of **293** followed by further transformations ultimately led to the brassinolide side chain (Scheme 77).

BS unsubstituted at C-24 can be obtained from the hemiacetals also (Zhou, W.-S. et al., 1993). The reaction sequence includes protection of **294** with trimethyl orthoformate, conjugate addition of lithium dimethylcuprate, stereoselective reduction of the 23-ketone **295**, and transformation of **296** into the open-chained structure **297** (Scheme 78).

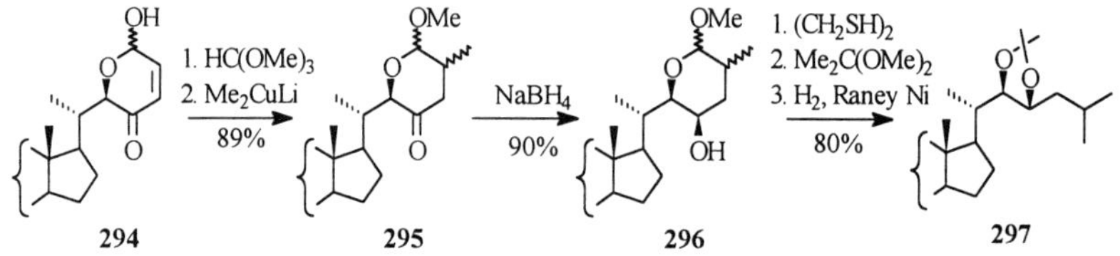

SCHEME 78

6) Via hydroxy lactones. In some approaches the control of the stereochemistry at C-22 and C-23 in one chemical step is possible. For instance the reaction of enolate **298** at 0 °C (kinetic conditions) with aldehyde **219** leads to the preferential formation of the 22R,23S-hydroxy butenolide **299** (Scheme 79), whereas the same reaction at -78 °C mainly gives the 22R,23R intermediate **300** (Donaubauer et al., 1984; McMorris et al., 1991). In both cases the preferred stereochemistry at C-22 is in accordance with Cram's rule.

Catalytic hydrogenation of the intermediate **300** afforded in quantitative yield a mixture (78:22) of saturated lactones that are epimeric at C-24. Reduction of the main isomer **301** led to opening of the lactone ring with formation of the triol **302**, which could be converted into derivative **305** with a brassinolide side chain as indicated.

A shorter route to the acetonide **304** was based on the reaction of aldehyde **219** with anion **306** and is described in Scheme 80 (Donaubauer et al., 1984).

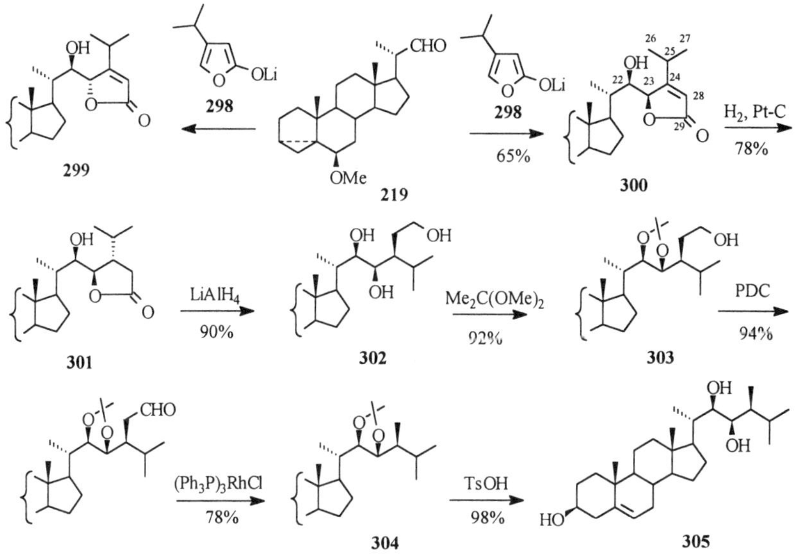

SCHEME 79

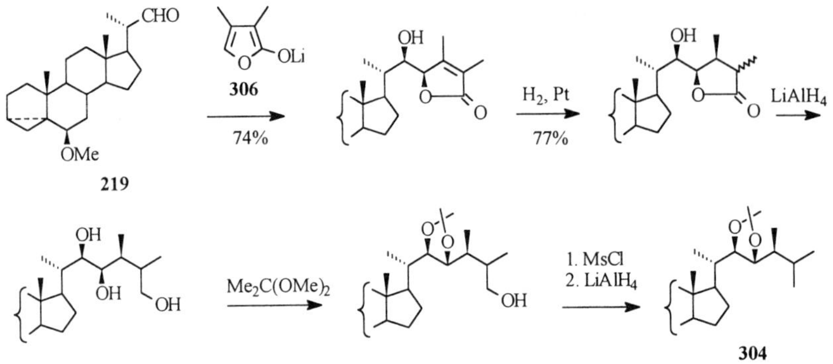

SCHEME 80

A third variant of this approach used the aldol reaction of the aldehyde **273** with the enolate **307** (Zhou, W.-S., and Tian, W.-S., 1987). The addition gave a

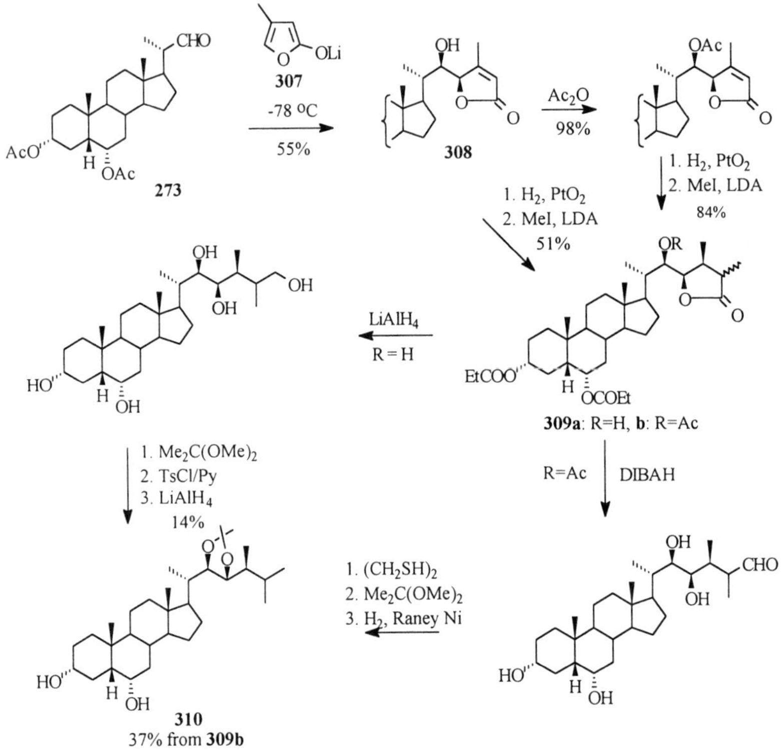

SCHEME 81

mixture (64:36) of 22-alcohols in 86% overall yield. Two routes have been realized for the transformation of the main 22R-alcohol **308** into the acetonide **310** as is depicted in Scheme 81.

7) Via 22-hydroxy-23-aldehydes. At first this approach was mainly used for the preparation of BS without a chiral center at C-24, e.g., 28-nor- or $\Delta^{24(28)}$-BS (Takatsuto and Ikekawa, 1983b,c; Takatsuto *et al.*, 1984a). The required intermediate **311** was prepared in a traditional way by addition of the lithium salt of 1,3-dithiane (Scheme 82). The 22-hydroxy group in this adduct had to be protected prior to the desulfurization step because of the instability of the α-hydroxy aldehyde. The acetoxy aldehyde **312** was suitable for transformation both into norbrassinolide (Takatsuto *et al.*, 1981) and into brassinolide precursors **313** (Hazra *et al.*, 1994). Better results have been achieved by MOM protection because chelation-controlled addition of the Grignard reagent gave the correct configuration at C-23 (Takatsuto and Ikekawa, 1983b,c; Takatsuto *et al.*, 1984a).

A similar chelation-controlled ene reaction (Scheme 83) of the aldehyde **314a** with isobutylene in the presence of a Lewis acid gave the ether **315** as the main product (Mikami *et al.*, 1990; Nakai and Mikami, 1989a; Ihara Chem. Ind. Co., Ltd., 1982b). Further advances in this field allowed achievement of the

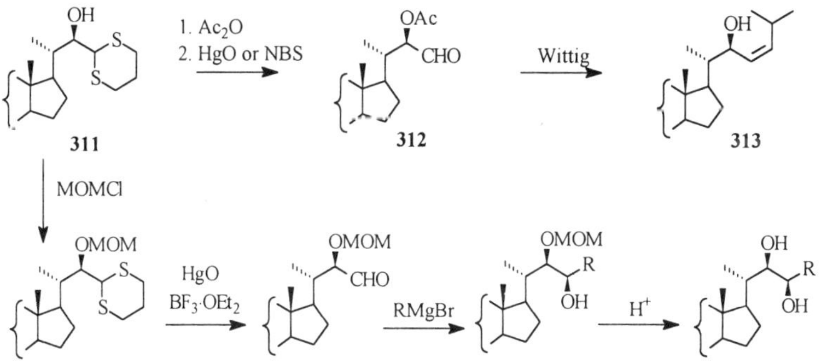

SCHEME 82

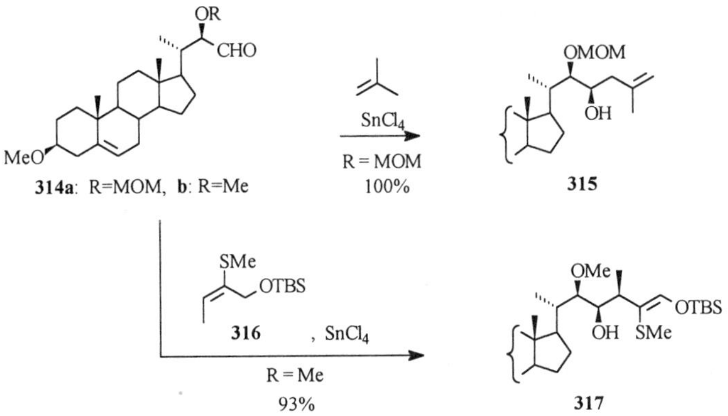

SCHEME 83

introduction of a C-24 substituent *via* addition of thiosilyl ether **316**. Compound **317** was isolated together with only a small amount of the corresponding C-23 epimer (97:3) (Nakamura *et al.*, 1993). Similar studies were carried out on tin-organic derivatives (Nakai and Mikami, 1986a,b).

Another possibility for controlling the stereochemistry of the substituent at C-24 has been shown on model compounds (Mikami and Sakuda, 1993). The key stage of the method was a chelation-controlled carbonyl ene reaction of the α-benzyloxy aldehyde **318** with vinyl sulfide **319** (Scheme 84). Dependent on the substituent R the threo-**320** or the erythro-isomer **321** could be obtained.

Synthesis of the brassinolide side chain has also been based on an α-silyl radical cyclization-protiodesilylation sequence as the key step for the

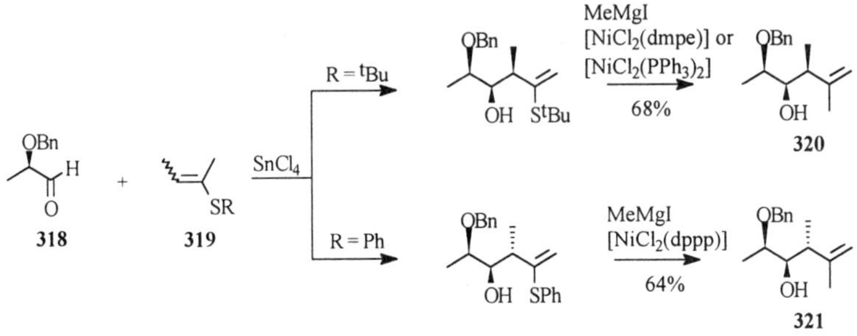

SCHEME 84

stereoselective introduction of the 24-methyl group (Koreeda and Wu, 1995). Treatment of the MOM-protected hydroxy aldehyde **322** with 1-lithio-2-methyl-1-propene gave the syn-diol mono MOM ether **323** (72%) in addition to its epimer (7%) (Scheme 85). The α-silyl radical cyclization proceeded exclusively in the 5-exo mode to furnish the trans-siloxane **324**. The heterocyclic compound **325** was subjected to protiodesilylation to afford 24-methyl derivative **326**. Treatment with acid led simultaneously to regeneration of the 3β-hydroxy-5-ene system and deprotection of the MOM group to give triol **305**.

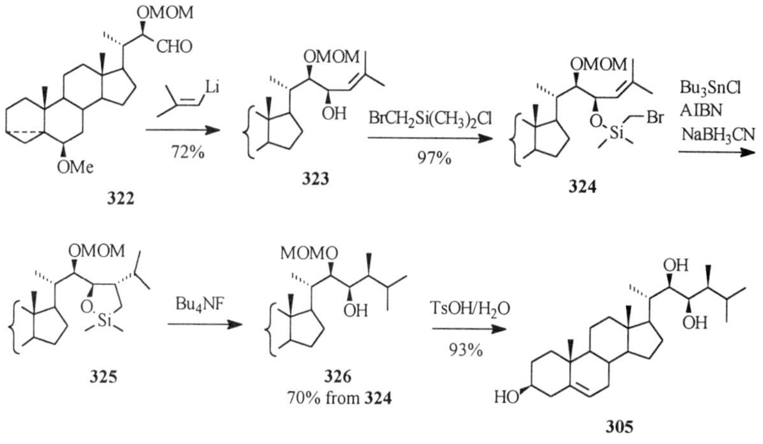

SCHEME 85

8) Via hydroxy ethers. An effective method for construction of the side chain of norbrassinolide is the addition of (*S*)-α-methoxy derivative **328** to 22-aldehyde **327** (Furuta and Yamamoto, 1992). The only product of the reaction was the monomethyl ether **329** (Scheme 86). A similar reaction with the corresponding (*R*)-organolead derivative proceeded very slowly and with partial racemization at the chiral center of the reagent. This allows the use of a racemate under the conditions of kinetic resolution. Thus, the reaction of aldehyde **327** with 4 equiv of racemic **328** gave compound **329** in 99% yield. A serious problem of the approach is the removal of the protecting group in the last stage.

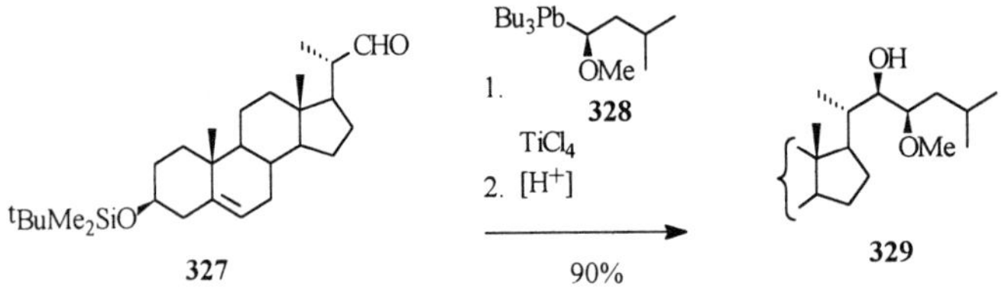

SCHEME 86

Another possibility for achieving the correct stereochemistry of both the C-22 and C-23 centers in one step is the addition of tin organic compounds **330** and **331** to 22-aldehydes (Scheme 87). It was shown that in both cases the threo-monomethyl ethers were the only products of the reaction (Koreeda and Tanaka, 1987).

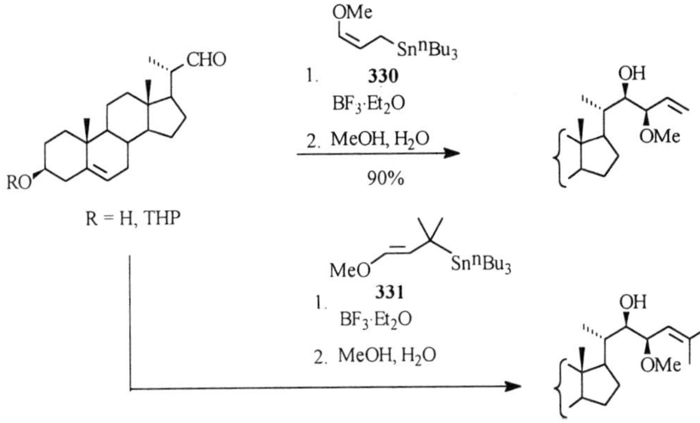

SCHEME 87

9) Via isoxazoles and isoxazolines. The 1,3-dipolar cycloaddition of nitrile oxides to alkenes and acetylenes followed by transformation of the obtained isoxazolines or isoxazoles has proven to be a useful method for the synthesis of many natural products (Torsell, 1988). Some of the developed methods could be successfully applied for BS synthesis (Khripach, 1990; Khripach *et al.*, 1991d; Baranovskii *et al.*, 1993). The addition of olefin **332** to the nitrile oxide generated from *N*-chlorosuccinimide and isobutyraldoxime gave a mixture of isoxazolines **333**, epimeric at C-22 (Khripach *et al.*, 1991a-c). Hydrogenolysis of

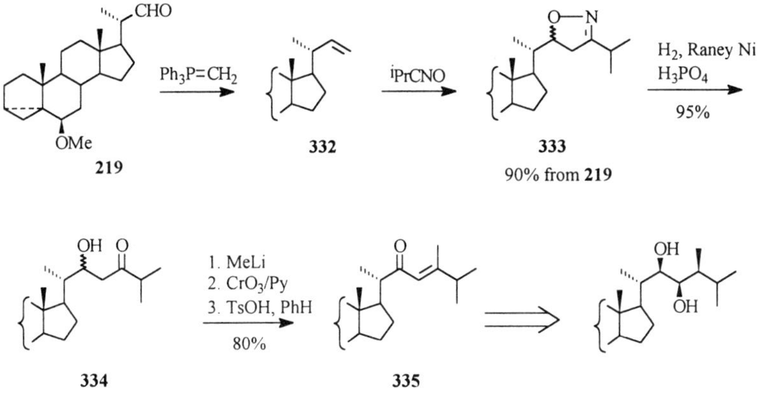

SCHEME 88

333 over Raney nickel led to the ketols **334** (Scheme 88). Introduction of the 24-methyl group, oxidation of the C-22 secondary alcohol group, and dehydration of the resulting ketols led to the enone **335** (Khripach *et al.*, 1990f), useful as an intermediate for brassinolide synthesis (see Scheme 46).

Intermediate enones such as **338** could be obtained from 20-isoxazolines **336** (Scheme 89) *via* base-catalyzed cleavage of the heterocycle followed by hydrolysis of the unsaturated oxime **337** (Khripach *et al.*, 1990e, 1993c). Transformation of Δ^{22}-23-ketones **338** into derivatives such as **335** has been described in Scheme 72.

The route toward BS side chains *via* this 1,3-dipolar addition is a very flexible one because the nitrile oxide can be prepared also from a 22-aldehyde,

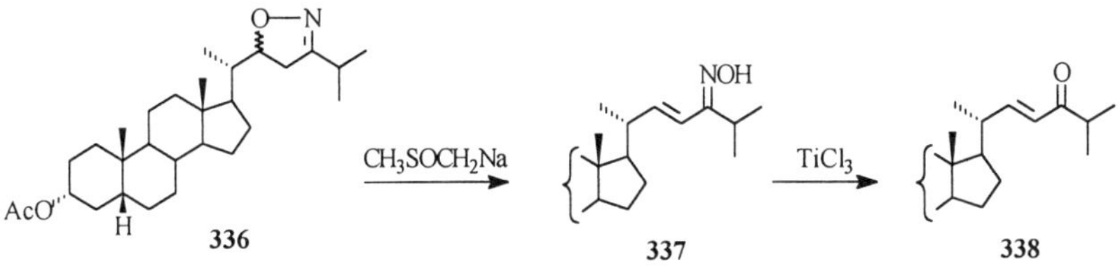

SCHEME 89

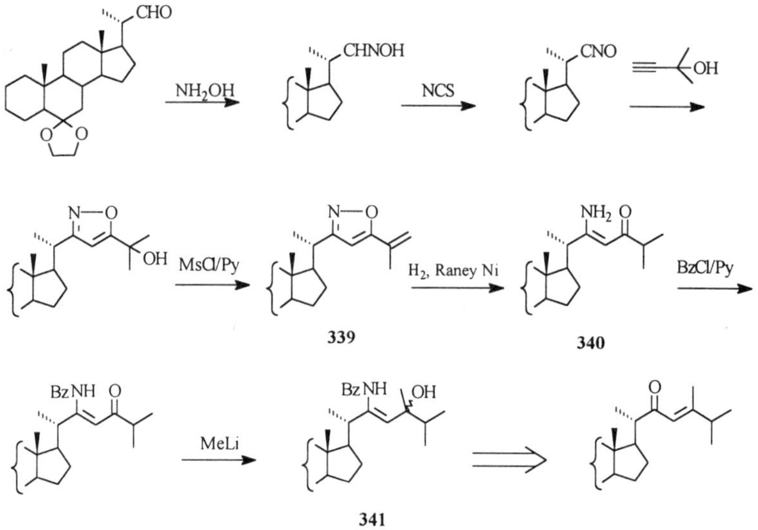

SCHEME 90

which allows the use of a variety of alkenes and alkynes for the preparation of natural and nonnatural side chains (Khripach et al., 1993e). The intermediate 22-nitrile oxides (Scheme 90) proved to be remarkably stable and could be isolated in a pure state (Khripach et al., 1993f). Hydrogenolysis of the obtained isoxazole **339** led to the enamino ketone **340**, which had to be benzoylated to ensure the 1,2-introduction of the 24-methyl group. Compound **341** could be transformed into a Δ^{23}-22-ketone by known methods (Khripach et al., 1993d).

A variation of the alkyne allowed the preparation of enone **344** (Khripach et al., 1998) as depicted in Scheme 91. The hydrogenolysis of bromoisoxazole **342** was accompanied by simultaneous debromination to give the enamino ketone **343**. Hydride reduction of the carbonyl group of the corresponding keto benzamide and acid-catalyzed hydrolysis gave the enone **344**. This compound can be transformed into the allylic alcohol **249**, which is a suitable intermediate for the construction of the brassinolide side chain (see Scheme 62).

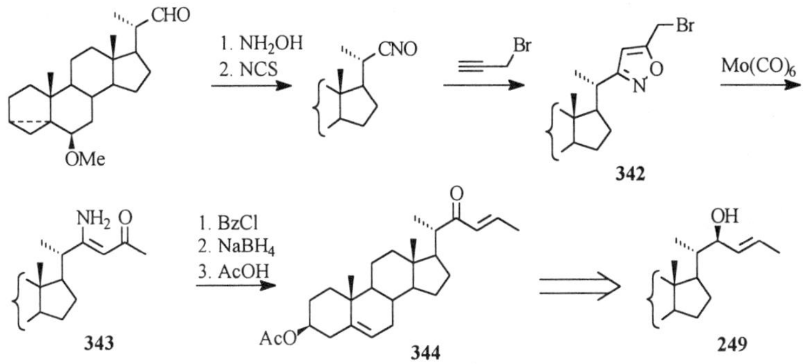

SCHEME 91

1,3-Dipolar cycloaddition of acetonitrile oxide to the allylic alcohol **345** (Scheme 92) led to a mixture (4:1) of two hydroxy isoxazolines epimeric at C-22, with compound **346** as the main isomer (Litvinovskaya *et al.*, 1995). Hydrogenolysis of this hydroxy isoxazoline **346** over Raney nickel gave ketol **347**, which proved to be rather unstable under basic conditions (Wittig reaction). Addition of methyllithium to triol **348** proceeded smoothly. The same approach could be applied for the preparation of the side chain of 28-norbrassinolide (Litvinovskaya *et al.*, 1996a).

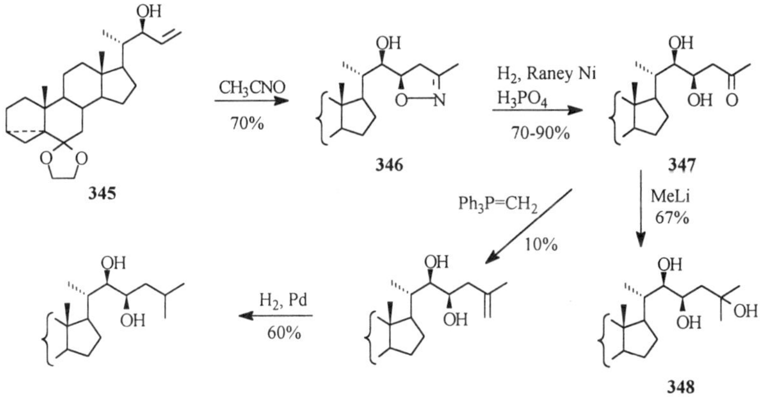

SCHEME 92

Additional possibilities, especially for synthesis of BS analogs, offer the 1,3-dipolar addition of nitrile oxides to Δ^{22}-olefins followed by introduction of the hydroxyl group at C-22 (Litvinovskaya et al., 1994, 1997b).

d. Starting with 24-esters. Most methods that used this approach were based on derivatives of hyodeoxycholic acid. Decarboxylation of the side chain and transformations of intermediate Δ^{22}-olefins into 22-aldehydes are the synthetic routes that have been followed mostly (Zhou, W.-S., 1989) and these approaches have been discussed in the previous section.

Some approaches used the carboxylic group of the starting material to construct the side chain as is indicated for 25-methyl-BS in Scheme 93 (Zhou, W.-S., and Huang, L.-F., 1992). Treatment of **349** with *tert*-butyllithium gave ketone **350**, which upon dehydrogenation afforded the enone **351**.

A similar approach has been used for synthesis of the brassinolide side chain from **349**. Reaction of isopropylmagnesium chloride in the presence of lithium borohydride furnished a mixture of C-24 epimeric alcohols **352**, which were oxidized to ketone **353**. Dehydrogenation of this ketone proved to be a problem, because both possible enones **354** and **355** (2:1) were obtained in this reaction.

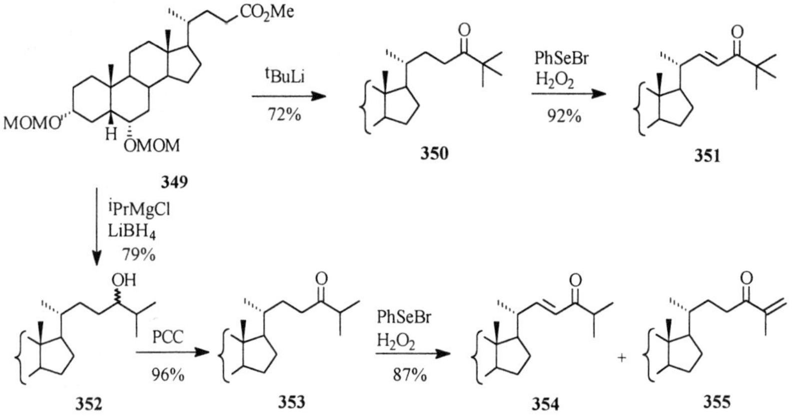

SCHEME 93

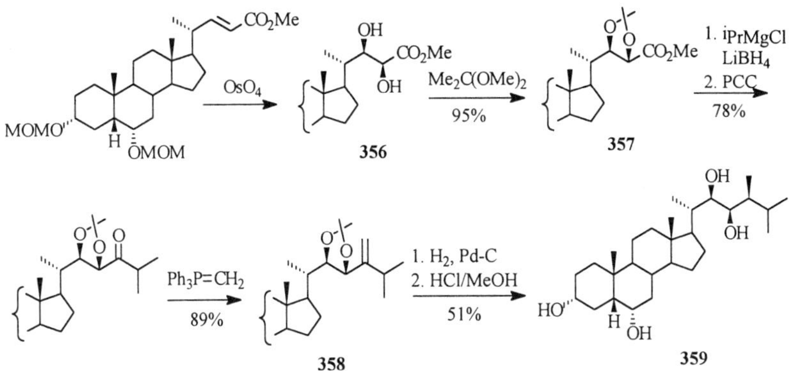

SCHEME 94

Transformation of compounds such as **351** and **355** into those with a BS side chain has been described above (see Scheme 72).

Derivatives of hyodeoxycholic acid that contain a Δ^{22}-double bond offer another possibility for construction of the side chain of BS (Zhou, W.-S. *et al.*, 1991a, 1992a). Catalytic hydroxylation of this double bond with OsO_4 in the presence of dihydroquinidine *p*-chlorobenzoate gave the diol **356** in addition to its isomer (4:1) (Scheme 94). Treatment of the diol **356** with 2,2-dimethoxypropane led to the corresponding acetonide **357**, which was converted into the $\Delta^{24(28)}$-olefin **358** as indicated. Its hydrogenation in the presence of palladium on charcoal followed by acetonide deprotection gave a mixture of diol **359** and its C-24 epimer (4:1), which could be separated by column chromatography.

The material presented in this chapter basically covers all methods of BS synthesis which have been published to date. Some of these methods originate from research on synthesis of ecdysteroids or vitamin D, whereas others have been developed especially for the synthesis of BS and undoubtedly will be useful for the preparation of related compounds.

CHAPTER VII

SYNTHESES OF NATURAL BS

Nowadays chemists have to and can rely on partial synthesis of BS as a source for these compounds. Neither isolation from natural sources (Tokuda, 1986; Saimoto et al., 1989; Yamaji et al., 1992) nor total synthesis (Berthon et al., 1994; Diziere et al., 1994; Zoller et al., 1997) can provide BS in sufficient amounts for practical purposes.

In this chapter partial syntheses of natural BS are described starting from available steroids. The presentation is focused on the synthetic strategy of each approach. It is necessary to note that for synthesis of compounds containing many functional groups such as BS even a change in reaction sequence can lead to considerable improvement of the overall yield. This is an important aspect especially when larger quantities of BS have to be produced for practical application.

A. SYNTHESES OF BRASSINOLIDE

Brassinolide is currently considered to be the most important synthetic target among BS. First of all this is due to its extremely high biological activity. In addition, preparation of brassinolide is a relatively difficult task and a synthetic challenge. The first brassinolide syntheses were not the most effective ones but more a demonstration of capability and speed or meant to prepare brassinolide and congeners in sufficient amounts for biological testing.

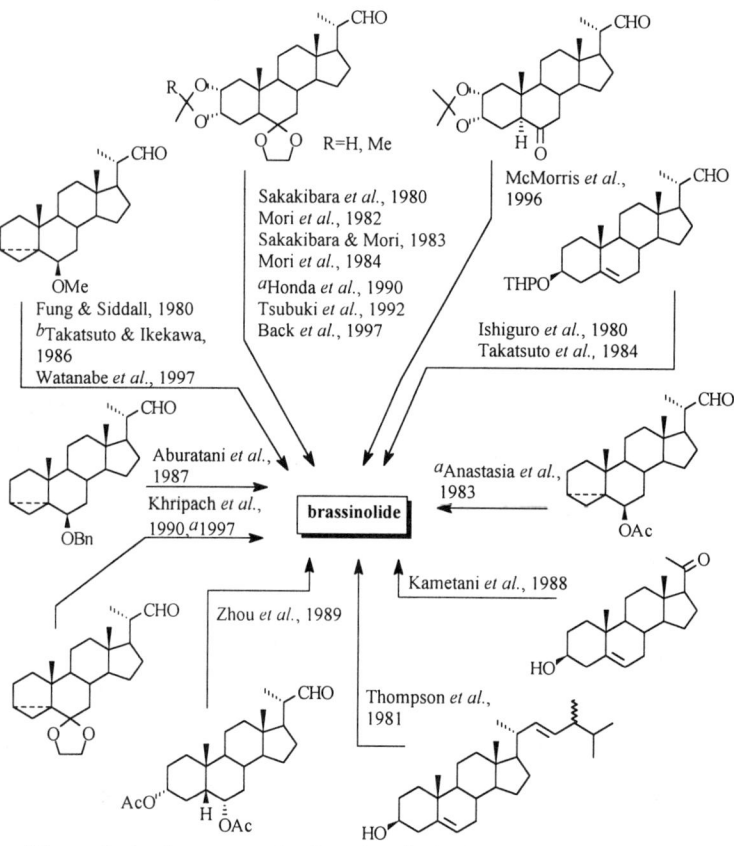

[a] The synthesis of castasterone has been described only.
[b] Preparation of labeled brassinolide.

SCHEME 95

Brassinolide syntheses have been accomplished starting from pregnenolone, from hyodeoxycholic acid, from a mixture of sterols containing crinosterol, and especially from stigmasterol. Although stigmasterol is not included in Scheme 95, all of the 22-aldehydes depicted there are in fact derived from this sterol. Since by far the majority of the reported syntheses started from these 22-aldehydes, this approach will be treated first, followed by syntheses starting from other sterols.

1. Starting with Stigmasterol

As has been mentioned above, syntheses described in this group make use of 22-aldehydes as key intermediates, all of which are derived from stigmasterol. As usual in complicated syntheses, several strategies can be employed to reach one's goal. In the previous chapter already the obvious division of the molecule into a cyclic part and the side chain was used to describe the subject. Also in the syntheses of BS, strategies have been followed in which first the cyclic part of the molecule was constructed and properly protected, followed by elaboration of the side chain, or the other way around. Both approaches have been applied for the synthesis of brassinolide and will be shown in the following schemes.

When the synthesis starts with the cyclic part of the molecule, it is useful to demonstrate how the starting 22-aldehydes can be obtained. Scheme 96 indicates that stigmasterol tosylate **360** can be solvolyzed to give the rearranged ethers, which can be ozonolyzed directly to the 22-aldehydes. The 6-hydroxyl compounds can be oxidized, protected, and ozonolyzed to give the 22-aldehyde **244**. Introduction of the $2\alpha,3\alpha$-diol function may be often desirable prior to synthesis of the side chain and this can be achieved *via* isomerization of ketone **361** followed by regiospecific hydroxylation. Before the addition of nucleophiles to 22-aldehydes the $2\alpha,3\alpha$-diol function and the 6-keto group have to be protected. Dioxolane derivatives give the best possibility to do this and after ozonolysis 22-aldehyde **362** is obtained. Oxidative cleavage of the Δ^{22}-double

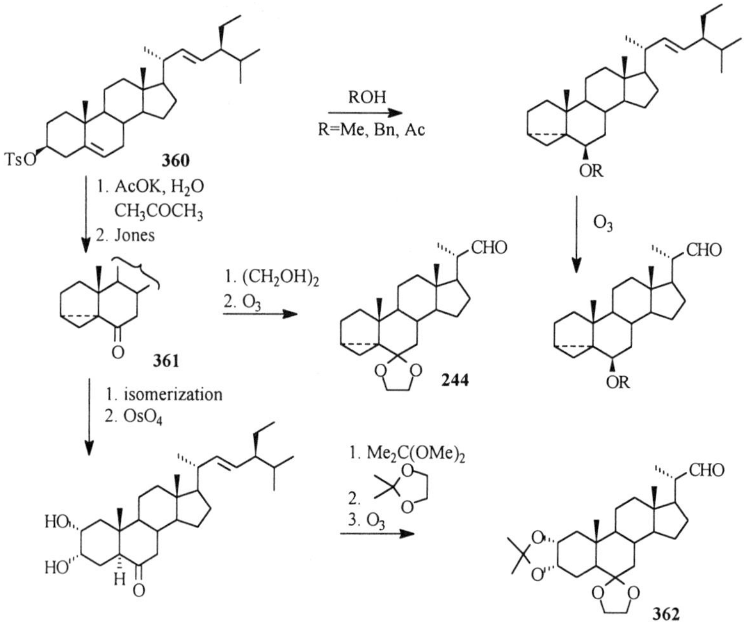

SCHEME 96

bond *via* hydroxylation and H_5IO_6 oxidation sometimes gives better results than ozonolysis (McMorris *et al.*, 1996).

22-Aldehydes such as **362** are considered to be the most attractive intermediates for the synthesis of BS because they have a complete functionalized cyclic part that is characteristic for many representatives of this class of compounds. When these functional groups are protected in a suitable way, the construction of the side chain can be undertaken as the second and most difficult part of the synthesis.

The first synthesis of brassinolide was completed by the American chemists Fung and Siddall (1980). Alkylation of the aldehyde **219** with the alanate **363** proceeded with formation of the allylic alcohol **364** as the main product, isolated in 46% yield after column chromatography (Scheme 97). The epoxidation of the Δ^{23}-double bond was investigated using MCPBA or ^{t}BuOOH in the presence of $VO(acac)_2$. In both cases a mixture of the two possible epoxy alcohols was

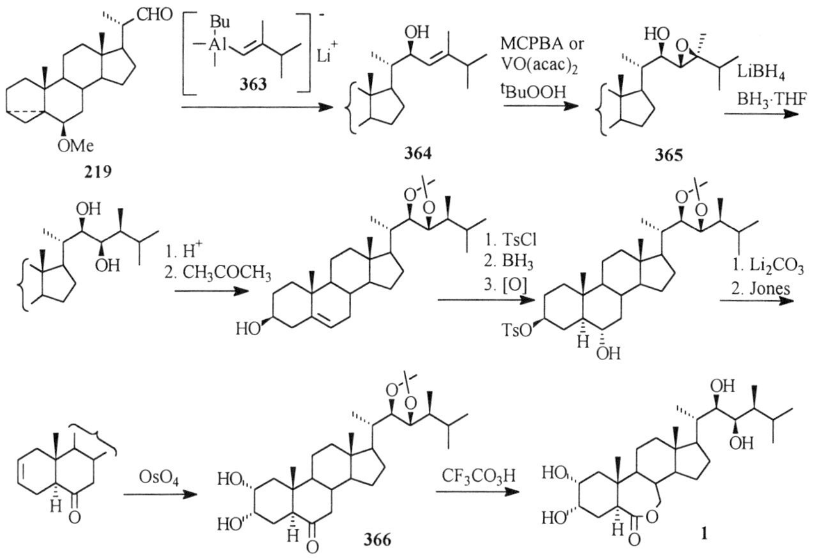

SCHEME 97

formed. The first method gave a slightly better ratio (95:5) of the desired epoxy alcohol **365** to its erythro isomer compared to 85:15 for the oxidation with ^{t}BuOOH. Anti-Markovnikov reduction of the epoxy alcohol **365** afforded a mixture of the 1,2- and 1,3-diols in a ratio of 3:1. The 1,2-diol was protected as its acetonide and then the cyclic part of the molecule was prepared by standard methods. As the last step the monoacetonide **366** was treated with CF_3CO_3H, which led to lactonization of ring B with simultaneous deprotection of the diol function in the side chain to give brassinolide **1**.

Practically simultaneously Japanese scientists reported their synthesis of brassinolide (Ishiguro et al., 1980; Takatsuto et al., 1984b; Suntory, Ltd., 1980). They used aldehyde **367** as a starting compound, which was treated with (2-methylbut-1-ynyl)lithium to give a mixture (1:1) of 22-alcohols (Scheme 98). The desired 22R-alcohol **368** was isolated from the mixture by single crystallization. Its partial hydrogenolysis followed by epoxidation led to the epoxy alcohol **369**. All attempts to introduce the 24-methyl group directly with

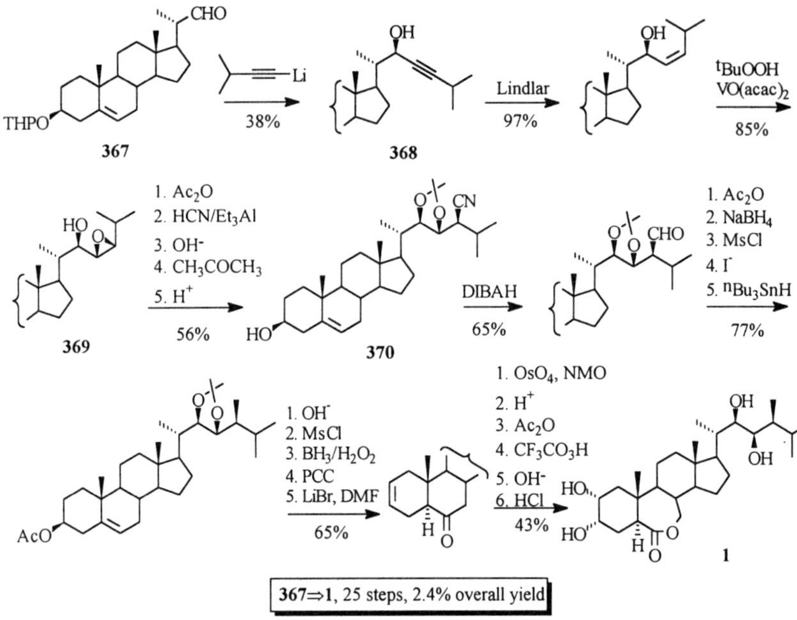

SCHEME 98

367⇒1, 25 steps, 2.4% overall yield

various reagents (Me$_2$CuLi, MeMgBr-CuI) failed however. The epoxide ring could be opened with an excess of HCN-Et$_3$Al. It was noted that protection of the 22-OH group as its acetate improved the regioselectivity of the reaction. The nitrile **370** was reduced to a methyl group in several steps. At this point the side chain of brassinolide was ready in a suitably protected form and the transformation of the cyclic part could be undertaken to give brassinolide **1**.

Several approaches to the synthesis of brassinolide were reported by Mori's group. The first one was based on the Julia olefination reaction as a method for construction of the carbon skeleton of the side chain followed by introduction of the diol function *via* the 22,23-epoxides (Sakakibara *et al.*, 1982; Mori *et al.*, 1982). The sulfone **282** had been prepared from optically pure (*R*)-(+)-citronellic acid (see Scheme 74) and in this way the stereochemistry at C-24 was predetermined. The Julia olefination of the aldehyde **362** with sulfone **282** followed by acetylation produced a mixture of the acetoxy sulfones **371** (Scheme

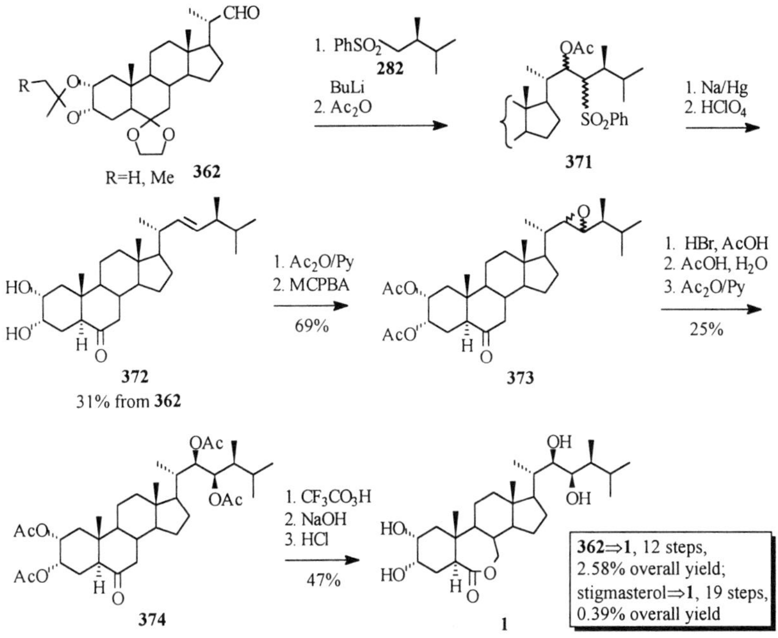

SCHEME 99

99). Treatment with sodium amalgam gave after deprotection the enediol **372**. The diol function in the side chain was introduced by epoxidation followed by transformation of the 22R,23R-epoxide **373** into the corresponding acetate **374** (see also Scheme 37). The further transformation into brassinolide **1** is indicated in Scheme 99.

A comparable approach to brassinolide was realized by Khripach *et al.* (1990k). Optimization of the reaction conditions at all stages led to a considerable improvement of the total yield. The racemic sulfone **375** was used in the olefination process (Scheme 100). An inseparable mixture of C-24 epimers **376** was obtained after sodium amalgam reduction and deprotection of the 6-keto group. Cyclopropane ring opening followed by epoxidation of the Δ^{22}-double bond and dehydrobromination gave the epoxy enone **377**. The isomeric pair of 22R,23R- and 22S,23S-epoxides could not be separated at this stage, but

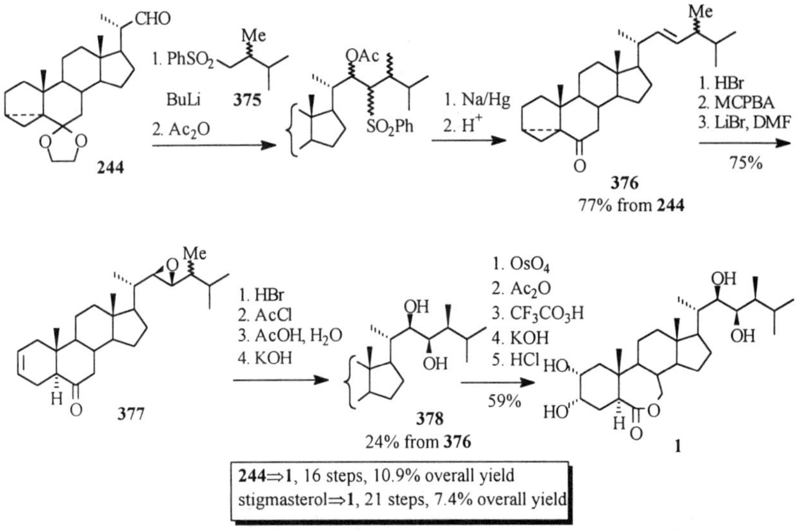

SCHEME 100

this problem was solved after transformation of the mixture into the corresponding diols. The desired diol **378** was purified by column chromatography. Direct hydroxylation of **378** with osmium tetroxide afforded castasterone, which was used then for the preparation of brassinolide **1**.

Sulfone **379** and the same aldehyde **244** were used in another variant of this approach (Zhabinskii *et al.*, 1997). The addition of the anion derived from

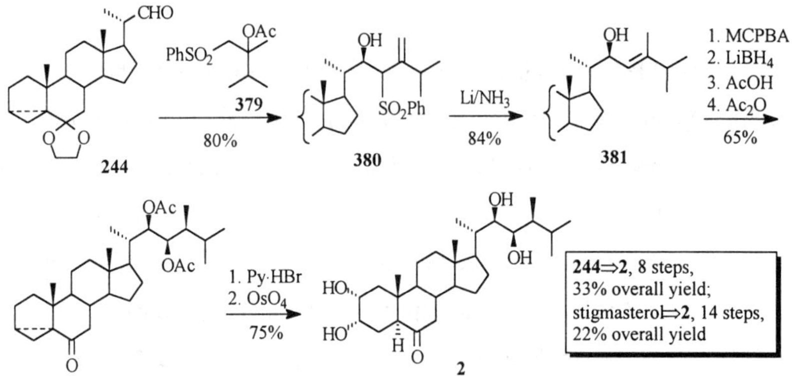

SCHEME 101

379 and BuLi to aldehyde 244 followed by desulfonization of the intermediate β-hydroxy sulfone 380 led to the allylic alcohol 381 (Scheme 101). The shift of the double bond during the desulfonization was fortunate and made this approach a rather short one. The remaining part of this castasterone synthesis was straightforward.

A low total yield prompted Mori to develop another approach to the synthesis of brassinolide (Scheme 102). The aldehyde 362 (R=Me) (Sakakibara and Mori, 1983a; Mori et al., 1984) or later 362 (R=H) (Aburatani et al., 1985b) was used as a key intermediate. Its reaction with the anion derived from 1,1-dibromo-3-methylbutene gave a separable mixture of the alkynyl alcohols 382. The unwanted isomer 382b could be transformed into the 22α-alcohol 382a either by an oxidation-reduction sequence or by treatment of the corresponding mesylate with KO_2 followed by reductive workup. However, a mixture of both alcohols was used in practice. Partial hydrogenation of 382 gave the allylic alcohols 383, which on epoxidation afforded a mixture (1:1) of the less polar 22S,23S,24R- and the more polar 22R,23R,24R-epoxy alcohols 384. The alkylative cleavage of the epoxide followed by a deprotection step gave castasterone 2.

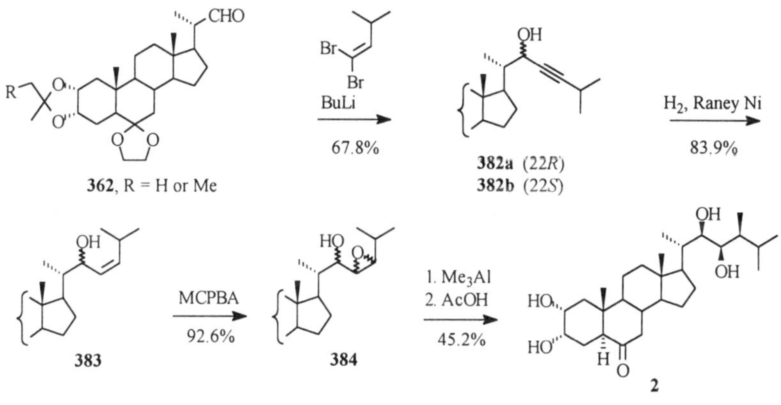

SCHEME 102

When the aldehyde **362** was used as a pure compound (R=H) (Aburatani *et al.*, 1985b), the alkynyl alcohols **382a** and **382b** could be separated easily. In this way the allylic alcohol **383** and the epoxy alcohol **384** were obtained as pure compounds. Another improvement of the approach proved to be the use of a mixture of hexane-cyclohexane (4:1) for cleavage of the epoxide ring. This was a crucial step in the whole route, and this improvement led to a considerably (from 69 h to 4 h) shorter reaction time. With this modification, approximately 30 g of brassinolide could be prepared in 7% overall yield starting from stigmasterol.

The same method of side chain construction was applied in the brassinolide synthesis reported later by Aburatani *et al.* (1986, 1987b). Both castasterone and brassinolide were available from the diacetoxyacetonide **385** (Scheme 103).

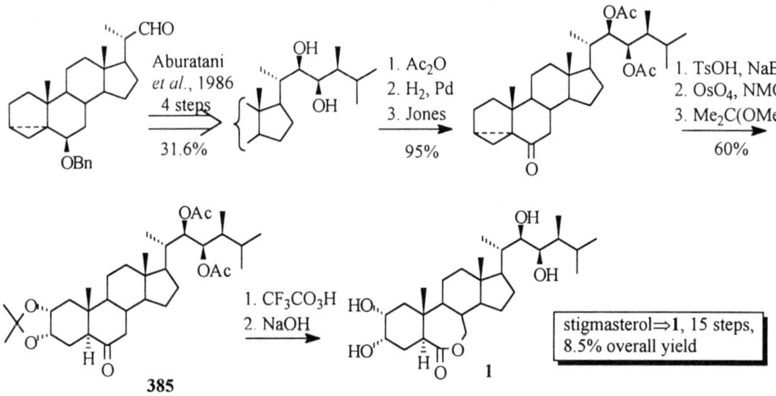

SCHEME 103

The preparation of castasterone **2** has been described starting from the aldehyde **362** *via* addition of a lithio silyl propyne derivative (Back and Krishna, 1991). The construction of the side chain was done as described in Scheme 55. The remaining part of castasterone synthesis was rather straightforward and included epoxidation, anti-Markovnikov reduction, and deprotection stages as indicated in Scheme 104.

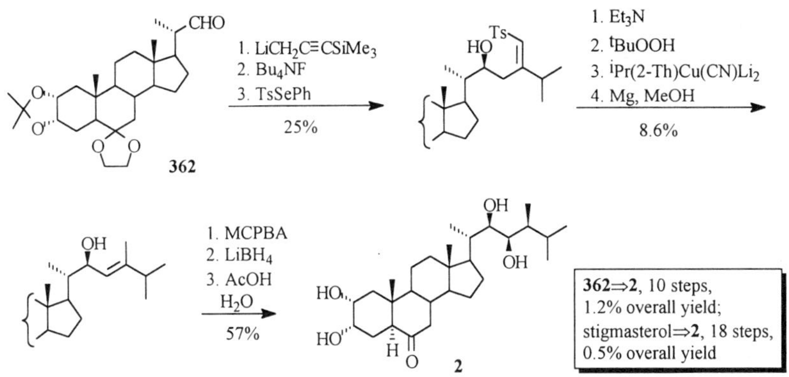

SCHEME 104

An improved synthesis for brassinolide has been published also by Back *et al.* (1997a,b). Addition of the anion **386**, with a bulky substituent in the α-position to the anionic center, resulted almost exclusively in the formation of the desired Cram product **387** (Scheme 105). Although the hydroxy selenides **387** could be isolated and separated at this stage, it was not necessary because both of them, upon oxidation with hydrogen peroxide, produced the same *E*-allylic alcohol **388**. The best method for its transformation into the epoxy alcohol **389**

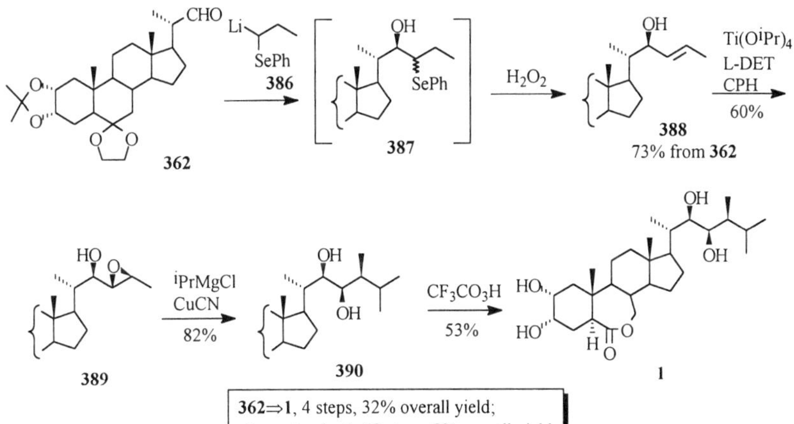

SCHEME 105

was found to be by Sharpless epoxidation with cumene hydroperoxide at -20 °C in the presence of (+)-(L)-diethyl tartrate. The reaction produced an inseparable mixture of the epoxy alcohol **389** and its erythro isomer (70:30) in 85% yield. Treatment of the mixture with isopropylmagnesium bromide in the presence of a catalytic amount of cuprous bromide afforded the diol **390**. The Baeyer-Villiger oxidation proceeded with simultaneous deprotection to give brassinolide **1**. Some intermediate products including the threo epoxy alcohol and the selenide derivatives could be recycled and in this way an additional improvement of the total yield could be achieved.

Scheme 106 demonstrates an example of the application of hydroxy furans for the synthesis of brassinolide (Honda, 1990; Tsubuki *et al.*, 1992a). The synthetic route was based on the approaches described in Scheme 76. The lactone **391** available in 6 steps from aldehyde **362** possessed all the necessary stereocenters characteristic for the brassinolide side chain. Synthesis of the latter

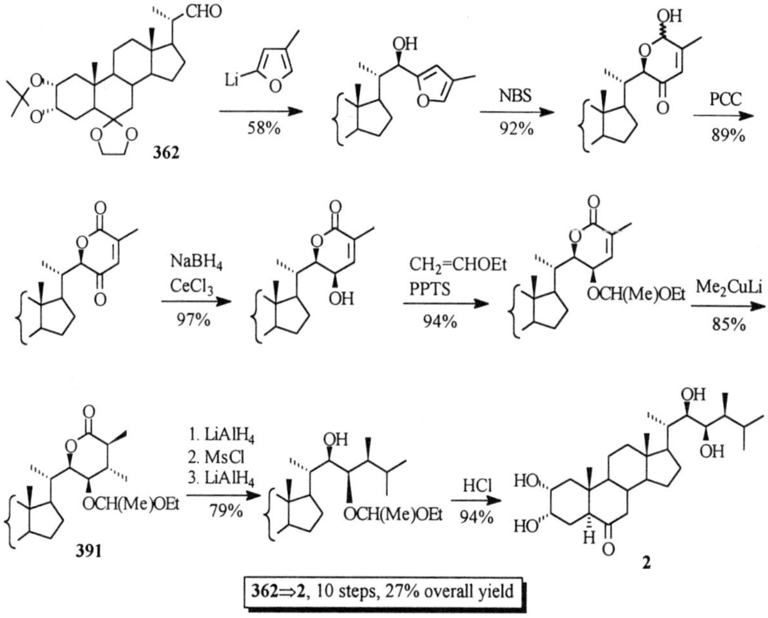

SCHEME 106

was realized by hydride reduction of the lactone **391** to the corresponding C-29 alcohol. The hydroxyl group was reduced and finally acid hydrolysis, to remove all protecting groups, afforded castasterone **2**.

An improved variant of the brassinolide synthesis based on the aldol reaction of butenolide **298** with 22-aldehyde **392** has been reported by McMorris *et al.* (1996) (Scheme 107). Compound **392** could be used without prior protection of the carbonyl group at C-6 and the synthetic route followed the side chain formation method described in Scheme 79. Opening of the lactone ring by hydride reduction was accompanied by formation of a mixture (6:1) of 6β- and 6α-alcohols. To avoid the formation of a hemiacetal with the C-29 hydroxyl group, protection of the 22,23-diol function was carried out with anhydrous acetone in the presence of camphor-10-sulfonic acid. Oxidation of the intermediate 29-alcohol with Collins reagent followed by decarbonylation afforded the methyl group in diacetonide **393**. This compound could be

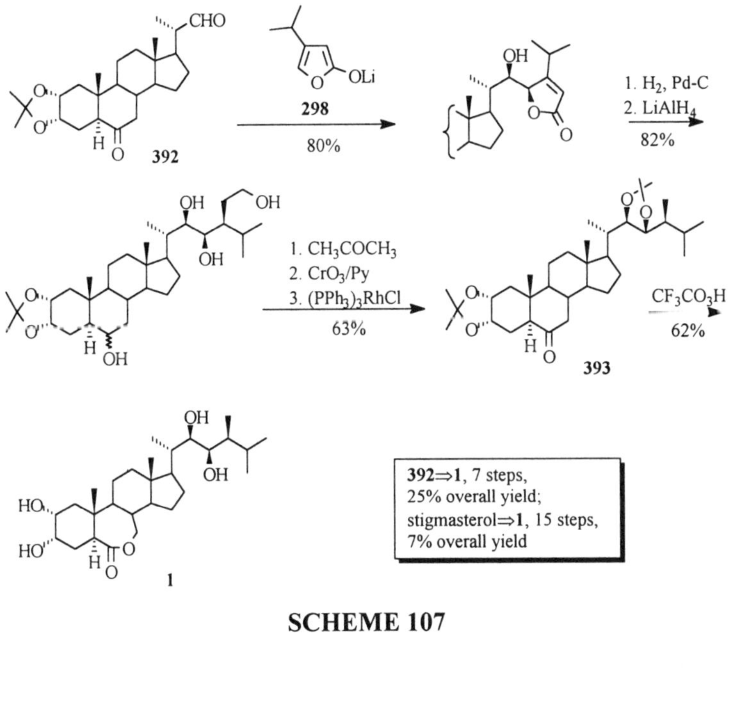

SCHEME 107

deprotected to give castasterone or transformed into brassinolide **1** by Baeyer-Villiger oxidation.

The Claisen rearrangement was the key step in the castasterone synthesis developed by Anastasia *et al.* (1983d). Addition of (3-methylbutynyl)magnesium bromide to aldehyde **394** gave a mixture (3:2) of two acetylenic alcohols in 65% overall yield (Scheme 108). The minor isomer **395** was partially hydrogenated to give the allylic alcohol **396**, which was subjected to the Claisen rearrangement with triethyl orthoacetate. The reaction proceeded stereoselectively to afford the unsaturated ester **397** with the correct configuration at C-24. Removal of the ester group was achieved by reduction of the ester **397** into the corresponding 28-aldehyde followed by decarbonylation over Wilkinson's catalyst. The remaining steps included transformation of the alcohol **398** into the dienone **399** and hydroxylation of the latter with OsO_4. The last step gave only 6.8% of castasterone **2**, but now this yield can be considerably improved by employment of chiral catalysts (Marino *et al.*, 1996).

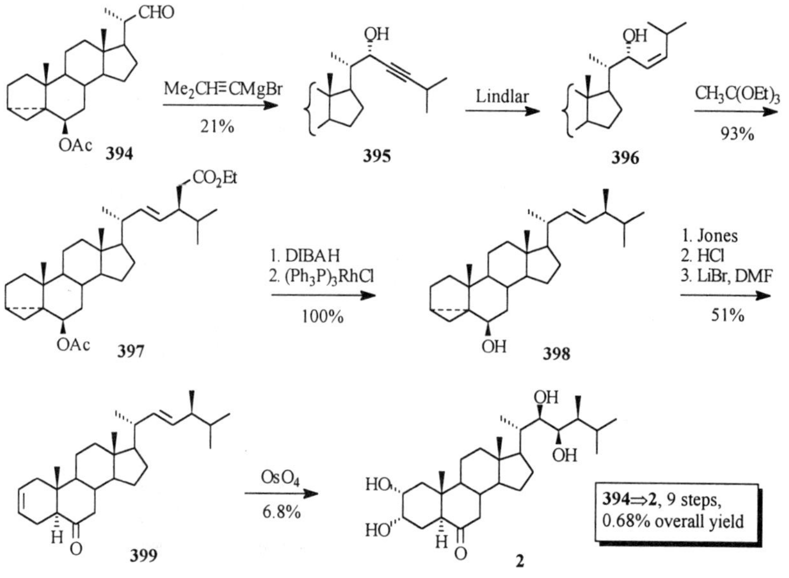

SCHEME 108

2. Starting with Crinosterol

Crinosterol would have been an ideal starting material for the synthesis of brassinolide if it were available. Unfortunately, pure crinosterol can be obtained only *via* chemical synthesis (Anastasia *et al.*, 1983a, see Scheme 68; Zhou, W.-S. *et al.*, 1992a,b). A number of BS, including brassinolide, have been prepared starting from a natural mixture of crinosterol **70** (R=α-Me) and brassicasterol **71** (R=β-Me) (40:60) obtained from oysters (Thompson *et al.*, 1981). Tosylation of the mixture followed by isosteroidal rearrangement and Jones oxidation led to the corresponding 6-ketones (Scheme 109). Further transformations of the cyclic part included cyclopropane ring opening followed by saponification of the intermediate 3β-acetates to give the corresponding 3β-alcohols, which were converted into the Δ^2-6-ketones *via* a tosylation-dehydrotosylation sequence. Equimolecular oxidation led to a separable mixture of four isomeric tetraols. Brassinolide and other BS were prepared in 4 steps *via* Baeyer-Villiger oxidation of the individual tetraacetates.

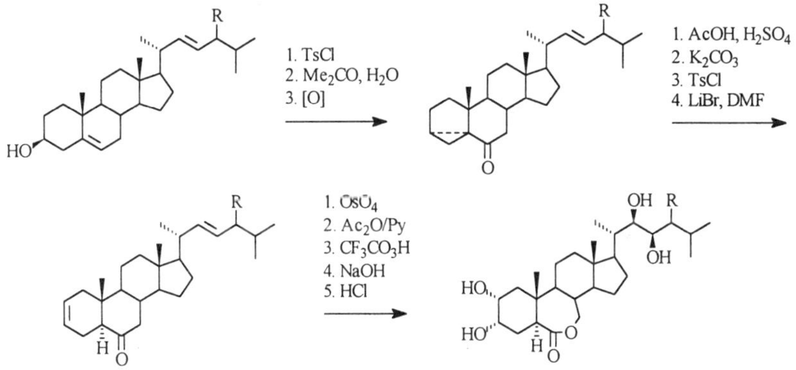

SCHEME 109

3. Starting with Pregnenolone

A synthesis of brassinolide starting from pregnenolone has been reported by Kametani *et al.* (1988a). All the necessary carbon atoms of the side chain were introduced *via* a chelation-controlled addition of the anion obtained from 3-isopropyltetronic acid **178** to ketone **400** (Scheme 110). Treatment of the adduct with methoxymethyl chloride gave **401** and dehydration of the tertiary hydroxyl group *via* the corresponding trifluoroacetate led to olefin **402**. The hydrogenation of this compound afforded the lactone **403** with remarkable selectivity; all four contiguous chiral centers of the brassinolide side chain could be established in one step. Hydride reduction of the lactone **403** led to the opening of the ring to give diol **404**. Removal of the 28-hydroxyl group was achieved as indicated and further deprotection of **405** afforded castasterone **2**.

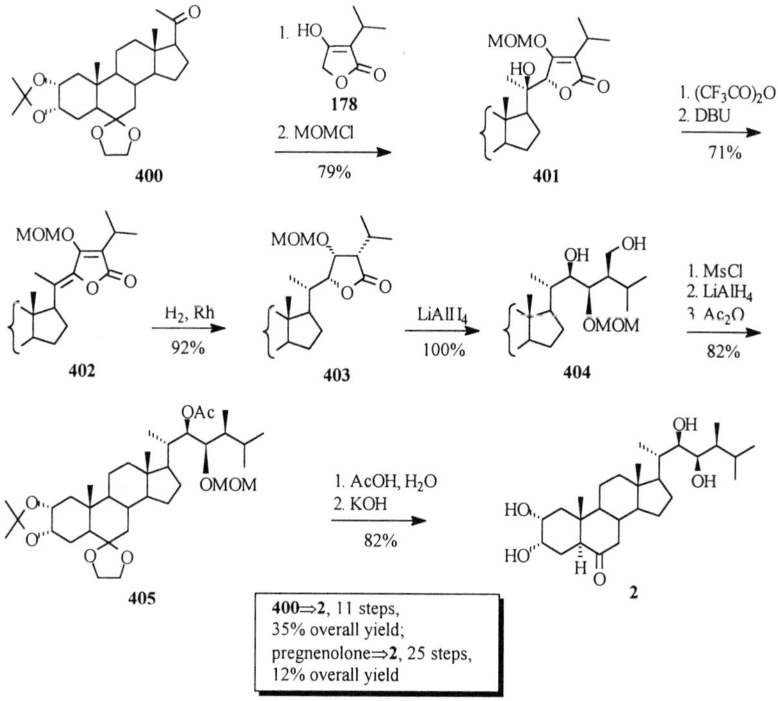

SCHEME 110

4. Starting with Hyodeoxycholic Acid Derivatives

A synthesis of brassinolide starting from hyodeoxycholic acid has been elaborated by Zhou, W.-S. *et al.* (1989a). The Wittig olefination of the aldehyde **273** by an arsonium ylide (see Scheme 72) followed by a number of side chain transformations described in led to the lactone **406** (Scheme 111). Its conversion into the acetonide **407** was achieved by reductive opening of lactone **406**, which was accompanied by simultaneous deacetylation at C-3 and C-6. Protection of the 22,23-diol function, decarbonylation of the C-29 aldehyde, oxidation of the hydroxy groups in **407**, and deprotection gave the diketone **408** in which the keto group at C-3 could be selectively transformed into the Δ^2-olefin **409** *via* zinc reduction of the corresponding silyl enol ether. The remaining part of the brassinolide synthesis included osmylation to give castasterone and the Baeyer-Villiger oxidation of the latter to give brassinolide **1**.

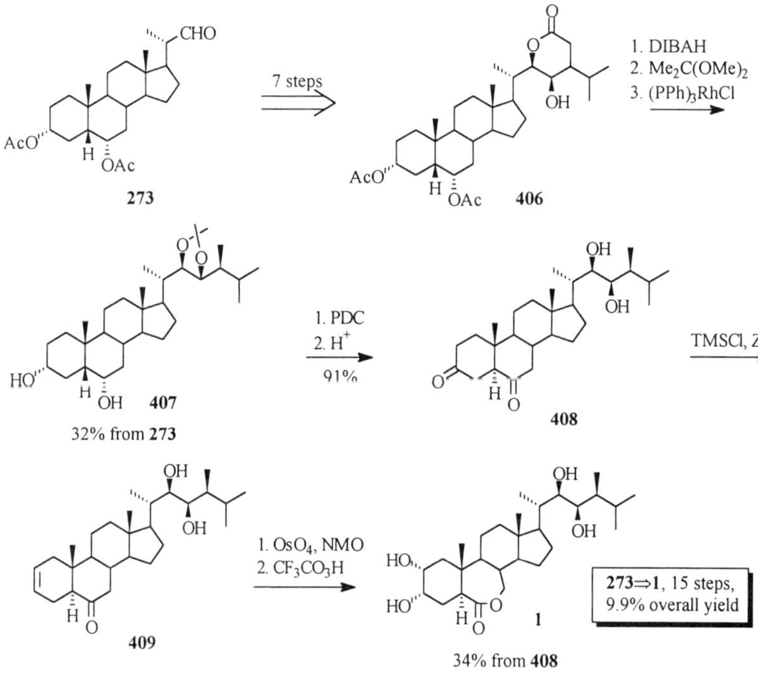

SCHEME 111

B. SYNTHESES OF EPIBRASSINOLIDE

Epibrassinolide is considered to be the most probable candidate for practical application of BS in agriculture. It combines important features such as high biological activity and relatively good availability. In many tests its activity is comparable with that of brassinolide. Furthermore, it can be synthesized from ergosterol, which is the starting material also for the industrial production of vitamin D_2. Most of the epibrassinolide syntheses have started from ergosterol (Scheme 112). Brassicasterol, which is available as a component (10-20%) in the sterol fraction of rapeseed oil, can be used for the preparation of epibrassinolide also; in both cases reconstruction of the carbon side chain is not necessary.

The first synthesis of epibrassinolide was reported in 1979 (Thompson et al., 1979, 1980a), but it was not until 1988 that this compound was identified as a natural hormone in pollen of the broad bean *Vicia faba* L. (Ikekawa et al., 1988). The synthesis relied mainly on classical methods known from ecdysteroid chemistry and is depicted in Scheme 113. Epibrassinolide was obtained in 4.6%

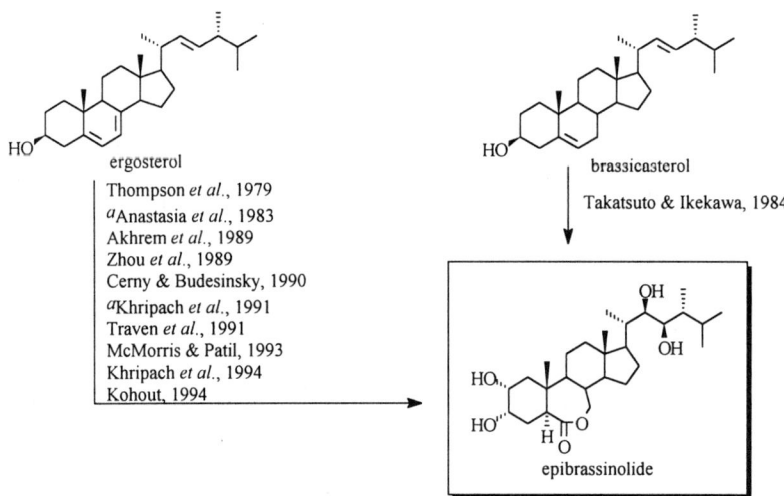

[a]The synthesis of epicastasterone has been described only.

SCHEME 112

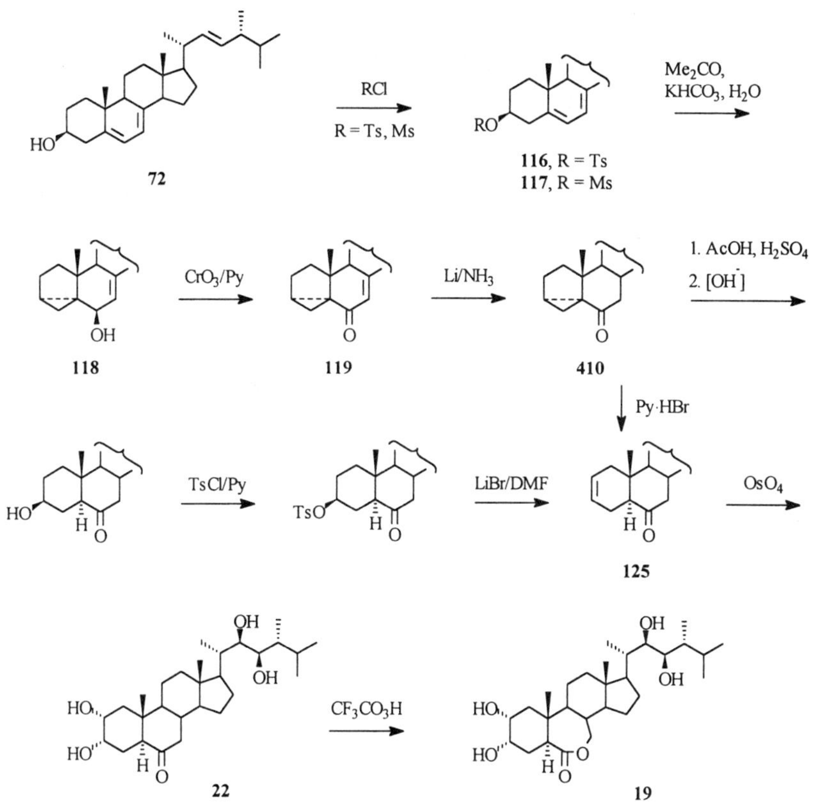

SCHEME 113

overall yield in 13 steps.

Many alternative approaches to the synthesis of epibrassinolide have been elaborated since, but currently the synthesis described in Scheme 113 remains the best with respect to practical aspects and total yield. Several improvements have been introduced during the past years (Traven et al., 1991a; McMorris and Patil, 1993; Khripach et al., 1994c). The use of mesylate **117** instead of the tosylate **116** in the isosteroidal rearrangement allowed a considerable improvement of the yield of alcohol **118** (Khripach et al., 1991f). When the reduction of enone **119** with lithium in liquid ammonia was carried out at low temperature, overreduction could be avoided (Traven et al., 1991b). A one-step transformation of the isosterone **410** into the dienone **125** was elaborated using pyridinium chloride and lithium bromide in *N,N*-dimethylacetamide (McMorris and Patil, 1993) or pyridinium bromide in DMF (Khripach *et al.*, 1994c).

The cis hydroxylation of the dienone **125** according to the Sharpless procedure made it possible to improve the ratio of (22*R*,23*R*)-epicastasterone **22** to its 22*S*,23*S* isomer up to 10:1. The Baeyer-Villiger oxidation could be done directly with epicastasterone, thus avoiding protection-deprotection steps. These improvements have led to a synthesis of epibrassinolide **19** in 7 steps from ergosterol **72** in 26% overall yield (McMorris and Patil, 1993). Similar results have been obtained by Traven *et al.* (1991a), Khripach *et al.* (1994c), and Kohout (1994b).

Synthesis of epibrassinolide **19** has been accomplished by Anastasia *et al.* (1983c) starting from the adduct **111** of ergosterol acetate with 4-phenyl-1,2,4-triazoline-3,5-dione (Scheme 114). Its reduction with lithium in ethylamine gave a mixture (3:2) of brassicasterol **71** and its Δ^7-isomer **112** (see Scheme 26). Only brassicasterol tosylate reacted under the conditions of isosteroidal rearrangement

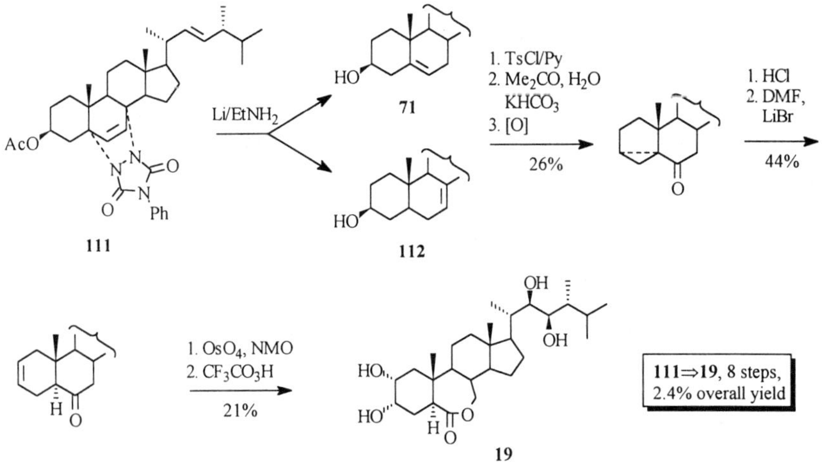

SCHEME 114

to give the 3α,5-cyclo-6-alcohol, whereas the tosylate of the Δ^7-isomer was left unchanged and could be separated by column chromatography.

A similar synthesis of epibrassinolide starting from ergosterol was realized by Akhrem et al. (1989c). A mixture of brassicasterol 71 and its Δ^7-isomer 112 in this case was obtained via direct reduction of ergosterol with lithium (see Scheme 26).

Brassicasterol 71 as starting material for the preparation of epibrassinolide 19 was used only once by Takatsuto and Ikekawa (1984a). Treatment of brassicasterol mesylate with sodium acetate in dimethyl sulfoxide gave the isosterone 410 in a moderate yield (Scheme 115). It was transformed into dienone 125 by heating with TsOH in sulfolane. Catalytic hydroxylation with OsO_4 in the presence of N-methylmorpholine N-oxide followed by the Baeyer-Villiger oxidation of epicastasterone 22 furnished epibrassinolide 19.

Nowadays the Sharpless hydroxylation is considered to be the method of choice for the construction of the epibrassinolide side chain starting from the corresponding Δ^{22}-precursors. In the beginning of BS research, when this method was not yet known, an approach via epoxidation was elaborated to

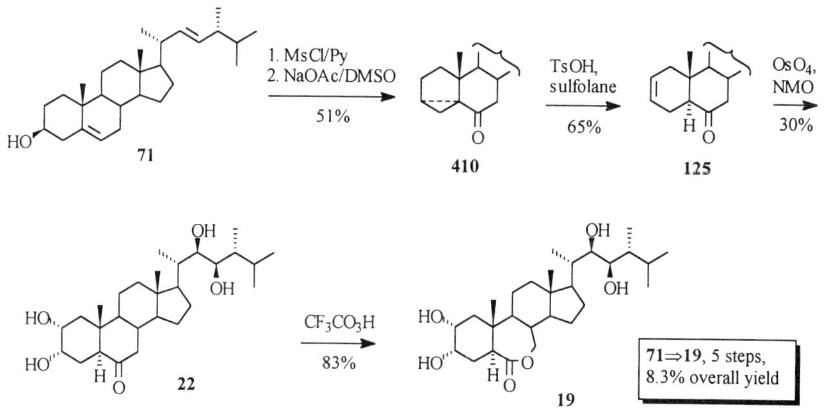

SCHEME 115

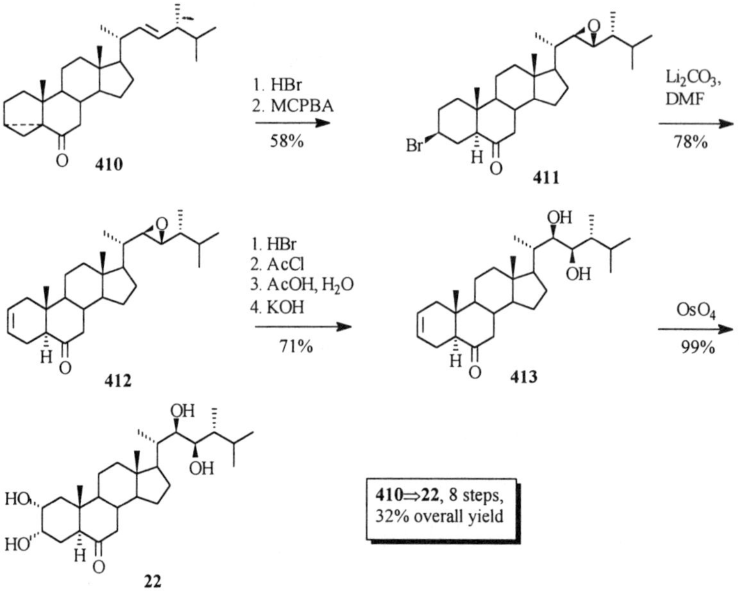

SCHEME 116

circumvent the unfavorable stereochemistry of the OsO$_4$ oxidation of the Δ^{22}-double bond. An example of such an approach is demonstrated in the synthesis of epibrassinolide (Khripach et al., 1991g) via isosterone **410**, which was prepared from ergosterol in 4 steps according to procedures described above. Hydrobromination of **410** followed by epoxidation gave the bromo epoxide **411** (Scheme 116). Dehydrobromination of **411** after chromatographic separation from the minor 22S,23S isomer was effected by heating in DMF in the presence of lithium carbonate. A four-step transformation of the epoxide **412** (see Scheme 37) afforded the diol **413** and hydroxylation of the latter gave epicastasterone **22**.

This methodology can be applied also for the construction of both diol functions, that in the cyclic part and that in the side chain, simultaneously (Cerny and Budesinsky, 1990). Epoxidation of the dienone **125** led to formation of two diepoxides which differed with respect to the configuration of the epoxides in the side chain (Scheme 117). Treatment of the epoxide **414** with hydrobromic

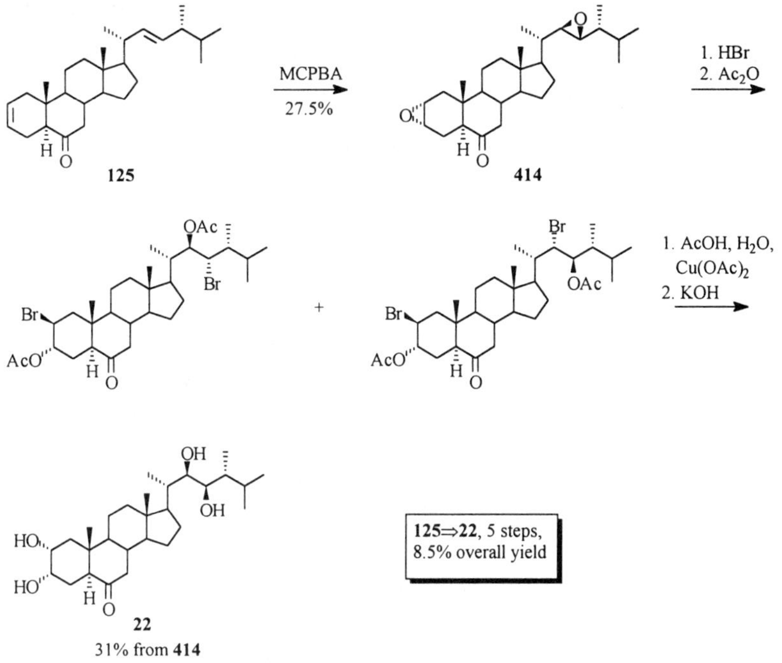

SCHEME 117

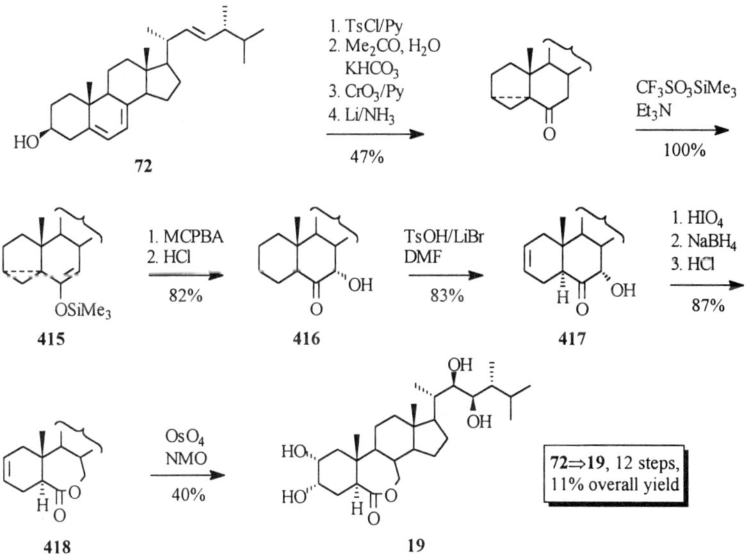

SCHEME 118

acid gave a mixture of the bromohydrins, which could be processed further without separation, because all the isomers led after acetylation, nucleophilic substitution, and saponification to epicastasterone **22**.

The characteristic feature in the approach developed by Zhou, W.-S. *et al.* (1989c) was the absence of the Baeyer-Villiger oxidation step for the construction of the lactone in ring B (Scheme 118). The 7-oxalactone ring was formed regioselectively in a few steps by oxidation of the silyl enol ether **415**; the resulting ketol **416** was transformed directly into the Δ^2-ketol **417**. Oxidative cleavage of the ketol **417** followed by reduction of the intermediate secosteroid (see Scheme 25) and acidification afforded the lactone **418**. Hydroxylation gave epibrassinolide **19** in addition to its 22S,23S isomer.

C. SYNTHESES OF HOMOBRASSINOLIDE

There are no problems with the choice of an appropriate starting material for the synthesis of homobrassinolide. Stigmasterol isolated from Soya bean oil is a

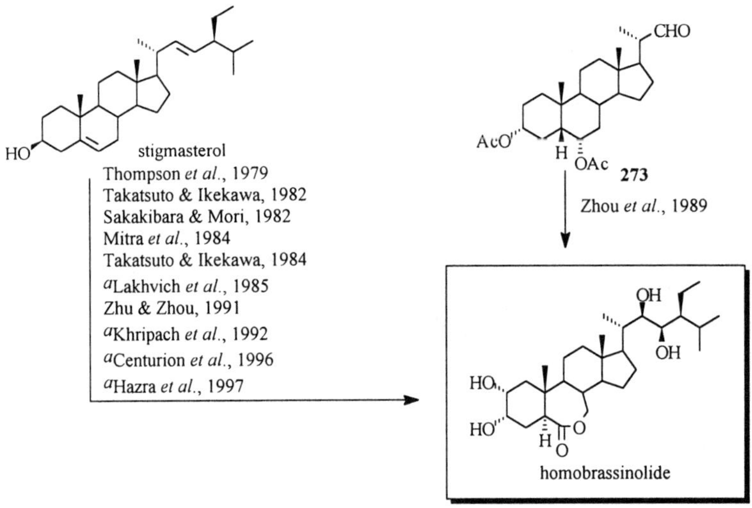

stigmasterol
Thompson *et al.*, 1979
Takatsuto & Ikekawa, 1982
Sakakibara & Mori, 1982
Mitra *et al.*, 1984
Takatsuto & Ikekawa, 1984
[a]Lakhvich *et al.*, 1985
Zhu & Zhou, 1991
[a]Khripach *et al.*, 1992
[a]Centurion *et al.*, 1996
[a]Hazra *et al.*, 1997

273
Zhou *et al.*, 1989

homobrassinolide

[a]The synthesis of (24S)-ethylbrassinone has been described only.

SCHEME 119

commercial product, it is relatively cheap, and it is available in bulk quantities. Therefore it is natural that practically all homobrassinolide syntheses have started from stigmasterol (Scheme 119). The only exception was the preparation of homobrassinolide from the hyodeoxycholic acid derivative **273** (Zhou, W.-S. et al., 1989a), which was a demonstration of the synthetic methodology.

Many homobrassinolide syntheses are very similar to each other and the differences concern details. Ketone **361**, acetate **419**, and especially dienone **420** (Scheme 120) are the most evident intermediates for the preparation of homobrassinolide.

The first attempt of homobrassinolide synthesis was unsuccessful (Mori, 1980a,b; Mori et al., 1981). A mixture of (22S,23S)- and (22R,23R)-homobrassinolide was obtained; however, later it was shown (Mori et al., 1982) that the oxidation of the Δ^{22}-double bond with OsO_4 and purification of the final product by means of crystallization had led inevitably to (22S,23S)-homobrassinolide as the sole product.

The first synthesis of (22R,23R)-homobrassinolide was accomplished by Thompson et al. (1982). The synthetic route was very similar to that used for brassinolide (see Scheme 109). A few years later homobrassinolide was

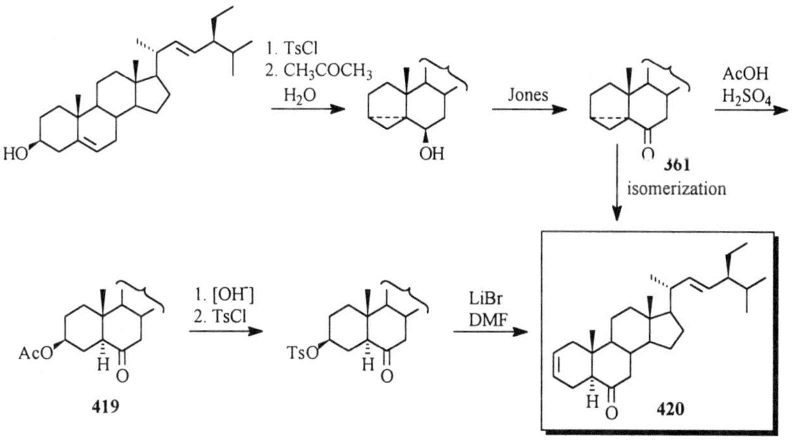

SCHEME 120

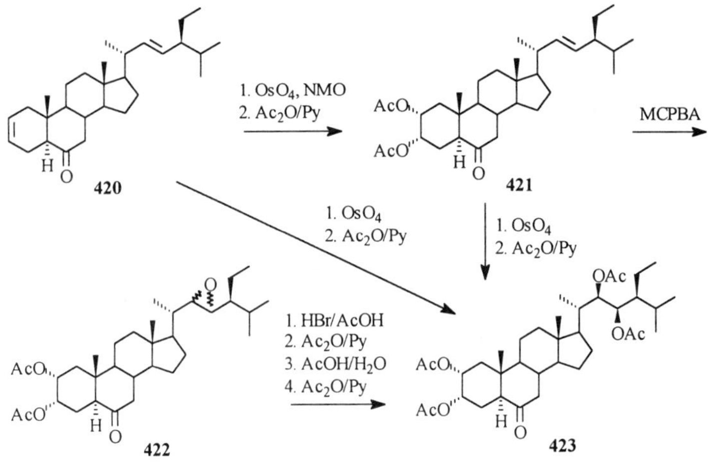

SCHEME 121

synthesized practically simultaneously by two Japanese groups (Takatsuto and Ikekawa, 1982; Sakakibara and Mori, 1982) (Scheme 121). Direct oxidation of the olefin **420** with OsO$_4$ followed by acetylation gave the tetraacetate **423** in 2.2% yield only. An alternative approach *via* the 22,23-epoxides **422** gave the tetraacetate **423** in 10.5% yield from **421**.

A considerable improvement could be achieved by employment of the Sharpless hydroxylation for construction of the side chain (Brosa *et al.*, 1992; Huang, L.-F. *et al.*, 1993; McMorris *et al.*, 1994). The best results here were obtained by McMorris *et al.* (1994), and homobrassinolide was prepared in 6 steps in an overall yield of 21%.

Hydroxylation of the Δ^{22}-double bond of stigmastane derivatives even in the presence of chiral ligands gave a considerable amount of the unwanted 22S,23S isomers. That allowed consideration of an alternative epoxide route to the 22R,23R-diols as being competitive. The last example of this approach is outlined in Scheme 122 (Hazra *et al.*, 1997b). A number of other improvements such as the use of tetradecyltrimethylammonium permanganate for hydroxylation of the Δ^2-double bond of the dienone **420**, a one-step epoxidation and Baeyer-

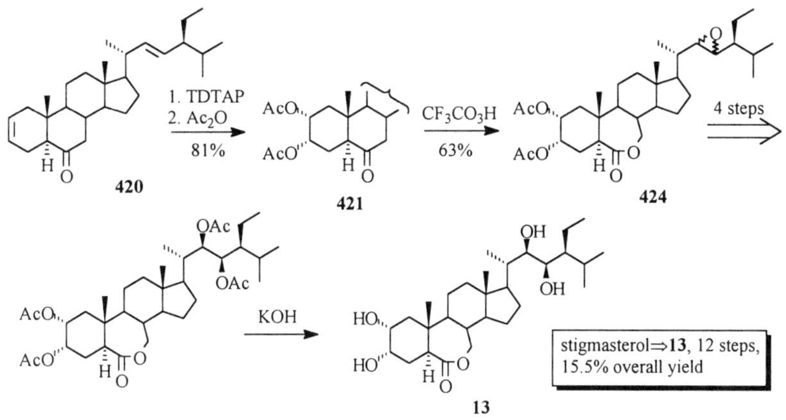

SCHEME 122

Villiger oxidation of **421**, and the epoxide **424** opening to the corresponding bromohydrins with LiBr and Amberlyst-15 resin have been accomplished also. A slightly different reaction sequence was applied by Takatsuto and Ikekawa (1984b) in their synthesis of homobrassinolide **13**.

Another synthesis of homobrassinolide based on the employment of 22R,23R-epoxides for construction of the side chain was realized by Lakhvich *et al.* (1985a-c). It was similar to that described for preparation of epibrassinolide (see Scheme 116) and gave homobrassinolide in 15 steps in 6.8% overall yield starting from stigmasterol. The formation of the 22R,23R- and 2α,3α-diol groups could be combined in this synthesis also (Khripach *et al.*, 1992c).

Stereoselective epoxidation of the Δ^5-double bond in stigmasterol acetate was the key step of the method developed by Centurion *et al.* (1996). The reaction led to formation of the 5β,6β-epoxide **425** (Scheme 123), which after reductive opening afforded the diol **426**. Attempts to carry out selective Jones oxidation failed, but the equatorial 3β-hydroxy group could be selectively silylated, thus allowing the necessary transformations at C-6. After that, compound **427** was dehydrated and the obtained dienone **420** was subjected to catalytic asymmetric

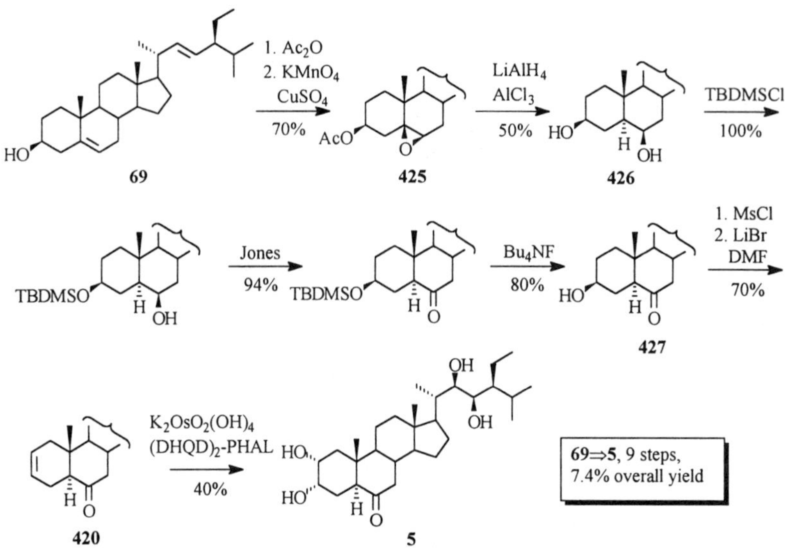

SCHEME 123

hydroxylation. (24*S*)-Ethylbrassinone **5** and its 22*S*,23*S* isomer were obtained in equal amounts in this case.

The approach to homobrassinolide described by Zhu, J.-L. and Zhou, W.-S. (1991) was based on the previous experience of the authors in the synthesis of epibrassinolide (see Scheme 118). A characteristic feature in this approach is the formation of the ring B lactone *via* the corresponding seco derivative, generated by oxidative cleavage of hydroxy ketone **428** (Scheme 124).

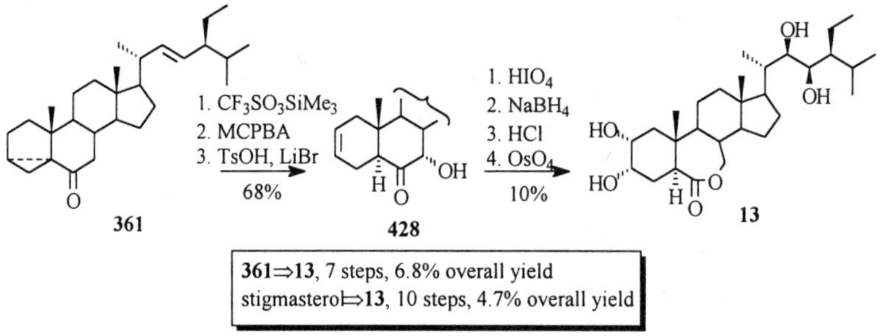

SCHEME 124

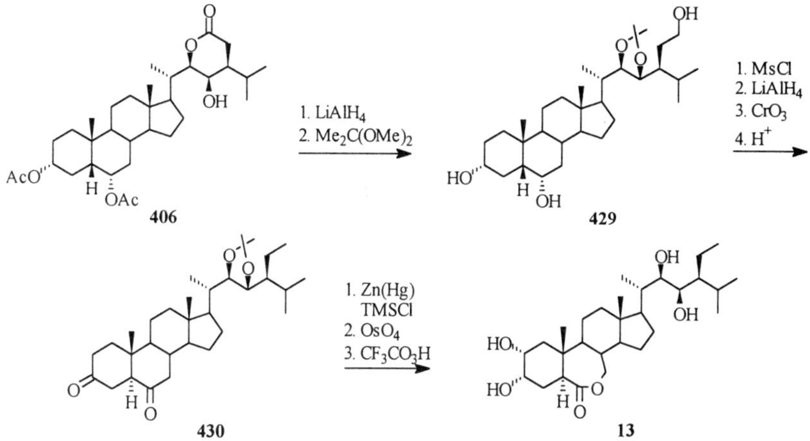

SYNTHESES OF NATURAL BS

SCHEME 125

The only synthesis of homobrassinolide that included construction of the carbon skeleton of the side chain was performed by Zhou, W.-S. *et al.* (1989a) (Scheme 125). The hydroxy lactone **406** (also an intermediate in the synthesis of brassinolide; see Scheme 111) was used as the starting compound. Reductive opening of the lactone ring followed by selective protection of the 22,23-diol function as its acetonide gave the triol **429**. Removal of the C-29 hydroxyl group was achieved *via* successive mesylation and hydride reduction. The secondary methanesulfonyl groups in the cyclic part were reconverted to the original hydroxyl functions in this reduction. Oxidation with chromic acid afforded the diketone **430** and selective reductive elimination of the 3-keto group furnished the corresponding Δ^2-derivative, which was hydroxylated to homocastasterone. Baeyer-Villiger oxidation then led to homobrassinolide **13**.

D. SYNTHESES OF 28-NORBRASSINOLIDE

The characteristic feature of norbrassinolide is the absence of a chiral center at C-24 and this simplifies the construction of the side chain considerably. On the other hand, 22-dehydrocholesterol, which is the best starting material for

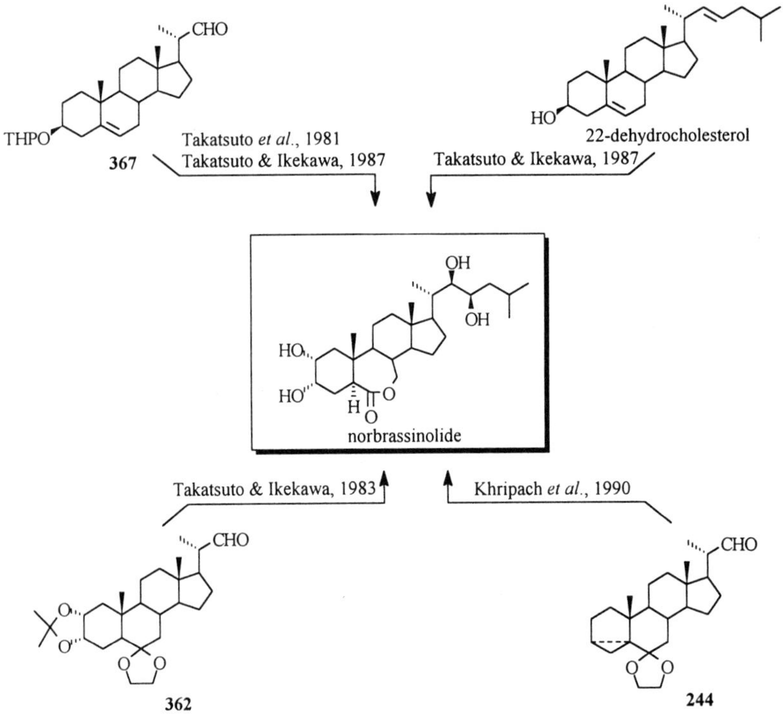

SCHEME 126

norbrassinolide synthesis, is not readily available. That means that other sterols (usually stigmasterol) have to be used as starting compound and partial replacement of the carbon skeleton of the side chain is an obligatory element of each synthetic approach in this case. The 22-aldehydes **367**, **362**, and **244**, available from stigmasterol, have been used for preparation of norbrassinolide (Scheme 126).

The most significant contribution in this field has been made by Japanese scientists. The first synthesis of norbrassinolide was performed by Takatsuto *et al.* (1981), starting from aldehyde **367** (Scheme 127). Its reaction with the lithium salt of 1,3-dithiane followed by deprotection of the ether group at C-3 gave the adduct **431**. Treatment of the corresponding diacetate with HgO led to the 23-aldehyde **432**. The Grignard reaction followed by saponification

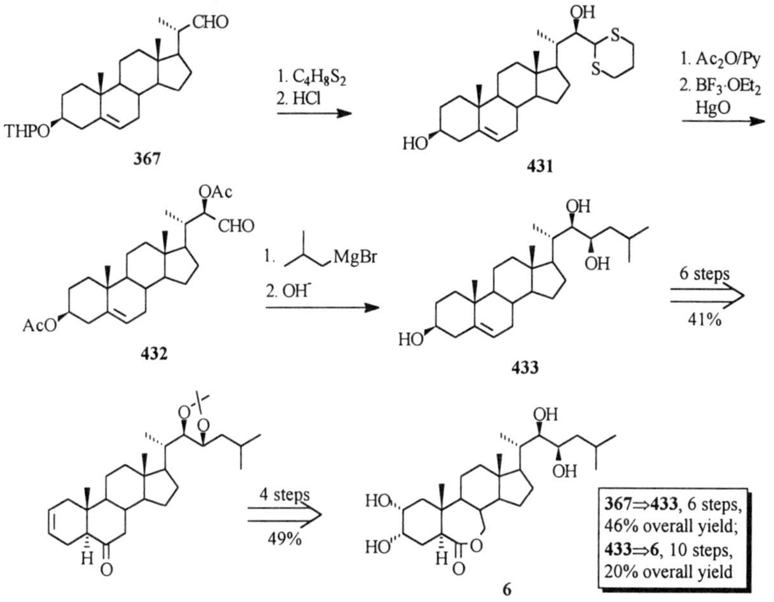

SCHEME 127

furnished the triol **433**, which proved to be a useful intermediate for syntheses of norbrassinolide that were performed by other authors also (see below).

Later on, the synthesis of the triol **433** was improved (Takatsuto and

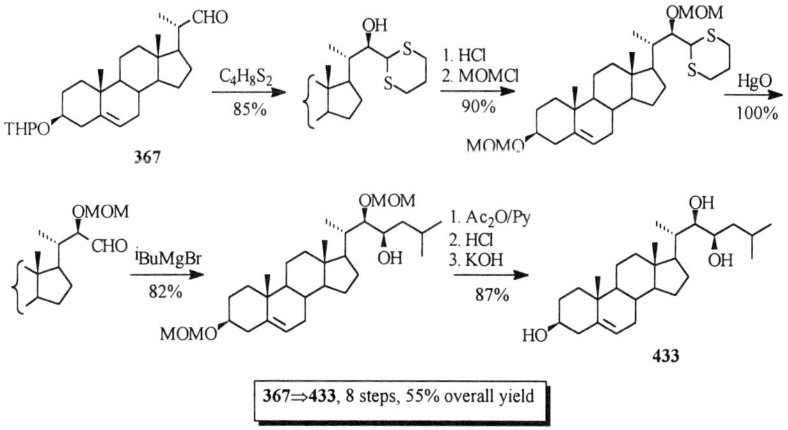

SCHEME 128

Ikekawa, 1987a) and the methoxymethyl group was used instead of the acetyl group for protection of the hydroxyl at C-22 (Scheme 128). This gave a considerable improvement of the stereoselectivity of the isobutylmagnesium bromide addition. After protection of the 22,23-diol function as its acetonide the AB rings were functionalized as required using standard procedures.

A variant of the synthesis of norbrassinolide starting from the C-22 aldehyde with functionalized AB rings and the same method of construction of the side chain were reported also (Takatsuto and Ikekawa, 1983c). The triol **433** was prepared also from the epoxide **369** (Scheme 129) by Takatsuto *et al.* (1981).

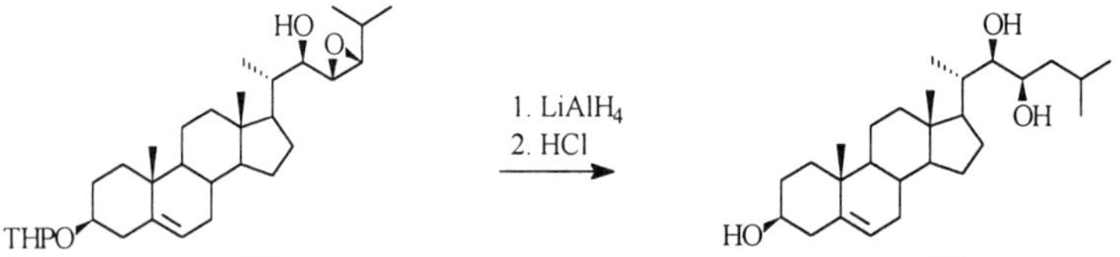

SCHEME 129

A synthesis of norbrassinolide from 22-dehydrocholesterol **434** was reported by Thompson *et al.* (1982) and by Takatsuto and Ikekawa (1987a). The carbon skeleton of the side chain does not need any adaptation in this case and only functional groups have to be introduced. This situation is similar to that for the

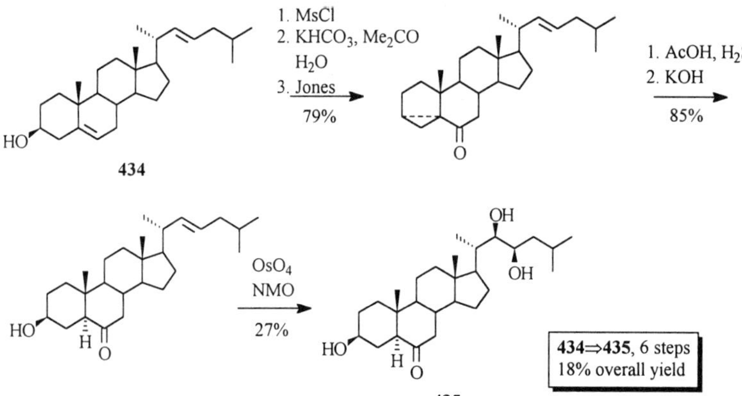

SCHEME 130

synthesis of homobrassinolide from stigmasterol and these two approaches are practically the same as is demonstrated in Scheme 109 (Thompson et al., 1982). Another synthetic route to norbrassinolide starting from 22-dehydrocholesterol **434** involved formation of the triol **435** as is shown in Scheme 130 (Takatsuto and Ikekawa, 1987a).

The synthesis of norbrassinolide has been achieved also from the aldehyde **244** (Khripach et al., 1990j). Coupling with the anion derived from phenyl isoamyl sulfone followed by acetylation led to a mixture of the β-acetoxy sulfones **436** (Scheme 131). Treatment of **436** with sodium amalgam gave, after deprotection of the C-6 carbonyl group, the olefin **437** and this could be transformed by conventional methods into brassinone **4** and norbrassinolide **6**.

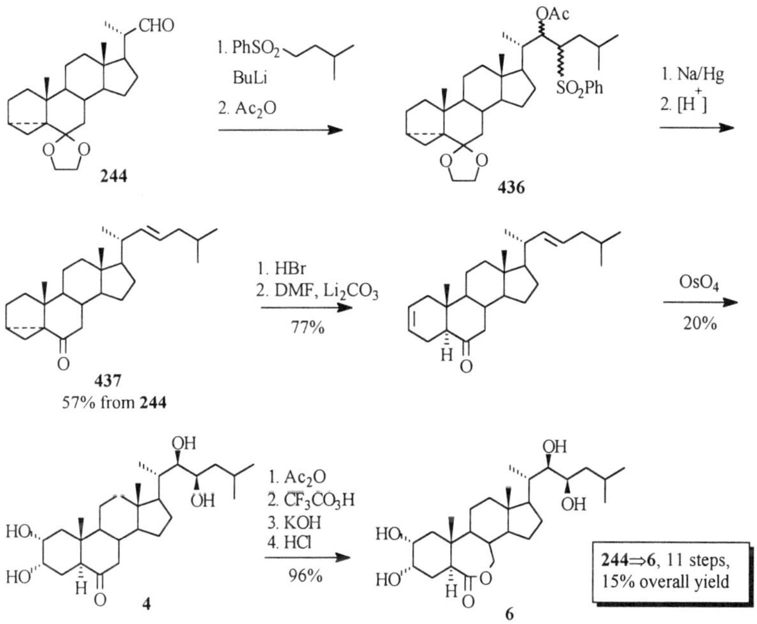

SCHEME 131

E. SYNTHESES OF BS CONTAINING A $\Delta^{24(28)}$-DOUBLE BOND

The presence of the $\Delta^{24(28)}$-double bond in the side chain of BS necessitates special methods for their synthesis. Since this double bond is the most labile element, the side chain has to be constructed as the last part of the molecule. The $\Delta^{24(28)}$-double bond is not compatible with the conditions of the Baeyer-Villiger oxidation and therefore the lactone has to be introduced prior to the construction of the double bond. A number of 22-aldehydes with fully functionalized AB ring, brassicasterol and stigmasterol were used as starting material for the preparation of $\Delta^{24(28)}$-BS (Scheme 132).

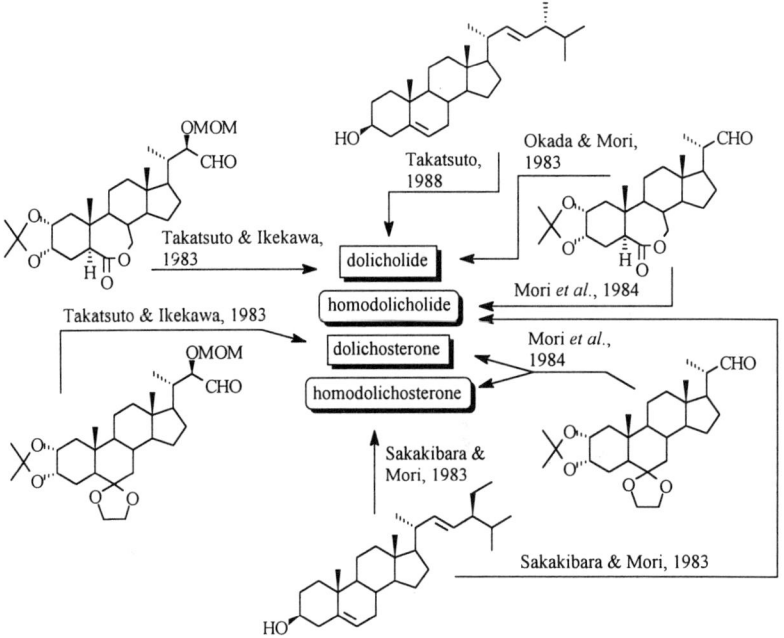

SCHEME 132

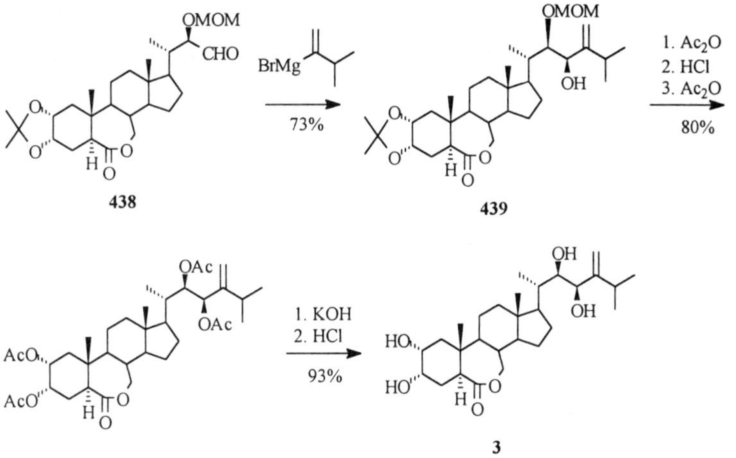

SCHEME 133

Synthesis of dolicholide **3** has been accomplished starting from the aldehyde **438** by Takatsuto and Ikekawa (1983b,c). Chelation-controlled addition (see Scheme 82) of the Grignard reagent derived from 2-bromo-3-methylbut-1-ene to the aldehyde **438** led to 23-alcohol **439** as the sole product. The remaining part of the dolicholide synthesis included protection-deprotection steps of the hydroxyl groups as indicated in Scheme 133.

A very similar approach (Scheme 134) was applied to the synthesis of dolichosterone **7** from the aldehyde **440** *via* adduct **441** (Takatsuto and Ikekawa, 1983b).

A number of syntheses of $\Delta^{24(28)}$-unsaturated BS were based on employment of aluminum organic reagents for extension of the side chain. An example of

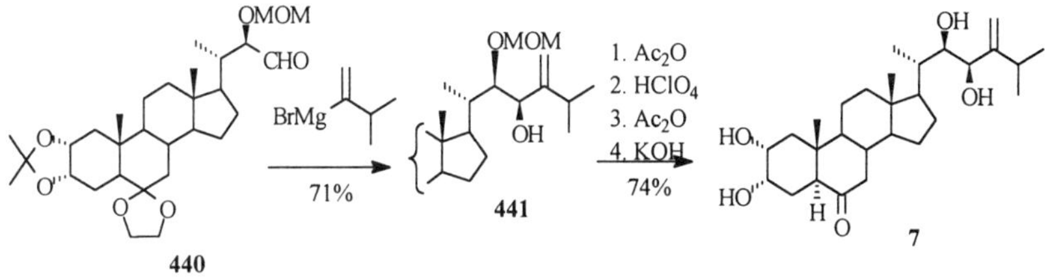

SCHEME 134

such an approach for the synthesis of dolicholide **3** is shown in Scheme 135 (Okada and Mori, 1983b; Mori *et al.*, 1984). Treatment of the aldehyde **442** with the alanate **363** gave the allylic alcohol **443**. Epoxidation followed by heating of the latter in toluene in the presence of aluminum isopropoxide furnished the corresponding $\Delta^{24(28)}$-allylic alcohol, which after removal of the 2α,3α-acetonide protecting group gave dolicholide **3**.

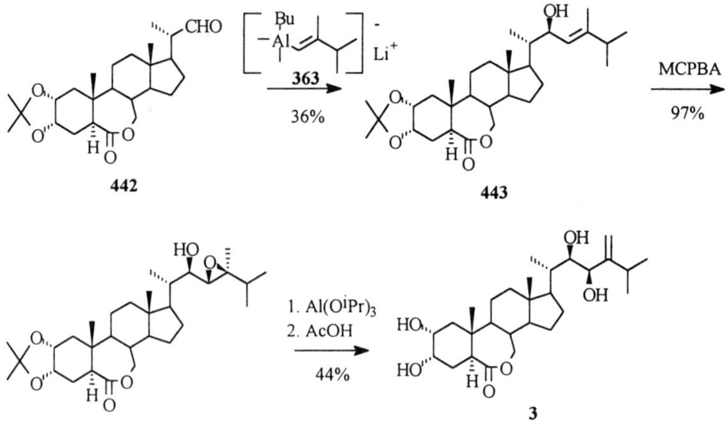

SCHEME 135

Essentially the same method of side chain construction was used by Mori *et al.* (1984, 1987b) for the preparation of dolichosterone **7** from aldehyde **362** *via* the allylic alcohol **444** (Scheme 136). The aldehydes **442** and **362** were also employed for the synthesis of homodolichosterone and homodolicholide (Mori *et al.*, 1984).

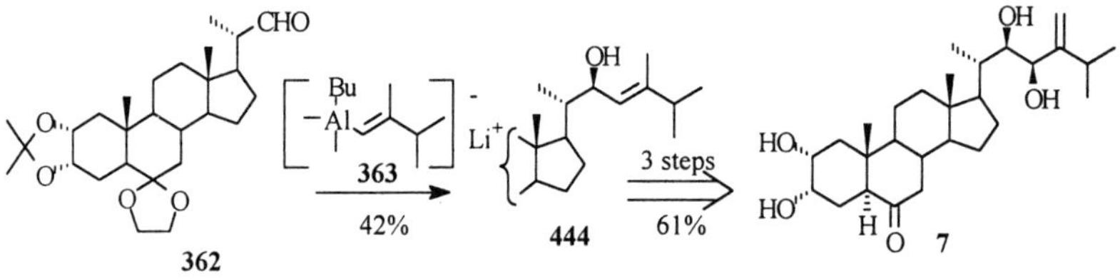

SCHEME 136

Dolicholide **3** was synthesized from brassicasterol *via* dienone **125** (Sumimoto Chem. Co., Ltd., 1983a; Takatsuto, 1988) as described in Scheme 137. In this approach the native carbon skeleton of the starting material was used for the construction of the dolicholide side chain. Hydroxylation of dienone **125** followed by 2α,3α-diol group protection led to the acetonide **445**. Treatment of this compound with an excess of MCPBA gave the epoxy lactone **446**. Opening of the epoxide with the phenyl selenide anion followed by oxidation of the intermediate hydroxy selenide led to a mixture that contained the allylic alcohol **447**. Its epoxidation furnished the epoxide **448**, which was transformed into the allylic alcohol **449**. Deprotection of the latter gave dolicholide **3**.

The same methodology of side chain construction was applied for the preparation of homodolichosterone (Sakakibara and Mori, 1983c; Takatsuto and

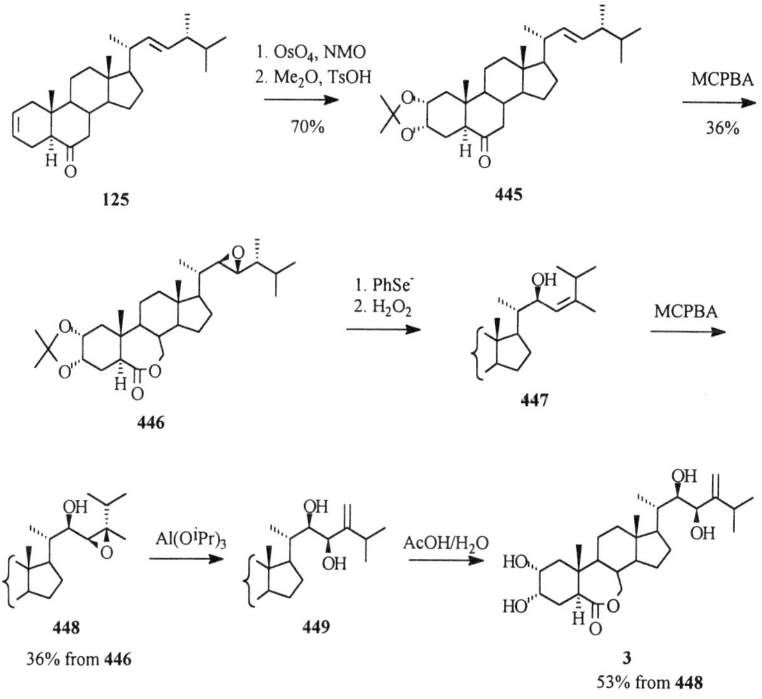

SCHEME 137

Ikekawa, 1987b) and homodolicholide (Sakakibara and Mori, 1983b; Sumimoto Chem. Co., Ltd., 1983b).

Transformation of homodolicholide **9** into dolicholide **3** was reported as described in Scheme 138 (Yuya and Takeuchi, 1986). The diacetonide **450** was ozonized to give the 24-ketone **451**, which was used for the introduction of the methylene group *via* a Wittig reaction.

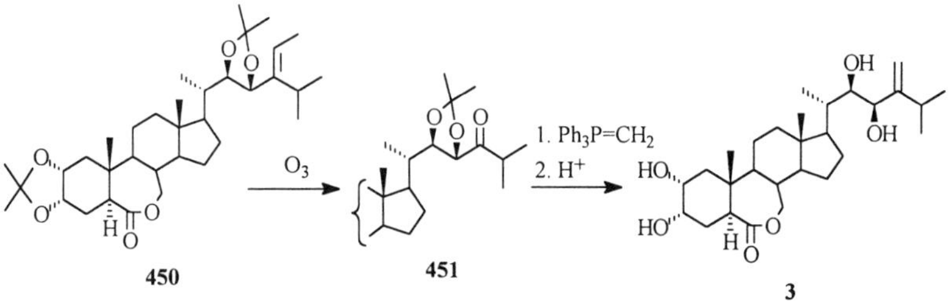

SCHEME 138

F. SYNTHESES OF 6-DEOXO-BS

The most straightforward approach to 6-deoxo-BS proved to be the direct reduction of 6-ketones to the corresponding 6-deoxo derivatives (Scheme 139). Thus the Wolff-Kishner reduction of castasterone **2** gave 6-deoxocastasterone **10** (Mori *et al.*, 1984).

Another reduction was applied for the synthesis of 6-deoxo-24-epicastasterone by Spengler *et al.* (1995). Treatment of epicastasterone **22** with ethanedithiol and boron trifluoride etherate followed by radical desulfurization of the thioketal **452** afforded the 6-deoxo-BS **40** (Scheme 140).

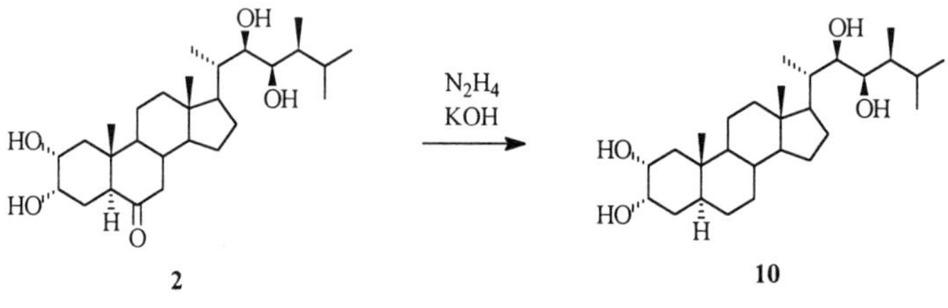

SCHEME 139

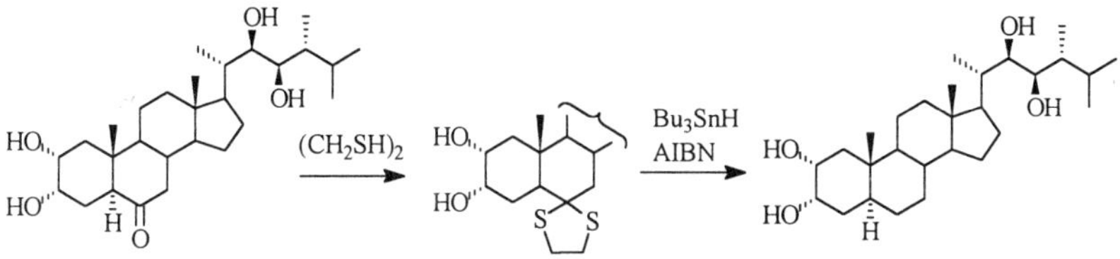

SCHEME 140

6-Deoxodolichosterone **11** was synthesized from triacetate **453** as described in Scheme 141 (Mori et al., 1984). Removal of the 6-keto group was achieved again via its thioacetal **454** by desulfurization with Raney nickel. The remaining steps were similar to those described in Scheme 135.

In the method of Takatsuto and Ikekawa (1986d) the saturated ring B was formed first (Scheme 142). Oppenauer oxidation of stigmasterol **69** to the enone **455**, reduction with lithium in liquid ammonia, elimination of the hydroxyl group, and catalytic hydroxylation of the olefin **456** followed by protection of the $2\alpha,3\alpha$-diol as the acetonide gave **457**. Its epoxidation furnished two epoxides, which could be separated chromatographically. Construction of the side chain was achieved as described in Scheme 137 and is depicted in Scheme 142.

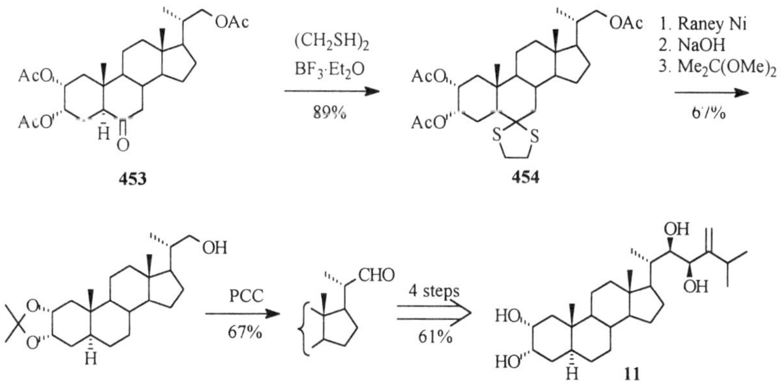

SCHEME 141

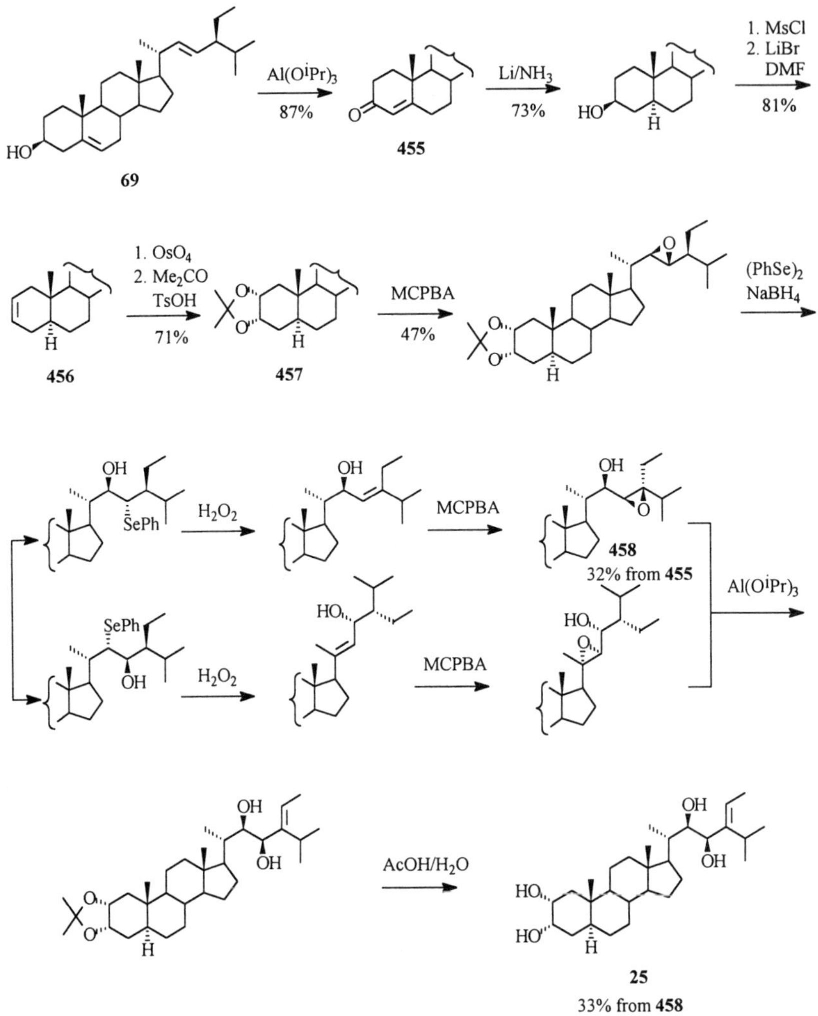

SCHEME 142

A number of 6-deoxo-BS have been prepared starting from aldehyde **219** (Takatsuto *et al.*, 1997a,b). The construction of the side chain involved addition of a Grignard reagent, orthoester Claisen rearrangement, and the Sharpless hydroxylation as indicated in Scheme 143. Acid treatment of the isosteroidal derivative **459** resulted in regeneration of the 3β-hydroxy-5-ene system.

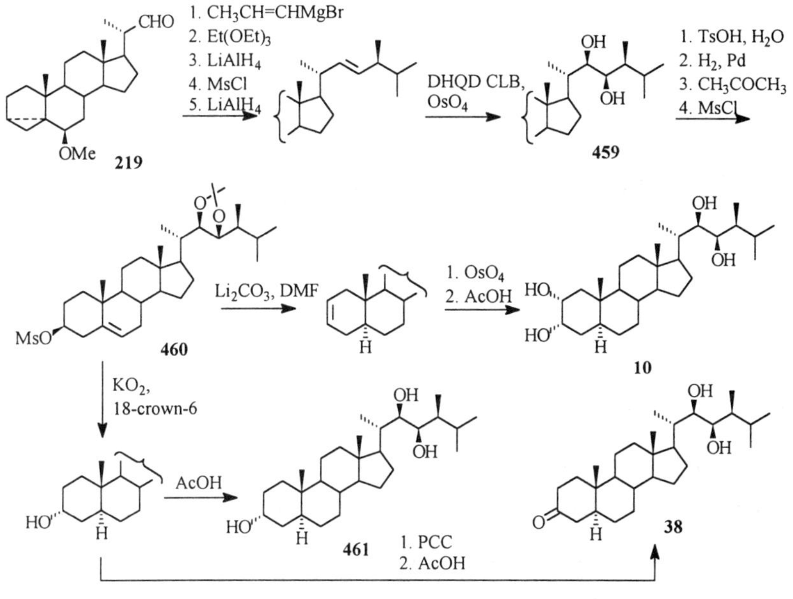

SCHEME 143

Hydrogenation of the Δ^5-double bond, protection of the diol function in the side chain, and mesylation afforded the mesylate **460**. This intermediate could be transformed into 6-deoxocastasterone **10** or, after inversion at C-3, into 6-deoxoteasterone **461** and 3-dehydro-6-deoxoteasterone **38** (Takatsuto *et al.*, 1997b).

G. SYNTHESES OF BS WITH MODIFICATIONS IN RING A

This group of BS consists of compounds with functionalities in ring A other than the 2α,3α-diol group, namely 2β,3β-, 2α,3β-, and 2β,3α-diols, 3α- and 3β-alcohols, or 2β,3β-epoxides. Such BS are necessary first of all as reference compounds for identification of minor BS in plants, which is essential for a better understanding of the biosynthesis of these hormones.

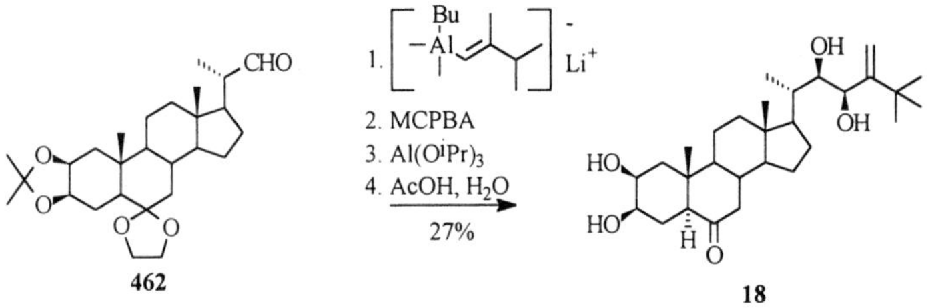

SCHEME 144

Synthesis of (2*S*,3*R*)-25-methyldolichosterone **18** (Scheme 144) was performed by Mori and Takeuchi (1988) *via* the aldehyde **462**, which had been obtained from stigmasterol in the course of a large-scale preparation of brassinolide. Aluminum organic reagents were used for the construction of the side chain as described in Scheme 135.

Synthesis of the tetraol **17** has been realized by Levinson and Traven (1996) starting from the diol **463**, which is available from ergosterol. Treatment of diol

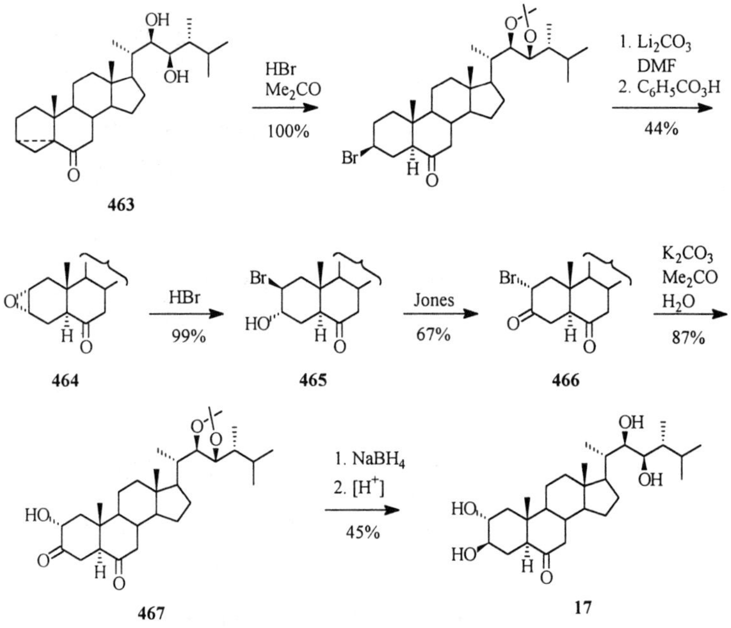

SCHEME 145

463 with hydrobromic acid in acetone led to protection of the glycol system as its acetonide and to opening of the cyclopropane ring (Scheme 145). Dehydrobromination and epoxidation of the intermediate Δ^2-olefin furnished the epoxide 464, and acid-catalyzed cleavage of this epoxide provided the trans-diaxial bromohydrin 465. Jones oxidation of 465 proceeded with simultaneous epimerization at C-2 to give the bromo ketone 466. Heating of this compound in aqueous acetone in the presence of potassium carbonate led to hydroxy ketone 467, which was converted into 17 by selective reduction and deprotection of the diol system in the side chain.

Synthesis of secasterone 44 has been performed (Voigt et al., 1995) from the Δ^2-olefin 468 (Scheme 146). Treatment of 468 with N-bromosuccinimide in aqueous dimethoxyethane gave bromohydrin 469. Acid-catalyzed deprotection of the acetonide 469 followed by formation of the epoxide gave secasterone 44.

The first synthesis of typhasterol 12 (Scheme 147) was accomplished by Takatsuto et al. (1984b) together with the preparation of brassinolide (see Scheme 98), starting from mesylate 470. Treatment of 470 with lithium carbonate in DMF gave the 3α-formate 471 in a low yield. Acetonide deprotection followed by hydrolysis of the formate ester gave typhasterol 12.

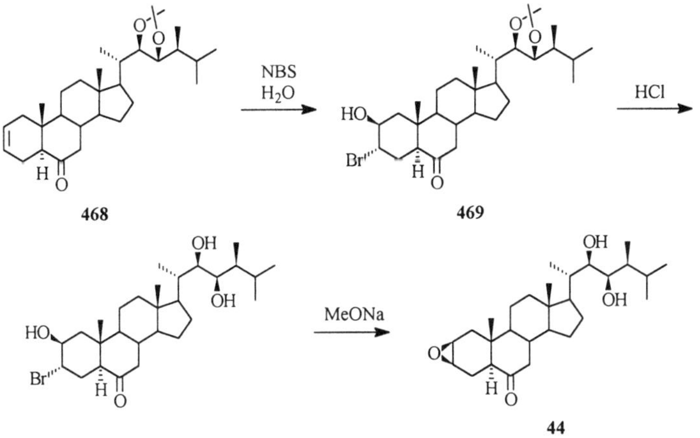

SCHEME 146

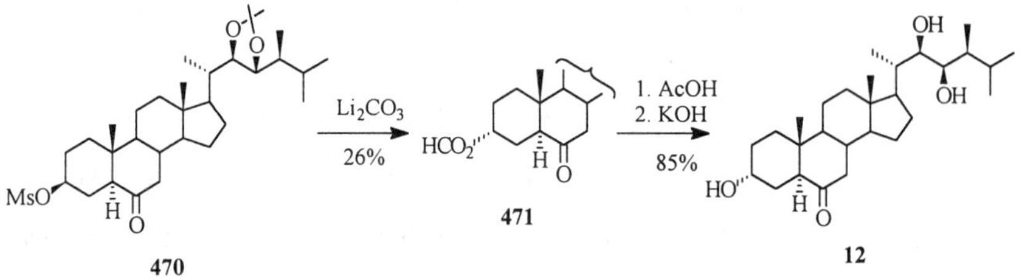

SCHEME 147

Better results have been achieved with ketone **472** (Scheme 103) as a starting material. Treatment with hydrobromic acid (Scheme 148) followed by nucleophilic substitution of the bromide atom and subsequent saponification gave the desired typhasterol **12** (Aburatani et al., 1987b). Ketone **472** proved to be suitable also for the preparation of teasterone **14** via cyclopropane ring opening followed by acetate deprotection.

Synthesis of teasterone and typhasterol from crinosterol has been described also (Takatsuto, 1986). This synthesis was based on reactions as described above for the preparation of typhasterol.

Synthesis of 3-oxo-BS is a rather simple task, provided that the corresponding derivatives with a suitable side chain are available. Thus 3-

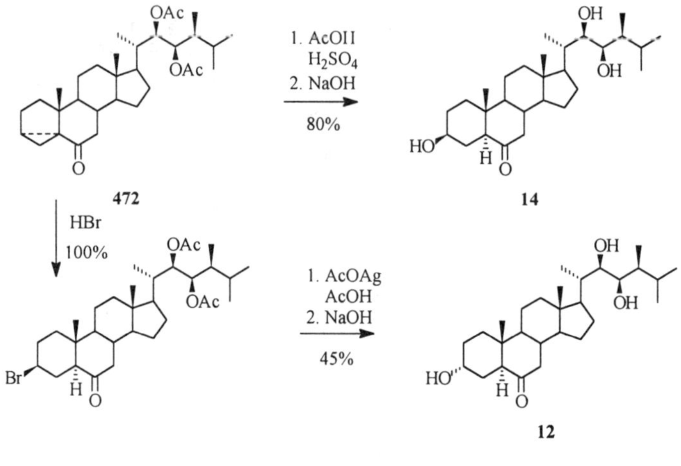

SCHEME 148

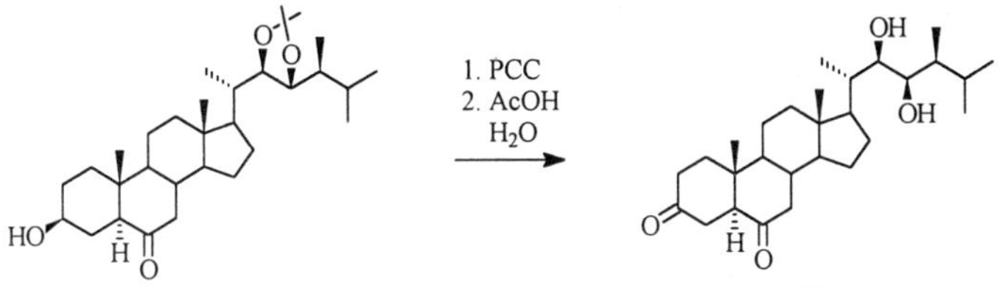

473 35

SCHEME 149

oxoteasterone **35** was synthesized from teasterone *via* the acetonide **473** (Scheme 149) by Abe *et al.* (1994). Oxidation of the 3-hydroxyl group followed deprotection of the diol gave the target compound **35**. Essentially the same method was applied for the preparation of 3-oxoteasterone **35** starting from typhasterol (Yokota *et al.*, 1994).

H. SYNTHESES OF ISOTOPICALLY LABELED BS

Research on the biosynthesis and metabolism of BS can be done efficiently only when radioactively labeled compounds are available. The easiest way for

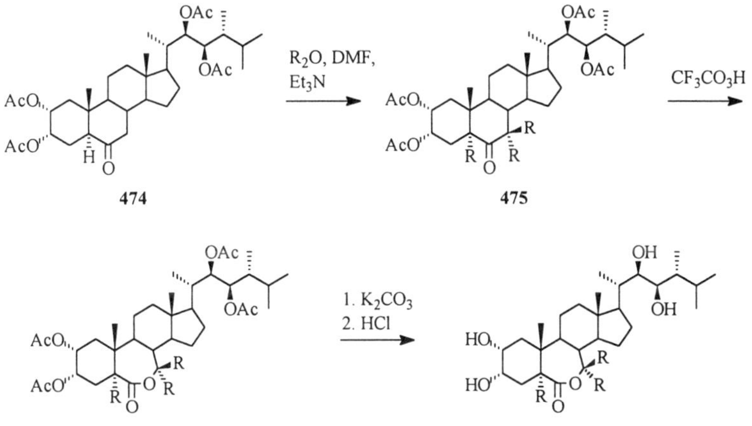

474 475 476

R = ^{2}H or ^{3}H

SCHEME 150

182 BRASSINOSTEROIDS: A NEW CLASS OF PLANT HORMONES

the preparation of such compounds proved to be the base-catalyzed exchange of hydrogen atoms by deuterium or tritium at C-5 and/or C-7. An example of this approach is shown in Scheme 150 (Kolbe *et al.*, 1992). Treatment of tetraacetate **474** with ^{2}H$_2$O or ^{3}H$_2$O in the presence of a base led to the tri-deuterated or -tritiated compounds **475**. The Baeyer-Villiger oxidation followed by deprotection and relactonization afforded the labeled epibrassinolides **476**.

An improved procedure for the tritiation of epibrassinolide was based on employment of its diacetonide (Kolbe *et al.*, 1992). Syntheses of 5,7,7'-trideuterated BS of this series were reported also by Khripach *et al.* (1995f) *via* labeling of the Δ^2-6-ketones by MeOD and D$_2$O.

Labeling of other positions of the steroid skeleton is a more difficult problem. On the other hand, such compounds (especially when labeled with ^{14}C)

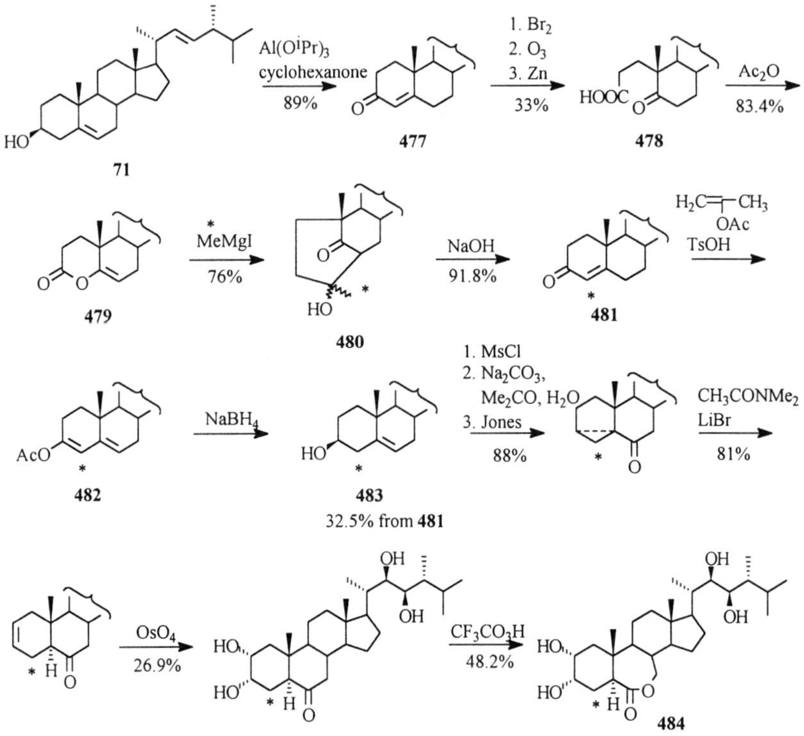

SCHEME 151

are more stable to metabolic loss. Synthesis of [4-^{14}C]-epibrassinolide **484** was performed by Seo *et al.* (1989) starting from brassicasterol **71**. Oppenauer oxidation led to brassicasterone **477** (Scheme 151). Cleavage of ring A after protection of the Δ^{22}-double bond as its dibromide followed by ozonolysis and debromination with zinc dust afforded the keto acid **478**. Dehydration to the enol lactone **479** followed by a Grignard reaction with [^{14}C]-methylmagnesium iodide led to the bridged ketone **480**, which was transformed into the enone **481** by alkaline treatment. The regeneration of the 3β-hydroxy-Δ^5-system was achieved by hydride reduction of the enol acetate **482**. [4-^{14}C]-Brassicasterol **483** was transformed into [4-^{14}C]-epibrassinolide **484** by methods described above (see Scheme 115).

A number of deuterio-labeled 28C-BS including [26,28-^{2}H$_6$]-brassinolide, [26,28-^{2}H$_6$]-castasterone **490**, [26,28-^{2}H$_6$]-typhasterol **489**, and [26,28-^{2}H$_6$]-teasterone were prepared from aldehyde **219** (Takatsuto and Ikekawa, 1986b,c,e). Addition of lithium acetylide to the aldehyde **219** led to a mixture (2:1) of the 22*R*-alcohol and its 22*S* epimer (Scheme 152). Protection of the hydroxyl group of the main isomer followed by alkylation of the acetylene moiety with iodomethane-d$_3$ gave, after removal of the silyl group, the d$_3$-acetylenic alcohol **485**. Partial hydrogenation over a Lindlar catalyst afforded the allylic alcohol **486**. The Claisen rearrangement with triethyl orthopropionate in the presence of propionic acid provided the ester **487** with the desired configuration at C-24. Transformation of the ester group into a d$_3$-methyl group followed by regeneration of the cyclic part of the molecule furnished [26,28-^{2}H$_6$]-crinosterol **488**. The remaining steps of the synthesis were similar to those described above for preparation of a normal BS as indicated in Scheme 152. Synthesis of [24,28-^{3}H$_2$]-castasterone was realized *via* catalytic reduction of dolichosterone over platinum with tritium gas (Yokota *et al.*, 1990a,b).

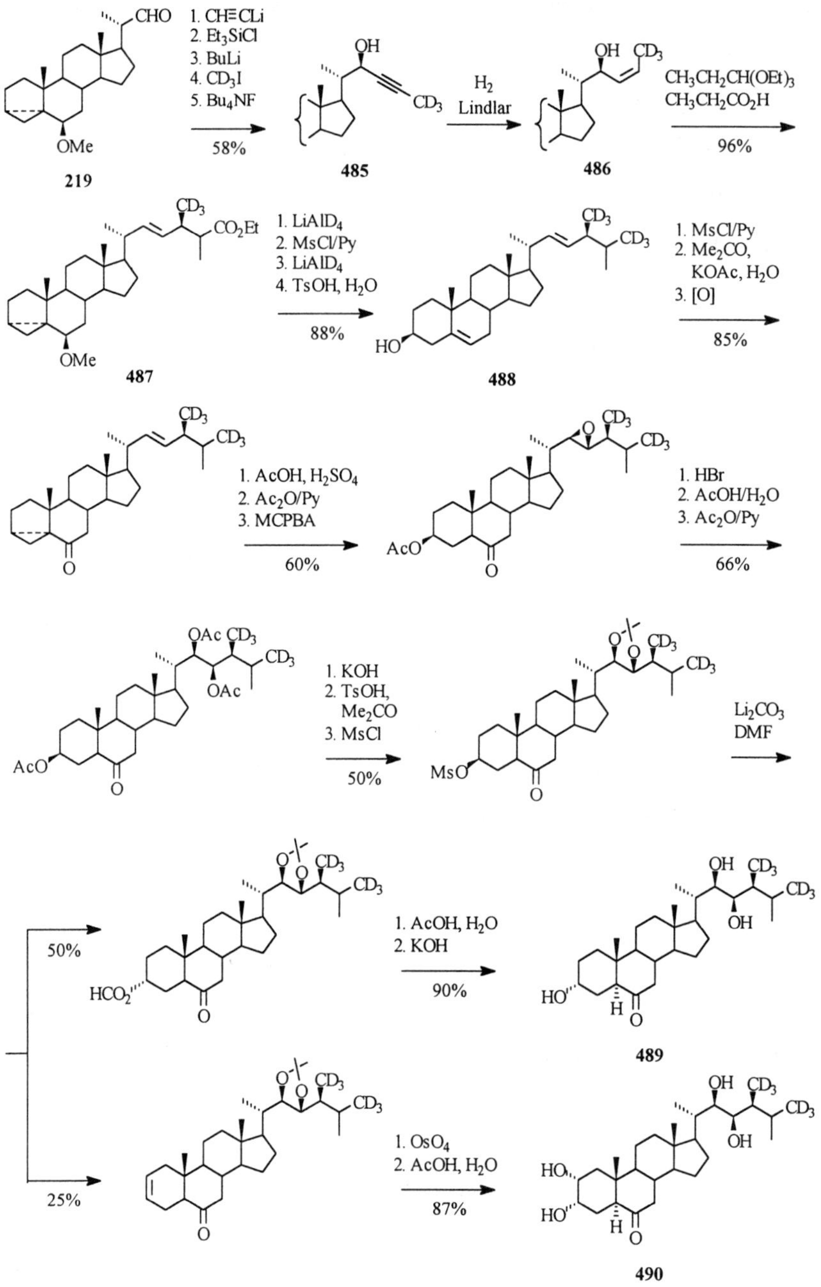

1. CH≡CLi
2. Et3SiCl
3. BuLi
4. CD3I
5. Bu4NF
58%
OMe
219
485
486
H2
Lindlar
CH3CH2CH(OEt)3
CH3CH2CO2H
96%
CO2Et
1. LiAlD4
2. MsCl/Py
3. LiAlD4
4. TsOH, H2O
88%
487
488
1. MsCl/Py
2. Me2CO, KOAc, H2O
3. [O]
85%
1. AcOH, H2SO4
2. Ac2O/Py
3. MCPBA
60%
1. HBr
2. AcOH/H2O
3. Ac2O/Py
66%
1. KOH
2. TsOH, Me2CO
3. MsCl
50%
Li2CO3
DMF
50%
25%
1. AcOH, H2O
2. KOH
90%
489
1. OsO4
2. AcOH, H2O
87%
490

SCHEME 152

CHAPTER **VIII**

SYNTHESES OF BS ANALOGS

The preparation of analogs has started simultaneously with the isolation of BS from plant sources and with the synthesis of natural BS. From earlier experiences with natural products, and in particular with steroids, it is known that frequently analogs can exhibit a high and selective type of activity that makes them mostly more suitable for practical application. Analogs are necessary also for scientific research, for example for the study of structure-activity relationships or for investigation of the biosynthesis and the metabolism of the natural product. Often analogs are obtained in the course of the chemical synthesis of the natural compounds as byproducts. For BS this is especially the case for the 22S,23S isomers which are formed in the oxidation of the Δ^{22}-double bond in the side chain along with the natural 22R,23R compounds. A set of analogs can also be obtained from synthetic research on model compounds. Very often however analogs are obtained from special research programs in which sets of compounds are designed to investigate structure-activity relationships, or to test some hypothesis about the mode of action. It is necessary to note that the concept "analog" has to be taken rather wide. Also some BS were synthesized first as analogs and only later identified in plant sources as an authentic natural products. In this chapter the syntheses of many analogs with structural variations in various parts of the molecules will be described. In the next chapters their biological activity will be discussed.

A. ANALOGS WITH THE SIDE CHAIN OF A STARTING MATERIAL

The synthesis of analogs with the side chain of the starting compound is simple from a chemical point of view. Only the A and B rings have to be functionalized in the proper way. This means that such derivatives are relatively cheap and that explains the interest in these, in general not very active, substances. To date, practically all commercially available steroids that contain appropriate functional groups in the AB rings, first of all the 3β-hydroxy-5-enes, have been used for the preparation of the corresponding analogs. Moreover, mixtures of phytosterols proved to be suitable for the preparation of compositions useful as plant growth regulators (Mitra et al., 1984b,c). The synthetic routes in most cases follow methods described earlier (see Scheme 18). As a rule the side chain is not subjected to additional transformations and sometimes, when no vulnerable functional groups are present, rather drastic methods can be employed (Liang et al., 1989).

A number of analogs have been prepared starting from cholesterol (Thompson et al., 1982; Akhrem et al., 1983a; Kondo and Mori, 1983; Mitra et al., 1984a; Kohout and Strnad, 1986; Liang et al., 1989; Kovganko and Netesova, 1991; Kohout et al., 1991) and β-sitosterol (Mitra et al., 1984a;

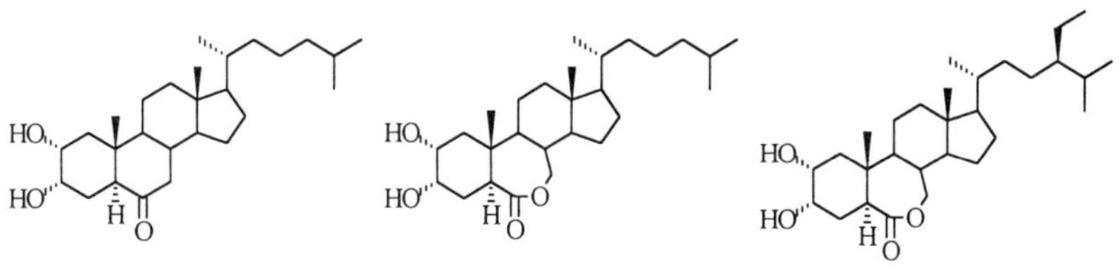

Fig. 26. Examples of analogs with unfunctionalized side chains.

Kovganko and Kashkan, 1990). Some structures of this type are shown in Fig. 26.

Steroids of the pregnane type were used also for the synthesis of the corresponding analogs (Kohout et al., 1991) (Fig. 27).

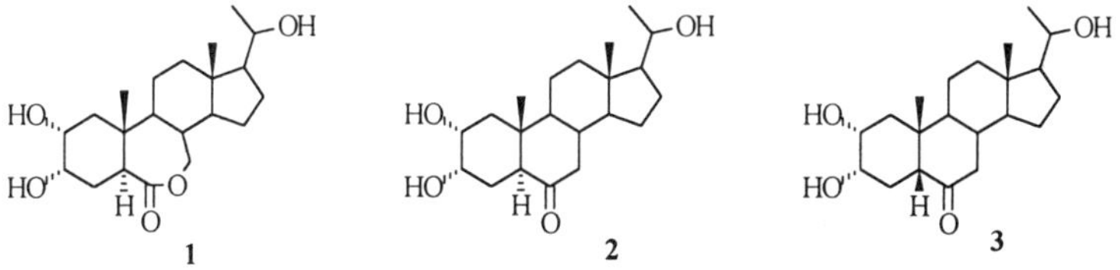

Fig. 27. Examples of analogs of the pregnane series.

The Baeyer-Villiger oxidation of diketone **491** (Scheme 153) was accompanied by a transformation in the side chain leading to triacetate **492** (Kondo and Mori, 1983; Kohout et al., 1986, 1987a,b, 1991), which could be transformed into the analog **493** with the androstane skeleton.

Similar types of derivatives were also available from androstanes and in this way analogs without a side chain were prepared (Kohout and Strnad, 1989a; Kohout et al., 1991) (Fig. 28).

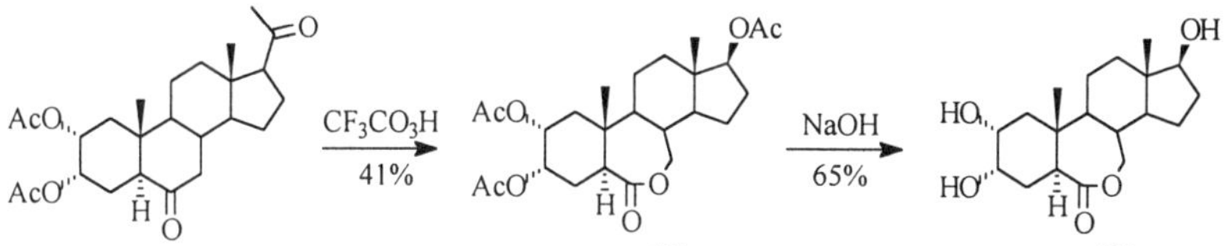

SCHEME 153

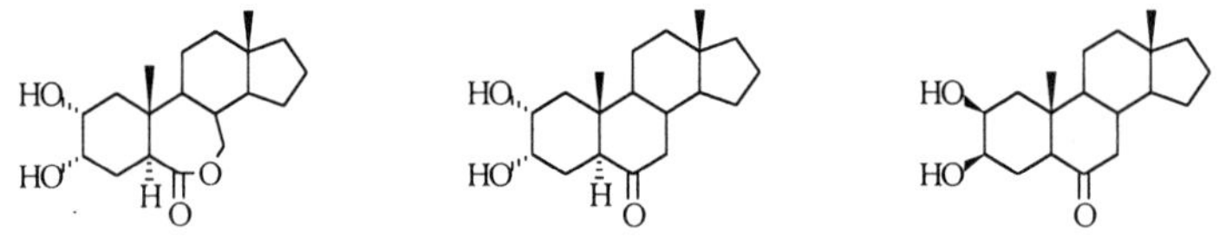

Fig. 28. Examples of analogs without a side chain.

Many analogs of spirostanes have been prepared by Marquardt *et al.* (1988), Adam *et al.* (1991), Tian (1992), Coll *et al.* (1992, 1995), and Iglesias *et al.* (1996, 1997) (Fig. 29).

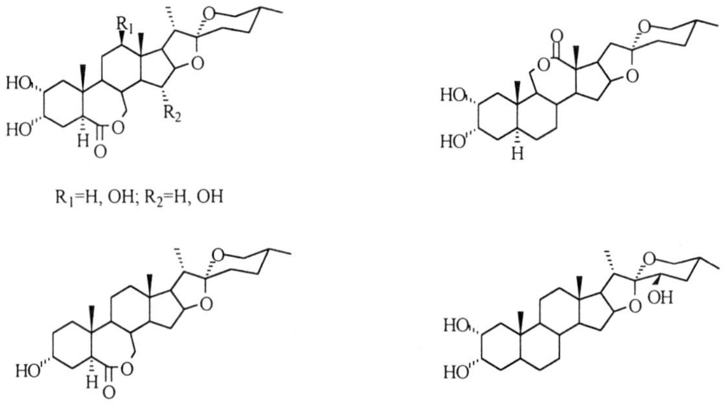

R_1=H, OH; R_2=H, OH

Fig. 29. Examples of analogs of the spirostane series.

The steroid alkaloids solasodine (Quyen *et al.*, 1994a) and solanidine (Quyen *et al.*, 1994b) were also used for the preparation of BS analogs (Fig. 30).

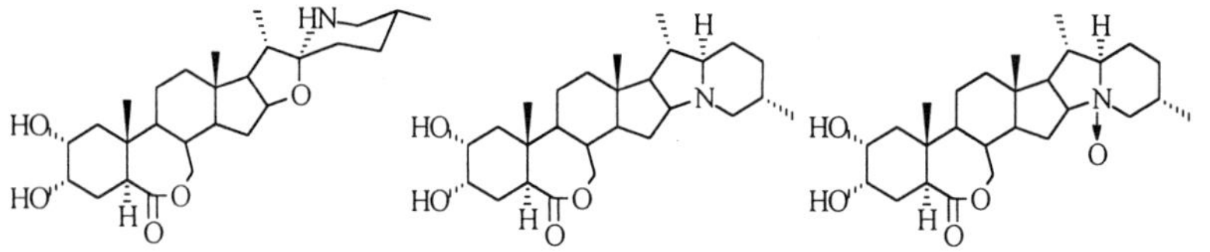

Fig. 30. Examples of solasodine and solanidine nitrogenous analogs.

B. ANALOGS WITH DIFFERENT FUNCTIONAL GROUPS IN THE SIDE CHAIN

In this type of analog the carbon skeleton of the side chain that is characteristic for natural BS is maintained but different functional groups or other stereoisomers are present. First of all the 22,23-diol should be mentioned, with the configuration at C-22 and/or C-23 opposite to that in the natural BS. Furthermore, many 22,23-epoxides and compounds with additional or without hydroxyl group(s) have been prepared, purposely or in the course of other syntheses of BS.

1. 22,23-Dihydroxy Derivatives

Compounds with the 22S,23S-diol function in the side chain (Fig. 31) are often obtained as BS analogs along with the natural 22R,23R compounds. This is the result of the usual hydroxylation methods of the Δ^{22}-steroids with osmium tetroxide which lead to the 22S,23S-diols as the major products. Even in oxidations with a chiral catalyst, considerable amounts of the 22S,23S-diols are formed. Probably (22S,23S)-homobrassinolide is the most interesting among these analogs because of its relatively high activity and good availability (Mori, 1980a,b, 1982).

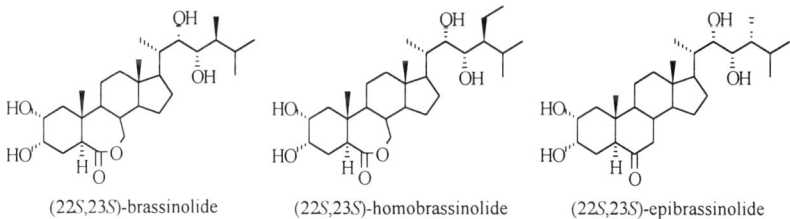

(22S,23S)-brassinolide   (22S,23S)-homobrassinolide   (22S,23S)-epibrassinolide

Fig. 31. Examples of 22S,23S analogs.

Very often (22S,23S)-BS were isolated and tested along with their natural counterparts (Thompson et al., 1981, 1982), and sometimes these BS were the actual target compounds (Akhrem et al., 1984a, 1985b, 1987b; Takatsudo and Hayashi, 1987). Scheme 154 shows some variants of (22S,23S)-BS analogs **494** and **495** that were synthesized in the stigmastane series (Akhrem et al., 1984a, 1987b).

A synthetic route to analogs with a 22R,23S- or 22S,23R-diol function is shown in Scheme 155. The Δ^{22}-double bond was epoxidized, followed by opening of the epoxy ring to afford the tetraols **496** and **497** (Fuendjiep et al., 1989).

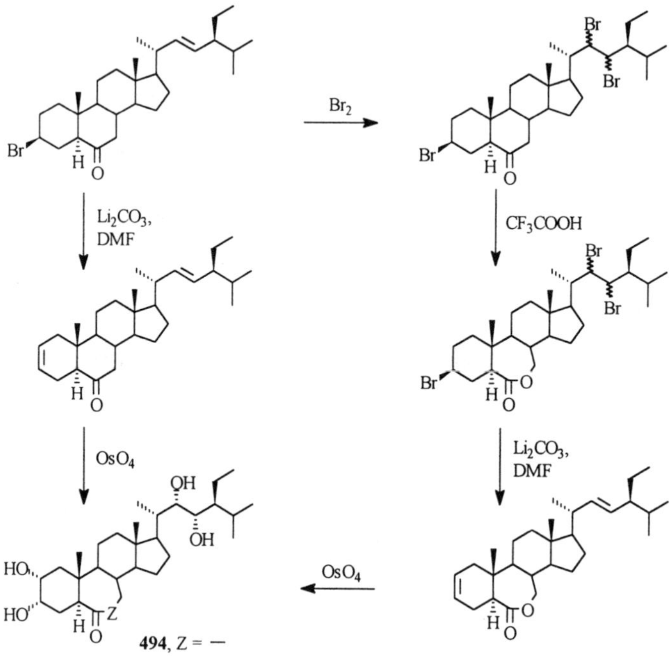

SCHEME 154

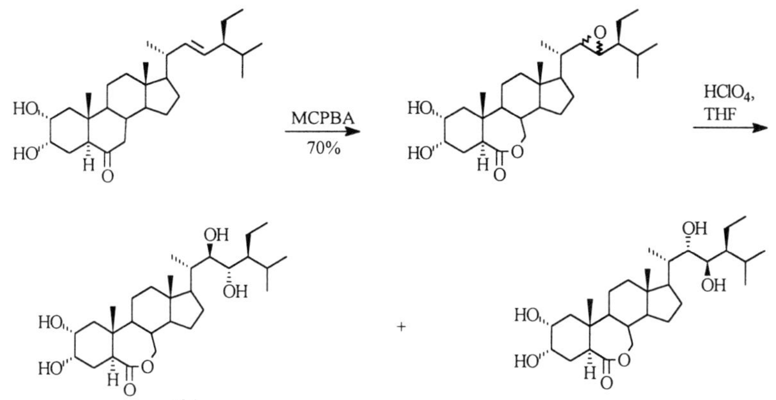

496
40%

497
40%

SCHEME 155

2. 22,23-Epoxides

Compounds with a 22,23-epoxy group in the side chain reveal a high level of biological activity (Suntory, Ltd., 1980; Sumitomo Chem. Co., Ltd., 1981; Kamuro *et al.*, 1987, 1988a,b, 1993, 1996, 1997; Kamuro and Takatsuto, 1991; Takatsuto *et al.*, 1989a, 1996c) and are considered to be very promising for practical application in agriculture because of their long-lasting biological activity under field conditions.

Synthesis of epoxy ester **499** (TS303) has been achieved from stigmasterol in 26% overall yield *via* dienone **420** and diester **498** as shown in Scheme 156 (Takatsuto *et al.*, 1996c).

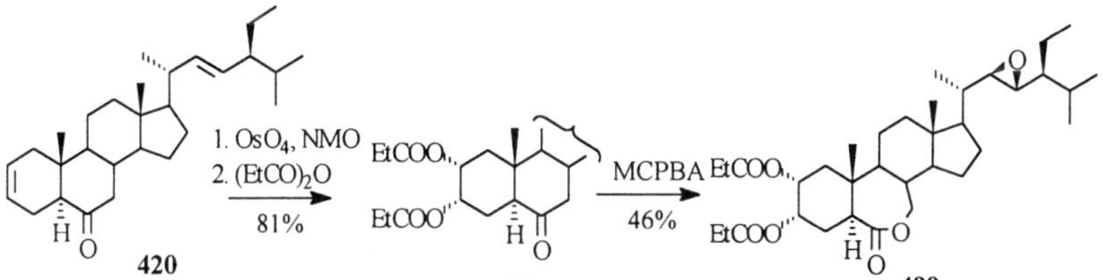

420 **498** **499**

SCHEME 156

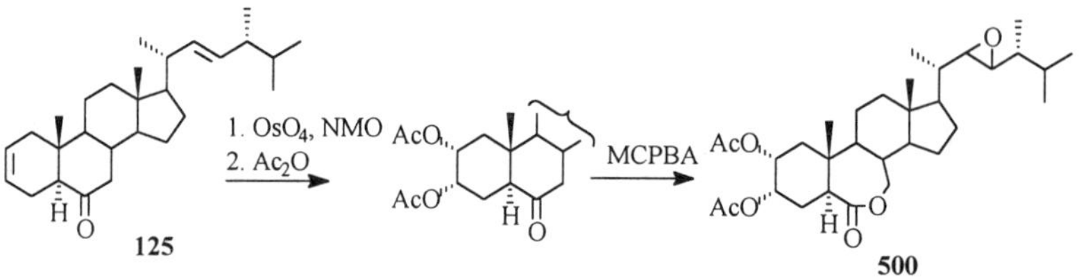

SCHEME 157

A similar epoxy ester **500** with an ergostane skeleton has been prepared by Takatsuto *et al.* (1987b, 1989a) as described in Scheme 157 and also by Khripach *et al.* (1995e).

A number of epoxy analogs have been synthesized by Brosa and Miro (1997). Treatment of dienone **420** with MCPBA or molecular oxygen and benzaldehyde proceeded with initial epoxidation of the Δ^2- and Δ^{22}-double bonds, followed by a Baeyer-Villiger oxidation of the 6-ketone to give a set of derivatives (Scheme 158). Probably, the diepoxide **501** and diepoxy lactone **502** are the most interesting analogs.

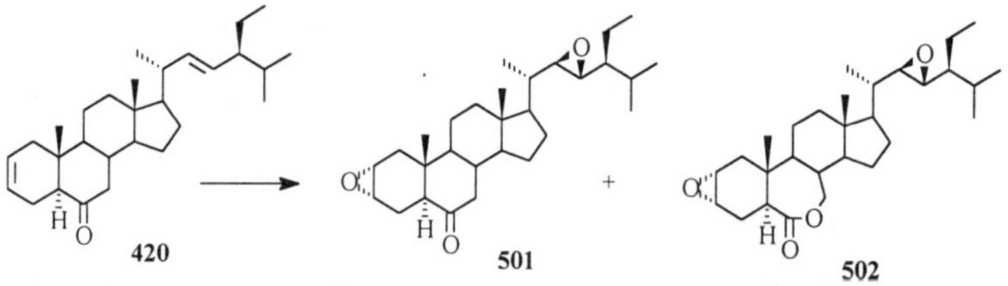

SCHEME 158

3. Analogs with Additional or without Hydroxy Group(s)

Synthesis of 23-deoxy analogs is a relatively simple task, and an example is shown in Scheme 159 (Ihara Chem. Ind. Co., Ltd., 1982a). Addition of a Grignard reagent to the C-22 aldehyde **219** followed by acetylation gave 22-

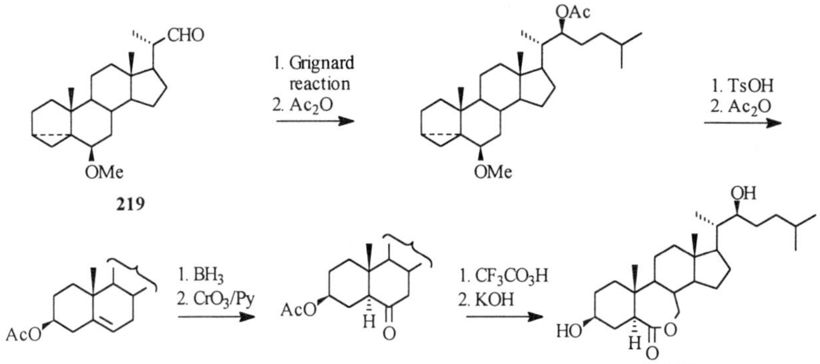

219

503

SCHEME 159

acetate, which in a few steps was converted into dihydroxy lactone **503** by methods described in the previous chapters.

Several new monohydroxy analogs such as 24-epicathasterone **504** and 22-deoxy-24-epiteasterone **505** (Fig. 32) have been prepared by Voigt *et al.* (1997).

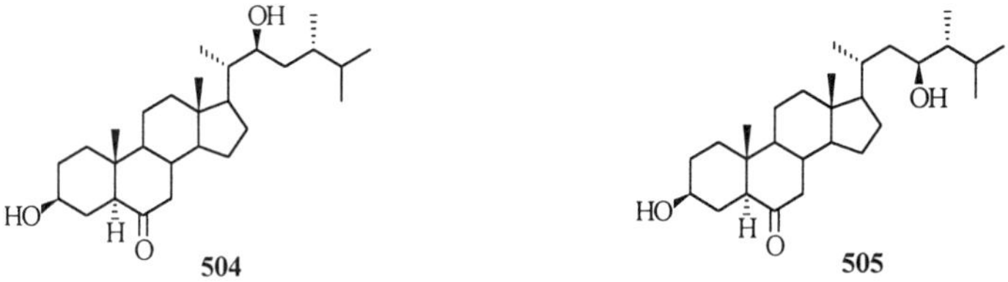

504 505

Fig. 32. BS analogs prepared starting from ergosterol *via* the 22,23-epoxides and the 22,23-bromohydrins.

To date, there is no firm evidence for the occurrence of 25-hydroxy-BS in plants. However, identification of the 25-glycoside of 25-hydroxy-24-epibrassinolide as a metabolite of exogenously applied epibrassinolide in

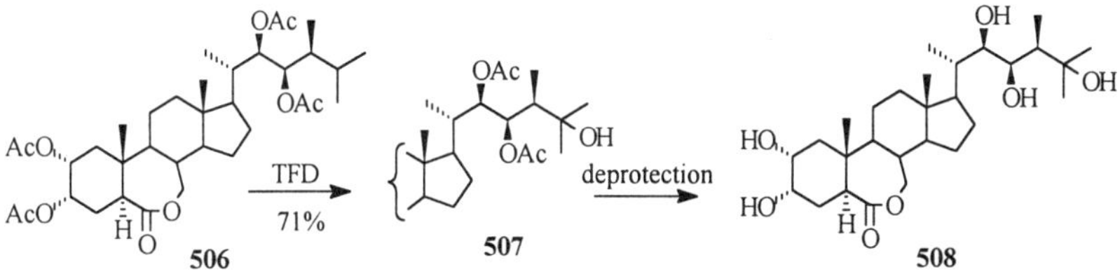

506 507 508

SCHEME 160

cultured cells of *Lycopersicon esculentum* (Hai *et al.*, 1995) suggests the existence of similar compounds in plants. The easiest way to prepare these compounds is the direct oxyfunctionalization of C-25 by regioselective C-H insertion using methyl(trifluoromethyl)dioxirane (Voigt *et al.*, 1996a,b). The reaction proceeds smoothly for brassinolide tetraacetate **506** to give the 25-alcohol **507** (Scheme 160), which upon basic hydrolysis afforded **508**.

A similar oxidation of the diacetoxy acetonide **509** (Scheme 161) followed by Baeyer-Villiger oxidation afforded 25-hydroxyepicastasterone **510**.

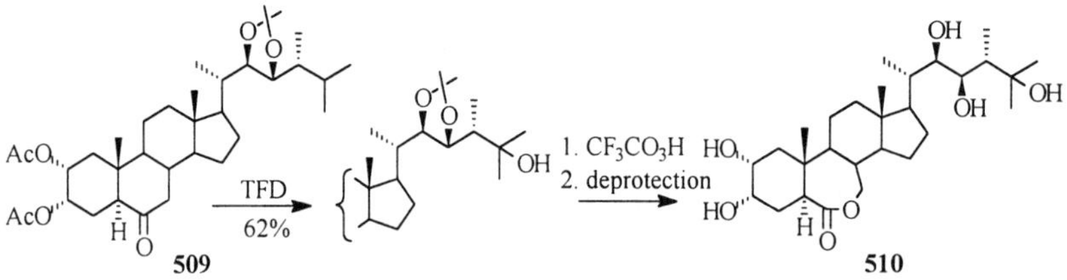

SCHEME 161

A number of analogs containing hydroxyl groups at C-20 and/or C-28 have been prepared (Kametani and Honda, 1986, 1987a; Kametani *et al.*, 1988a). Some of them are shown in Fig. 33.

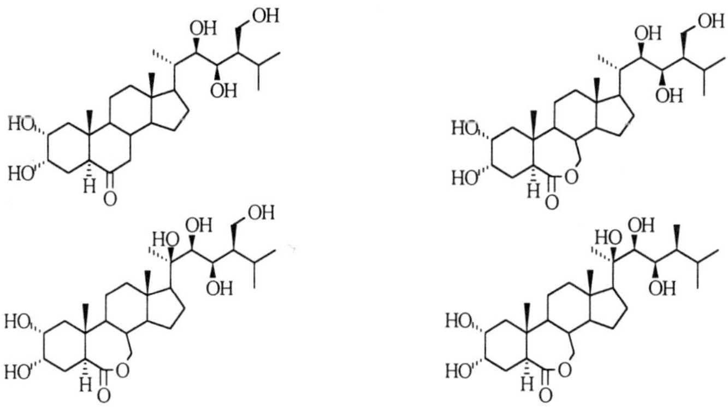

Fig. 33. Examples of C-20/C-28-hydroxy analogs prepared from pregnenolone. The synthetic route was similar to that described for the synthesis of brassinolide, which is outlined in Scheme 110.

C. ANALOGS WITH A MODIFIED CARBON SKELETON IN THE SIDE CHAIN

Modification of the carbon skeleton of a compound mostly makes its construction more complicated and consequently more expensive. The chances of such analogs for practical application in agriculture are therefore rather low, but for the study of structure-activity relationships they are interesting and useful.

1. With Side Chains Epimeric at C-20

The center at C-20 normally does not participate in metabolic transformations resulting in its inversion. For this reason practically all natural steroids, including BS, possess the same configuration at C-20. To elucidate the importance of this center for the biological activity of BS, a number of analogs with the opposite configuration at C-20 have been prepared.

The brassinolide analogs **514** and **515** were synthesized from the C-20-alcohol **511** by Kametani *et al.* (1988a). Dehydration of alcohol **511** gave the exo-olefin **512**, which upon catalytic reduction afforded a mixture (51:49) of

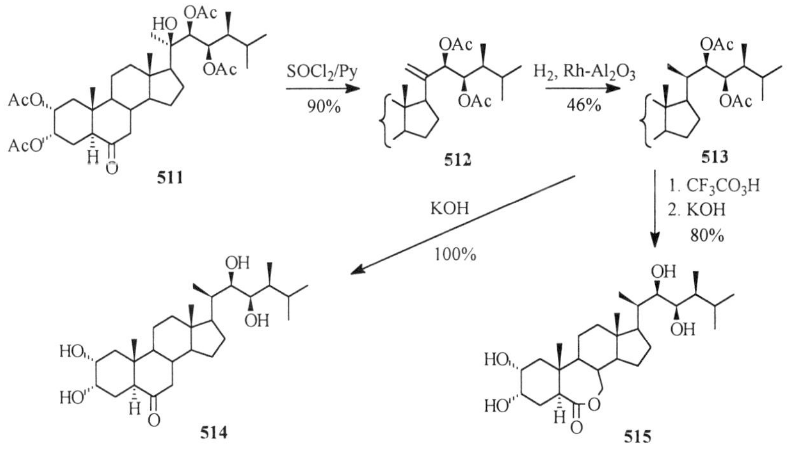

SCHEME 162

tetraacetates epimeric at C-20 (Scheme 162). Hydrolysis of epimer **513** furnished 20-epicastasterone **514**. The Baeyer-Villiger oxidation of tetraacetoxy ketone **513** followed by acetate hydrolysis gave 20-epibrassinolide **515**.

The synthesis of the (20R)-brassinones **519** and **520** was based on the ozonolysis of the Δ^{22}-olefin **516** in the presence of a mild base (Khripach, 1990). Under these conditions isomerization at C-20 took place to give a mixture (2:1) of both 22-aldehydes (Scheme 163). The equilibrium was shifted toward

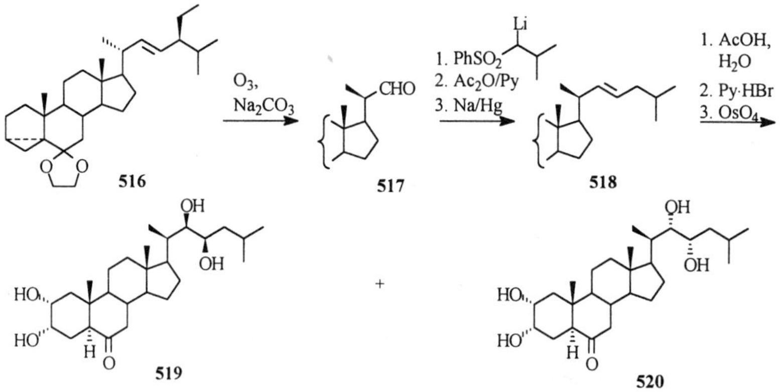

SCHEME 163

formation of the nonnatural (20R)-aldehyde **517**. It was transformed into the Δ^{22}-olefin **518** using the sulfone approach (see Scheme 131) and further into (20R)-brassinone **519** and (20R,22S,23S)-brassinone **520**.

2   With Side Chains Epimeric at C-24

To date, the isolation of 24α-ethyl-BS has not been described, although sterols with the corresponding carbon skeleton in the side chain are known as natural products. Probably these BS will be found in the future, but now such compounds are considered as analogs. Two syntheses of (24R)-homobrassinolide **522** have been published (Scheme 164). The first one was based on poriferasterol **521** with the 24R-carbon skeleton already in the side chain (Takatsuto *et al.*, 1983a). The synthetic route was identical to that

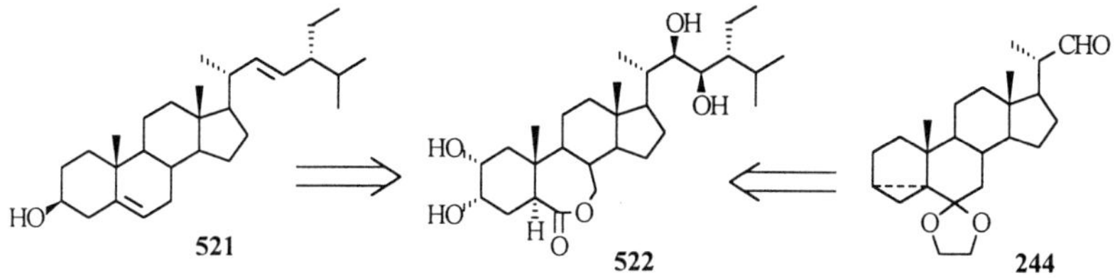

SCHEME 164

described for homobrassinolide (see Scheme 121). The second synthesis started from the C-22 aldehyde **244** (Khripach *et al.*, 1994b). Construction of the side chain was achieved *via* addition of the corresponding sulfone to **244** followed by hydroxylation of the obtained Δ^{22}-olefin as described in Scheme 131.

3   Analogs with a Shortened Side Chain

The side chain of BS with four contiguous chiral centers is a difficult element in their synthesis. It is natural that attention has been paid to the preparation of analogs with a "shortened" simple side chain.

Besides it was found that 26,27-bisnorbrassinolide showed a high biological activity (Ihara Chem. Ind. Co., Ltd., 1983; Takatsuto *et al.*, 1984a; Kametani *et al.*, 1987; Shen, Z.-W., and Zhou, W.-S., 1992). The synthesis of these analogs,

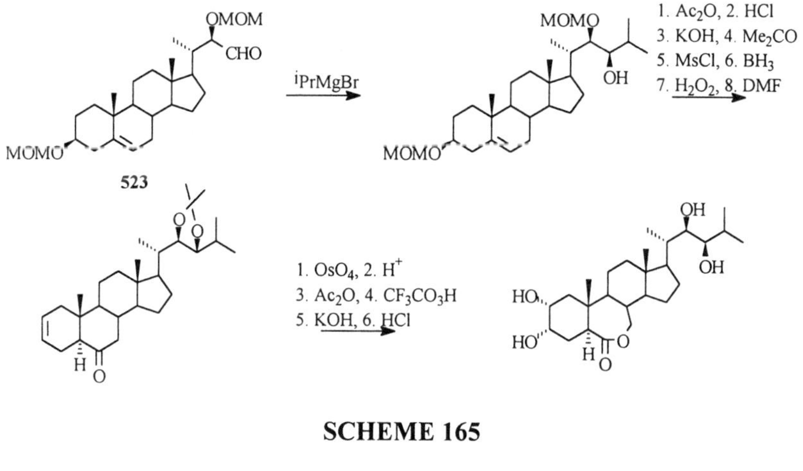

SCHEME 165

starting from the aldehyde **523**, was performed by Takatsuto *et al.* (1984a) and is shown in Scheme 165. The side chain in this compound was constructed using a chelation-controlled Grignard reaction.

A formal synthesis of 26,27-bisnorbrassinolide, reported by Kametani *et al.* (1987), is outlined in Scheme 166. Stereoselective reduction of the 5-ylidenetetronate **524** was used to control the stereochemistry of the chiral centers in the side chain (see also Scheme 45). Transformation of the lactone **525** led to triol **526**, which was described in the previous scheme as an intermediate for the synthesis of 26,27-bisnorbrassinolide.

Hazra *et al.* (1997a) prepared several 20,22-dihydroxy-BS analogs starting from 16-dehydropregnenolone **527** (Scheme 167). The first approach *via* hydroxylation of the intermediate $\Delta^{20(22)}$-olefin led to a mixture (1:1) of diols **528** epimeric at C-20. The stereoselective synthesis of the desired isomer, in the form of its 20,22-diethyl ether, was achieved in 16 steps. The stereochemistry at C-20 was established *via* addition of the lithium salt of 1,3-dithiane to the C-20 ketone. The poor yield obtained for the transformation of **529** into the final compound **530** was due to instability of the 20,22-diether function under the conditions of the acid-catalyzed rearrangement of the 3α,5-cyclo-6-ketone.

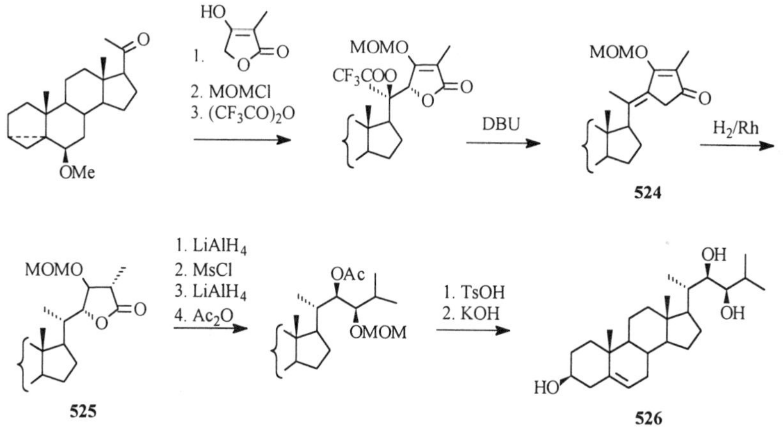

SCHEME 166

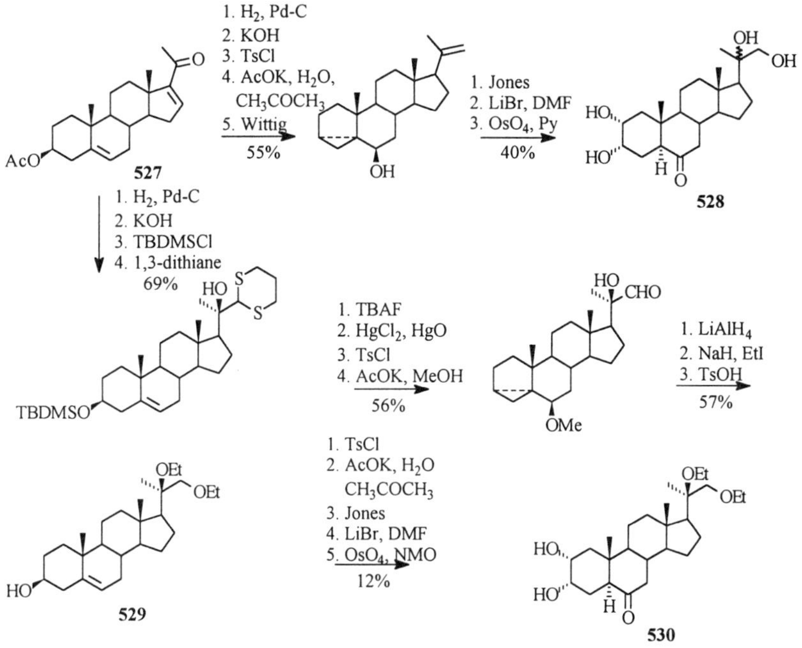

SCHEME 167

4 Analogs with an Ester, Ether, or Amide Function in the Side Chain

A number of BS analogs containing an ester or amide function in the side chain were prepared by Cerny *et al.* (1984, 1986, 1987) as shown in Scheme 168. The acid **531** was the key intermediate in this approach. Treatment with oxalyl chloride led to the chloro anhydride **532**, which was further subjected to reaction with an amine or an alcohol to give the corresponding amide **533** or ester **534**.

The reaction of simple acyl chlorides with androstane type (Kohout, 1989a-f; Kohout and Strnad, 1989b; Kohout *et al.*, 1991) and pregnane type (Kohout *et al.*, 1987b, 1991; Kohout and Strnad, 1992) steroid alcohols was used for the

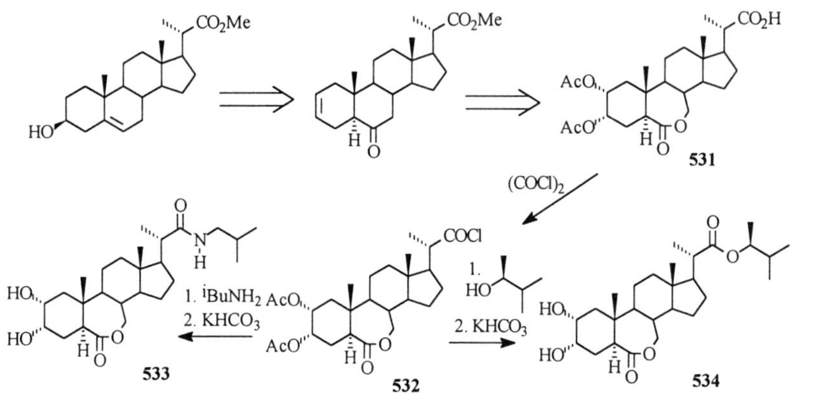

SCHEME 168

preparation of analogs with an ester function with five carbon atoms at C-17 or C-20, respectively. Several 2,3-diacyl and -alkylidene derivatives of these esters have been prepared by Kohout (1989a,c,d). A new type of analog containing a lactone in ring D (Kohout, 1997) was prepared as well. Structures of some of these analogs are shown in Fig. 34.

Compounds with a 22-ether function in the side chain have been prepared by Kerb *et al.* (1983a,b,d, 1985, 1986). Treatment of diketone **535** with chlorotrimethylsilane and zinc produced the Δ^2-6-ketone, which was converted into the tosylate **536** *via* acetate hydrolysis and tosylation (Scheme 169).

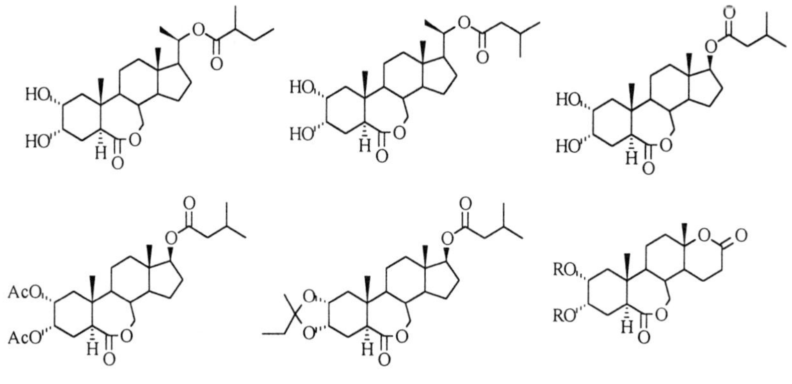

Fig. 34. BS analogs (esters) of the androstane and pregnane series.

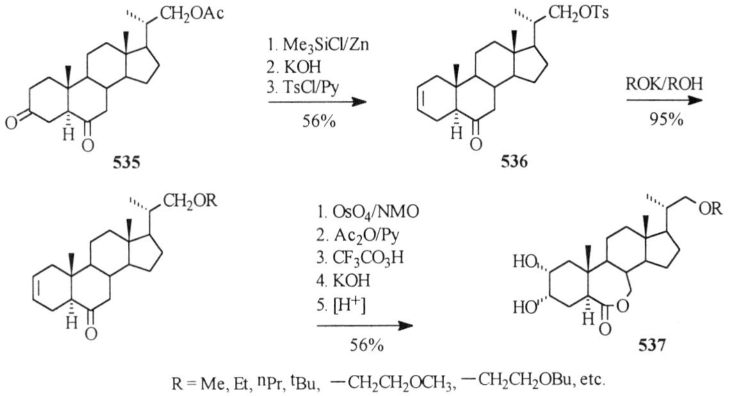

SCHEME 169

Introduction of the 22-ether group was achieved by treatment of this tosylate with potassium alcoholates. The remaining steps were traditional for BS synthesis and led to the hexanorbrassinolide 22-ethers **537**.

5. Analogs with a Heterocyclic Moiety in the Side Chain

Isoxazolines and isoxazoles were repeatedly used as intermediates for the construction of several polyfunctionalized steroid side chains (Khripach, 1990; Khripach et al., 1990a-h, 1991a-d, 1992a, 1993b-f; Litvinovskaya et al., 1994, 1995, 1996a, 1997a,b). It was natural to evaluate the corresponding analogs with these heterocyclic moieties in the side chain for their plant growth promoting activity. In general, the synthesis includes olefin formation followed by 1,3-cycloaddition of the appropriate nitrile oxide. Synthesis of the isoxazoline analog **538** starting from the aldehyde **244** demonstrates the methodology as shown in Scheme 170 (Khripach et al., 1990h, 1993b). The heterocyclic ring proved to be stable enough to withstand the further transformations of the AB rings to dihydroxy lactone **538**.

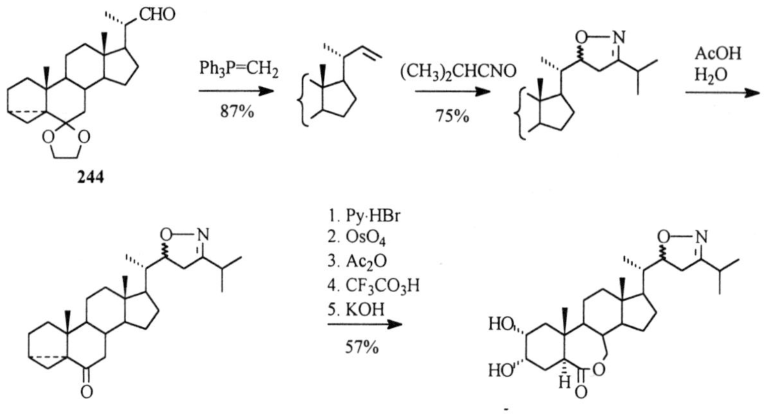

SCHEME 170

A similar heterocyclic analog **540** with an additional hydroxyl group at C-20 has been prepared from the pregnenolone THP ether **182** (Litvinovskaya et al., 1997a). Formation of the side chain was achieved by addition of vinylmagnesium bromide to ketone **182** followed by the reaction of the allylic alcohol **539** with isobutyronitrile oxide. Synthesis of the cyclic part was performed in a usual manner, as shown for preparation of another isoxazoline analog **541** in Scheme 171.

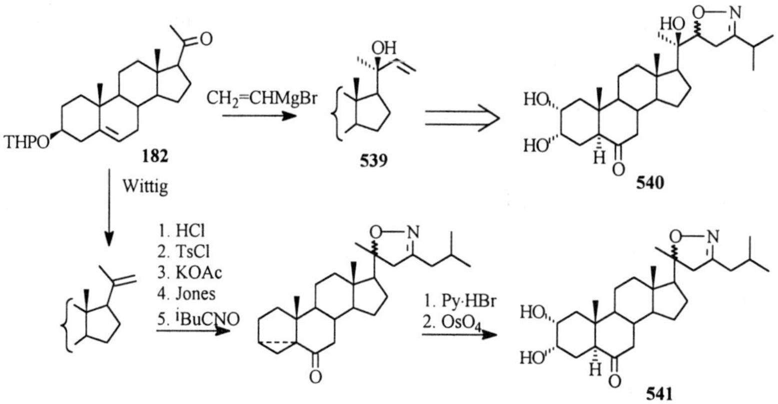

SCHEME 171

6. Other Analogs

A high plant growth promoting activity has been demonstrated for BS analogs containing an aryl substituent instead of a terminal alkyl group in the side chain (Hayashi et al., 1987). A number of 23-aryl-BS have been prepared using the Heck arylation for the introduction of the aromatic moiety (Huang, L.-F., and Zhou, W.-S., 1994). Treatment of olefin **542** with iodobenzene in the presence of palladium acetate and triphenylphosphine gave the phenyl olefin **543** (Scheme 172). Functionalization of the cyclic part was followed by asymmetric hydroxylation of the Δ^{23}-double bond and afforded the analogs **544**, **545**, and **546**.

A fluorinated analog **550** has been prepared by Jin et al. (1993) via an aldol reaction of aldehyde **273** with a 2,2-difluoro silyl enol ether in the presence of a Lewis acid. The difluoro alcohol **547** was protected as its MOM ether **548** (Scheme 173) and a Wittig reaction with an excess of phosphorane at low temperature followed by reacetylation afforded the olefin **549**. Catalytic

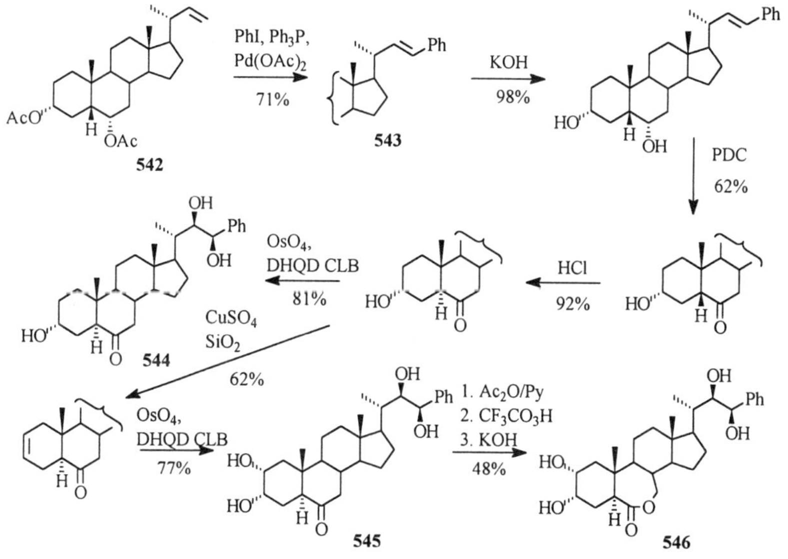

SCHEME 172

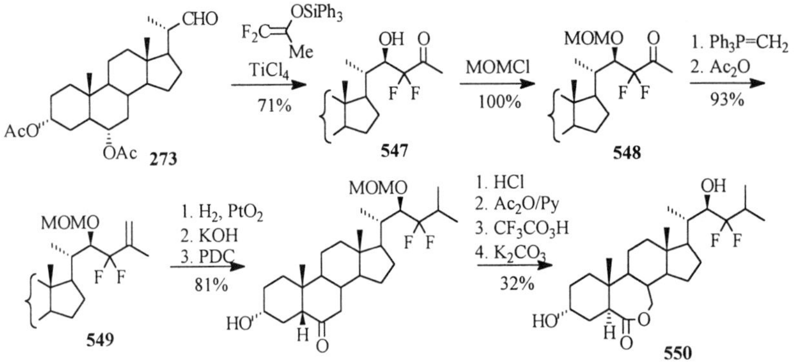

SCHEME 173

hydrogenation of the double bond and construction of the cyclic part furnished lactone **550**.

D. ANALOGS WITH A MODIFIED CYCLIC PART

1. Analogs with an Additional Double Bond in the Cyclic Part

From a structural point of view BS are close to ecdysteroids and it was natural to investigate the biological activity of compounds with structural elements common to both groups of steroids. BS analogs with a Δ^7-double bond could be prepared either from the corresponding 6-ketones *via* bromination-

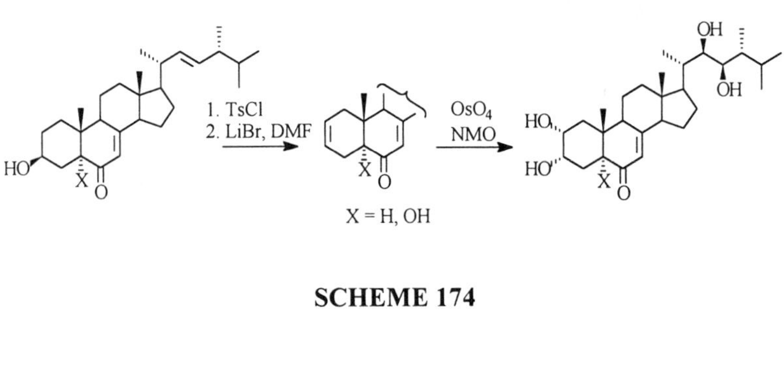

SCHEME 174

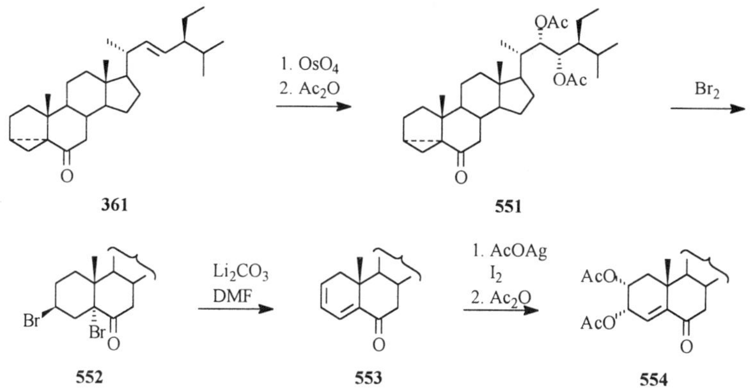

SCHEME 175

dehydrobromination (Akhrem et al., 1982) or from ergosterol derivatives (Takatsuto et al., 1987a) using the structural features of the starting material as is shown in Scheme 174.

Another type of BS analogs with an additional double bond in the cyclic part are the Δ^4-derivatives (Akhrem et al., 1981, 1983b,c, 1988a, 1989d, 1990a). The synthesis of one of these is shown in Scheme 175. Hydroxylation of the olefin **361** followed by acetylation led to the diacetate **551**. Opening of the cyclopropane ring by treatment of **551** with bromine gave the dibromide **552**. Heating in DMF led to elimination of HBr to afford the dienone **553**, which upon Prevost hydroxylation furnished the Δ^4-enone **554**.

Some Δ^5-oxygenated and $\Delta^{5,7}$-unsaturated BS analogs have been prepared by Hellrung et al. (1997) from isostigmasterol. Hydroxylation followed by acid-catalyzed isomerization and acetylation provided the Δ^5-triacetate **555** (Scheme 176). Treatment of **555** with chromic acid in dichloromethane resulted in allylic oxidation to give the enone **556**. The hydride reduction of the latter with sodium borohydride in the presence of cerium chloride followed by alkaline hydrolysis furnished the 7β-hydroxylated analog **557a** in 74% yield. The corresponding 7α-epimer **557b** was prepared in 76% yield by reduction of enone **556** with L-Selectride and subsequent deacetylation. The $\Delta^{5,7}$-unsaturated analog **559** was obtained from the enone **556** *via* reductive elimination of tosylhydrazone **558** followed by alkaline hydrolysis.

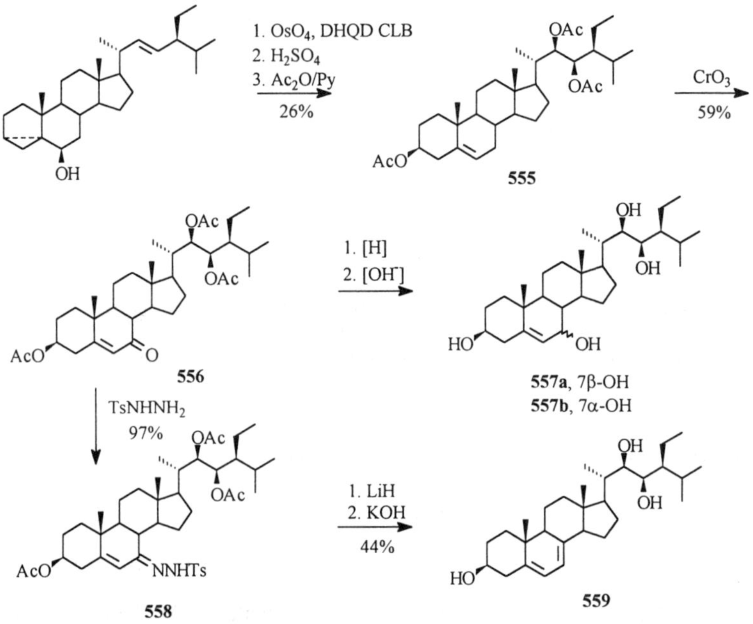

SCHEME 176

2. Ring B Hetero Analogs

Considerable efforts were directed toward the synthesis of 7-aza analogs with a lactam function in ring B. The most obvious method for their preparation seemed to be a Beckmann rearrangement; however, attempts to apply this reaction for the preparation of the 7-aza analogs proved to be unsuccessful and led to the 6-azalactams as the sole products (Scheme 177) (Ahmad *et al.*, 1981; Okada and Mori, 1983a; Anastasia *et al.*, 1984a, 1986b). Similar results were obtained for the Beckmann rearrangement of the Δ^4-6-ketoximes (Ahmad *et al.*,

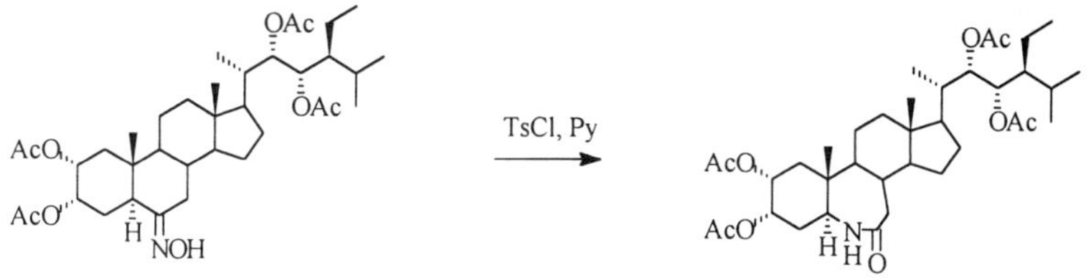

SCHEME 177

1981; Anastasia et al., 1986b) and for the of several 6-ketones (Okada and Mori, 1983a; Anastasia et al., 1984a).

The first successful synthesis of 7-aza analogs of BS was accomplished by Anastasia et al. (1986a). Kinetically controlled enolsilylation of dienone **420** followed by hydroxylation and protection of the diol groups led to the hydroxy ketone **560** (Scheme 178). Its oxidative cleavage with periodic acid gave, after treatment with diazomethane, the aldehydo ester **561**. Further transformation into

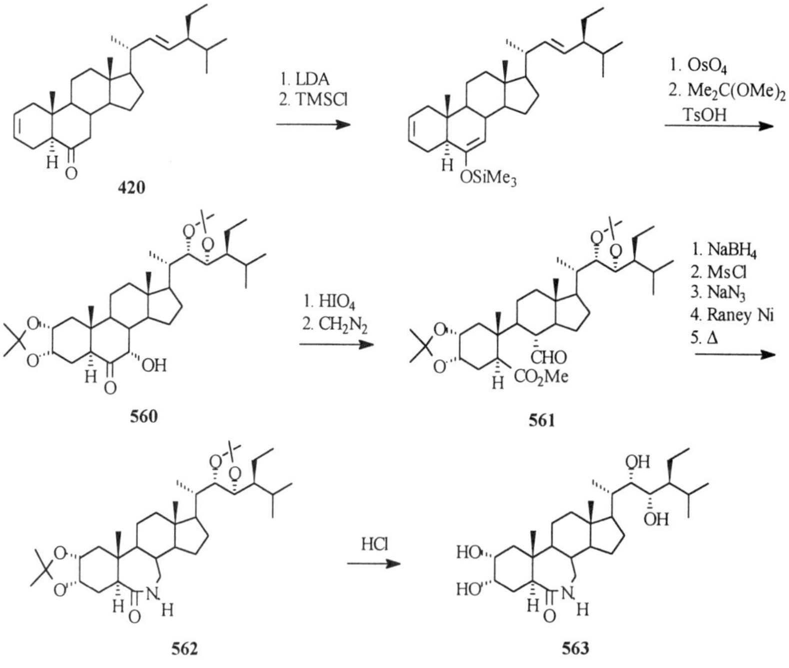

SCHEME 178

the lactam **562** was effected in several ways. Reductive amination of the aldehydo ester **561** with hydrogen in the presence of Raney nickel afforded the desired product **562** in a low yield. Much better results were obtained *via* catalytic reduction of the corresponding hydroxyimino derivative with Adam's catalyst. Alternatively, lactam **562** could be prepared by reduction of the aldehydo ester **561** with sodium borohydride in the presence of ammonium acetate followed by cyclization of the intermediate amino ester. A third possible way was based on a multistep transformation of aldehydo ester **561** with the mesyl ester and the azide ester as intermediates (Scheme 178). Deprotection of the diacetonide **562** furnished the 7-aza anolog **563**.

The synthesis of a number of 7-aza analogs including 7-azahomobrassinolide, (22*S*,23*S*)-7-azahomobrassinolide, and the corresponding analog with a cholestane side chain was reported by Kishi *et al.* (1986). Bromination of homocastasterone tetraacetate **423** followed by base hydrolysis of the intermediate bromide led to hydroxy ketone **564** (Scheme 179), which was cleaved to the seco acid **565**. Reductive amination of **565** with sodium borohydride in the presence of ammonium acetate to lactam **566** and saponification afforded the tetrahydroxy lactam **567**.

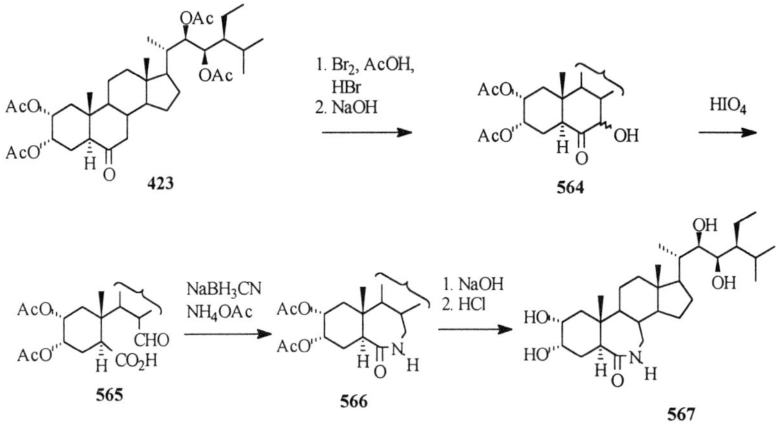

SCHEME 179

The intermediate seco acid **565** proved to be useful also for the preparation of 7-thiahomobrassinolide **574** (Mori and Kishi, 1985; Kishi et al., 1986). Esterification of seco acid **565** with diazomethane produced the corresponding aldehydo ester **568**, which was reduced to alcohol **569** and further transformed into the mesylate **570** (Scheme 180). Heating of the latter with sodium thiosulfate in aqueous ethanol followed by treatment of the resulting S-alkylthiosulfate with iodine afforded the disulfide **571**. Removal of the protecting groups first by lithium iodide and then by alkaline hydrolysis furnished the tetraol **572**. Reduction with zinc gave the thio acid **573**, which was converted into the desired 7-thiahomobrassinolide **574** by dehydration with dicyclohexylcarbodiimide.

6-Deoxo analogs could be easily prepared from the corresponding lactones (Kishi et al., 1986). Hydride reduction of homobrassinolide **13** followed by acid-catalyzed cyclization of the intermediate alcohol **575** led to 6-deoxohomobrassinolide **576** (Scheme 181). Preparation of the 7-aza, 7-thia, and 6-deoxo analogs of brassinolide was also reported by Back et al. (1997b).

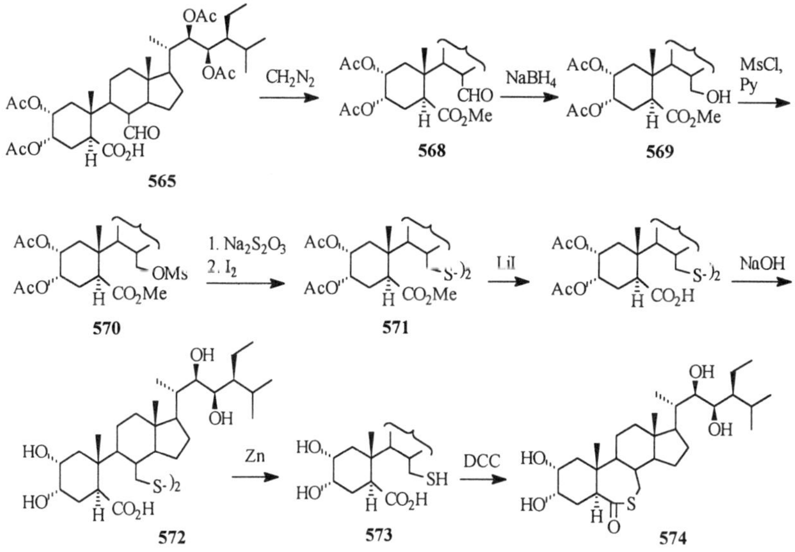

SCHEME 180

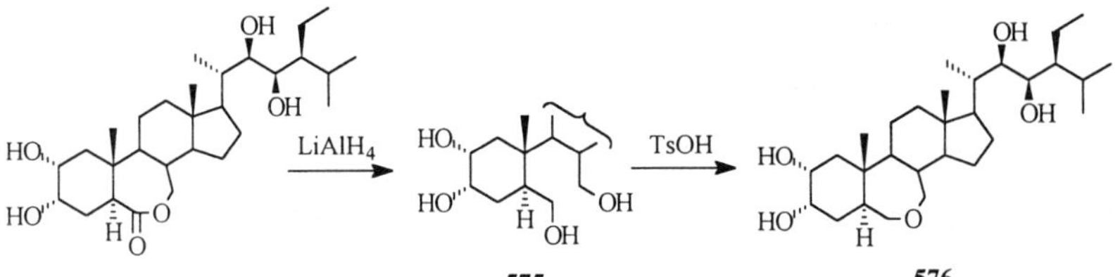

SCHEME 181

Several seco analogs were synthesized by Adam et al. (1991). The 2,3-seco derivatives were obtained from diol **577** (Scheme 182). Periodate cleavage of **577** gave an unstable dialdehyde, which was reduced to afford the corresponding seco diol. Hydroxylation of the Δ^{22}-double bond and some deprotection-protection steps led to the tetraacetate **578**. Baeyer-Villiger oxidation followed by saponification gave a mixture (3:1) of seco lactone **579** and the 5-membered lactone **580**.

Irradiation of 6-ketones was the key step in the synthesis of 5,6-seco analogs (Adam et al., 1988, 1991). In this way (22S,23S)-homocastasterone **494** gave ketene **581** (Scheme 183), which reacted with the solvent to give the ester **582**. Saponification afforded the 5,6-seco carboxylic acid **583**.

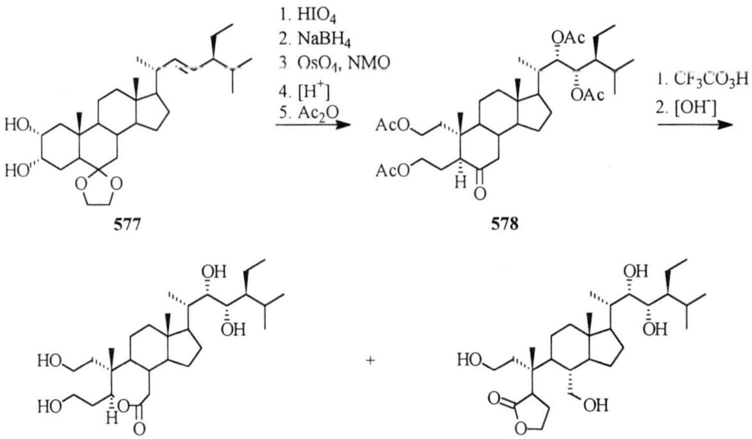

SCHEME 182

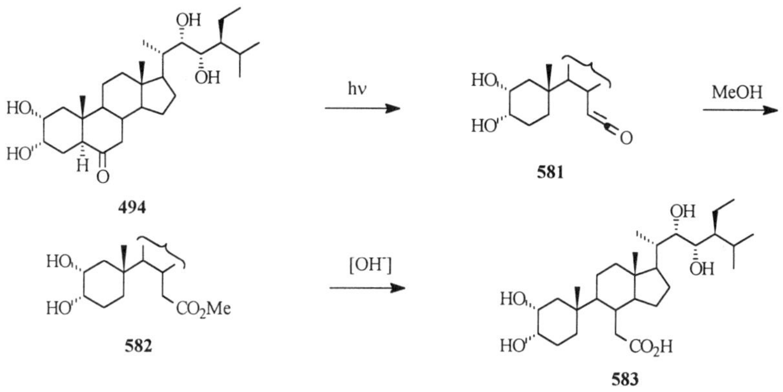

SCHEME 183

The synthesis of analogs with a 7a-oxa-7-keto functionality in ring B was performed by Anastasia *et al.* (1985b). Reaction of ergosterol **72** with 4-phenyl-1,2,4-triazoline-3,5-dione followed by tosylation gave the adduct **584** (Scheme 184). Elimination of the tosylate and reduction of the adduct **585** with lithium in

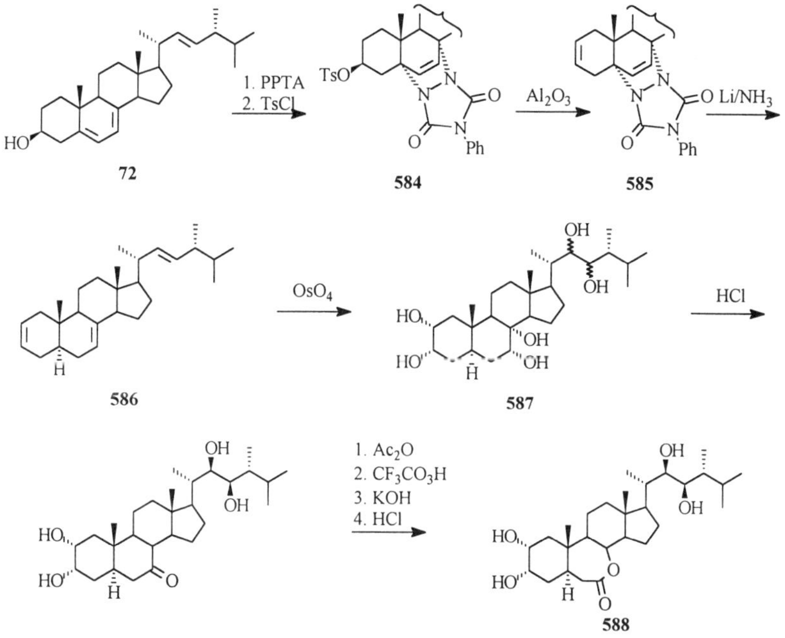

SCHEME 184

liquid ammonia proceeded smoothly to afford triene **586**. Hydroxylation with osmium tetroxide in pyridine for 5 days furnished a mixture of hexaols **587**. Treatment of **587** with acid gave a pinacol rearrangement of the 7α,8α-diol group in ring B, whereas the 2α,3α- and 22,23-glycols were stable under these conditions. The Baeyer-Villiger oxidation of the appropriate tetraacetate followed by acetate deprotection and relactonization gave lactone **588**. Preparation of similar derivatives was reported by Takatsudo *et al.* (1986).

Cholestane analogs have been obtained starting from the 7-keto derivative **589** by Kohout (1994a). Heating of the corresponding tosylate in 2,4,6-collidine led to a mixture of the Δ^2- and Δ^3-olefins **590** and **591** (Scheme 185). The isomeric olefins were separated by column chromatography and subjected to oxidation with osmium tetroxide. Two diols **592** and **593** could be isolated in the case of olefin **590**; both were converted into the lactones **594** and **595** by Baeyer-Villiger oxidation. The 3α,4α-dihydroxy analog **596** was prepared in a

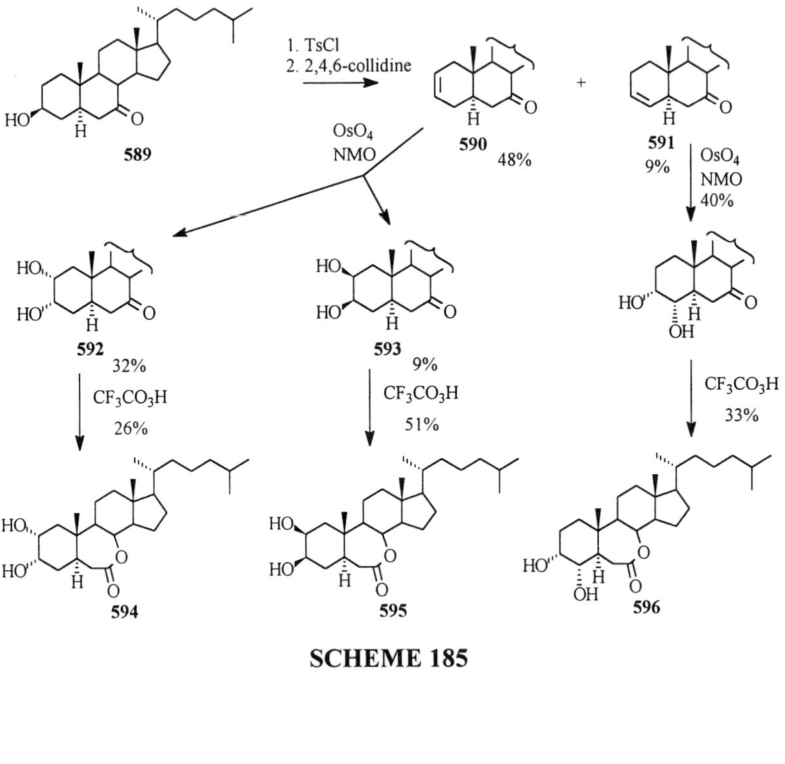

SCHEME 185

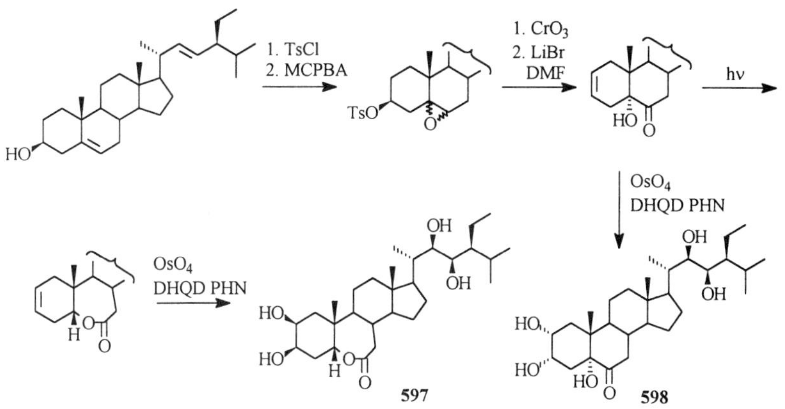

SCHEME 186

similar way.

The 5α-hydroxy analog of (24S)-ethylbrassinone **598**, the lactone **597**, and the corresponding 22S,23S isomers have been prepared from stigmasterol by Brosa et al. (1996c) as described in Scheme 186. Synthesis of the 5α-hydroxy analogs of BS has been performed also by Kovganko and Ananich (1991).

One of the obligatory stages in the synthesis of BS is the Baeyer-Villiger oxidation. This reaction proceeds with formation of both possible regioisomers. The amount of the nonnatural 6-oxa-7-ketones normally does not exceed 15-20%, but it may be increased by the introduction of an appropriate substituent in ring A. In the majority of publications where the preparation of isomeric lactones was described, the synthesis of the natural products was the main purpose. However, in some synthetic approaches formation of a considerable amount of the 6-oxa-7-ketones was used for the preparation of the corresponding analogs (Takatsuto and Ikekawa, 1984b). This was the case in the Baeyer-Villiger oxidation of triacetate **599** (Scheme 187). After chromatographic separation of the "natural" regioisomer, a considerable amount of **600** was obtained as well. Saponification of **600**, protection of the 22,23-glycol as the acetonide, and mesylation gave the mesylate **601**. Dehydromesylation of **601** led to formation of

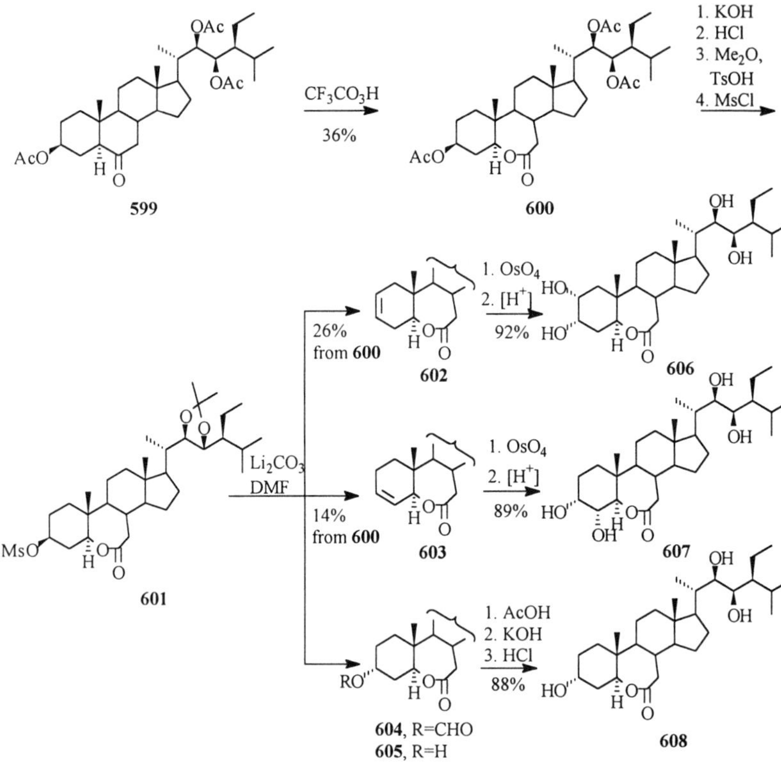

SCHEME 187

the Δ^2- and Δ^3-lactones **602** and **603** and a mixture of the 3α-derivatives **604** and **605**. Hydroxylation of the olefins **602** and **603** followed by removal of the acetonide protecting group afforded the corresponding tetraols **606** and **607**. A mixture of **604** and **605** was converted into triol **608**.

3. Analogs with Modifications in Ring A

A number of (22*S*,23*S*)-homobrassinolide analogs of this type have been prepared by Wada and Marumo (1981). A standard set of reactions was used for preparation of the key intermediate lactone **609** (Scheme 188). It could be easily reduced to the saturated lactone **610**. The epoxy lactone **611**, the trans-diol **612**,

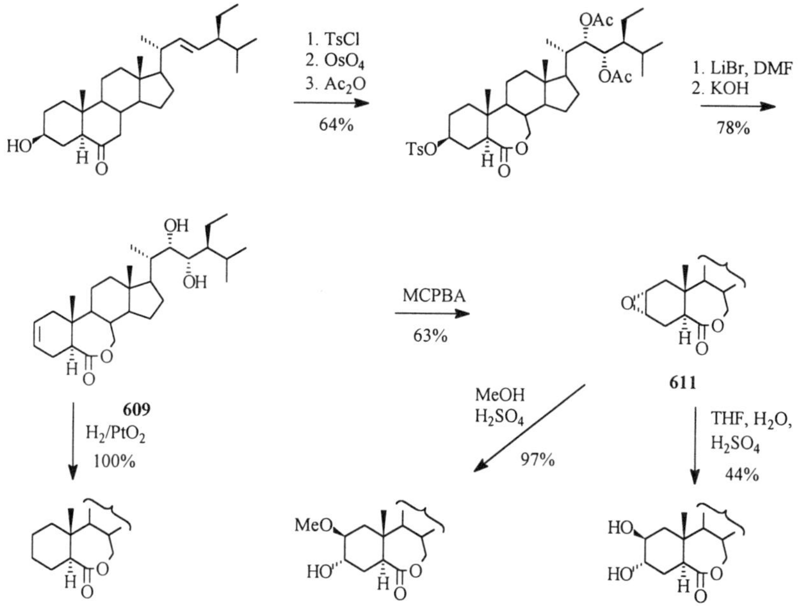

SCHEME 188

and the corresponding ether **613** could be obtained from **609** as well. Compounds such as epoxy lactone **611** proved to be prospective as long-lasting agents for agricultural and horticultural applications (Takatsudo et al., 1993).

Treatment of dienone **420** with silver acetate and iodine in acetic acid followed by alkaline hydrolysis gave a mixture of the two 2β,3β-diols **614** and **615**, which are epimeric at C-5 (Scheme 189). Both could be transformed into the tetraols **616** and **617** or into lactones such as **618** (Brosa et al., 1994). Hydroxylation of the olefins **614** and **615** in the presence of DHQD PHN provided an easy access to the corresponding 22R,23R analogs (Brosa et al., 1996a). Synthesis of some 2β,3β analogs in the cholestane series was described by Gao et al. (1993).

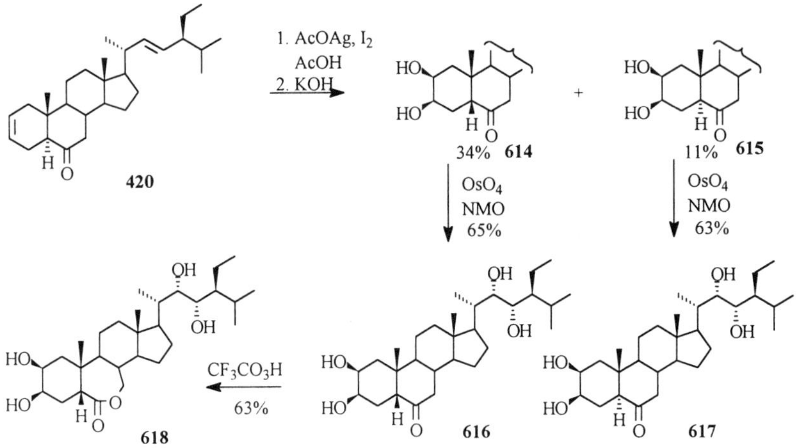

SCHEME 189

The synthesis of 1α-hydroxy analog **619** was performed by Khripach *et al.* (1995b) as indicated in Scheme 190.

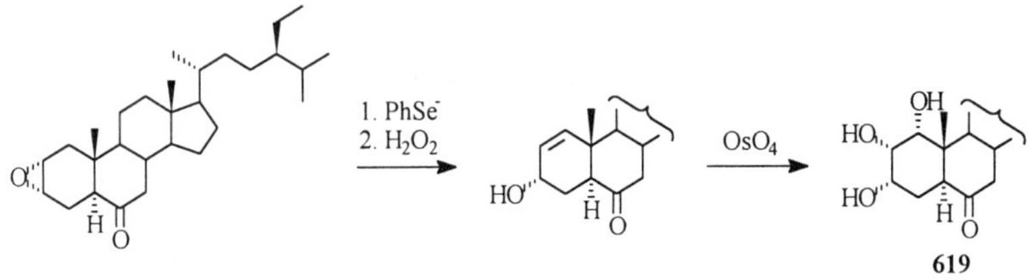

SCHEME 190

Acid-catalyzed opening of a 2α,3α-epoxide was used for preparation of (2*S*,3*S*)-epicastasterone (2,24-diepicastasterone) by Levinson *et al.* (1994).

E. ESTERS AND ETHERS

It has been repeatedly shown for mammalian steroids that derivatization (such as acetylation or etherification) results in new compounds that are more useful for practical application than the natural hormones. The reason is mostly a rapid inactivation of the natural product; normally their level can be effectively controlled by the living organism. It is reasonable to expect a similar effect for

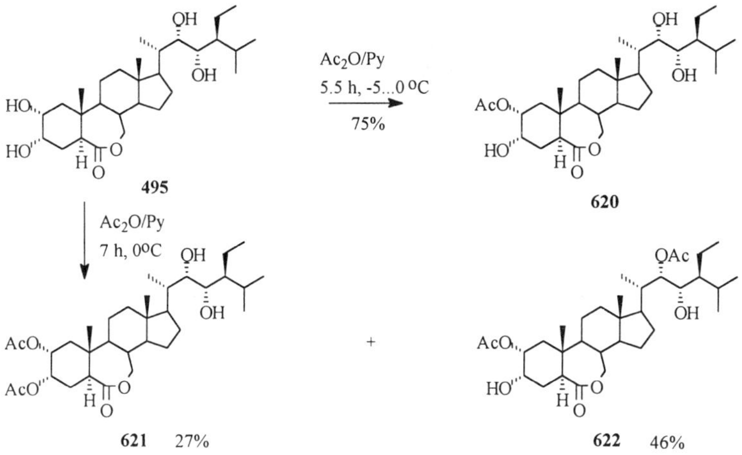

SCHEME 191

BS, and it was shown for example that teasterone 3-palmitate revealed a greater activity in the rice lamina joint test than teasterone itself (Abe and Yuya, 1992).

The first attempts to synthesize acyl derivatives of BS suitable for practical application were made by Kerb *et al.* (1982a-e, 1983c,e). Since BS contain mostly several hydroxyl groups, knowledge of their relative reactivity is necessary. Experiments with (22S,23S)-homobrassinolide **495** revealed (Scheme 191) that only the equatorial 2α-monoacetate **620** could be obtained in a reasonable yield *via* direct acetylation. Prolonged acetylation led to the formation of the diacetates **621** and **622** as the main products. Synthesis of the 2α-monoacetate of brassinolide has been performed by Yuya *et al.* (1984a).

Compound **621** showed a relatively high biological activity, but the direct acetylation gave only a 27% yield and a tedious chromatographic purification was necessary. An alternative route to diacetate **621** (Takatsuto and Muramatsu, 1988) could be achieved *via* initial protection of the 22S,23S-diol function followed by functionalization of the cyclic part (Scheme 192). The same approach was applied for the preparation of the corresponding 22R,23R isomer (Takatsuto and Hayashi, 1987).

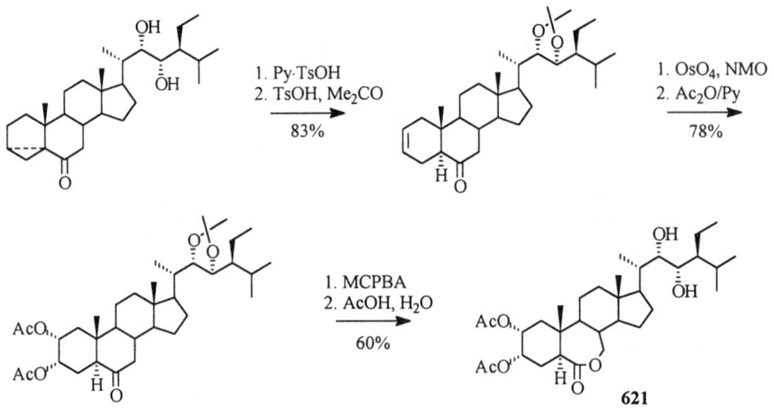

SCHEME 192

Chemical synthesis of BS can be considered as one of the first steps in the understanding of the role and the place of brassinosteroids. Synthesis first of all can provide for sufficient material for biological testing but especially the synthesis of analogs can contribute to give insight into structure-activity relationships and the physiological mode of action of BS. This ultimately may lead to discovery of such simple analogs that the practical application of BS comes close to reality.

CHAPTER **IX**

PHYSIOLOGICAL MODE OF ACTION OF BS

Specific physiological action in plants has triggered the interest and curiosity of many researchers to new substances which later became known as BS. Although this subject has been widely discussed in many reviews (Gamburg, 1986; Mendt and Nes, 1987; Mandava, 1988; Marquardt and Adam, 1991; Takeuchi, 1992; Sakurai and Fujioka, 1993; Baiguz and Czerpak, 1995; Prusakova and Chizhova, 1996; Brosa, 1997; Sasse, 1991b, 1996, 1997) and in a previous monograph (Khripach *et al.*, 1993a), new developments in this field are proceeding quickly and regularly need reconsideration. This and the following chapters are a new attempt to analyze the situation in the area at this time taking into account all the data published in the literature to date. Special attention is paid to the data obtained by investigators from Eastern Europe, which have never been reviewed before and which are accessible with difficulty for international readership.

Nowadays the wide spectrum of biological activity of BS and the complex character of their mode of action are well documented. These features are the result of a cascade of biochemical reactions which can be initiated *via* direct action of BS on the genome or by an extragenetic route. Both routes assume the participation of a system of secondary messengers and can act together. This complexity and interactions of the bioeffects of BS make a classification of these effects rather difficult. The current level of knowledge in this area does not yet allow a systematic approach for the presentation of data based on the mechanism of action of BS, and a combined variant is used. For easier reading the whole plant effects and the effects on physiological processes will be discussed separately. For example, plant growth initiated by BS is discussed separately from the effects on processes such as cell wall loosening, enzyme activation, and protein biosynthesis. A mixed scheme is used for the explanation of phenomena when effects of different levels are involved.

A. GROWTH AND DEVELOPMENT

The best known and the most widely studied biological effect of BS is their ability to stimulate plant growth in a variety of systems such as whole plants, excised segments, cuttings, and seedlings. Research on the physiological mode of action of BS in plants had already begun prior to the structural elucidation of Bl, the first representative of this group of phytohormones (Mitchell *et al.*, 1970; Mitchell and Gregory, 1972; Worley and Mitchell, 1971; Worley and Krizek, 1972; Milborrow and Pryce, 1973; Mandava *et al.*, 1978). The first data were obtained from investigations of the brassin complex. It was shown that when brassins were applied in a concentration of 10 µg per plant, they induced remarkable elongation of the second and third internodes of intact bean plants. For example, the second internodes grew at an average of 155 mm four days after treatment, while controls grew only 12 mm (Mitchell et al., 1970). Brassins typically caused elongation of all parts of the bean plants, and increased the length of stem, shoots, roots, weight of pods, and quantity of buds (Worley and Krizek, 1972; Mitchell and Gregory, 1972; Gregory, 1981). A study of the effect of brassins on the retardation of the bean hypocotyl hook opening and the reversal of the light inhibition of hypocotyl elongation demonstrated that they exhibit both "auxin"- and "gibberellin"-like activities (Yopp et al., 1979). The results of the research till 1979 have however some restrictions, because a complex mixture of compounds was used. The situation changed after pure Bl was isolated, its structure determined, and its synthesis carried out.

For the investigation of the physiological mode of action of BS many test systems were used that were developed earlier for other phytohormones. It was shown that BS have a wide spectrum of biological action and are capable of influencing various physiological processes in plants. Thus, the observed growth acceleration of bean internodes after treatment with brassins was the result of stimulation of cell elongation and cell division. Later, similar responses, related to cell growth and accumulation of biomass, were found to be typical for BS and were observed in various plants. It was found that Bl induced elongation of pea epicotyls, apical segments of dwarf pea (Sasse, 1988, 1990), mung bean epicotyls (Kamuro and Inada, 1987), segments of azuki bean epicotyls (Mandava et al., 1981), and hypocotyls of cucumber (Katsumi, 1985), sunflower (Mandava et al., 1981), and radish (Choi, C. et al., 1986). Also stimulation of wheat leaves and root growth and growth of mustard seedlings were observed (Braun and Wild, 1984a). These data correlate well with those (Gregory, 1981) obtained for barley plants cultivated from seeds treated with BS. They had larger leaves and stems and grew faster in comparison with control plants. Similar results were obtained in rice (Zhou, A.-Q., 1987; Hirai and Fujii, 1985), barley, lettuce

(Meudt et al., 1983), and celery (Wang, Y. et al., 1988) plants when sprayed with a BS solution. Growth acceleration after BS treatment was observed also for some woody species (Worley and Krizek, 1972) and fungi (Gartz et al., 1990). Experiments with mung bean seedlings (Zhao and Wu, 1990) revealed that stimulation of stem growth after treatment with EBl was mainly the result of cell expansion and connected with enhanced water absorption.

Among all growth responses to BS, the influence on the root system is the least studied and unequivocal due to its variability in different conditions. In contrast to the well-documented growth-promoting effects on aerial tissues, BS caused either growth inhibition or growth promotion in roots. This depends on the part of the plant to which BS are applied, the mode of treatment, the dose, the time of exposure, and so on. In early experiments with mung bean hypocotyls (Morishita et al., 1983) an inhibition of adventitious root growth under the influence of BS was shown, whereas other authors (Romani et al., 1983; Cerana et al., 1985), using cuttings of maize roots, reported growth promotion. A stimulating effect on adventitious root development in terms of number and length was obtained with a very low concentration of EBl when applied to soybean hypocotyl segments (Sathiyamoorthy and Nakamura, 1990).

An opposite effect was obtained in epicotyl and hypocotyl cuttings of mung bean (Guan and Roddick, 1988a). The number and mean length of adventitious roots were decreased after treatment with EBl and significant reduction (approximately 50%) in the total root length was observed with 0.01 µM BS. A similar effect was obtained with tomato cuttings (Guan and Roddick, 1988b). In cultured tomato roots EBl (Roddick et al., 1993) and three other BS (Bl, (22S,23S)-EBl, and HBl) with different side chains caused an inhibitory effect in concentration range between 10^{-6} and 10^{-10} M. Bl was the most active and evoked inhibition at 10^{-10} M, compared to HBl at 10^{-6} M (Roddick, 1994). The highest response was observed in the main axis. This differed from the situation with intact seedlings of wheat, mung bean, and maize, where inhibition first was seen on the lateral root development (Roddick and Ikekawa, 1992). In excised rice roots, the action of Bl was shown to depend on the mode of treatment; when supplied to the scutellum, it stimulated growth, whereas application to the root tip caused an inhibitory effect (Radi and Maeda, 1988).

An interesting result was obtained with respect to root formation in cuttings from the *Norway spruce* tree after treatment with synthetic (22S,23S)-HBl (Rönsch *et al.*, 1993). Cuttings were treated with 60 ppm of BS analog after being harvested in March, stored in darkness, and planted in May. In the controls 50% of the cuttings rooted; in the treated cuttings this number increased to 92%. A promoting effect was observed when rice seeds were soaked in a solution of 10^{-5} and 10^{-3} ppm BS before germination. Root weight and rooting ability were significantly increased also when rice plants at tillering were fed with BS through the roots or by foliar spray (Wang, S.-G., and Deng, 1992). Growth promotion of root explants of young tobacco seedlings was observed when cultured in a medium with 0.01 or 0.05 ppm EBl (Chen *et al.*, 1990). BS were found to promote the growth and rooting of potato cuttings when added to the nutritient solution. HBl, Bl, and EBl (0.02-0.2 mg/l) caused a twofold and higher increase of the number of adventitious roots (Bobrik, 1995).

The variable influence of BS on root development was discussed in comprehensive reviews by Roddick and Guan (1991) and Sasse (1994). Analyzing the responses of different root systems to BS, some common peculiarities can be indicated. For cultured excised roots mostly growth reduction is observed, whereas in cuttings and seedlings where shoot tissues are present, promotion takes place more often, especially when treatment is made *via* aerial tissue. Although the physiological background of root responses to BS is still not clear, investigations in this field are of great value for understanding the role and mechanism of action of both endogenous and exogenous BS in plants. The observation of root growth promoting ability of BS in intact seedlings and plants is important for potential practical application.

In this context special attention should be paid to reports about the enhancement of resistance to stress conditions of plants treated by BS. An example is the increase of root activity, plant growth, and weight of roots and shoots of gram plants (*Cicer arietinum*) under water stress (Singh *et al.*, 1993) after application of 0.001 ppm EB1 as a foliar spray. The effects of salt stress can be decreased by soaking seeds in a BS solution before germination. An enhancement of the length and number of roots of rice plants cultivated under salt stress conditions from seeds soaked in a HB1 solution was reported (Uesono *et al.*, 1985; Takematsu and Takeuchi, 1989).

Along with growth alteration, BS usually influence other aspects of plant development. In particular, their effect on reproduction, maturation, senescence, and seed yield has been noticed. For example, B1 induced the formation of bisexual and pistillate flowers in *Luffa cylindrica* staminate inflorescences and slightly promoted flowering in nonvernalized, but not vernalized, radish plants (Suge, 1986). EB1 strongly retarded maturation and senescence in hypocotyl segments of mung bean seedlings (Zhao *et al.*, 1987), and it efficiently enhanced cell reproduction and colony formation of Chinese cabbage mesophyll protoplasts (Nakajima *et al.*, 1996). The biosynthesis of BS was suggested to be essential for the differentiation of cells into tracheary elements in isolated *Zinnia* mesophyll cells (Iwasaki and Shibaoka, 1991) and for inducing entry into the final stage of differentiation (Yamamoto *et al.*, 1997). Recently, an exclusive role of BS in plant development was demonstrated in mutants defective in B1 synthesis or response to it, which showed dramatically changed phenotypes (Clouse, 1996a, 1997). The increase of BS content in plant pollen with maturity also confirms this role (Asakawa *et al.*, 1996).

The data on the influence of BS on seed yields are quite extensive and will be discussed in detail in Chapter XI. These effects were observed not only for natural BS but also for their synthetic analogs, which sometimes appeared to be very active. One of them, (22S,23S)-EB1, evoked the growth of leaves, increased the weight of plants by 30-75%, and increased seed weight per plant by about 45% after treatment of string bean seedlings with a lanolin solution (Meudt *et al.*, 1983).

The numerous results that show the similarity in action of natural BS and their synthetic analogs, combined with the higher availability of the latter, have initiated extensive application of these analogs as model compounds in the investigations of the mode of action of BS. The present knowledge about the occurrence of BS in the plant kingdom is not exhaustive and some, nowadays still "nonnatural", BS might be found in natural sources in the future. Such a situation took place with EBl, described and investigated first as an accessible synthetic analog of Bl (Thompson *et al.*, 1979) and then isolated from *Vicia faba* L. pollen (Ikekawa *et al.*, 1988). At the same time, by using available synthetic analogs as a model for natural BS, one should keep in mind the possible restrictions, which are connected with the dependence of the activity on variations in the structure.

The growth-promoting activity of BS usually takes place only after treatment of plants in the appropriate phase of development and within a certain concentration range, which is different for each kind of plant and type of BS. Such a result is probably caused by the influence of BS on the total plant hormone status. The ratio between some components in the hormonal spectrum could also be important as it is known for other phytohormones, in particular for cytokinins (Galston *et al*, 1983).

Noncorrespondence between the actual conditions of treatment and the optimal conditions for treatment can result in an absence of response or in inhibition of growth and decrease of crop yield. An example of such a behavior is given by Luo (1986b), who investigated the influence of concentration and conditions of BS treatment on the growth and crop yield of soybean. It was shown that a concentration of 0.1 ppm Bl enhanced the growth and dry weight of plants, whereas an increase of the concentration up to 1 ppm resulted in some delay of leaf growth and reduction of weight in comparison with the control. In another case an increase of crop yield was reached after treatment of the plants in the period prior to flowering, but when treatment was carried out after flowering, the crop yield decreased. Probably, these effects have stimulated studies on the applicability of Bl in agents for restriction of growth (Hata et al., 1986). It is probable that a significant part of the contradictions in the literature on the influence of BS on plant growth and development is connected with the indicated peculiarities of the realization of their effects. That is why special attention should be paid to the planning and execution of experiments and to a detailed analysis of all conditions in order to obtain reliable and comparable results and arrive at correct conclusions.

A comparative analysis of the action of different BS in an early stage of plant growth and on further productive development, which was done recently using wheat cv. Chinese Spring, euploid, and 18 related ditelocentric lines (DT-lines) with fixed genetic differences as a model system, showed a complex picture of effects (Mazets, 1997). The marked lines of hexaploid soft summer wheat cv. Chinese Spring differ from the euploid by the absence of one of the arms in one pair of homologous chromosomes. Some of the tested genotypes were highly sensitive to the action of BS, some of them showed medium

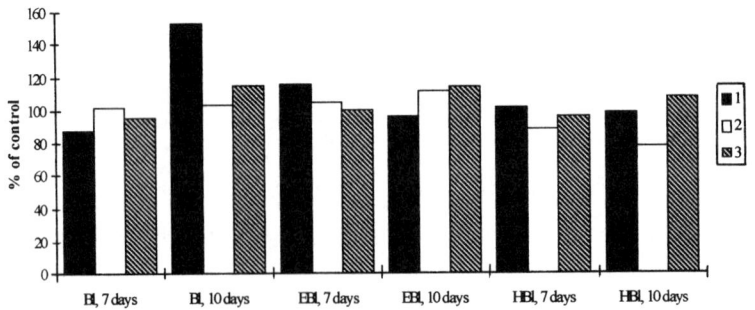

Fig. 35. Effect of BS on the fresh weight of 7-day-old and 10-day-old seedlings grown from seeds soaked in a 0.01 ppm solution of BS (1 - euploid, 2 - $5B^L$-line, and 3 - $5B^S$-line).

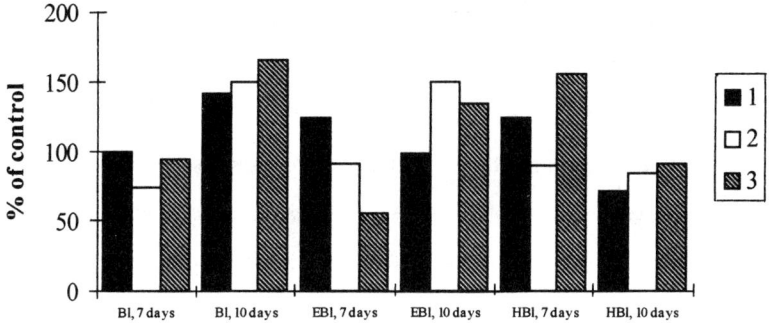

Fig. 36. Effect of BS on the fresh weight of roots of 7-day-old and 10-day-old wheat plants (1 - euploid, 2 - $5B^L$-line, 3 - $5B^S$-line).

sensitivity, and others were not sensitive at all. The effect of Bl, EBl, and HBl in two different stages of plant development in the euploid and in two DT-lines that are characterized by the absence of a short ($5B^L$) or a long arm ($5B^S$) in the chromosome 5B is shown in Fig. 35.

Although the data obtained for the treated plants are not much different in this period from the control, a high variability in plant response dependent on the structure of the BS and plant system was observed. Figure 36 illustrates the changes in root growth in the same plant system when BS were applied by soaking of the seeds with a 0.01 ppm solution of BS.

These data add another example to the previously described discrepancies in the literature data on root responses. Figure 36 shows a wide spectrum of effects within a short period of time. They strongly depend on the plant genotype and range from inhibition to stimulation. An interesting detail is the higher stimulation of the $5B^S$-line, which is characterized by the highest deficiency in the chromosome apparatus.

This tendency found in BS-influenced root growth became more obvious when the plant productivity was measured. Figures 37 and 38 show the data on two elements of crop structure in comparison with the control. These data were obtained for two years, which had rather different weather conditions. Although the final effect of BS on the grain yield for these years is different and the ratio between the parameters has changed, in general a higher stimulation of the $5B^S$-line in comparison with the euploid and $5B^L$-line is seen. These data indicate that there is no clear correlation between the growth responses at early stages of plant development and the crop yield. They also lead to the conclusion that minor structural differences in BS, which usually are considered as unimportant for their activity, can be critical for the effect and the same can be said about the minor shifts in the genetic structure of the plant.

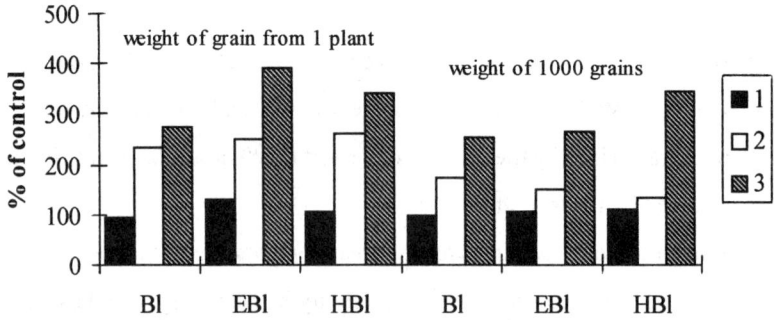

Fig. 37. Effect of BS on grain weight per plant and the weight of 1000 grains in 1991 (1 - euploid, 2 - $5B^L$-line, 3 - $5B^S$-line).

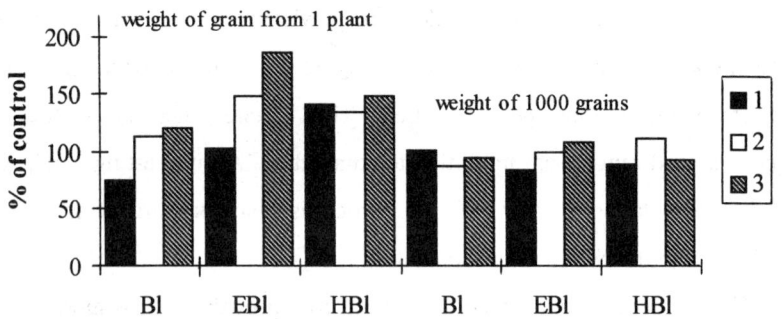

Fig. 38. Effect of BS on grain weight per plant and the weight of 1000 grains in 1995 (1 - euploid, 2 - $5B^L$-line, 3 - $5B^S$-line).

Additional confirmation of the complexity of the relation between the mode of application of BS and their effects on plant growth and development was obtained recently with new plant model systems and new methodological approaches. A characteristic influence of BS on plant development that depends on the time of BS application was found in investigations of the effects of EBl on large cranberry plants, *Oxycoccus macrocarpus* Pers. (Volodko and Zelenkevich, 1998). A special feature of the generative development of these plants is that the flower buds are formed on the top of uprights in the period from the second half of July to the first half of August in the year preceding the year of fruiting. The process of morphogenesis of the generative sphere continues till October and then it starts again in April. In May, when the average day temperature comes above the level of 5-7 °C, the development of the generative organs becomes visible. The flower bud starts to swell and then opens, and a new axial shoot with an inflorescence appears. As a rule, one top bud gives only one inflorescence with a variable amount of flowers (from 1 to 9-11, but usually 3-5). The 5-year-old plants were sprayed with a 10^{-5}% solution of EBl at a dose equivalent to 12.5 mg/ha at the beginning of August. Analysis of the crop parameters in the year of plant treatment showed no visible changes in comparison with the control. At the same time, the treatment acted efficiently on the development of the generative sphere, which at that time was in the beginning of its formation. In the year after treatment it was observed that in many plant buds more than one axial shoot with inflorescences appeared simultaneously from the top. The morphological structure and development of these shoots were not different from those of the control. On the inflorescences the normal buds were formed, opened, and flowered. The flowers were normally fertilized and gave ovaries. The percentage of top generative buds with this type of development was about 10-15% depending on the variety (Table XII). It should be pointed out that the amount of buds on such a shoot was the same as in the control.

TABLE XII

Effect of EBl on Cranberry Generative Development

Variety	Treatment	Number of generative uprights, % of total amount			
		1 Infloresc.	2 Infloresc.	3 Infloresc.	4 Infloresc. or more
Stevens	control	100.0	0.0	0.0	0.0
	EBl	84.6	7.6	4.3	3.5
Ben Lir	control	100.0	0.0	0.0	0.0
	EBl	88.8	5.7	3.1	2.4
Franklin	control	100.0	0.0	0.0	0.0
	EBl	89.1	6.0	3.2	1.7

An enhancement of the number of inflorescences was reflected in a larger crop yield of berries. Although the average weight of a berry was similar to that of the control, their larger amount was the reason for the increase of plant productivity by 9-15% (Table XIII). The treatment of plants with EBl did not affect the development of vegetative buds; in all shoots with vegetative buds only one new shoot was formed.

TABLE XIII

Effect of EBl on the Productivity of Cranberry Plants

Variety	Treatment	Number of berries from 1 m^2	Weight of one berry	Yield of berries, g/m^2	Crop increase	
					g/m^2	%
Stevens	control	2870	0.91	2583	-	-
	EBl	3343	0.89	2975	392	15.1
Ben Lir	control	2994	0.80	2395	-	-
	EBl	3321	0.81	2690	295	12.3
Franklin	control	3214	0.75	2410	-	-
	EBl	3519	0.75	2649	239	9.9

The discussed effect was observed only in the first year after plant treatment. The effect may be explained by a partial release of apical dominance in the generative organs after the application of EBl by regulation of the hormonal status in the apical part of the plant, which resulted in the enhancement of the number of inflorescences.

This conclusion is in agreement with the data obtained from the studies of the effect of EBl on the apical dominance in barley plants. Because this hormone-regulated phenomenon determines the character of interconnections between the shoots in cereal plants and strongly influences their productivity, an attempt to reduce it by the application of EBl has been done recently (Laman *et al.*, 1997; Vlasova *et al.*, 1998).

The final effect of BS is strictly dependent on the stage of plant development at the time of BS application; therefore the morphophysiological correlation between shoots in barley plants was studied for different times of EBl application, e.g., in the phase of full unfolding of the 2nd leaf, the 3rd leaf, and the 4th leaf in the main shoot. The best result in apical dominance release was achieved when the plants were treated with a dose that was equivalent to 50 mg of EBl per hectare in the phase of the 2nd leaf; when the main shoot apex becomes generative. The data were similar for two varieties of barley plants, cv. Prima Belarusi and cv. Visit. The final effect of EBl became apparent in the appearance of a larger amount of reproductive shoots and in a higher grain yield (Table XIV).

The observed effect is directly connected with a higher synchrony in the development of tillering shoots initiated by the application of BS. A high synchrony is important for the proportional distribution of assimilates in cereal plants and for a better grain production. The synchrony of shoot development was estimated for five varieties of barley plants treated with EBl in the phase of the 2nd leaf. The calculations were performed using a number of approaches (Paroda, 1971; Dahiya *et al.*, 1976; Faris and Klink, 1982) based on counting the

TABLE XIV

Effect of EBl on Tillering Shoot Formation and Crop Yield

Phase of development	Treatment	Number of productive shoots	Grain from one plant, g
two leaves	control	4.2	4.759
	EBl	6.0	5.372
three leaves	control	4.2	4.759
	EBl	4.9	4.049
four leaves	control	4.2	4.759
	EBl	3.9	3.551

number of days after planting till the development of the main shoot and the shoots of primary tillering had reached stage 50 (Zadoks et al., 1974). These data were used for the calculation of the intraplant variance (σ^2), the regression coefficient (b_{xy}), the synchrony measure (SM), and the synchrony range (SR). The higher the synchrony of the development of the main and the side shoots is, the lower are the values of the mentioned parameters corresponding to it. As shown in Table XV, the treatment of plants with EBl significantly increased the synchrony of shoot development.

An additional quantitative characteristic of the ability of BS to stimulate growth processes in plants is illustrated by Fig. 39. It shows the reduction of phyllochrone value (Ph) for EBl-treated plants. This value is determined as the number of growing degree days (GDD) that are necessary for the unfolding of each next leaf. Ph and GDD are indicated for different stages of development in accordance with Haun (1973). Barley plants were treated at the 2nd leaf stage with EBl or gibberellic acid (GA).

TABLE XV
Effect of EBl on Synchrony of Plant Development

Variety	Treatment	Parameters of synchrony			
		SR	SM	b_{xy}	σ^2
		1995			
Prima Belarusi	control	21.5	31.4	6.8	10.1
	EBl	11.5	18.8	4.4	3.7
		1996			
Visit	control	7.0	16.4	2.2	11.8
	EBl	4.5	10.1	1.5	3.9
Lipen	control	11.1	18.3	3.5	22.5
	EBl	5.3	13.6	1.7	5.5
Maladzik	control	3.7	5.3	1.3	3.0
	EBl	3.5	4.5	1.0	2.3
		1997			
Visit	control	7.5	16.4	2.5	10.7
	EBl	3.9	8.4	1.3	2.8
Lipen	control	5.5	13.3	1.8	5.8
	EBl	3.5	6.3	1.2	2.5
Zazersky	control	6.8	12.2	2.3	9.4
	EBl	4.6	8.7	1.6	4.2

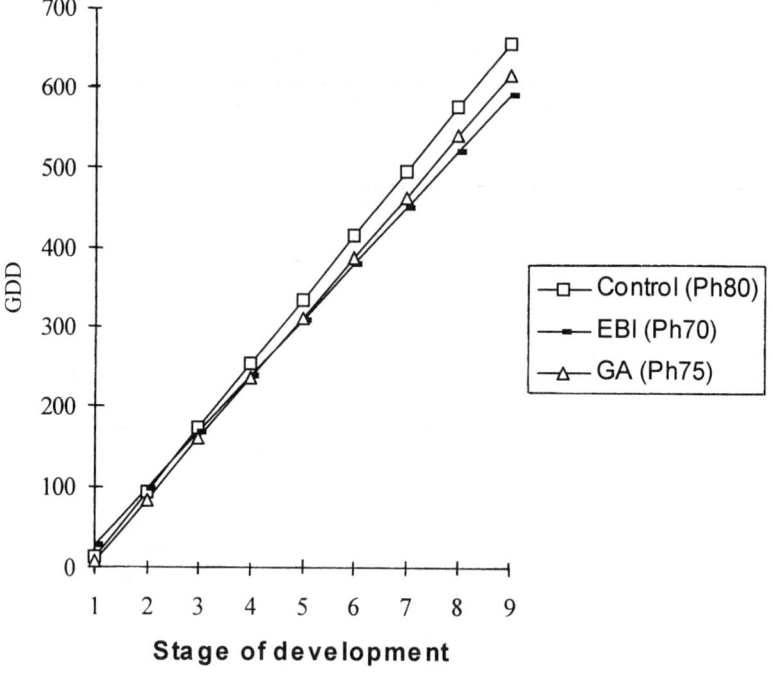

Fig. 39. The reduction of phyllochrone value (Ph) for EBl-treated plants.

A new illustration of the ability of BS to initiate a better realization of plant genome resource than usually takes place in normal plant development was found in studies of the effect of EBl on the ratio of morphotypes of barley plants that differed in the number of leaves in the main shoot (Laman et al., 1997; Vlasova et al., 1998). Treatment with EBl induced the development of plants with the maximum amount of leaves that is characteristic for the variety (Table XVI). For example, in the cv. Visit only 53% of the control plants came to full development with 9 leaves, which was the maximum number for this variety. After treatment with EBl, 100% of the plants showed full development. Similar effects were observed in the other two varieties.

In recent years several attempts to clarify the origin of the influence of BS on plant growth were undertaken. To explain the phenomenon of cell expansion under the action of BS, different hypotheses were developed and examined experimentally. It was found that the effect of BS is genetically determined and that BS are probably involved in all steps of cell growth regulation. Alterations in the mechanical properties of the cell walls (Tominaga et al., 1994) and their relaxation and expansion may be regulated by BS *via* activation of specific genes controlling these processes (Clouse et al., 1992; Wang, T.-W. et al., 1993; Zurek and Clouse, 1994; Zurek et al., 1994). Minor structural changes in plant tissues *via* the initiation of shifts in the orientation of microtubules within the cell followed by the corresponding shifts in the orientation of cellulose microfibrils are involved in the process of elongation and may be induced by BS (Mayumi and Shibaoka, 1995). The mechanistic aspects of BS-regulated plant growth will be discussed below.

TABLE XVI

Number of Leaves in the Main Shoot of Barley Plants Treated with EBl

Variety	Treatment	Number of plants, % of total amount			
		6 Leaves	7 Leaves	8 Leaves	9 Leaves
Visit	control	0	0	47	53
	EBl	0	0	0	100
Lipen	control	27	73	0	0
	EBl	0	80	20	0
Zazersky	control	0	0	46	54
	EBl	0	0	7	93

B. INTERACTION WITH OTHER PHYTOHORMONES

One of the first questions evoked by the determination of the high plant growth promoting activity of BS was the behavior of these new phytohormones in the tests that were supposed to be specific for auxins, gibberellins, and cytokinins. Also the mutual interaction with other plant hormones was a subject of interest. In tests (Yopp *et al.*, 1981) on elongation of etiolated maize mesocotyl segments, elongation of pea epicotyl and azuki bean epicotyl segments, and retardation of hook opening of etiolated bean hypocotyls, a similarity in the effects of Bl and IAA was shown. However, in a number of other tests such as inhibition of cress root elongation and inhibition of lateral bud growth on decapitated pea plants, Bl appeared to be inactive. It did not promote, but elicited retardation, of mung bean hypocotyl rooting.

Strong synergism between auxins and BS was observed by several investigators (Yopp *et al.*, 1981; Takeno and Pharis, 1982; Arteca *et al.*, 1983; Meudt and Thompson, 1983; Choi, C.-D. *et al.*, 1990b; Cao and Chen, 1995). It proved to be dependent on the sequence of treatments of plants and it appeared only in cases where BS treatment preceded the auxin treatment. Typical auxin inhibitors, such as (*p*-chlorophenoxy)isobutyric acid, suppressed the BS action and the joint influence of both hormones. One of the most sensitive tests for BS is based on the enhancement of the auxin-induced curvature of the first bean internode. The response to Bl of vertically placed hypocotyls was dependent on exogenous auxin. In contrast, the response of gravistimulated sections to Bl was observed in the absence of auxin (Meudt, 1987). The assumption (Takeno and Pharis, 1982) about the mediating of the effects of BS in plants *via* an increase of the auxin level was checked and it was proposed that Bl did not depend on auxin as a mediator but could interact with it in a complex manner (Sasse, 1990, 1991a). This is in agreement with previous data obtained with EBl, which did not affect the level of IAA in plant tissue, the rate of IAA transport, or its metabolism (Cohen and Meudt, 1983).

Further evidence for synergism between BS and auxin was obtained in studies of ethylene production by etiolated mung bean (*Vigna radiata*) hypocotyl segments (Arteca *et al.*, 1983, 1985; Arteca, 1984; Schlagnhaufer *et al.*, 1984a; Arteca and Schlagnhaufer, 1984; Tsai and Arteca, 1985; Schlagnhaufer and Arteca, 1985b). The combined treatment of plants with Bl and IAA caused a production of 43.6 nl/h ethylene; in the control the corresponding value was 0.41 nl/h. After separate treatments, these values were 10.5 nl/h for Bl and 17 nl/h for IAA. It was shown that Bl and IAA both affect ethylene production in the stage of transformation of S-adenosylmethionine to aminocyclopropane-1-carboxylic acid (Schlagnhaufer *et al.*, 1984b) and that their effects on ethylene production could be inhibited by the application of fusicoccin (Arteca *et al.*, 1988b), (aminooxy)acetic acid (Arteca *et al.*, 1988a), or some other auxin antagonists (Arteca *et al.*, 1991). Increase in ethylene production was found in rice plants treated several days before ear formation with Bl or EBl together with plant growth retardants, when they were controlled at the milk-ripe stage (Saka *et al.*, 1992). When Bl and ethylene were used in combination, an additive growth-stimulatory effect was observed in rice coleoptiles (Zhou, A.-Q., 1987).

The cooperative action of Bl with gibberellin A_3 was also investigated (Gregory and Mandava, 1982). The treatment of young mung bean seedlings with both growth promoters separately resulted in elongation of epicotyls, but BS showed this effect at a lower concentration and gave higher elongation (up to 1000%). Also epinastic curvature of the epicotyls and petioles was observed for BS whereas this was not the case for gibberellic acid. Similar behavior was noticed for tomato plants (Schlagnhaufer and Arteca, 1985a,c). Treatment with BS and GA_3 together did not show any synergism, but the two hormones acted in an additive manner (Gregory and Mandava, 1982). A retardant of the elongation effect for GA_3, e.g., ancymidol [α-cyclopropyl-α-(4-methoxyphenyl)pyrimidine-methanol], did not interfere with the BS response, and this fact was taken as evidence for an independent role of both phytohormones in plants. A similar conclusion can be drawn from experiments on pea segments (Sasse, 1988) and rice root cuttings (Radi and Maeda, 1988). BS were reported to affect to some extent the endogenous gibberellin level (Skorobogatova, 1991).

It was shown that BS changed the composition of cytokinins in plant leaves (Kislin and Semicheva, 1991; Kozik and Kislin, 1991). Treatment of barley plants with EBl in a dose of 10-50 mg/ha resulted in a 6-12-fold increase of zeatin-riboside (ZR); a reduction of zeatin was noticed simultaneously. Dihydrozeatin-riboside, dihydrozeatin, and isopentyladenine were found in leaves of treated plants, but not in the control plants.

Inconsistent data on the influence of BS on the level of abscisic acid (ABA) were obtained (Eun et al., 1989; Kozik and Kislin, 1991; Korablyova and Sukhova, 1991; Kuraishi et al., 1991). Thus, treatment of barley plants with EBl in the beginning of flowering resulted in a 10-fold decrease of endogenous ABA analyzed in the milk phase in comparison with the control (Kozik and Kislin, 1991; Kurapov et al., 1992). In the range of the tested doses (10-50 mg/ha) no dose-effect relationship was found; evidently, the threshold of sensitivity was lower than 10 mg/ha. When barley plants were treated in the booting phase, a significant increase was observed only for IAA (Kurapov, 1996). A simultaneous increase of the levels of IAA and ABA was observed in *Gossypium hirsutum* plants after their treatment with Bl (Luo et al., 1988), and GA_3 and ABA in cucumber hypocotyls after treatment with EBl (Xu et al., 1990). An increase of the ABA level was found in potato tubers treated with EBl before storage (Korablyova and Sukhova, 1991).

A comparative investigation of the changes in levels of IAA and ABA and in the evolution of ethylene in squash hypocotyls after treatment with Bl showed its stimulative effect on the levels of IAA and a tendency to decrease the level of

ABA. The production of ethylene was not much influenced by Bl in these experiments (Eun et al., 1989).

An analysis of shifts in the content of three phytohormones (IAA, ABA, and ZR) under the action of BS showed a significant difference in the effects of Bl, EBl, and HBl in wheat cv. Chinese Spring and its DT-lines (Mazets, 1997). The data obtained for two of them, the $5B^L$- and $5B^S$-lines, and for the euploid are shown in Fig. 40. The 10-day-old wheat plants grown from seeds treated with BS (soaked in 0.01 ppm solution) were analyzed and compared with an untreated control.

These data show the high specificity of the interaction of each BS with the genotypes. Strong stimulation of the ABA level was observed for the plants of the $5B^L$-line that were obtained from seeds treated with EBl or HBl (850% and 680% of control, respectively). The plants of the $5B^S$-line also gave a good response (363% and 298% of control, respectively). A reasonable increase of the IAA level was registered only for EBl in the euploid (187% of the control), and an increase of the ZR level was found in the same plants when treated with HBl (180% of control). A number of these measurements showed a suppression in phytohormone accumulation in this period of plant development under the action of BS.

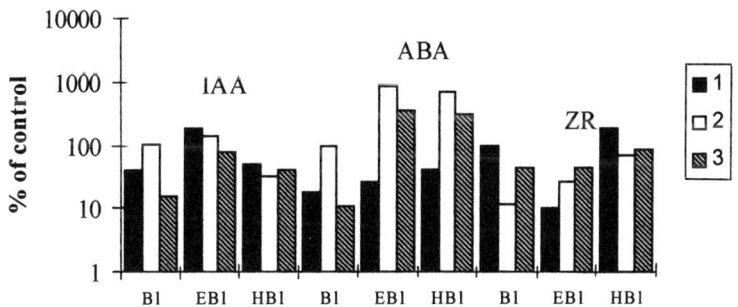

Fig. 40. Effects of BS on content of different phytohormones in wheat cv. Chinese Spring and in its DT-lines (1 - euploid, 2 - $5B^L$-line, 3 - $5B^S$-line).

The evidence mentioned above about the influence of BS on the function and content of other hormones in plants evoked the suggestion that the observed macroeffects could be mediated in this way. Similar considerations created the basis for a hypothesis about the managing role of BS with respect to other hormones (Takematsu, 1988), but this hypothesis still requires confirmation. As for the place of BS in the total hormone spectrum, it was shown that the maximum sensitivity to EBl during plant development occurred after that to GA and cytokinin and began before that for auxin in isolated wheat coleoptiles. In dwarf pea segments, the maximum sensitivity to BS laid between those of GA and auxin (Sasse, 1985).

C.   Effect on Cell Membranes

1. H^+-Pump Activation and Electrical Properties

The biochemical changes that take place in the cell membranes play a key role in the growth responses of plants to the application of exogenous BS. Membranes also define, to a high degree, many other aspects of the physiological action of BS. The data obtained to date show that BS are capable of influencing the electrical properties of membranes, their permeability, and the structure, stability, and activity of membrane enzymes. One of these effects, connected with the activation of the membrane-bound proton pump, showed similarity in the action of BS and IAA. This fact allows one to assume that the initial stage of cell elongation in both cases is connected with loosening of the cell wall as a result of its acidification by H^+ ions and activation of polysaccharide hydrolases. Increase of the cell size takes place as a consequence of reduction of the cell wall resistance to the pressure of the intracellular contents followed by water uptake.

In an early study of BS-stimulated acid secretion, a pronounced effect of EBl in azuki bean epicotyls, in shoots, and in apical and subapical root segments of maize at concentrations of 10^{-7}-10^{-5} M was shown (Cerana et al., 1983a,b; Romani et al., 1983). A significant stimulation of growth associated with increased acid secretion was accompanied by an early hyperpolarization of the transmembrane electric potential. These effects of BS, and those of IAA, were suppressed by inhibitors of RNA and protein synthesis (Cerana et al., 1983a). BS-induced acid secretion was enhanced by the presence of K^+ ions in the medium (Romani et al., 1983). Additivity of the BS and IAA effects was found for azuki bean epicotyls (Cerana et al., 1983a). A difference in the behavior of two phytohormones under some experimental conditions (Romani et al., 1983) suggested different pathways for the action of BS and IAA on the proton pump. The suggestion that BS affects the cytoplasmic metabolism through acid secretion is based on data on the enhancement of malate content and rate of CO_2 fixation in maize root segments together with activation of H^+ extrusion. Study of the same segments under conditions when IAA decreased acid secretion, malate content, and rate of dark CO_2 fixation led the authors to the conclusion that activation and inhibition of the rate of operation of the H^+ pump was connected with a change in phosphoenolpyruvate carboxylase activity (Cerana et al., 1983b).

BS-induced acid secretion showed a close correlation with the electrical parameters of the membranes and with BS-stimulated plant growth. Quite characteristic in this respect is the influence of Bl and EBl on growth and proton extrusion in the green alga *Chlorella vulgaris* (Baiguz and Czerpak, 1996). Although some of the natural BS, including EBl, have been found in *Chlorophyta*, there are few data on the biochemical effects of exogenous BS in algae. The results indicate an intense stimulation of growth of the algae under the influence of BS at concentrations of 10^{-15}-10^{-8} M. Bl showed slightly more activity in comparison to EBl. Between the 12th and the 36th hour of the experiment BS induced an intense growth of the alga that was two to three times higher than in the control. The intensity of H^+ secretion correlated with the dynamics of cell growth and was dependent on the concentration of BS. A slightly higher activity of Bl as H^+ secretion stimulant was observed here also.

As already mentioned above, many studies on the biological activity of BS were based on the employment of easily available nonnatural BS analogs such as (22S,23S)-HBl. When only these analogs are used to find out how the real hormone is functioning in plants, there is a great probability that the outcome will not be correct. The structural incompatibility of the analog with the putative hormone receptor together with the multifunctionality of steroids in plants, which usually will cause some plant response, easily could lead to false conclusions. Experiments with nonnatural BS analogs are, however, very useful for determination of the experimental boundaries of BS test systems.

An example of this approach is the study on proton extrusion and electric cell properties in *Egeria* leaf cells under the action of (22S,23S)-HBl and 2α,3α-dihydroxy-5α-stigmast-22-en-6-one, two BS analogs with a different level of similarity to Bl (Dahse *et al.*, 1990). Their effects were compared with the activity of stigmasterol and the nonsteroidal plant growth regulator fusicoccin and the structural requirements for their action were estimated. BS analogs and fusicoccin in the light showed similar effects in hyperpolarization and proton extrusion whereas stigmasterol was less effective. In darkness, the effects of the three steroids were comparable. The fact that all steroids caused acidification of the medium indicates that no special structural requirements are needed for the stimulation of acid secretion. The ability of all steroids to hyperpolarize the cell potential in accordance with the theoretical view on the components contributing to the plasma membrane potential strongly suggests the stimulation of an electrogenic pump. Rather characteristic changes in the membrane potentials caused by different steroids, in comparison with the less specific data on medium acidification at the same conditions, suggest that the membrane potential reflects proton pumping more sensitively than medium acidification. In contrast to fusicoccin, the effect of the steroids on the membrane potential and on medium acidification was reversible. A different behavior of (22S,23S)-HBl and fusicoccin was observed in an experiment on Fe^{3+} reduction and also on sucrose and amino acid uptake. These facts indicate that the mechanism of steroid action on proton pumping is different from that of IAA and fusicoccin. These data together with earlier results (Cerana *et al.*, 1984) on the plasmalemma proton pump activation by different sterols show that the most probable mode of action of BS is an indirect effect on the lipid environment of the enzyme; however, shifts in metabolic processes cannot be excluded.

New data on the effect of BS on the membrane properties in root cells have been obtained from studies of the action of EBl on young barley and triticale plants (Kalituho *et al.*, 1997b; Kalituho and Kabashnikova, 1998). The seeds were soaked in an EBl solution (10^{-8} M) for 6 h, then grown in tap water for 4 days, and cultivated in a solution of 10^{-3} M KCl + 10^{-4} M $CaCl_2$. Investigation of 7-day-old plants showed a significant decrease of the H^+ concentration in culture medium in comparison with the control (Table XVII). No essential changes in fresh weight or in growth of roots, coleoptiles, and the first leaves were observed except for a significant root elongation in one of the two varieties of the triticale plants (cv. Dar Belarusi).

Along with the diminishing of the total H^+ concentration ([H^+]), the H^+ concentration related to the root biomass ([CH^+]) was decreased also. These data, obtained from experiments with intact seedlings, are opposite to the data previously discussed on the activation of acid secretion by roots that were treated with EBl. The main reason for this difference could be the mode of treatment of the plants, application of EBl by addition to the nutrient medium or by soaking the seeds. The last variant reflects a long term effect of EBl: from the moment when the seeds were treated to a time after 7-8 days when the plants were studied for the results.

TABLE XVII

Effect of Epibrassinolide on Growth and H^+-Pump Activity

Treatment	Length, % of control			Fresh weight, % of control		pH	$[H^+]$		$[CH^+]$	
	Root	Coleop-tile	First leaf	Root	Leaf+Co-leoptile		M/g	% of control	M/g	% of control
Barley, cv. Honar										
control	100	100	100	100	100	6.12	$0.75 \cdot 10^{-6}$	100	$0.80 \cdot 10^{-6}$	100
EBl, 10^{-8} M	104	100	101	103	101	6.19	$0.66 \cdot 10^{-6}$	88	$0.69 \cdot 10^{-6}$	86
Winter triticale, cv. Malno										
control	100	100	100	100	100	4.70	$0.20 \cdot 10^{-6}$	100	$0.16 \cdot 10^{-4}$	100
EBl, 10^{-8} M	105	101	104	93	100	4.86	$0.13 \cdot 10^{-6}$	65	$0.12 \cdot 10^{-4}$	75
Winter triticale, cv. Dar Belarusi										
control	100	100	100	100	100	4.52	$0.50 \cdot 10^{-6}$	100	$0.43 \cdot 10^{-4}$	100
EBl, 10^{-8} M	133	100	102	97	97	5.04	$0.11 \cdot 10^{-6}$	22	$0.10 \cdot 10^{-4}$	23

The obtained data can be considered as an additional indirect confirmation of the existence of a concerted mechanism between BS-initiated growth and acid secretion. Rather promising for further clarification of this relation is an estimation of proton pump activity under the conditions of growth inhibition, which are well documented in the literature (Roddick and Guan, 1991).

Because the activity of the proton pump is closely related to the transmembrane transport of different ions and of organic substances, an attempt to estimate the functioning of membranes has been done based on the excretion by cells of some neutral metabolites (Kalituho and Kabashnikova, 1998). Some products of the nucleotide metabolism have a characteristic UV absorption at 260 nm and this wavelength was used to follow spectrophotometrically the changes in excretion of these metabolites. Usually this excretion is rather low under normal conditions but becomes much higher under stress conditions. The results summarized in Table XVIII show that at 20 °C and under heat stress (50 °C) the excretion of these metabolites was lower for the plants grown from treated seeds than in the control.

Similar to the results on the deactivation of the proton pump, these data suggest a stabilizing effect of EB1 on cells *via* modification of the plasma

TABLE XVIII

Effect of EB1 on the Secretion of Nucleotide Metabolites

Treatment	Leaves				Roots			
	t=20 °C		t=50 °C		t=20 °C		t=50 °C	
	D_{260}	% of control	D_{260}	% of control	D_{260}	% of control	D_{260}	% of control
control	0.099	100	0.114	100	0.138	100	0.176	100
EB1	0.096	96	0.106	93	0.094	68	0.144	81

membrane activity during plant development. Such an effect could play a role in the maintenance of plant cell homeostasis and contribute to the adaptation mechanism. This also might be an explanation for the antistress action of BS in plants.

A comparison of the data in Table XVIII on the effects of EB1 on the roots and on the first leaf leads to the conclusion that the stabilizing effect in the leaves is much weaker than in the roots, at least in this stage of plant development. The tissue specificity of EB1 action is reflected also in the metabolite excretion in the roots and in the leaves under heat stress. In the roots the relative level of BS inhibition was lower under stress than in normal conditions (32% and 19%, respectively), and in the leaves the degree of BS-affected excretion inhibition was higher under stress conditions (4% and 7%, respectively).

Of considerable interest for better understanding the correlation between growth processes and changes in the activity of membrane enzymes influenced by BS are the results of studies of the effect of EB1 on growth, acid secretion, and H^+-ATPase activity in roots of maize (Palladina and Simchuk, 1993; Palladina *et al.*, 1995). Hydroponically grown, 8-day-old maize seedlings were examined for changes produced by the application of EB1 *via* three different ways: (1) soaking the seeds in an EB1 solution for 24 h; (2) exposure the roots in a solution containing EB1; (3) spraying the upper part of the seedlings with an EB1 solution.

A 24-h exposure of the 7-day-old seedlings in EB1 solution (10^{-7}-10^{-9} M) increased the mass of roots, acid secretion, and K^+ uptake, which was in agreement with the earlier results of Cerana *et al.* (1983a) discussed above; most efficient was the concentration of 10^{-7} M EB1. A dependence of the acid secretion on the functioning H^+ pump was illustrated by the inhibition of medium acidification in the presence of orthovanadate, an inhibitor of transport H^+-

ATPase, the enzyme that is responsible for the functioning H^+ pump in plasma membranes.

An attempt was made to estimate the effect of EBl on the activity of H^+-ATPase in plasma membrane preparations obtained from maize roots. It was found that all three different treatments with EBl *in vivo* increased the H^+-ATPase activity. A kinetic analysis showed that the application of EBl decreased the K_M value but significantly increased V_{max}. In *in vitro* experiments with plasma membranes, EBl showed no effect on the H^+-ATPase activity within an interval of concentrations from 10^{-7} to 10^{-9} M at pH 6.5. On the contrary, all the employed concentrations of EBl strongly inhibited of enzyme activity at pH 7 and pH 7.5.

A difference in the effects of EBl on the H^+-ATPase activities of plasmalemma and cytoplasmic components was found in 5-day-old seedlings of buckwheat (*Fagopyrum esculentum* Moench.), where total enhancement of the enzyme activity was fully controlled by cytoplasmic components and took place along with the enzyme activity decrease in the plasmalemma (Deeva *et al.*, 1996a).

The data of the studies (Palladina and Simchuk, 1993; Palladina *et al.*, 1996) indicated a possible role of BS-induced changes in the lipid environment of the proton pump protein as one of the mechanisms of BS control of the H^+-pump activity. It is well known that sterols play an important role in the structural organization of plasma membranes; therefore the effect of EBl on the sterol compositions and their quantity in membranes was investigated. It was shown that exposure the maize roots in a 10^{-7} M EBl solution led to a diminution of the sterol/phospholipid and sterol/protein ratios. Similar changes also took place with seedlings grown from EBl-treated seeds. Some variation in the contents of different phytosterols was also found.

This result logically leads to the detailed analysis of the BS-influenced changes in membrane composition, which can be considered to be an important element in the mode of action of BS.

2. Effect on Chemical Composition. Membrane-Protective Action

The BS-initiated changes in the phytosterol composition of membranes discussed above were mainly connected with the ratio of three sterols: cholesterol, sitosterol, and stigmasterol. The first and second ones have a saturated side chain, and the third one is the main plant sterol with an unsaturated side chain. In treated plants the contents of stigmasterol was increased, and the contents of sitosterol and cholesterol were decreased. In the case of cholesterol the decrease was about 50% in comparison with the control. The reduction of the total sterol content may be expected if there is a similarity in sterol biosynthesis regulation in plants and in mammals. In such a case, a mechanism in which 3-hydroxy-3-methylglutaryl coenzyme A reductase could be blocked *via* binding of oxygenated steroids, such as BS, with a cell component similar to the oxysterol receptor of mammals (Zeelen, 1990) has to be present in plants.

Along with the changes in sterol content, the alteration in protein and phospholipid components is very important in membranes, due to the direct connection of these parameters with the functional properties of membranes. Especially the chemical nature of the fatty acid composition of membranes is an important factor in the regulation of permeability because of their influence on the surface properties of phospholipids, on lipid-protein and lipid-lipid interactions, and on the activity of lipolytic enzymes.

The early findings on changes in fatty acid composition of membrane lipids showed clearly an enhancement of the unsaturated acids in comparison with the saturated ones after BS treatment (Zhao and Wu, 1990; Katsumi, 1991). Further investigation with new plants and experimental conditions confirmed the previous results and gave a better insight into the phenomenon (Vedeneev et al., 1997b).

Two varieties of barley plants, cv. Roland (R) and cv. Zazersky (Z), and their isoplasmatic lines R(Z), bearing in cells the nucleus of R and the cytoplasm of Z, and Z(R), bearing in cells the nucleus of Z and the cytoplasm of R, were treated with EBl solution (0.01 ppm) in the beginning of the tillering phase. After 72 h plant leaves were investigated for their fatty acid composition using GC analysis. It was found that the application of EBl significantly affected the content of fatty acids and resulted in an enhancement of the unsaturated acids and in a reduction of the saturated ones (Table XIX).

The Zazersky variety, which initially had the lowest unsaturation index (I) {I =$([C_{16:1}] + [C_{18:1}] + 2[C_{18:2}] + 3[C_{18:3}])/([C_{16:0}] + [C_{18:0}] + [C_{20:0}])$}, showed the highest changes after EBl treatment; the unsaturation degree became about three

TABLE XIX

Effect of EBl on the Fatty Acid Composition in the Fraction of Saponifiable Lipids for Different Genotypes of Barley Plants (72 h after Treatment), %

Plants	Treatment	$C_{16:0}$	$C_{16:1}$	$C_{18:0}$	$C_{18:1}$	$C_{18:2}$	$C_{18:3}$	Satur/ unsatur	Index
R	control	4.14	0.29	1.13	1.46	92.02	0.06	0.056	35.28
	EBl	2.48	0.24	0.36	1.78	94.88	0.26	0.029	67.80
Z	control	7.55	0.40	1.79	5.08	84.76	0.42	0.103	18.87
	EBl	2.93	0.15	0.47	2.10	94.19	0.16	0.035	56.21
R(Z)	control	4.32	0.57	0.54	1.99	92.46	0.12	0.051	38.56
	EBl	2.49	0.27	0.25	1.50	95.38	0.11	0.028	70.39
Z(R)	control	3.80	0.70	0.60	1.38	93.40	0.12	0.046	43.00
	EBl	3.09	0.06	0.51	1.38	94.48	0.48	0.037	53.29

times higher. The weakest response was exhibited by the cytoplasmic line Z(R), which already had the highest unsaturation degree in the control. The Roland variety and the cytoplasmic line bearing its nucleus had quite similar parameters. This experimental approach gave no clear indications for the role of the nuclear or cytoplasmic genome in the mediation of the effect of EB1. The obtained data do not allow one to exclude a short-distance mechanism, implying a regulation by EB1 of some membrane enzymes that are responsible for the biotransformation of fatty acids.

The shifts in composition of fatty acids are closely connected with the changes of lipid content and with the dynamics of lipid peroxidation, two additional parameters which are usually used for the specification of the status of the membrane. The influence of EB1 on the lipid content in barley plants (cv. Roland) showed that all treatments gave a pronounced enhancement of total lipids. The data for a 4-day period in the tillering phase are shown in Table XX.

Application of EB1 by spraying gave a higher effect than treatment of the seeds, and the combination of both treatments gave an efficiency similar to that of spraying alone. The observed effect of lipid increase was rather stable and

TABLE XX

Effect of EB1 on the Lipid Content in Barley Plants

Treatment	mg/g of fresh weight					mg/g of dry weight				
	time, days					time, days				
	0	1	2	3	4	0	1	2	3	4
control	12.5	12.2	12.0	11.5	12.5	68.3	67.0	66.0	65.3	66.5
EB1, 0.01 ppm (seeds)	12.0	12.5	14.2	14.0	14.2	70.2	72.8	75.0	77.8	78.9
EB1, 0.01 ppm (spraying)	12.5	12.5	15.5	16.0	17.1	68.7	70.2	92.7	88.9	92.0

was preserved in the plants for a longer time.

A characteristic of the influence of BS on the membrane lipid components is not complete without information about changes in the fractions of saponifiable (fatty acid lipids) and unsaponifiable (sterols, glycosides, terpenoids, etc.) lipids because of their importance for the membrane fluidity. The data of Table XXI illustrate the changes of these components in four genotypes of barley plants under the influence of EBl, applied by spraying at a concentration of 0.01 ppm in the tillering phase.

At first glance, a diminution of the total content of lipids is rather unexpected, but taking into account that the total lipid fraction (Table XX) also included a growing amount of plant pigments, mainly chlorophyll, that are not present in the fractions of saponifiable and unsaponifiable lipids, the data of Table XXI become understandable. The most important result shown here is the

TABLE XXI

Effect of EBl on Saponifiable and Unsaponifiable Lipids in Different Genotypes of Barley Plants

Plants	Treatment	Content of lipids		
		Both fractions, mg/g of fresh weight	Saponifiable, %	Unsaponifiable, %
Z	control	8.18	49.51	50.49
	EBl	5.58	69.53	30.47
Z(R)	control	7.20	50.13	49.87
	EBl	6.24	58.33	41.67
R	control	9.42	68.79	31.21
	EBl	8.58	75.29	24.71
R(Z)	control	9.10	39.56	60.44
	EBl	6.96	48.85	51.15

pronounced change in the ratio of the two groups of lipids, reflecting a significant increase of the saponifiable lipids, which results in an increase of the membrane fluidity.

The dynamics of lipid peroxidation in cell membranes was studied (Vedeneev and Deeva, 1997) using the accumulation of malonic dialdehyde as the end product of the lipid peroxidation in plants (Table XXII).

The plants were sprayed with a 0.01 ppm solution of EB1 in the tillering phase and then monitored for 4 days after the treatment. A significant retardation of the malonic dialdehyde accumulation was found starting from the first day after EB1 application.

Since the chemical composition of the membranes is an important factor affecting the behavior of plants in stress conditions, further studies were carried out to clarify the relationship between higher stress resistance of BS-stimulated plants and the shifts in the membrane chemical composition. The peroxidation of

TABLE XXII

Effect of EB1 on Lipid Peroxidation in Barley Plants (Malonic Dialdehyde Content, $\mu M/g$ of Fresh Weight)

Plants	Treatment	Exposure, days				
		0	1	2	3	4
Z	control	33.53	42.05	28.21	58.55	50.03
	EB1	33.53	39.39	26.61	55.04	49.50
Z(R)	control	31.40	45.25	35.96	82.50	69.19
	EB1	31.40	33.00	25.02	63.34	50.56
R	control	44.18	44.71	46.57	48.97	63.87
	EB1	44.18	43.11	43.11	43.65	44.18
R(Z)	control	53.22	53.22	58.54	53.22	61.21
	EB1	53.22	43.11	46.84	44.71	45.24

lipid components of membranes occurs normally as part of the lipid catabolism. In stress conditions destructive processes in membranes become more intense, and this phenomenon is considered to be connected with lipid peroxidation. Taking into account the stress-protecting properties of BS, the lipid peroxidation was chosen as an approach for better understanding the mode of action of BS on cell membranes in different conditions (Ershova and Khripach, 1996).

The oxidative lipid degradation influenced by EBl was studied in 2-week-old pea seedlings (*Pisum sativum* L., cv. Ramonskii 77) at normal aeration, under oxygen deficiency, and in a CO_2-enriched atmosphere. The two last models are known to activate the degradation of phospholipids in biological membranes. Kinetin was used as a comparison because of its ability to maintain the elevated level of phospholipid desaturation under the applied conditions. Both EBl and the kinetin solution were introduced into the shoots by transpiration flow in the darkness. As criteria for the degree of degradation, the contents of products of lipid peroxidation, such as conjugated dienoic acids and malonic dialdehyde, were determined. The content of conjugated dienoic acids was shown to decrease in seedlings treated with EBl, and this effect was stronger than the lipid peroxidation inhibition caused by kinetin (Fig. 41). The

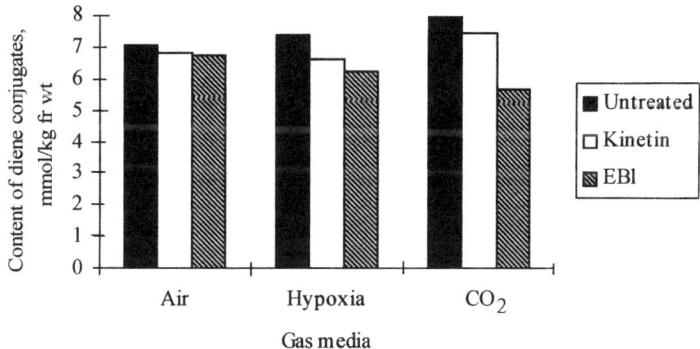

Fig. 41. Conjugated dienoic fatty acid content in pea seedlings treated with EBl or kinetin in different gas media.

inhibition was more efficient in plants under hypoxia or in the CO_2-enriched atmosphere, and the content of conjugated dienes for these cases was lower in treated plants by 13 and 21%, respectively, in comparison with air-grown seedlings.

The accumulation of malonic dialdehyde in plants in normal conditions was considerably decreased by EBl but not by kinetin (Fig. 42). Both hypoxia and elevated CO_2 inhibited malonic dialdehyde formation. Under hypoxia, this inhibition was higher when the seedlings were treated with the phytohormones, but in the CO_2-enriched atmosphere EBl did not show such an effect.

Taking into account the data (Leshem, 1984) on the lipoxygenase blocking mechanism of kinetin action, a similar influence on the lipoxygenase activity could be assumed to explain the mode of action of EBl, because its direct antioxidative activity is less probable. Finally, the inhibition of the lipid peroxidation by EBl contributes to a better maintenance and stability of biomembranes, and this is probably one of the ways in which the stress resistance of plants can be improved by BS.

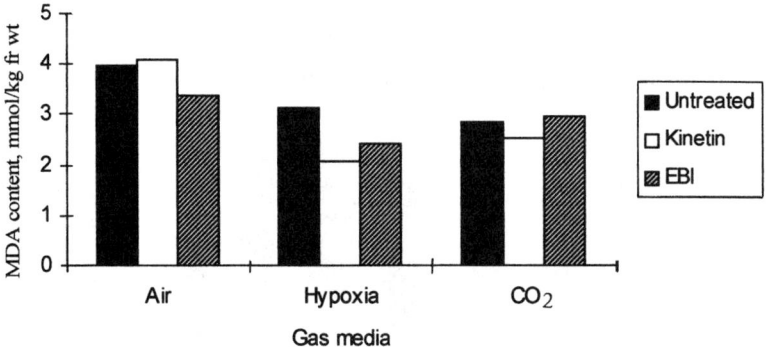

Fig. 42. Content of MDA in pea seedlings treated with kinetin or EBl in different gas media.

3. Membrane Permeability and Transport

The plasma membrane plays an important role among the different membrane structures of the cell because of its location as a peripheral external surface, bordering the cell content from the environment. Along with this mechanical barrier function, there are two other important functions of the plasma membranes: (1) providing conditions for exchange of information with the environment, mainly by reception and transmission of chemical signals in both directions, and (2) maintaining the cell homeostasis *via* the transport of ions and biologically important substances.

These functions are strongly related to the membrane bilayer permeability and the status of the membrane phospholipid matrix. As shown above, the latter can be regulated efficiently by BS.

The effect of BS on cell membrane permeability and transport has been reported by several groups. Some of the data were mentioned above in the discussion about the H^+-pump regulation and proton transport. Investigations on their correlation with transport and assimilate uptake by plant tissues showed that these processes are closely interconnected (Dahse *et al.*, 1990, 1991). Treatment of *Egeria densa* plants with (22S,23S)-HBl, along with hyperpolarization of the plasma membrane in light and in dark conditions, promoted the uptake of ^{14}C-labeled sucrose by the leaf cells. The changes in uptake rate for plants treated with 1 µM (22S,23S)-HBl were 110 and 121% compared to untreated controls for the light and for the dark, respectively. When a 10 µM concentration of (22S,23S)-HBl was applied, the corresponding results were 118 and 129% compared to the control. These values were even higher than that of fusicoccin, a known initiator of similar effects. (22S,23S)-HBl also stimulated the uptake of α-aminoisobutyric acid. At the most efficient concentration (0.1 µM) its uptake rate was 138 and 127% compared to an untreated control in the light and in the dark, respectively.

With respect to stimulation of the membrane transport of amino acids, it should be pointed out that a similar effect of human steroidal hormones, in particular glucocorticoids, on human cells has been reported in the literature (Sergeev, 1984).

This and other similarities between BS and mammalian steroids suggest that it may be useful to look a bit further than just to plants alone to gain knowledge about their mode of action. This means that in the study of a new BS phenomenon in plants, a look at the analogous situation with steroids in mammals or in insects might be more fruitful than the traditional comparison with effects of other plant hormones. Nevertheless, the traditional approach is also necessary because the role of BS in the total hormonal spectrum is not fully clarified yet.

An important aspect of BS-regulated transport in plants has been found in studies on the partitioning of ^{14}C-labeled photosynthates in *Vicia faba* under the action of (22*S*,23*S*)-HBl (Petzold *et al.*, 1992). This compound, when applied exogenously to the source leaves, was found to increase the retention of [^{14}C]sucrose and to activate the uptake in discs of the source leaves during a 4-h treatment. IAA and GA$_3$ showed similar behavior in such experiments. A comparison of the treated plants after 24 h revealed the differences between the effects of the applied phytohormones. Only (22*S*,23*S*)-HBl and GA$_3$ significantly enhanced the transport of a model assimilate to the apical sink region. The hormone-activated sucrose uptake is probably caused by modification of the H$^+$-ATPase activity, which was illustrated by enhanced V$_{max}$ values for sucrose uptake. The measured alteration in the kinetic properties of the uptake system correlates with the rapid effect (during 1 h) of BS on sucrose uptake in leaf discs. This work provides further support for hypothesis on the interrelation between the plasma membrane proton pump, energetization the proton-sucrose-coordinated transport, the sucrose uptake, and the phloem loading in leaf tissues.

A supposition about BS activation of phloem transport *via* initial promotion of the H$^+$-pump stimulated investigations of the effect of EBl on the dynamics of glucose accumulation in the organs of winter wheat (cv. Mironovskaya 808), barley plants (cv. Zazersky), and potatoes (cv. Nevskii) (Kurapov *et al.*, 1995). Application of [^{14}C]glucose on the leaves of control and EBl-treated plants was used to follow the assimilate transport. Treatment of cereal plants with EBl in the flowering stage activated the transport of labeled substance into the ear; the maximal intensity of this process was observed after 24 h. The application of EBl to potato plants activated the accumulation of the label in the lower part of the plants and in the tubers, and its content grew during the month.

Practically promising results concerning the regulation by BS of the cell permeability for ions were obtained recently using different model systems in experiments on the absorption of heavy metals and radionuclides by plants. The accumulation of metals (Cd, Zn, Pb, and Cu) under the influence of EBl has been studied for different agricultural plants such as barley, tomatoes, radish, and sugar beet (Voronina *et al.*, 1997; Mineev *et al.*, 1996).

It was found that the application of EBl in appropriate doses in a certain stage of development reduces the metal absorption significantly. For example, the results obtained for barley plants (cv. Zazersky) treated with EBl by spraying in the booting stage at a dose of 10 mg/ha showed that the diminution of metal content in the plants was 40-98% in comparison with the control (Table XXIII).

TABLE XXIII

Effect of EBl on Heavy Metal Absorption in Barley Plants

Treatment	Content of metal, mg/kg			
	Cd	Pb	Zn	Cu
control	18.30	1.47	17.20	2.20
EBl	0.38	0.70	10.40	1.30

TABLE XXIV

Effect of EBl on the Uptake of Heavy Metals from the Soil into Tomato Fruits

Treatment	Content of Cd, mg/kg		Content of Zn, mg/kg	
	Fruits	Soil	Fruits	Soil
Me^a	0.05	10.20	5.0	340
EBl^b+Me	0.02	9.25	3.3	320
EBl+Me+EBl	0.04	4.28	3.6	280

^{a} Addition of the corresponding salt to the soil. ^{b} Treatment with EBl.

Soaking tomato seeds (cv. Ranniy) for 12 h in a 10^{-8} M solution of EBl before sowing was more efficient in decreasing the content of Zn and Cd in tomato fruits than a double treatment consisting of soaking the seeds and spraying the plants in the budding stage with a 10^{-7} M solution of EBl (Table XXIV). However, in the last case, the atomic absorption spectroscopic analysis of the soil where the tomato plants were grown showed a lower content of the metals than in the case of the single treatment by soaking the seeds.

Sugar beet plants (cv. Bordo) were treated with EBl at a dose of 5 mg/ha in stage 3 true leaves. The content of Pb in beet roots was more than 50% lower and the content of Cd was 8% lower than in the control. It should be pointed out that the crop yield in this experiment was not higher than in the control. This means that this effect cannot be explained by dilution in the plant tissues but probably is connected with the regulation by EBl of the metal ion transport through the cell membranes.

Similar behavior concerning the accumulation of radionuclides under the influence of BS was found recently for different plant species. Experiments with barley plants (cv. Zazersky) on the absorption of cesium and strontium ions were carried out under laboratory conditions using modeled soil with added cesium and strontium salts. It was shown that changes in the ion content were influenced by EBl and depended on the stage of plant development. Table XXV illustrates the ion content in plants which were treated with 0.01 ppm of EBl solution in the

TABLE XXV

Effect of EBl on Cesium and Strontium Accumulation by Barley Plants

Treatment	Cs,Sr content in plants (2 weeks after treatment)			
	Cs		Sr	
	mg/g dry weight	% of control	mg/g dry weight	% of control
1. control	0.063	100	0.025	100
2. control + Cs - 27 mg/kg, Sr - 8.3 mg/kg	0.095	151	0.036	144
3. EBl, soil as in no. 2	0.068	108	0.030	120

booting stage and then were mowed after 2 weeks and dried (Khripach et al., 1995g).

The Cs content in EBl-treated plants was close to that in the unpolluted control and was different from that in EBl-untreated plants by 43%. The Sr content was 24% lower than that in the polluted control. The dynamics of changes in ion accumulation during the vegetation period (Table XXVI) was quite favorable with respect to Cs and Sr in grains (70 and 36% lower than for the untreated plants, respectively) and was less evident for straw.

In much the same way, a minimization of ^{137}Cs accumulation after EBl treatment (0.01 ppm solution, spraying) was observed for Timothy plants (*Phleum pratense*) grown in an area with a high level of radioactivity (13.8 Ci/km^2). The radioactivity of treated plants was 27% lower than that of untreated plants grown under the same conditions: 27.0 and 37.0 Bq/kg, respectively.

TABLE XXVI

Effect of EBl on Cesium and Strontium Absorption by Barley Grain and Straw

Treatment	Cs,Sr content							
	Grain				Straw			
	Cs		Sr		Cs		Sr	
	mg/kg	%	mg/kg	%	mg/kg	%	mg/kg	%
1. control	0.016	100	0.003	100	0.069	100	0.062	100
2. control + Cs -27mg/kg, Sr - 8.3 mg/kg	0.038	238	0.008	236	0.181	262	0.082	132
3. EBl, soil as in no. 2	0.027	168	0.006	200	0.174	252	0.075	120

A comparison of the efficiency of two methods of EBl application to prevent ^{137}Cs accumulation in lupine plants (*L.luteus*, cv. Zhemchug) showed that the treatment of plants by spraying with a 10^{-9} M solution of EBl in the flowering stage gave better results than soaking seeds in a 10^{-9} M EBl solution (Zabolotny et al., 1997). The data summarized in Table XXVII were obtained from an experimental area with a 72.5 Ci/km^2 level of radioactivity for ^{137}Cs. The first method of EBl application gave a significant diminution of ^{137}Cs content both in vegetative and in reproductive organs. When the seeds were treated, the content of ^{137}Cs in the plants at the flowering stage was even higher than in the untreated control. Fully maturated plants showed some decrease of ^{137}Cs content, especially in the vegetative organs, probably as a result of vegetative dilution.

TABLE XXVII

Effect of EBl Treatment on ^{137}Cs Content in Plants of *L. luteus* (cv. Zhemchug), Bq/kg of Dry Weight

Treatment	Flowering stage	Full maturity stage		
	Vegetative weight	Vegetative weight	Bean shells	Seeds
control	4038	3338	3687	4842
EBl, seeds	5956	2771	3634	4763
EBl, plants	-	2054	2528	3355

An application of EBl by treatment of the seeds as described above to the lupine plants (*L. angustifolius,* cv. Helena) and some cereal plants showed no effect on ^{137}Cs accumulation.

Additional confirmation of the ability of BS to regulate radionuclide absorption in plants can be concluded from the data obtained for *Calamagrostis epigeios* L. plants grown under natural conditions with a high level of radioactivity (59.9 Ci/km^2). The experimental plants were sprayed with a 10^{-9} M EBl solution either at the beginning of the growth or in the heading stage. γ-Radiometric analysis of the plants was first performed at the heading stage and then was done again at the end of the vegetation period.

Summarized data, including both plant and soil radioactivity, together with the transfer factors (TF) are shown in Table XXVIII. The TF is a ratio of radioactivity of the weight of dried plants in Bq/kg to the radioactivity of the weight of dried soil in Bq/kg, and it reflects the transfer of radioactivity from the soil to the plants.

The data in Table XXVIII show that EBl application caused a significant reduction of the transfer of radioactivity from soil to plants; the TF for treated plants was about 2.4-4 times lower than for untreated plants.

TABLE XXVIII

Effect of EBl Treatment on ^{137}Cs Accumulation in Plants of *Calamagrostis epigeios* L., Bq/kg of Dry Weight

Treatment	Sampling, dates and activity					
	3 June		TF	14 August		TF
	Plant	Soil		Plant	Soil	
control	3263	3126	1,04	3513	1710	2,05
EBl, 16 April (growth beginning)	3811	5626	0,68	2991	5626	0,53
EBl, 3 June (heading stage)	-	3754	-	3166	3754	0,84

An explanation of the observed BS-induced alterations in metal ion uptake from the environment to plants might be the influence of BS on the competition between ions in K^+-Cs^+ and Ca^{2+}-Sr^{2+} pairs for their uptake by the plant. This was checked under laboratory conditions using barley plants (cv. Roland). After growing as water culture, 7-day-old seedlings were put in a nutrient medium with equivalent amounts of the mentioned ions (Vedeneev *et al.*, 1997a). Half of the experimental plants were sprayed with an EBl solution (0.1 and 0.01 ppm), and the rest was treated by addition of EBl to the nutrient medium at the same concentration. EBl-untreated plants grown in the same nutrient medium were used as an experimental control in addition to a pure control without added Cs and Sr salts.

Five days of culturing in the presence of Cs^+ and Sr^{2+} led to the inhibition of seedling growth and development and to diminution of biomass. The seedlings that were treated with EBl grew better. The presence of EBl in the nutrient medium reduced the Cs^+ absorption by a factor 1.3 in comparison with the K^+ absorption, but the ratio of absorbed Ca^{2+} and Sr^{2+} was not changed. This effect was found to be independent of the EBl concentration. For the seedlings sprayed with EBl, Cs^+ absorption was 1.6 times lower than K^+ absorption, and Sr^{2+} was absorbed 1.5 times slower than Ca^{2+}.

In the soil culture barley plants were treated in the tillering stage with an EBl solution (0.01 ppm). This had little effect on plant growth and development. Two weeks after the EBl treatment the enhancement of uptake of Cs^+ and Sr^{2+} by the plants was not significant, but for untreated plants it became 1.5 times higher. These data suggest the existence of a K^+-Ca^{2+}-dependent mechanism of regulation of radionuclide absorption by plants, which is influenced by BS.

D. EFFECT ON PROTEIN AND NUCLEIC ACID METABOLISM. STRESS RESPONSE

The obligatory precondition of growth, cell elongation and cell division, is an active biosynthesis of proteins. The increase in the rate of protein synthesis has to be preceded by activation of transcribing nuclear DNA into RNA, which is catalyzed by the DNA-dependent RNA polymerases. Increases in their activity ultimately will result in an increased number of ribosomes and in activation of protein synthesis. Indeed, the study of the nucleic acid-protein metabolism in plants treated with BS showed an activation of DNA and RNA polymerases and an increased level of DNA, RNA, and protein biosynthesis.

The first model plant systems, pinto bean and mung bean, were chosen because of their known high response to BS action and the ability to differentiate the BS-induced response from that of other plant hormones (Kalinich *et al.*, 1985, 1986). In pinto beans, excised swollen and split internodes were compared with the excised internodes of untreated controls. In mung beans, hypocotyls and epicotyls from Bl-treated and untreated seedlings were compared.

It was found that treatment with Bl resulted in increased RNA polymerase activity in both experimental models. As could be expected, in swollen tissue of pinto beans the RNA polymerase activity was the greatest. The increase in enzyme activity led to a higher protein production, which directly linked to cell elongation. The RNA polymerase activity was also increased together with the protein production in split internodes but not to the same extent as in swollen tissue. It should be pointed out that these effects were strictly correlated with the data on elongation and weight changes of tissue sections obtained for the same plant systems.

In mung beans, Bl treatment resulted in cell division and elongation of the epicotyls while the morphology of hypocotyls remained relatively unaffected. An interesting feature of this system was that, nevertheless, the RNA polymerase activity was increased in both types of mung bean tissue, but to a greater extent in epicotyls. The parameters of protein synthesis followed the activities of RNA polymerases.

Similar alterations for both plant systems were found in the activity of DNA-dependent DNA polymerases after treatment with Bl. Because the enhanced activities of RNA and DNA polymerases could affect the nucleic acid level, RNA and DNA contents in the excised bean sections were also measured, and an enhancement was found in the tissues of all treated plants. A characteristic detail was the higher increase in DNA polymerase activity in split tissue than in the swollen tissue of pinto beans. This caused a greater increase in DNA content and a higher level of cell division. Thus, a close correlation between the morphological changes resulting from Bl treatment, on one side, and increased activities of the RNA and DNA polymerases and an enhanced level of proteins, RNA, and DNA, on the other side, was observed. Both enzyme systems were enhanced by Bl treatment, although one system was predominant in each type of tissue. The system that was enhanced most determined the tissue response. Further investigation (Mandava *et al.*, 1987) using selective inhibitors of RNA and protein synthesis on EBl- and (22*S*,23*S*)-EBl-induced responses in mung bean confirmed the previous results and showed that the decline in epicotyl growth caused by the inhibitors can be overcome by the effect of BS.

These results led to the assumption that Bl acted on the genome in the form of a hormone-receptor complex. The action of Bl resulted in genome activity regulation and could be responsible for the effects of Bl on transcription and DNA-replication. This hypothesis was examined further to reveal the main sites of BS action inside the cell and to learn the details of its molecular mechanism. If the hypothesis is realistic, specific genes that are transcriptionally regulated by BS must exist. Thus, investigations of the effect of BS on gene expression, mostly *via* examination of a variety of specific gene products, became one of the directions in studies of the mechanism of action of BS. The studies on the effect of Bl on gene expression using a molecular-genetic approach (Clouse *et al.*, 1992; Zurek and Clouse, 1994; Zurek *et al.*, 1994) led to characterization of the BS-regulated gene from soybean (Zurek and Clouse, 1994). Although, the problems formulated initially when this program was started (Clouse and Zurek, 1991) are not fully solved, this approach did bring progress in the understanding of the mechanism of action of BS. The main difficulties for further developments are caused by the extreme multifunctionality of BS in plants, which is not connected with the genetic level only.

An attempt to localize a starting point in the chain of BS-initiated signal transduction was undertaken in studies on the action of BS on RNA and protein biosynthesis in genetically different barley (Deeva *et al.*, 1996b) and wheat plants (Mazets and Deeva, 1996; Deeva *et al.*, 1997a; Mazets, 1997).

To distinguish the effects of the nuclear and the cytoplasmic genome, two varieties of barley plants, Roland (R) and Zazersky (Z), and their isoplasmatic lines R(Z), bearing the nucleus of the Roland variety and the cytoplasm of the Zazersky variety, and Z(R), bearing the nucleus of the Zazersky variety and the cytoplasm of the Roland variety, were chosen as experimental plant systems to study the effect of EBl application on the qualitative and quantitative composition of RNA and of soluble and structural proteins. The Roland variety is characterized by a higher rate of nucleic acid and protein biosynthesis, by a higher heterogeneity of the polypeptide composition of soluble proteins in organelles, and by an increased metabolic activity of some enzymes, e.g., alanine and aspartate-aminotransferases. Such differences are very important for the plant response to external influences.

Two variants of EB1 application were used: (1) soaking the seeds in a 0.01 ppm EB1 solution and (2) spraying the plants with the same solution in the tillering stage. It was found that EB1 did not affect the qualitative composition of the RNA fraction but changed the content of some its components. In the Zazersky variety EB1 enhanced the quantity of tRNA and rRNA, and in the isoplasmatic line Z(R) their synthesis was slower, especially that of tRNA. In the Roland variety EB1 enhanced the content of tRNA more efficiently than in the Zazersky variety. In the line R(Z) the synthesis of (28S)- and (18S)-rRNA was also increased.

Qualitative and quantitative changes were found in the composition of soluble and insoluble proteins in the EB1-treated plants. The character of these changes was dependent on the genotype and on the time of observation. In the case of the Roland variety and in the line R(Z), new polypeptides in the region of 67 and 30-20 kDa appeared in the electrophoretic spectra of soluble proteins, 5 h after the plants were sprayed with EB1. In the line R(Z), some components in the high-molecular-mass region were decreased additionally. After 72 h in the Roland variety, treated plants showed only changes in the low-mobility components, and in the R(Z) line, in some components with low and middle mobility. The changes in the high-molecular-mass region in both genotypes were similar after treatment with EB1. Polypeptides with REM (relative electrophoretic mobility) 0.13 and 0.17 were induced and two components with MW 80 and 43 kDa were degraded. An interesting detail of the study (Deeva *et al.*, 1997b) was the comparison of the EB1 effect with the changes in the protein spectrum induced by the nonnatural synthetic plant growth regulator "kvartazin" [N,N-dimethyl-N-(2-chloroethyl)hydrazine chloride]. Under the same conditions it led also to a change of intensity of some components in the protein spectrum, but not to the appearance of new components.

In the Zazersky variety a new polypeptide with REM 0.40 was found only 48 h after the application of EB1. The intensity of the component with a MW of 43 kDa was decreased, and the component with a MW of 30-20 kDa was increased. Similar changes in the electrophoretic spectra of soluble proteins were observed in the line Z(R); however, *de novo* synthesis of the corresponding polypeptides had not occurred.

The most significant changes in the composition of soluble proteins in the Zazersky variety were found 72 h after the application of EB1. The intensity of the already existing components with REM 0.31 and 0.34 became higher for the first one and lower for the second one. Besides, the components with REM of 0.22 and 0.24 disappeared, and a new polypeptide with REM of 0.26 was found. The synthesis of polypeptides with medium mobility was activated, and a new highly mobile component (REM 0.92) was induced (Fig. 43). The changes in the line Z(R) after the application of EB1 mostly followed the behavior of the nuclear donor (Fig. 44).

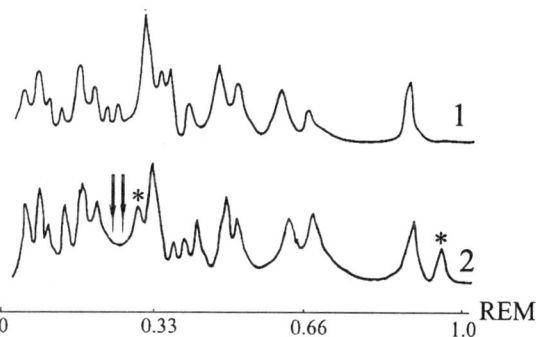

Fig. 43. Densitogram of the soluble proteins in barley plants cv. Zazersky: 1 - control, 2 - treatment with EB1; * - induction, ↓ - degradation.

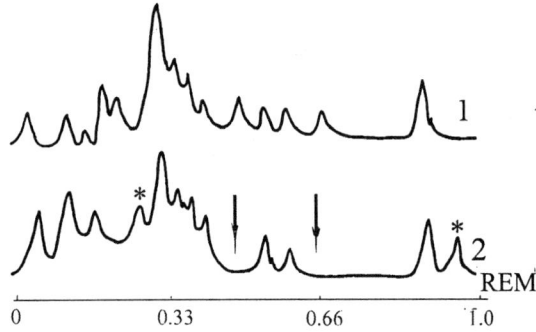

Fig. 44. Densitogram of the soluble proteins in barley plants cv. Zazersky (Roland): 1 - control, 2 - treatment with EBl; * - induction, ↓ - degradation.

The treatment with EBl had a different effect on the composition of structural proteins in the different genotypes and it came more to expression 72 h after treatment. The Roland variety and the line R(Z), bearing its nucleus, showed similar changes in protein content: a relative diminution of the high-molecular-mass polypeptides and an increase of the middle- and low-molecular-mass polypeptide components. In both cases a new peptide with a REM of 0.90 was found. A similar tendency in changes of the protein spectrum under the influence of EBl was found also for the Zazersky variety and for the line Z(R) bearing its nucleus.

A conclusion that can be drawn from the obtained results is that the effect of BS on the protein biosynthesis is mainly controlled by the nuclear genome, although the role of cytoplasmic structures cannot be excluded.

In another study, based on the employment of experimental plant systems with fixed genetic differences, the effect of EBl on the protein biosynthesis in wheat cv. Chinese Spring was investigated. The euploid and two related DT-lines, which differed from the euploid by the absence of a short ($5B^L$) or a long ($5B^S$) arm in their chromosomes, were compared with respect to their reaction to a treatment with EBl (Mazets and Deeva, 1996; Mazets, 1997; Deeva et al., 1997a). A specific feature of the approach was the parallel analysis of shifts in growth, hormonal spectrum, and protein synthesis. The mentioned lines were chosen out of 18 initially studied DT-lines because of their characteristic BS-dependent behavior.

Six-day-old plants were treated with a 10^{-7}% solution of EBl by spraying. The absence of one of the chromosome arms led to changes in plant growth rate, reduction of phytohormone accumulation, and deviations in protein synthesis resulting in changes in the quantitative and qualitative composition of soluble proteins. The effect of EBl in treated plants was visible even in the early stages of plant growth. EBl reduced the shoot weight of the euploid and to a lesser degree that of the DT-line $5B^L$ 3 days after treatment. The shoot weight was slightly increased at the end of the experiment (Fig. 45). The shoot weight of the line $5B^S$ was increased in the beginning, but lowered till 120 h after the treatment. The length of shoots was changed to a lesser extent in the same period.

The changes in shoot growth were accompanied by differences in accumulation of IAA. The quantity of IAA was noticeably decreased shortly

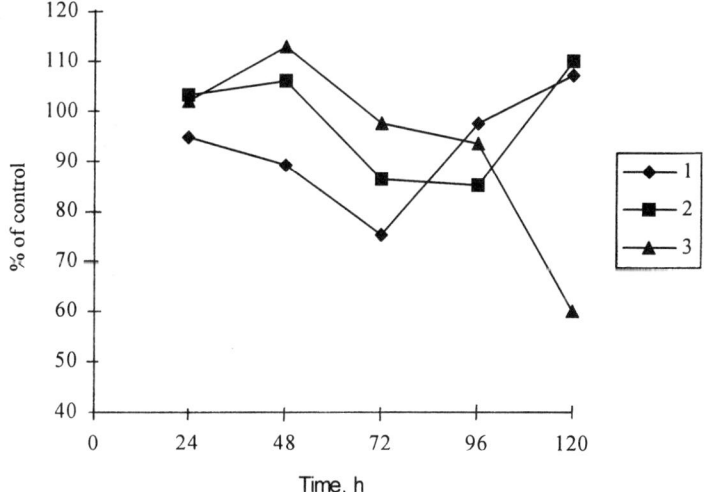

Fig. 45. Effect of EBl on shoot weight of euploid and DT-lines of wheat cv. Chinese Spring (1 - euploid, 2 - $5B^L$-line, 3 - $5B^S$-line).

after EBl treatment of the euploid, but later it increased and stabilized after 96 h (Fig. 46).

The IAA content in the $5B^L$-line showed an enhancement in two steps and was closer to that in the euploid than in the $5B^S$-line. The last one demonstrated an extremely high level of IAA in the beginning of the experiment and a dramatic reduction of this content in the final period. This tendency of higher initial EBl activation of the plants with more deficiency in genetic material was found also in the shoot development, as shown above, and in the EBl-induced regulation of the soluble protein synthesis (Fig. 47). Such a correlation between the effect of EBl and the genome looks like a compensatory action of EBl, which could take place when EBl-regulated genes are not present in the lacking part of the chromosome.

The quantitative and qualitative compositions of RNA and proteins were affected by EBl, and the *in vivo* synthesis of soluble proteins was activated in the euploid and in the lines, except for the $5B^L$-line on the first day after the treatment.

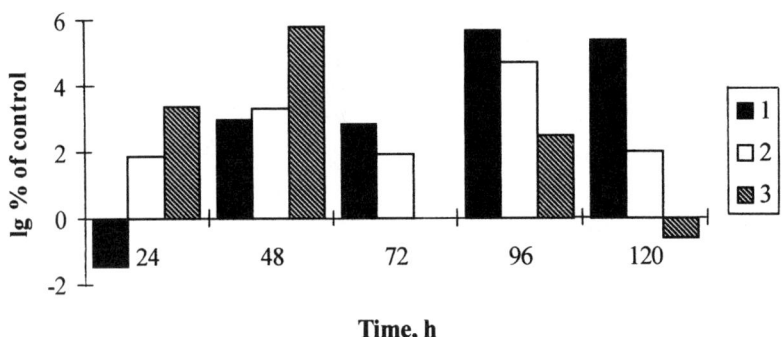

Fig. 46. Effect of EBl on the content of free IAA in the euploid and in the DT-lines of wheat cv. Chinese Spring (1 - euploid, 2 - $5B^L$-line, 3 - $5B^S$-line).

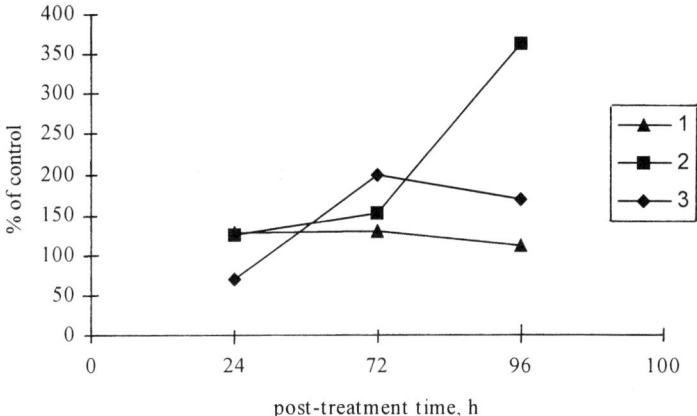

Fig. 47. Effect of EBl on the synthesis of soluble proteins in the euploid and the DT-lines of wheat cv. Chinese Spring (1 - euploid, 2 - 5B^S-line, 3 - 5B^L-line).

Many data published in the literature during the period of BS investigations indicate, directly or indirectly, a strong influence of BS on the protein metabolism. Such data were obtained for wheat and mustard plants (Braun and Wild, 1984a), celery plants (Wang, Y. et al., 1988), rice (Mai et al., 1989), radish (Beinhauer et al., 1990), sugar beet (Schilling et al., 1991), and some other crops. As a rule, this influence activates the protein synthesis and enhances the total protein yield. As discussed above, the changes in protein biosynthesis can be observed at different levels, in the separate components of the protein spectrum, in the ratio of different components, and in the total protein yield. The last effect is well documented, for example for lupine plants that showed a positive effect of BS on the protein production in all experiments (Mironenko et al., 1996, 1997; Kandelinskaya and Khripach, 1990). The results obtained in field experiments for two lupine varieties, L. luteus cv. BSKA and L. angustifolius cv. Omega, after spraying with two concentrations of EBl in the beginning of flowering are shown in Table XXIX. Although the changes in protein content for two varieties were quite different, a positive trend in protein accumulation by the seeds was observed.

TABLE XXIX

Effect of EBl on Protein Content in Seeds of *L. luteus* and *L. angustifolius*

Treatment	Protein content, % of dry mass	
	L. luteus	*L. angustifolius*
control	40.26	37.72
EBl, 10^{-9} M	43.13	40.22
EBl, 10^{-7} M	42.00	40.13

The treatment of seeds with Bl was less efficient with respect to the total protein accumulation, but it gave an opportunity to study the Bl-influenced protein synthesis in the early stages of plant development and to estimate the range of active concentrations (Fig. 48).

Together with quantitative shifts in total protein, some qualitative changes were also registered. It was shown, for example, that the legumine fraction of the total proteins of the seeds obtained from treated plants was about 1.5 times higher than in the control. In addition, the study of the activity of some enzymes

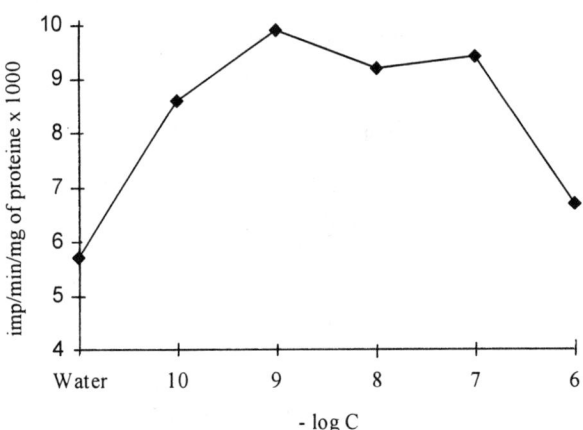

Fig. 48. Effect of Bl on the rate of protein synthesis in embryos of *L. luteus* after treatment of seeds with different concentrations of Bl (rate of incorporation of [^{14}C]leucine).

involved in the regulation of the protein metabolism showed that Bl and EBl did not affect the neutral protease activity.

A comparison of the behavior of growing seedlings and ripening seeds showed that both Bl and EBl act as anabolics in relation to chlorophyll, DNA, RNA, and protein synthesis in cotyledons and embryos in growing seeds. The character of action of the hormones was similar, but in heat stress conditions EBl was more efficient. Similar effects took place also in maturated seeds. The total nucleic acid content was increased in treated plants till the 26th day after EBl application, and it mainly resulted from enhancement of the DNA content. In this period the activity of hydrolytic enzymes, DNAases, and RNAases was decreased and started to regenerate gradually till the 37th day after treatment (Fig. 49).

The obtained results could be interpreted as an anabolic effect of BS in the growing plant and in the maturating seeds created by initiation of synthetic processes, by partial inhibition of catabolic processes, and by preservation in this way of tissue functions as centers of assimilate attraction.

An interesting aspect of the influence of BS on protein metabolism is the change found in the amino acid composition of the total protein. The data indicate significant shifts in the content of some amino acids in the total protein

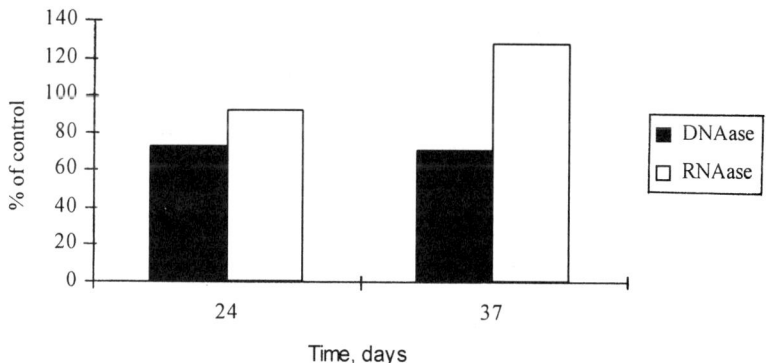

Fig. 49. Nuclease activity in ripening seeds of *L. luteus* after treatment with EBl (10^{-9} M/l).

TABLE XXX

Content of Some Amino Acids in the Total Protein Content of Seeds of *L. luteus*, %

Amino acid	Treatment	
	Control	EBl, 10^{-9} M
methionine	0.35	0.51
lysine	4.87	5.32
cysteine	3.14	4.35
glutamic acid	22.92	21.50
tyrosine	2.72	3.03

content of lupine seeds obtained from plants treated with EBl (Table XXX) (Mironenko *et al.*, 1997).

These data are important not just to indicate how the protein nutrient value is affected by BS but also as a possible characteristic for the plant stress resistance, because some amino acids are known to be important for plant stress response. Along with the amino acids, osmoregulating metabolites such as choline, betaine, and other tertiary ammonium compounds (TAC) are interesting in this respect. The balance of these substances in the cell regulates to a large extent the physicochemical status of the cell matrix and affects the mesomorphic membrane structure *via* quick alterations in the content of some components. Accumulation of these substances by plants increases their adaptive ability to external factors. Research on the effect of EBl on the accumulation of these compounds by barley plants was recently carried out (Deeva *et al.*, 1997b). The data shown in Fig. 50 were obtained in the tillering phase for barley plants (cv. Roland) grown from seeds treated with a 0.01 ppm EBl solution. Similar results were observed for the same plants when sprayed in the tillering phase.

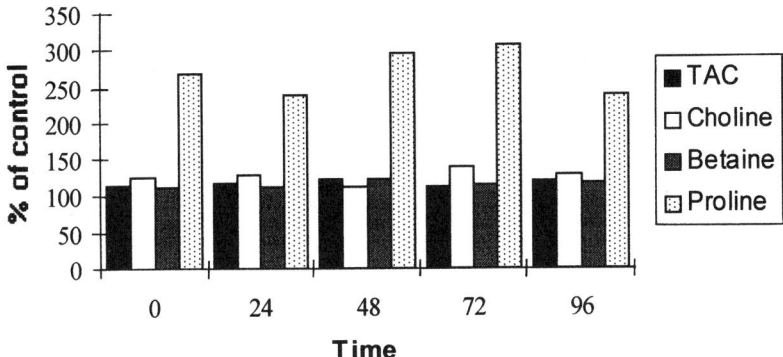

Fig. 50. Effect of EBl on the content of osmoregulating metabolites in barley plants.

The accumulation of all metabolites was activated in EBl-treated plants and was quite high during the studied period. The change of the proline content was most significant. It was about 3 times higher than the control in the period between 48 and 72 h after the beginning of the measuring.

A specific feature of this research was a simultaneous study of similar effects of some other nonsteroidal synthetic plant growth regulators under the same conditions applied either separately or as a mixture with EBl. It was found that in some cases the effect of these regulators was rather high and additive with EBl. After a short period that was comparable with the duration of their metabolic degradation, the total effect of the mixed treatment was decreased to the level of EBl treatment alone, which was quite stable and continuous.

These and other experiments gave additional information on the nature of the adaptogenic effect of BS, which may be connected with selective changes in specific links in metabolic pathways initiated *via* gene expression and synthesis of certain enzymes, proteins, and accompanying metabolites. An essential contribution to the understanding of the mechanism of the antistress activity of BS was made by investigations on BS-influenced protein synthesis in wheat leaves at normal and high temperature (Kulaeva *et al.*, 1989). High temperature was used as a stress factor causing heat shock protein synthesis, which is considered to be important in effecting of a protective response in plants. It was found that both EB1 and (22S,23S)-HB1 activated in stress conditions and in normal conditions the total protein synthesis and initiated essential shifts in the protein spectrum. An interesting feature was that EB1 and (22S,23S)-HB1 induced protein patterns similar to heat shock protein patterns not only under stress but also at normal temperature. This resulted in enhancement of the thermotolerance of protein synthesis in BS-treated plants, an effect that is connected with an increase of the thermostability of membranes.

The protection of cucumber plants against heat stress after treatment with BS was observed at high (40 °C) and at low (5 °C) temperature and it resulted in a significant increase of the germination ability of treated seeds, better growth, higher dry mass, and more chlorophyll accumulation (Katsumi, 1991). Similar data on the promotion of the chilling resistance of maize seedlings by B1 resulting in better growth recovery after rewarming were reported (He *et al.*, 1991). B1 also increased the ripening of rice plants under low-temperature conditions (Hirai *et al.*, 1991).

A comparison of the nature of the temperature stress protective action of BS with similar properties of ABA in bromegrass (*Bromus inermis*) as a plant model system showed different mechanisms by which these compounds exert antistress effects (Wilen *et al.*, 1995).

A protective effect of BS in salt stress conditions was shown for barley plants grown from seeds treated with (22S,23S)-HB1 or EB1 (Kulaeva *et al.*, 1991; Bokebaeva and Khripach, 1993). Electron microscopy showed that treatment with BS prevented the degradation of the cell ultrastructure, which was induced by a 0.5 M solution of NaCl. Some changes in ion content and an enhancement in chlorophyll accumulation were found in 7-day-old plants.

A decrease of the effect of water stress in mung beans treated with EB1 was shown to be connected with a higher ability of plants to assimilate water, which was confirmed by experiments with tritiated water (Zhao and Wu, 1990). At the same time, a significant increase of proline content was found in treated plants, which was interpreted by the authors as an indication for the increase of the resistance of the plants to stress conditions.

Application of (22S,23S)-HBl in rather high doses (1 g/ha) to sugar beet plants in conditions of mild water stress did not lead to a stress reaction while in the control a loss of 8% of biomass was observed (Schilling *et al.*, 1991). Treatment with BS activated the growth of the lateral roots by 25-30% in comparison with untreated plants under stress conditions.

E. RESISTANCE TO DISEASES

Although the stress-protective properties of BS have been known for a long time, systematic investigations of the potential of BS to enhance plant resistance to diseases, which can be considered also as a kind of stress, were started only recently. Among the results obtained to date in this area, the largest part is connected with the influence of BS on fungal phytopathogenesis. The early data obtained for potato plants as a model system showed that treatment of plants or sowing material with BS could significantly affect the development of fungal infection. The treatment of plants was found to be protective against fungal infection for all the applied doses of BS, but the treatment of tubers sometimes activated the disease. Thus, investigation on the interaction between *Phytophthora infestans* (Mont.) de Bary and potato showed that EBl and HBl induced a higher susceptibility of potato tuber tissues in the concentration range of 10^{-8}-10^{-16} M (Vasyukova *et al.*, 1993, 1994). This effect was caused by stimulation of the growth of mycelia of the fungi and enhancement of spore formation. This was confirmed by a direct experiment where fungi were grown in media that contained EBl in different concentrations. The highest concentration, 10^{-5} M, induced lysis of zoospores. Lower concentrations did not affect the growth of fungi, but starting from a concentration of 10^{-10} M, the stimulation of hyphae growth increased to its maximum at 10^{-14} M and then it decreased gradually, although even at 10^{-20} M some stimulation could be seen. A specific feature of the EBl action was the influence on the vegetative growth of fungi, but not on the reproductive functions. The last mode of action is typical for some phytosterols that activate the process of spore formation in sterol-depending fungi. An additional explanation for the observed phenomenon of pathogen initiation as a result of weakening of the immune status of plant tissues was suggested based on data on the inhibition of wound tissue reparation under the action of BS in the absence of phytopathogen.

The data (Korableva *et al.*, 1991, 1992, 1995; Platonova *et al.*, 1993) suggest that the mentioned results can be considered from different points of view. It was shown that treatment of potato plants or tubers after harvesting led to a prolongation of the period of deep dormancy of tubers and to enhancement of their resistance to phytophthora infection and other diseases. Under the influence of EBl the production of ethylene by tuber tissues increased and the biosynthesis of protective substances of phenolic and terpenoid nature was activated. These effects took place when the intact tubers were treated with EBl at a concentration of 0.1-0.01 mg/l, which was higher than the stimulative concentration for pathogen development. This means that the protective or deprotective type of activity of BS depended on the method and time of BS application and was connected with the different stimulating points of either the plant or the pathogen. In such cases the desirable result can be achieved by choosing the optimal conditions.

As mentioned above, the treatment of potato plants with BS essentially decreased the level of phytophthora development. It was shown in field experiments that most efficient in this respect was spraying plants with BS (Bl, EBl, and HBl) solutions in doses of about 10-20 mg/ha in the beginning of the budding stage (Khripach *et al.*, 1996c, 1997) (Table XXXI).

TABLE XXXI

Effect of BS on Potato Plant Resistance to Phytophthora Infection

Treatment	Tubers affected by disease		
	number/1000 tubers	% of total amount	% of control
cv. Orbita			
control	19	1.9	100
HBl	16	1.6	84
EBl	17	1.7	89
cv. Sante			
control	25	2.5	100
EBl	19	1.9	76
cv. Rosinka			
control	29	2.9	100
EBl	19	1.9	66

In some cases the efficiency of BS in protection against fungi was even higher than for plants treated with standard fungicides. Usually, in conditions with a higher level of pest development, a higher expression of protective properties of BS was observed. The data shown in Fig. 51 were obtained for

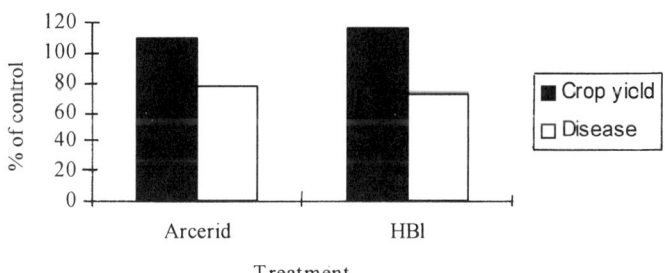

Fig. 51. Effect of HBl on the yield of potato tubers and resistance of plants to phytophthora infection in comparison with the standard fungicide arcerid (cv. Adretta).

potato plants treated with HBl at a dose of 20 mg/ha in the beginning of the budding stage. The protective effect against fungi produced by this single treatment with BS was similar to the effect produced by a double treatment with the standard fungicide arcerid (composition of ridomil and polycarbacine) at a dose of 2 kg/ha.

Later, a study on the mechanism of the protective effect of BS was carried out with barley plants as a model system in field and laboratory conditions. It was found that spraying plants in the tillering phase with a solution of EBl significantly decreased the extent of leaf diseases induced by *Helmintosporium teres* Sacc. under laboratory conditions. The effect of EBl on the resistance of barley plants to leaf diseases induced by mixed fungi infection in field conditions is illustrated by Fig. 52; the development of disease is indicated for the heading stage. This effect was accompanied by an increase in grain yield that was significant even at a dose of 5 mg of EBl per hectare (Pshenichnaya *et al.*, 1997; Volynets *et al.*, 1997a,b).

The highest level of disease suppression took place at a dose of 15 mg of EBl per hectare and it was comparable with the effect induced by the standard fungicide Bayleton when applied in the usual dose. An attractive feature of the

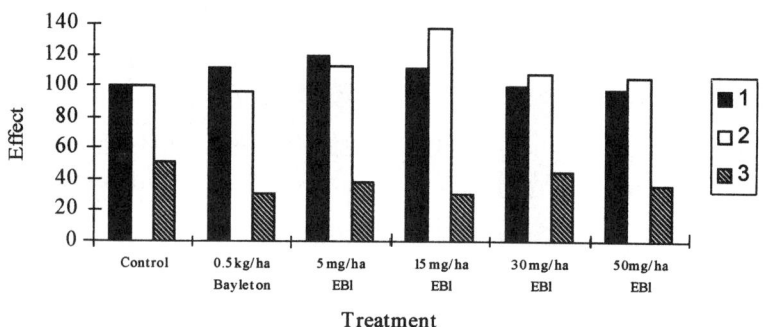

Fig. 52. Effect of EBl on barley plant resistance to leaf diseases and on productivity (cv. Prima Belarusi) (1 - bushiness, % of control; 2 - weight of grains, % of control; 3 - development of disease, %).

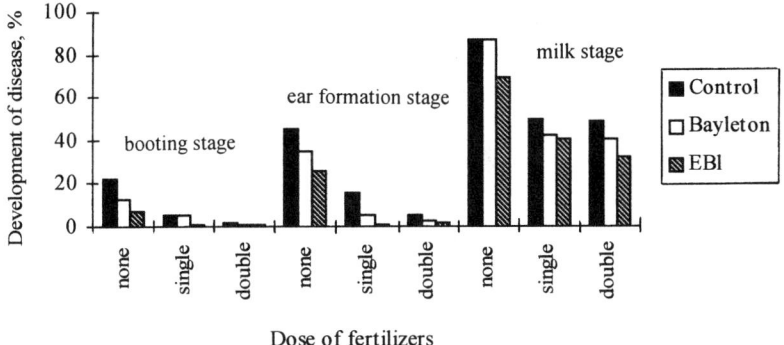

Fig. 53. Effect of EBl on barley leaf diseases at different doses of mineral fertilizers (cv. Prima Belarusi).

action of EBl was a simultaneous stimulation of the plant fungi protective properties and the higher productivity that was observed for all doses.

An interesting result on BS-regulated resistance of barley plants to leaf diseases was obtained during the study of the effect of EBl at different doses of mineral fertilizers. All experiments showed a protective effect of EBl, which was especially significant at a high dose of fertilizers (Fig. 53).

The data shown in Fig. 53 were obtained for a 5 mg/ha dose of EBl and a usual (0.5 kg/ha) dose of Bayleton. They illustrate the higher capacity of EBl as a protective factor against fungi under these conditions in comparison with the traditional fungicide.

In separate experiments (Pshenichnaya *et al.*, 1997b) it was shown that the fungistatic activity of EBl, measured in cultures of the fungus *H. teres*, was rather low (the active concentration was higher than 20 mg/l). So, it was concluded that only a hormonal effect, leading to activation of the internal mechanism of plant resistance, could be responsible for the observed results. Since regulation by a hormonal signal has to include a series of biochemical shifts, some metabolic parameters of plants infected with *H. teres* Sacc. after treatment with BS were investigated (Volynets *et al.*, 1997a).

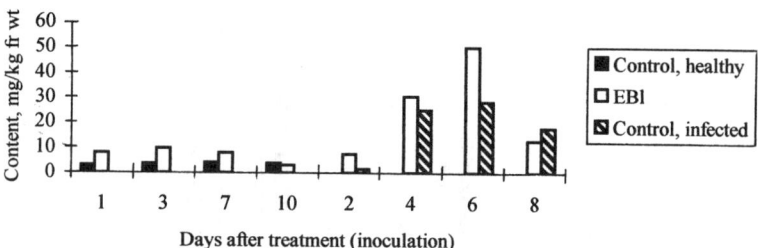

Fig. 4. Content of auxin in leaves of barley (cv. Zazersky) infected with *H. teres* Sacc.

The treatment of barley plants with EBl promoted the accumulation of auxin, and the same tendency was observed with the treatment of infected barley plants (Fig. 4).

It has been shown that the main reason for changes in the auxin level was the inhibition of auxin-oxidase activity under the influence of EBl in the infected barley plants (Fig. 55).

Simultaneously, the plants treated with EBl showed a higher peroxidase activity and an increased level of some phenolic components in comparison with the control, for both the healthy and the infected variant of the experiment. The total activity of gibberellins was not influenced under these conditions.

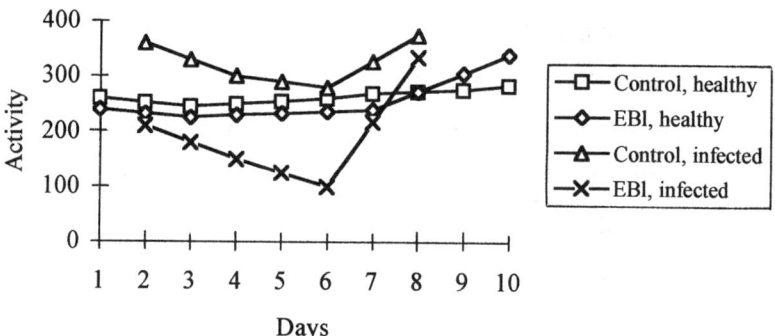

Fig. 5. Activity of IAA oxidase in leaves of barley plants (cv. Zazersky) infected by *H. teres* Sacc. and treated with EBl.

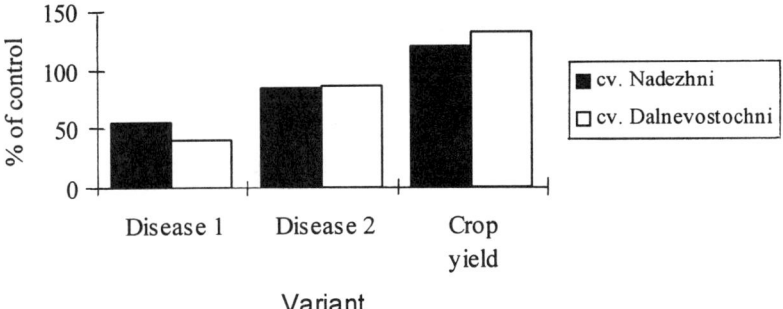

Fig. 56. Effect of EB1 on the yield of cucumbers and the resistance of the plants to peronosporosis (disease 1 - middle of vegetation period; disease 2 - the end of vegetation perion).

A protective effect of EB1 against fungi was established in field trials using cucumber plants as a model system (Churikova and Vladimirova, 1997). Figure 56 illustrates the results on suppression of peronosporosis in cucumber plants under the influence of EB1. In these experiments EB1 was applied twice. First, seeds were soaked in a 0.1 mg/l solution of EB1 and then the plants were sprayed at a dose of 25 mg/ha in the flowering stage.

An increase in the activity of some enzymes (peroxidase, polyphenoloxidase) in the leaves of cucumber plants has been found also. Since these enzymes are involved in the metabolism of polyphenols, a change in their activity may be considered as one of the factors that are connected with the increase of plant resistance to infection.

A new recently discovered aspect of the protective action of BS on plants is related to their ability to stimulate resistance to virus infection (Bobrick, 1993, 1995; Rodkin et al., 1997). Potato starting material, produced from cuttings, was cultured in a medium that contained B1, EB1, or HB1. This resulted in a significant increase of all the parameters that characterized the growth and development of the cuttings in comparison with the control and led to a higher yield of plants suitable for propagation by cutting. One of the most important effects was found in the reduction of virus infection in the resulting starting material used for planting. This effect was found in all stages of plant development and was observed also in the first and in the second tuber generations produced from the starting plant material grown in BS-containing media. Figure 57 shows such a long-term EB1 effect on the resistance to virus infection and productivity of potato plants grown from the tubers of the first tuber generation. Along with significant lowering of the virus infection, the plants obtained from BS-influenced sowing material gave a higher crop yield, which differed from the control to a maximum of 56%. The highest efficiency in crop increase corresponded to the lowest level of disease development. This was the case for plants for which the previous generation was produced from starting material that was grown in a nutrient medium with EB1 at a concentration of 0.25 mg/l.

The presented data indicate that exogenous BS can act efficiently in plants as immunomodulators when applied in the appropriate dose and in the correct stage of plant development. As in other cases of BS-regulated stress response, the protective action of BS against pathogens is the result of a complex sequence of biochemical shifts such as activation or suppression of key enzymatic reactions, induction of protein synthesis, and stress substances of different nature. These

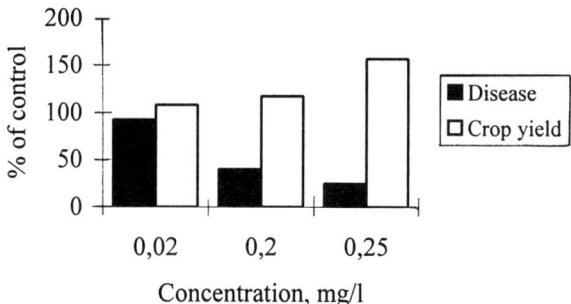

Fig. 57. Effect of EB1 on the resistance to virus infection and productivity of plants grown from the first tuber generation (cv. Temp).

relatively little investigated properties of BS are very promising from a practical point of view. They open up new approaches for plant protection based on the employment of very small amounts of environmentally friendly natural substances instead of the traditional pesticides, which often conflict with the environment.

F. EFFECT ON THE PHOTOSYNTHETIC APPARATUS

Plant ontogenesis depends on two integral processes, which are closely related to each other. These processes are growth and photosynthesis. Existing theories describe the growth-photosynthesis relationship from different points of view. One assumption is that the expression of genes responsible for growth is independent from genes that code the biosynthesis of the photosynthetic pigments. The interconnection of these processes during further plant development is realized *via* the regulatory systems of the plant, particularly *via* a light-dependent growth regulating system. Together with its role as a source of energy for photosynthesis, light is a source of information for plant photomorphogenesis. It is assumed that the transduction of a light signal from the photoreceptors can be mediated by alterations in the endogenous phytohormone balance. It is known, for example, that the content of some phytohormones in tissues of plants grown in the dark can be changed very quickly after a short irradiation with red light and that this effect is mediated by phytochrome.

A possible role of brassinosteroids in this process was a subject of interest from the beginning of research on BS. A characteristic feature of BS action, their ability to increase a crop yield, suggested such a role because the improvement of photosynthetic efficiency is an important prerequisite for an increase in productivity. Indeed, the importance of illumination with light of the correct spectral quality was shown for BS-activated growth of beans, which correlated with the accumulation of chlorophyll and photosynthetic assimilates (Krizek and Worley, 1973, 1981; Krizek and Mandava, 1983a,b; Kamuro and Inada, 1991; Kamuro and Takatsuto, 1991; Hirai *et al.*, 1985). Results on the enhancement of the photosynthetic capacity and the translocation of photosynthetic products after the application of BS were obtained in rice (Fujii *et al.*, 1991), in maize (He *et al.*, 1991), in wheat and mustard (Braun and Wild, 1984a,b), and in lupine plants (Mironenko *et al*, 1996). In the last case the stimulation of chlorophyll accumulation after application of Bl under normal conditions was accompanied by an increase of the thermoresistance of the photosynthesis under heat shock. Similar results with respect to chilling stress were obtained during investigations of the action of Bl on the growth of cucumber hypocotyls (Katsumi, 1991).

The data derived from these studies confirm that the action of BS is a light-dependent process. Some of these data are important for practical application of BS, showing for instance that the result of treatment is influenced by the photoperiod and can be different for long-day plants and short-day plants. Highly important from a mechanistic point of view were the data that illustrated a compensatory effect of BS in plant growth inhibited by specific light

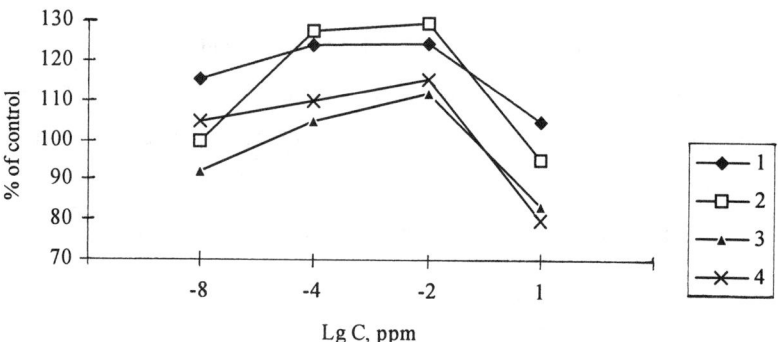

Fig. 58. Effect of different EBl concentrations on the chlorophyll accumulation in triticale seedlings: 1 - content related to area unit, leaf; 2 - content related to dry weight, leaf; 3 - content related to area unit, coleoptile; 4 - content related to dry weight, coleoptile.

conditions. This leads to the supposition of a relation between BS-induced growth and the action of phytochrome. To clarify this relationship some further experiments were carried out.

An investigation (Kalituho et al., 1996) of the action of EBl on the formation of the pigment apparatus in young triticale plants in a period when the leaf changes its functions from a sink to a source showed that soaking the seeds with EBl stimulated the accumulation of chlorophyll in 6-day-old first leaf and in the coleoptile over a wide range of concentrations (Fig. 58).

The analysis of green seedlings showed that the maximal effect of EBl in the enhancement of the chlorophyll content took place in the earliest stages of development and in the period of leaf emergence from the coleoptile (Fig. 59).

The period when the leaf came out from the coleoptile was found to be the most sensitive to the action of EBl also in postetiolated plants. At first, the chlorophyll content in these plants was lower than in the control. This last effect could be considered as a confirmation of the role of light in the realization of the stimulative influence of BS on the pigment apparatus and as an indication for their probable participation in mediation of the phytochrome regulatory functions.

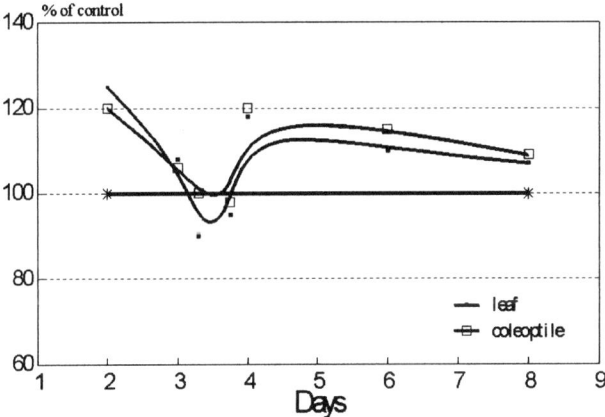

Fig. 59. Effect of EBl on the chlorophyll accumulation in leaves and coleoptiles of triticale seedlings grown under normal light conditions.

Further studies of the same plant model system under various light conditions, such as irradiation by white light, low-intensity flashing red (670 nm) light, far-red (730 nm) light, a combination of these, and dark, brought new support for this hypothesis (Kalituho, 1997; Kalituho *et al.*, 1997a). It was found that a short treatment of triticale seedlings with red or with a combination red + far-red light stimulated the chlorophyll and protochlorophyllide biosynthesis in the dark, while treatment with far-red light did not provoke chlorophyll synthesis in the dark and did not change the protochlorophyllide content significantly. During dark incubation the content of pigments increased till about 28 h of incubation, and after that the total pigment amount decreased. When the effects of successive irradiation by red and far-red light and those of only a red-light irradiation on pigment synthesis were compared, EBl-treated plants and untreated controls reacted differently. In red-light-irradiated plants the relative efficiency of pigment synthesis was higher in the EBl-untreated control; the red + far-red-irradiated plants showed a higher efficiency of pigment synthesis in the EBl-treated plants. This result probably means that the phytochrome-deactivating effect of far-red light was counterbalanced partially by the action of EBl, which could imply a phytochrome dependence of the action of EBl on the accumulation of photosynthetic pigments.

Further confirmation of this supposition was obtained from experiments with preilluminated plants which were exposed after 12 h of darkness to continuous white light. It was found that red light increased the rate of chlorophyll accumulation and that the combination of red + far-red light gave a lower effect in postetiolated triticale leaf. The effect of EBl on chlorophyll accumulation in red-light-pretreated plants was most significant in the early stages of greening.

Figure 60 illustrates the effect of red light and of EBl treatment, acting together or separately, on the chlorophyll accumulation in the first triticale leaf. The data were obtained for seedlings, which were postetiolated for 1 h. It was shown that the combined effect produced by light activation of phytochrome and

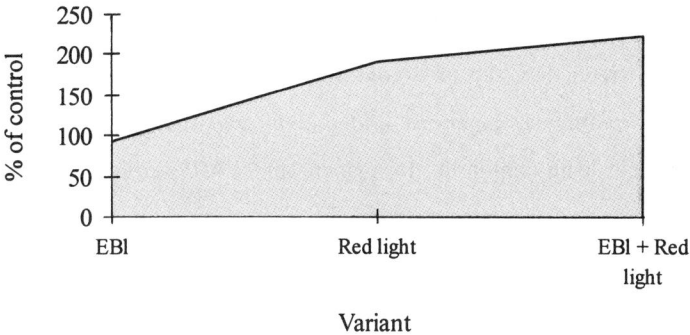

Fig. 60. Effect of red light and EBl treatment on the chlorophyll content in the first triticale leaf.

the action of EBl was higher than the effects of both factors separately; the effect of EBl alone was close to the control. These results suggest an essential role of phytochrome in the realization of the effect of EBl.

The mechanism of interaction of BS and the phytochrome system is unclear. It is possible that phytochrome$_{FR}$ formation under the flash of red light causes an alteration in the membrane permeability that could facilitate phytohormone transport or adhesion of BS by specific receptors. On the other hand, an interaction of BS and phytochrome at the genetic level cannot be excluded also, and their relationships in different plant systems should be investigated further.

A new intriguing aspect of the relationship between BS-growth-regulating and light-growth-regulating effects in plants is connected with the recent discovery of mutants in *Arabidopsis* that are considered to be defective in the biosynthesis of brassinolide (Ecker, 1997). These plants show in darkness a morphology that is very close to that of light-grown plants; for that reason defects in genes encoding signal substances involved in light regulation of growth was suggested. Treatment with Bl of these mutants restored the normal etiolated phenotypes grown in the dark and also initiated their normal wild-type growth in the light. When the genes responsible for this type of development were cloned, it was found that one of them, characteristic for the *deetiolated 2 (det 2)* mutant, encoded a protein that was similar to the mammalian enzyme that catalyzes the 5α-reduction step of steroid biosynthesis. Another one, related to the *constitutive photomorphogenesis and dwarf (cpd)* mutant, was found to encode a protein with similarity to cytochrome P450/steroid hydroxylase, responsible for one of the later steps in the biosynthesis of steroids. It was experimentally confirmed that both the *det 2* and the *cpd* mutants were defective in steroid biosynthesis but in different steps. The fact that exogenous Bl was able to restore the wild-type phenotype was interpreted as a confirmation of the deficiency of Bl in these mutants. These data are very important as new evidence for the role of BS in plant development.

G. MECHANISM OF ACTION, RECEPTION, AND TRANSPORT

The mechanism of action has been an attractive target for researchers since the elucidation of the structure of Bl, but for many reasons only recently has some significant progress in this direction been achieved. Although this problem is still rather far from its final solution, many important data have brought us closer to understanding the mode of regulatory action of BS.

Taking into account the high variability of the physiological effects of BS, it is probable that more than one molecular mechanism of their action exists. Two main aspects of the primary mechanism have to be considered first: an effect of BS on the biosynthesis of enzymes *via* an effect on genome expression and an effect of BS on membranes. The first effect is responsible for the slow reactions of plants on exogenous hormones, and the second one for the quick reactions. The data discussed in the previous sections of this chapter showed that the complicated character of BS-initiated processes is probably caused by both types of action with overlap and close interconnections between them.

The fact that not long before the discovery of Bl the main postulates of steroid hormone action in mammalians were formulated (Roy and Klark, 1980) was important for researchers who were interested in mechanisms of hormone action. Since the idea of hormone regulation in plants came from human endocrinology, these mechanisms were considered to be applicable also for plant hormones (Muromtsev *et al.*, 1987), and in particular for BS (Kalinich *et al.*, 1985, 1986). This point of view was purely hypothetical but recently, after obtaining evidence for BS-induced gene expression, this traditional mechanism of steroid hormone action in application to BS became more realistic.

The molecular genetic approach has played an important role in bringing the insight in the mechanism of action of BS to a higher level. While the first investigations in this direction led to the recognition of gene expression under the action of BS (Kulaeva *et al.*, 1989, 1991; Clouse and Zurek, 1991), further developments have resulted in the identification of the BS up-regulated gene (BRU1) from soybean. It was found that the expression of BRU1 was specifically initiated by BS that were structurally close to Bl, but not by other plant hormones or steroids (Clouse *et al.*, 1992; Zurek *et al.*, 1994). The significant sequence homology of BRU1 to xyloglucan endotransglycosylase (XET), an enzyme involved in the regulation of some cell wall components, indicated a possible role of BRU1 in cell expansion that is realized *via* an increase of the activity of XET and loosening of the cell wall (Clouse, 1996a).

If gene expression under BS action in plants is similar to the steroid-regulated gene expression in animals and a chain of events such as

BS + receptor → BS-receptor complex → nuclear DNA → mRNA → protein (enzyme)

exists, a detection of the Bl receptor is an actual challenge. The history of classical phytohormone receptor studies, which is much longer than the history of BS, does not allow us to suppose that a solution for this problem can be found easily, because even for these much better investigated hormones the traditional approach based on hormone-binding experiments led to the structure elucidation of only one auxin receptor.

Substantial progress in this direction has been achieved by application of a new methodology, developed after the isolation of *Arabidopsis* hormone-response mutants (Ecker, 1997). Some of these mutants grown in the dark or in the light, with phenotypes similar to those observed in hormone-deficient mutants, showed a nearly normal response to the usual phytohormones but were found to be insensitive to Bl (Clouse and Langford, 1995; Clouse *et al.*, 1996a; Kauschmann *et al.*, 1996a,b; Li and Chory, 1997). Study of the mutants led to the identification of only one Bl signaling gene that was assumed to be encoding a protein similar to the steroid receptor of animals. The cloning of this gene, called BRI1, and further elucidation of the protein structure showed its similarity to known receptor-like molecules, but not to those that were known previously to participate in steroid signaling events (Li and Chory, 1997). BRI1 is a member of a family of receptor-like transmembrane kinases that have structural similarity with a variety of proteins that are widespread from bacteria to human and are involved in protein-protein interactions. Although their interactions with nonprotein ligands are not yet known, some specific structural features of BRI1 suggest it has a role either in the direct binding of Bl or in binding of a complex of Bl with an intermediary protein. Intensive ongoing research to solve this and related problems seems very promising to get a confirmation of BRI receptive function in the very near future. In this way its real mode of action, membrane localization, and the targets to which its intracellular signals are transmitted can be elucidated.

To answer the questions about the action of endogenous BS in plants, knowledge about their transport and localization of biosynthesis is necessary. Unfortunately, there are very few data about these subjects in the literature now. Very useful for the clarification of the last question could be the application of highly sensitive analytical approaches based on immunochemical methods to study the subcellular localization of BS. Although such approaches are widely used for the microanalysis of several kinds of antigens as well as for phytohormones and a first attempt of their application to BS was done rather long ago (Horgen *et al.*, 1984), this methodology became accessible for the BS series just recently. It became possible after solving the problems of chemical synthesis of BS, their haptens, and conjugates and preparation of antibrassinosteroid antibodies (Yokota *et al.*, 1990b; Schlagnhaufer *et al.*, 1988, 1991; Naren *et al.*, 1996).

The approach using polyclonal antibodies against castasterone to study the localization of BS in germinating pollen of *Brassica napus* showed a specific binding of BS to nuclear components (Smith *et al.*, 1992). The detection of strong labeling in the amyloplasts suggested a role of plastids as storage organelles for BS (Sasse *et al.*, 1992). This supposition was further supported by the study of pollen of ryegrass (*Lolium perenne*) in different stages of development. It was found that BS were increasingly accumulated in starch granules during amyloplast maturation and this can be considered as an indication for the storage function of these organelles for BS (Taylor *et al.*, 1993). The detection of heavy labeling within starch granules and in the zone closely connected to it in the bicellular stage of pollen development, when the differentiation of proplastids is not yet finished and the stromal tissue is partially maintained in the starch granules, led to the conclusion that BS might be synthesized in the stroma, whose location close to the starch granule would allow the absorption of BS in these particles. An easy release of BS from starch granules was shown to be possible after hydration and that could make these compounds available to affect the process of pollen germination.

In the case of *Brassica napus* and *Lolium perenne* no specific binding of BS to any soluble proteins from pollen extract was found (Smith *et al.*, 1992; Taylor *et al.*, 1993). This might be interpreted as the absence in the cytoplasm of a specific cytoplasmic receptor as suggested earlier as a means of BS transportation to the nucleus (Kalinich *et al.*, 1985, 1986). In this connection, the preparation of antibodies to Bl and a comparative study of different types of conjugates based on differently modified Bl as possible antigens in anti-BS antibody induction will be important tasks for the future because it will increase the sensitivity and selectivity of the analysis.

A short-distance auto- or a paracrine-like way of interaction, in terms of animal endocrinology (Muromtsev and Danilina, 1996), between BS and their targets is probable also for the effects of endogenous hormones in the plant reproductive system and for exogenously applied BS. Endocrine interaction, which assumes long-distance transportation of the hormones *via* the xylem, has not yet been found to be important for the realization of the endogenous effects of BS. Nevertheless, such a long-distance transport probably *via* the xylem of exogenously supplied labeled BS that were translocated from the roots to the shoots was shown to take place in rice (Yokota *et al.*, 1992) and in cucumber and wheat (Nishikawa *et al.*, 1994). When labeled Bl and Bk were applied to the leaf surface of rice, only a slow transport of these compounds or their metabolites from the leaves to the roots was observed (Yokota *et al.*, 1992).

The question about BS transport proteins that might be similar to the corresponding ones of mammalian hormones is still open. The existence of special transport forms of BS similar to those in gibberellins is still uncertain. Data (Yokota *et al.*, 1992) on different rates of transportation obtained for natural BS with various structures of the cyclic part might be considered as an indication of such a possibility.

New possibilities for study of the mechanism of the action of BS became accessible due to the discovery of the first selective inhibitor of BS, which was found recently in the fungus *Drechslera avenae* (Kim, S.-K. *et al.*, 1994a, 1995).

H. OTHER EFFECTS OF BS

As mentioned in Chapter II, BS are very close structurally to ecdysteroids (ES), which are the moulting hormones (MH) of insects and other arthropods. ES are widespread in both the animal and plant kingdoms. This structural similarity and the corresponding possibility to bind ES receptor were the reasons which initiated the search for MH-like or anti-MH-like properties in BS series.

The first study of MH activity of BS was done using the test with imaginal disks isolated from the fly *Phormia terra novae* (Hetru *et al.*, 1986). Some BS showed a promoting effect on the evagination of imaginal discs but at concentrations 10-100 times higher than those of ecdysterone. An inhibition of ecdysterone action by BS was observed, and this agonistic effect of Bk and (22S,23S)-HBl was maximal at a concentration of $5 \cdot 10^{-5}$ M when the action of ecdysterone was fully inhibited. This fact indicated possible competition between BS and ES for the hormone receptor. It was confirmed in experiments on the competitive binding of two BS analogs [(22S,23S)-HBl and (22S,23S)-HBk] and radiolabeled ponasterone A to the ES receptor obtained from the blowfly *Calliphora vicina* larvae (Lehmann *et al.*, 1988). Ponasterone A is an ES ligand which is able to bind to the ES receptor with a high affinity (K_D 10^{-9} M). The affinity of (22S,23S)-HBk showed a K_D of about $5 \cdot 10^{-6}$ M and the binding of BS was reversible.

The antiecdysteroid activity was found also when BS analogs were applied to the cockroach *Periplaneta americana* (Richter *et al.*, 1987). It was shown that after feeding on a special diet containing (22S,23S)-HBl, a delay of 11 days took place in the moult of final-instar nymphs. (22S,23S)-HBk did not exhibit such action, but the extract of rape flowers, a natural source of brassinolide, also delayed the moult by 9 days. The finding of similarity in the action of natural extract and artificial BS analog was important in confirming the idea that the steric orientation of the hydroxy functions in the side chain of BS has no significant influence on their binding to the ES receptor, and both natural and synthetic hormones interact with the receptor in a similar way. However, this fact alone should not be overestimated because a significant amount of other steroid substances was contained in the extract, and the confirmation could be even more reliable if Bl or HBl was compared with (22S,23S)-HBl directly.

The fact that BS and their analogs possess anti-MH activity in insects stimulated a desire to learn whether they have other typical for ES types of activity, such as a neurotropic effect. This was investigated on isolated brains of last-instar larvae of the cockroach *Periplaneta americana* for (22S,23S)-HBl and (22S,23S)-HBk under *in vitro* conditions (Richter and Adam, 1991). The effect of both BS analogs was found to be similar to the effect of 20-OH-ecdysone. However, to reach the same level of response, a three fold higher concentration of (22S,23S)-HBk and a tenfold higher concentration of (22S,23S)-HBl were necessary. These results could be considered as an indication of the affinity of BS to neuronal ecdysteroid binding sites, which are the targets for 20-OH-ecdysone in insect brains. This means that BS possess ES agonist-like activity and their MH-inhibiting effect is realized probably not only at epidermal ES binding sites but also in the central nervous system. Similar results on the competition of BS with ES for the intracellular ES receptor from the epithelial cell line from *Chironomus tentans* were obtained with the same compounds (Spindler *et al.*, 1991), but they showed no pronounced ES-agonistic or ES-antagonistic properties in other model systems (Charrois *et al.*, 1996). A study of BS action on the viability and microfilarial production of adult *Brugia pahangi* cultured *in vitro* showed a behavior that was different from other studied compounds which apparently disrupted hormonally regulated processes in insects (Barker *et al.*, 1989). Similar to these compounds BS inhibited microfilarial production but did not exhibit filaricidal activity and did not influence the worm viability.

The analysis of BS effects in insects led to the conclusion that they are the first true antiecdysteroids found, and their properties and natural origin were considered to be very important for potential application in insect pest control as new safe substances, which represent the "third-generation pesticides" (Richter and Koolman, 1991).

Until recently, there were practically no attempts to investigate BS action in vertebrates except studies on the toxicology of BS. The obtained results reflected the very low toxicity of BS and their inactivity in tests for other negative influences on animals. Possible stimulating properties of BS similar to those known in the ecdysteroid series became the subject of interest only recently. The first study in this direction showed a pronounced toxicoprotective effect of EB1 on Russian sturgeon *Acipenser gueldenstaedti* fingerlings (Vitvitskaya *et al.*, 1997a). Treatment of fingerlings with EB1 solution at a concentration of 10^{-4} mg/l for 2 h before their 48-h exposure to solutions of toxicants such as $CuSO_4$ (0.004 mg/l), phenol (0.1 mg/l) and the detergent "Lotos" (0.2 mg/l) showed a significant decrease of the negative influence of the toxicants on the fingerlings in regard to swimming activity and ability to resist a current, reactivity to a sonic signal, and training. The effect of EB1 was higher than the effect of known protectants, which were compared with EB1 in the same experiments.

Since unfavorable factors such as pollutants, oxygen deficiency, high salinity, and others influence negatively the development and propagation of sturgeons, the revealed protective properties of EB1 were investigated further. It was found (Vitvitskaya *et al.*, 1997b) that the treatment of Russian sturgeon, white (Pacific) sturgeon *A. transmontanus*, and starred sturgeon *A. stellatus* eggs with EB1 at a concentration of 10^{-5}-10^{-9} mg/l significantly increased fecundation, hatching, and larvae survival (Fig. 61).

The data summarized in Fig. 61 show that the effect of EB1 on the eggs was higher in unfavorable conditions and especially in the case of the starred sturgeon. The fingerlings grown from EB1-treated eggs had better morphological

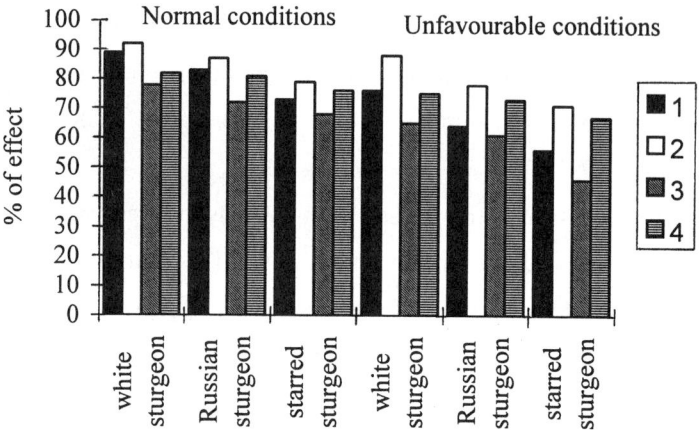

Fig. 61. Effect of EBl on sturgeon egg fecundation and hatching (1 - fecundation, control; 2 - fecundation, EBl; 3 - hatching, control; 4 - hatching, EBl).

characteristics and better resistance to stress factors, particularly to salt stress conditions.

The treatment of the sturgeon fish larvae with EBl increases the survival of the fingerlings. This effect for some species of sturgeons is illustrated by Fig. 62. Along with better survival of fingerlings, a tendency to increase of body weight was observed, and it was the most significant in the case of starred sturgeons, which were cultivated under worse hydrochemical and temperature conditions in

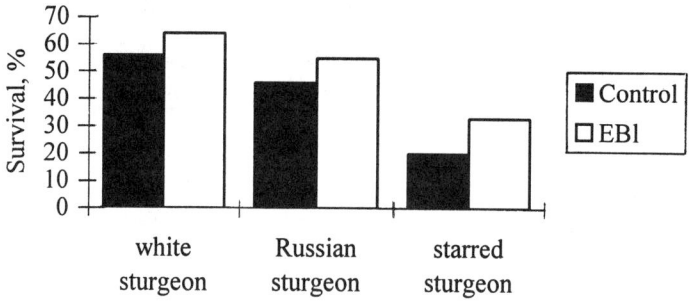

Fig. 62. Effect of EBl on the survival of sturgeon fingerlings.

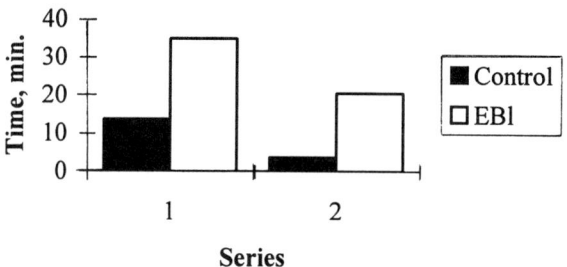

Fig. 63. Effect of EBl on spermatozoon viability time of unconserved (1) and cryoconserved (2) sperm in the Russian sturgeon.

comparison with other fishes.

The discovery of the influence of BS on the reproductive physiology of sturgeons stimulated investigations in this field, which showed prospects of wider employment of BS in aqua cultures. Thus, it was found that the application of EBl enhances efficiently the activity and viability of spermatozoons, which is very important in animal breeding. An example of such application is the activation of spermatozoons of the Russian sturgeon in normal conditions and after cryoconservation. Figure 63 illustrates the effect of EBl on the sperm from the males of Russian sturgeon (Vitvitskaya et al., 1997c). The activity and viability of spermatozoons in the samples of unconserved sperm treated with EBl at a concentration of 10^{-6} mg/l were about 250% of control. The treatment of cryoconserved sperm gave a corresponding value of about 550% of control.

The data on the action of EBl on fish physiology seem very valuable for clarifying the possible responses to BS in vertebrates. Although being the first clue, these data promise further findings that may become important for humans, if the spectrum of stimulative action of BS in vertebrates is as broad as in plants.

CHAPTER **X**

BIOASSAYS AND STRUCTURE-ACTIVITY RELATIONSHIPS OF BS

A. BIOASSAYS

Bioassays are very important for studies of plant growth substances. Since BS are present in plants in extremely low concentrations and exhibit high biological activity, their identification requires sensitive and specific bioassays that are able to discriminate BS from other endogenous plant growth-regulating hormones. Tests used for other plant hormones were applied for BS as well, but also new specific bioassays were developed.

The bean second-internode bioassay was the first test used for monitoring the biological activity throughout the isolation and purification of the "brassin-complex" (Mitchell et al., 1970, 1971; Worley and Mitchell, 1971; Mandava and Mitchell, 1971) and later for the isolation of pure Bl from rape pollen (Grove et al., 1979). The procedure of this bioassay is performed as follows (Thompson et al., 1981, 1982). Plants (*Phaseolus vulgaris* L., Pinto variety) are grown at 25-27 °C for 6 days. A dispersion of the test compound in lanolin is applied to the second internodes of seedlings. The length of the internodes has to be not more than 2 mm to ensure the optimum effect. Control plants are treated with lanolin only. After 4 days the increase in the length of the internodes of the treated plants is compared with that of the controls. Gibberellins are also active in this test but cause only elongation, whereas BS evoke also swelling, curvature, and splitting of the second internode. Such a response is often called "brassin activity" and can be observed only for BS. The morphological changes were rated by the authors on a 0-5 scale to evaluate the activity of different BS. The effects depend on the amount of BS, and swelling and curvature are visible at an application of 0.01 µg Bl per plant. The sensitivity of the test is increased by omitting nitrate and adding Ca^{2+} or Mn^{2+} to the solution in which the bean seedlings are grown (Kohout et al., 1991).

The bean first-internode bioassay is based on the enhancement of the auxin-induced elongation of the first bean internode (Meudt and Bennett, 1978). This bioassay is sensitive to auxins at picomolar concentrations. BS themselves have little effect on the internodes, but BS and IAA together induce a higher effect than the sum of their individual effects. In a typical procedure a solution of 100 ng of the test compound in ethanol is applied to the first internodes (4-5 cm) of etiolated 6-day-old plants followed in 1 h by the application of 0.1 nM auxin. Control plants are treated with auxin only. The elongation of the internode sections was measured after 1, 2, and 3 h. The BS promotion of the epicotyl growth was observed 1 h after the beginning of the experiment and in 3 h the test compounds [Bl, EBl, (22*S*,23*S*)-Bl and (22*S*,23*S*)-EBl] caused elongation of the epicotyls (5.28, 5.04, 4.80, and 6.82 mm, respectively), whereas control plants treated with IAA only showed an elongation of 3.4 mm (Thompson *et al.,* 1982). The sensitivity of the bioassay can be enhanced by application of the BS to the apical internode tissues (Strnad and Kaminek, 1985). The detection limit of the tested compounds in this improved method is $3 \cdot 10^{-13}$ M compared with $3 \cdot 10^{-10}$ M for the classical procedure.

The rice lamina inclination bioassay is one of the most popular and useful tests for the evaluation of BS activity. Originally it was developed by Maeda (1965) for gibberellins and IAA. For BS it was employed in two modifications: with intact plants (Takeno and Pharis, 1982) and with cuttings (Wada *et al.*, 1981). In the second variant the procedure is performed as follows. Seeds of dwarf rice (*Oryza sativa* L., cv. Kinmaze) are incubated in tap water in the light for 2 days. Germinated seeds are placed on the surface of 1% aqueous agar and incubated in the dark for 7 days with red light irradiation for 1-2 times per day. The leaf segments, each consisting of the second lamina (1 cm long), the lamina joint, and the sheath (1 cm long), are excised, floated on distilled water for 24 h, and incubated in 1 ml of 2.5 mM potassium maleate solution with a known amount of the test compound for 48 h at 30 °C in the darkness. The magnitudes of the induced angles between the laminae and the sheaths were measured after the incubation period; in the control segments the angles are about 90°. In the presence of 0.0005 μg/ml Bl, the angles between laminae and sheaths were about 140°; 0.005 μg/ml Bl caused a sharp lamina bending at the lamina joints to the abaxial side and some of them came into contact with the sheaths. HBl showed comparable activity and IAA evoked only a weak effect even at 50 μg/ml (Wada *et al.*, 1981). Gibberellins give also only a slight effect on the lamina inclination of rice seedlings at concentrations of about 100 μg/ml. The sensitivity of this test varied with the rice cultivars employed. Two cultivars, Arborio J-1 and Nihonbare, were selected as the most reliable ones out of sixty rice cultivars (Wada *et al.*, 1984).

In another modification of this test the bending of the second leaf lamina after treatment of intact seedlings of dwarf rice *Oryza sativa*, cv. Tan-ginbozu or Waito-C is measured. The effect is significant at 100 ng/plant and at higher dosages. Promotion of the second leaf sheath elongation is not so characteristic as for GA_3. The latter, however, has no significant effect on the bending of the second leaf lamina (Takeno and Pharis, 1982). Later, an improved dwarf rice (*Oryza sativa*, cv. Tan-ginbozu) lamina inclination bioassay was developed in which the detection of BS is made more sensitive by synergizing the response to BS with that to IAA. The minimal detectable amount of Bl is less than 0.1 ng/plant (Kim *et al.*, 1990). The rice cultivar Bahia was used in a similar manner for testing a large number of BS with different functionalities in the skeleton and in the side chain (Brosa, 1997).

The mung bean epicotyl bioassay has also been used for the determination of the biological activity of BS (Gregory and Mandava, 1982). For this test mung beans (*Phaseolus aureous* Roxb.) are grown for 9-10 days and plants with epicotyls of 25-35 mm long are selected. Cuttings are prepared with a hypocotyl of 3 cm, the epicotyl, two primary leaves, and the unexpanded terminal bud. These cuttings are placed in vials containing 2 ml of the test solution. The length of the epicotyls is measured at the beginning of the experiment and a final reading is taken after 48 h. The mean length of the epicotyls from the initial and final readings is calculated, and the average increase in epicotyl elongation of the treated plants is compared to that of the controls. The activity of Bl has been investigated in the concentration range from $2 \cdot 10^{-6}$ to $2 \cdot 10^{-12}$ M. The highest growth response (from 86% to 1000%) has been recorded in the range from $2 \cdot 10^{-9}$ to $2 \cdot 10^{-7}$ M. At concentrations of $2 \cdot 10^{-6}$ and higher the Bl activity levels off. This test system is responsive to gibberellic acid but insensitive to auxin. When Bl and GA_3 are applied together, an additive but not a synergistic effect can be detected at Bl concentrations below those for "maximal" growth. Like

BS-treated bean second internodes, mung bean epicotyls show elongation, curvature, and swelling.

The wheat leaf unrolling test was developed by Wada *et al.* (1985). Seedlings of wheat (*Triticum aestivum* L., cv. Norin no. 61) are grown in darkness at 26 °C for 6 days. Leaf segments of 1.5 cm length excised from the region of 1.5-3.0 cm from the leaf tip are incubated for 24 h at 30 °C in darkness in 1 ml of 2.5 mM dipotassium maleate solution containing the test sample. The unrolling of the leaf segments is determined by measuring their widths. The widths of the control segments are about 2.0 mm after incubation. Bl produced a complete unrolling of the leaf segments at concentrations of 0.01 µg/ml and higher. Zeatin had a quite strong activity also, but IAA showed a negative activity in this bioassay.

Radish (*Raphanus sativus*) and **tomato** (*Lycopersicon esculentum*, cv. Giant cherry) **bioassays** were found to be useful for the evaluation of the plant growth-promoting activity of BS and their analogs (Takatsuto *et al.*, 1983b). Plants are grown in wet sand for 7 (radishes) or 12 (tomatoes) days. Then they are transferred carefully into a 4 ml (radishes) or 2 ml (tomatoes) BS solution of known concentration. After incubation at 25 °C (radishes) or 30 °C (tomatoes) in darkness for 24 h, the elongation percentages of the cotyledon petioles and the hypocotyls with respect to the initial length of these organs are measured. Bl causes a significant elongation of the cotyledon petiole and the hypocotyl of young radish at a concentration of 0.01 ppm. Morphogenetic changes (curvature) of the cotyledon petioles and the hypocotyls are observed even at concentrations of 0.001 ppm. In the tomato test both responses take place at a concentration of 0.01 ppm of Bl. Gibberellins and auxins are active in these tests only at a level of 10 ppm and higher.

The above-described test systems are the commonly used bioassays. Examples of their application for monitoring the bioactivity during isolation and purification procedures of natural BS, for the clarification of their biological role, for the estimation of the activity, and for structure-activity relationships of BS and their synthetic analogs are given in the proper chapters. In Table XXXII some characteristic features of these bioassays are presented, such as the type of responses, the sensitivity to BS, and the testing time, so that the corresponding values can be compared.

Since the isolation of Bl, BS have been tested in bioassays, originally designed for auxins, gibberellins, and cytokinins, in order to evaluate the activity of BS, to study their interaction with other plant growth substances, and to estimate the position of BS among other plant hormones. In some cases the BS-elicited responses were similar to those of a particular hormone, but in other bioassays for the same hormone, BS proved to be inactive or gave a different response. Yopp *et al.* (1981) have screened Bl in a number of auxin tests. Like auxin, Bl retarded the hook opening of etiolated bean hypocotyls, induced the elongation of maize mesocotyl segments and segments of azuki bean epicotyls, and increased the fresh weight of slices of Jerusalem artichoke tissue.

The comparison of the responses caused by equimolecular amounts of phytohormones showed that auxin was about twice as active in all these tests, except for one with segments of the azuki bean, which was much more responsive to Bl than to IAA. An application of 10 µM Bl resulted in the same effect as 100 µM auxin. Little or no responses to BS were noticed in the cress root or decapitated pea lateral bud bioassays. Strong synergism was found between BS and IAA in the bioassays involving the elongation and fresh weight increase of dwarf pea hooks.

Bl was also tested (Mandava *et al.*, 1981) in bioassays for GA_3 and kinetin. In the gibberellin tests Bl caused effects similar to those of GA_3 in the elongation of dwarf pea epicotyls and etiolated bean hypocotyls and in the retardation of adventitious root formation in hypocotyls of mung bean and dwarf bean.

TABLE XXXII

Bioassays for BS

Bioassay	Plant system	Response	Limit, M	Time	Reference
bean second internode	seedlings	elongation of the second internode, morphogenetic responses (MR)	10^{-11} 10^{-14}	4 days	Thompson et al., 1981 Kohout et al., 1991
bean first internode	internode sections	increase of auxin-induced elongation	10^{-10} 10^{-13}	3 h	Thompson et al., 1982 Strnad and Kaminec, 1985
rice lamina inclination	cuttings	increase of angles between laminae and sheaths	10^{-13}	4 days	Wada et al., 1981
	seedlings	bending the second leaf laminae	10^{-10} 10^{-14}	3 days	Takeno and Pharis, 1982 Kim et al., 1990
mung bean epicotyl	cuttings	eongation of epicotyl, MR	10^{-12}	48 h	Gregory and Mandava, 1982
wheat leaf unrolling	leaf segments	unrolling the leaves	10^{-12}	24 h	Wada and Kondo, 1985
radish *Raphanus*	seedlings	elongation of cotyledon petiole and hypocotyl, MR	10^{-12}	24 h	Takatsuto et al., 1983
tomato	seedlings	same as above	10^{-11}	24 h	same as above

In cytokinin bioassays Bl was half as effective as kinetin in increasing the expansion of dark-grown cucumber cotyledons but had no effect on the expansion of dwarf pea epicotyl hooks. Bl was only weakly active at 0.1 µM concentration in the stimulation of the lateral expansion of pea apex. Hewitt et al. (1985) evaluated the influence of BS and other growth regulators on the germination and growth of pollen tubes of *Prunus avium*. The effect of Bl was 10-fold higher than that of auxin and GA_3; kinetin did not show any effect.

A test based on the brassinosteroid-induced ethylene production by etiolated mung bean (*Vigna radiata* L. Rwilcz, cv. Berken) segments was developed and used for the evaluation of the activity of BS and their relationships with other growth regulators (Arteca et al., 1983, 1984, 1985).

The synthesis of natural BS and new types of BS analogs created possibilities for extensive investigations and also stimulated the development of new tests for comprehensive and adequate evaluation of the biological activity of BS. Studies of the growth responses to BS application in dwarfed *Arabidopsis thaliana* mutants (Kauschmann et al., 1997) revealed a prominent hypocotyl elongation in the *dwf1-6* (*cbb1* and *cbb3*) mutants, which can be used as a novel highly specific and sensitive biotest for BS.

A new biotest for BS was developed based on the inhibition of the elongation of etiolated pea stems (Kohout et al., 1991). The procedure involves a treatment of young pea etiolated seedlings (cv. Amino) with a lanolin dispersion of the test compound, and as little as 10^{-14} M BS can be detected. The mechanism of the growth inhibition is connected with the ethylene production, which is mediated by BS. A similar effect was observed after application of auxin.

BS were tested in many other bioassays developed for plant growth substances and showed activity at a hormonal level. To obtain reproducible results strict fulfillment of all experimental conditions is necessary. The cultivar of the plants, the conditions of germination and growth, the phase of development, the light conditions, and the mode of treatment were all found to be of the utmost importance for each test.

B. STRUCTURE-ACTIVITY RELATIONSHIPS

Since the discovery of natural BS and the synthesis of their analogs, investigations have been carried out aiming at the evaluation of structure-activity relationships for these compounds. These data help to predict the activity of new analogs and are theoretically important for the elucidation of the mechanism of action and the biosynthesis of BS. Knowledge of the structural requirements for a molecule to possess biological activity creates a basis for the directed synthesis of bioactive analogs.

Structure-activity studies have been carried out using specific bioassays for a large number of natural BS and their analogs. The main modifications of the BS molecule concern ring A (variation in configuration, position, and number of hydroxyl groups, and 2,3-seco), ring B (6-keto, lactones, lactams and sulfo derivatives, exo- and endocyclic double bonds, unsubstituted, and 5,6-seco), *trans* or *cis* junction of AB rings, and the side chain (number of carbon atoms; number, position, and configuration of the hydroxyl groups; other substituents such as alkyl, carboxyl, ether, ester, amide, phenyl, and some other groups; and the double bond). From these studies it became clear that the hydroxyl groups in ring A and the oxygen functionality in ring B and in the side chain are a precondition for biological activity. The influence of a particular modification on the activity can be deduced from data for a group of compounds modified in only one part of the molecule. Such investigations were started soon after the elucidation of the structure of Bl and have continued until today.

The influence of the structure of the side chain for the $2\alpha,3\alpha$-dihydroxy-6-keto and the $2\alpha,3\alpha$-dihydroxy-6-keto-7-oxa BS derivatives on the activity has been studied by Thompson *et al.* (1981, 1982) with the bean second-internode and bean first-internode bioassays to monitor the activity. The obtained data are presented in Table XXXIII. In the tests all Bl stereoisomers (no. 1-4) showed a high activity. The 28-homobrassinolides (no. 5 and 6) showed a smaller activity and compounds without an alkyl group (no. 7-10) showed a weak activity. The order of activity for the substituent at C-24 was $CH_3 > C_2H_5 > H$. Compounds with a 22R,23R-cis-diol group evoked higher responses than their 22S,23S isomers. In the bean first-internode bioassay lactones proved to be more active than ketones.

The results of investigations of the dependence of BS activity on the structure of the side chain in the rice lamina inclination test (Takatsuto et al., 1983a), radish (*Raphanus*) test, and tomato test (Takatsuto et al., 1983b) are summarized in Table XXXIV.

The data presented in Table XXXIII and Table XXXIV show some similarities but also some differences in behavior of the compounds in different tests. Bl (no. 1) proved to be a highly active compound in all tests. In the bean second-internode bioassay the activity of HBl is equal to that of Bl, and (22S,23S,24R)-Bl is slightly less active. The results of the rice test reveal a strong dependence on the procedure employed. In the bioassay with rice segments only HBl (no. 4) has the same activity as Bl, and the other compounds are 2-10 times less active. In the bioassay with intact seedlings of dwarf rice the Bl isomers studied (no. 2 and 3) show comparable activity. (22S,23S,24R)-Bl (no. 3) has the same activity as Bl in the radish test but shows little activity in the tomato test. (22R,23R)-NBl, (no. 8), which shows about 5% of the activity of Bl in the rice lamina inclination test and a very low activity in the bean tests, is just as active as Bl in the radish test and has about 10% of the activity of Bl in the tomato test. However, (22R,23R,24S)-HBl (no. 4) and (22S,23S,24R)-HBl (no. 5), which possess high activity in the rice lamina inclination test, turn out to be much less active in the radish and tomato tests.

TABLE XXXIII
Effects of BS in the Bean First-Internode and the Bean Second-Internode Bioassays

No.	Side chain R	Bean first-internode test[a]		Bean second-internode test, lactone[b]		
		lactone	ketone	10[c]	1	0.1
1	(OH, OH)	547	273	136 (5)	136 (5)	156 (5)
2	(OH, OH)	198	216	237 (5)	178 (5)	228 (4)
3	(OH, OH)	189	171	163 (5)	136 (5)	152 (3)
4	(OH, OH)	249	35	257 (5)	129 (4)	116 (2)
5	(OH, OH)	180	75	411 (4)	317 (4)	278 (3)
6	(OH, OH)	141	9	274 (4)	26 (1)	22 (1)
7	(OH, OH)	25	17	94 (1)		
8	(OH, OH)	7	32	104 (1)		
9	(OH, OH)	33	35	9 (0)		
10	(OH, OH)	211	86	188 (2)		

[a] Growth compared to an auxin-treated control, %.

[b] Elongation (%) and growth responses. Numbers in parentheses mean: 0 - no elongation; 1 - elongation only; 2 - elongation with slight curvature and swelling; 3 - elongation with good curvature and swelling; 4 - elongation with excellent curvature and swelling; 5 - elongation with split internodes.

[c] Concentration of BS, μg/ml.

TABLE XXXIV

Relative Activity of BS in the Rice Lamina Inclination, Radish, and Tomato Tests

No.	Side chain R	Lactone			Ketone		
		Rice[a]	Radish	Tomato	Rice[a]	Radish	Tomato
1		100 (100)	100	100	50		
2		10 (87)	10	10	(75)		
3		5-10 (97)	100	3	(17)		
4		100 (87)	10	1	50 (97)	1	0.3
5		50 (14)	3	1	0.59 (17)	0.01	0
6		10	1	1	1.0	0.1	0
7		5	0.5	1	0.5	0.03	0
8		5	100	10	5.0	3	1
9		1.0	10	1	0.1	0.3	0.3
10		0.1	1	0.1			
11		0.05	1	0.5			
12[b]		100	100		50	10	

[a] Data in parentheses are taken from Brosa et al. (1996a).
[b] Data are taken from Takatsuto et al. (1984a).

For the latter tests, therefore, the replacement of the methyl group at C-24 by hydrogen gives a more active compound than replacement by an ethyl group. For the bean and rice bioassays the opposite situation is seen. With respect to the 22,23-vicinal diols, it is evident from Table XXXIV that cis-(22R,23R)-BS (no. 1, 2, 4, 6, and 8) are generally more active than the corresponding cis-(22S,23S) isomers (no. 3, 5, 7, and 9), and much more active than the trans-diols (no. 10 and 11). Analogs of BS lacking the 22,23-diol function exhibit weak or no activity in the radish and tomato tests. The introduction of a hydroxyl group at C-25 completely suppresses the plant growth-promoting activity. In the case of 6-keto steroids the responses are generally weaker than those of the lactones, although in the bioassay with intact seedlings the keto analog of EB1 and HB1 has the same activity as the corresponding lactones. (22R,23R)-28-Homo-6-keto steroid shows about 50% of the activity of B1 in the rice lamina inclination test but has only little activity in the radish and tomato bioassays. The stereochemical requirements observed for the side chain of BS with a lactone ring B are also true for the corresponding 6-keto analogs.

Several BS analogs were tested for their stimulative effect on ethylene production by etiolated mung bean hypocotyl segments (Arteca *et al.*, 1985). B1 and HB1 showed approximately the same activity. Changes of the configuration at C-22 and C-23 resulted in a decrease of activity at lower concentrations; however, at higher dosages there was no difference between B1, HB1, and (22S,23S)-EB1. The absence of hydroxyl groups in the side chain or the absence of the 7-oxa-function in ring B made the compounds inactive. Cholesterol and aldosterone failed to show any activity in this test.

A number of B1 methyl ethers were prepared in order to prevent glycosylation as a possible deactivating metabolic process that could decrease the activity of BS (Back *et al.*, 1997b). In the rice lamina inclination test tetramethoxy- and 2,3-dimethoxy-B1 proved to be inactive; 22,23-dimethoxy-B1 showed the same activity as EB1.

For natural steroids there are known metabolic routes leading to the breakdown of the side chain and it seems logical to assume that also for BS such metabolic pathways exist. For this reason it was interesting to investigate the bioactivity of BS with a shortened side chain. Especially 26,27-bisnorbrassinolide (Table XXXIV, no. 12) attracted attention due to its easy accessibility and high biological activity. In the radish test 26,27-bisnorbrassinolide stimulated the cotyledon petiole and hypocotyl elongation in the concentration range 10-0.01 ppm, with an effect equal to that of Bl. 26,27-Bisnorcastasterone showed only 10% of the activity of Bl in this test. In the rice test the activity of 26,27-bisnorbrassinolide was equal to that of Bl (Takatsuto *et al.*, 1984a).

A number of steroids with BS-like skeletons were prepared and tested in the bean second-internode bioassay. Compounds without a side chain showed little or no activity regardless of the orientation of the C-2 and C-3 hydroxyl groups or of the character of ring B (6-keto, 6-keto-7-oxa, or 6-oxa-7-keto). The activity of pregnane analogs with one hydroxyl group in the side chain was not high (Kohout *et al.*, 1991). A moderate activity in the mung bean epicotyl bioassay was shown by the 20,22-diethyl ether of hexanorcastasterone (Hazra *et al.*, 1997a). However, when the side chain was terminated at position 22 with a carboxyl group, the modification made compound **623** ten times more active then EBl in the bean first-internode bioassay (Fig. 64). The amides **533** and **624** showed a high activity in this test also. In the bean second-internode test these compounds showed a modest activity (Cerny *et al.*, 1986).

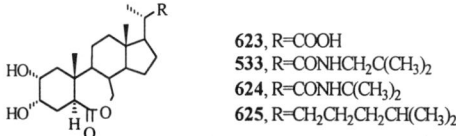

623, R=COOH
533, R=CONHCH$_2$C(CH$_3$)$_2$
624, R=CONHC(CH$_3$)$_2$
625, R=CH$_2$CH$_2$CH$_2$CH(CH$_3$)$_2$

Fig. 64. BS analogs with modified side chains.

A small elongation of the bean second internode was noticed for BS analogs with an unfunctionalized side chain, such as **625** (Thompson et al., 1982). In the bean first internode, the rice lamina inclination, the *Raphanus*, and the tomato bioassays it evoked practically no responses (Takatsuto et al., 1983b; Mandava and Thompson, 1984).

The influence of the functionality of AB rings on the biological activity of BS was examined with the rice lamina inclination test. Wada and Marumo (1981) investigated a number of steroids with the side chain of (22S,23S)-HBl (Table XXXV, no. 2-7). The most active compound proved to be (22S,23S)-HBl (no. 2), which showed 10% of the activity of Bl. The compound with a 3β-hydroxyl group in ring A (no. 3) was 10-20 times less active. Other analogs had very little activity, which confirmed the important role of the 2α,3α-diol group and the lactone ring. The same conclusion can be drawn from the results obtained by Takatsuto and Ikekawa (1984b) during investigations of HBl derivatives (Table XXXV, no. 20-25). These data show the importance of the B-homo-7-oxa-6-keto ring in the molecule for the biological activity. The compound with an isomeric lactone ring proved to be less active (no. 21); the compound with a 3α,4α-diol group (no. 23) showed, however, a rather high activity.

A large set of HBl and (22S,23S)-HBl derivatives with different configurations of chiral centers in the AB rings were tested in the rice lamina inclination bioassay with the cultivar Bahia in order to obtain comparable data about their activity (no. 8-14 and 26-32) (Brosa et al., 1996a). Of all compounds studied the highest activity, equal to that of HBl, was observed for the isomer with an AB *cis* junction and a 2β,3β configuration of the diol group in ring A (no. 27); the corresponding 6-ketone (no. 29) was slightly less active. All HBl analogs proved to be more active than the (22S,23S)-HBl derivatives.

TABLE XXXV

Relative Activity of (22S,23S)-HB1 and HB1 Derivatives in the Rice Lamina Inclination Test

No.	Compound	Relative activity	No.	Compound	Relative activity
1	Bl	100			
2[a]		10	20[c]		100
3[a]		1-0.5	21[c]		1
4[a]		0.1	22[c]		50
5[a]		0.1	23[c]		50
6[a]		0.1	24[c]		10
7[a]		0.1	25[c]		10
8[b]		7	26[b]		6

(continues)

TABLE XXXV (continued)

No.	Compound	Relative activity	No.	Compound	Relative activity
9[b]		6	27[b]		87
10[b]		11	28[b]		27
11[b]		7	29[b]		51
12[b]		11	30[b]		66
13[b]		0	31[b]		0
14[b]		6	32[b]		0
15[d]		0.1	33[d]		1
16[d]		0.1	34[d]		0.1

(continues)

TABLE XXXV (continued)

No.	Compound	Relative activity	No.	Compound	Relative activity
17[e]	(structure: HO, HO, O)	8	35[d]	(structure: HO, HO, O, H)	1
18[e]	(structure: HO, HO, O, O)	10	36[d]	(structure: HO, HO, HO, N, H)	0.01
19[e]	(structure: HO, HO, COOMe)	5			

Data are taken from: [a]Wada and Maruto, 1981; [b]Brosa et al., 1996a; [c]Takatsuto and Ikekawa, 1984b; [d]Kishi et al., 1986; [e]Adam et al., 1991.

2,3-Seco-BS (no. 17 and 18) showed about 80-100% of the activity of 22S,23S-HB1, while the 5,6-seco derivative (no. 19) was less active, probably owing to the more substantial conformational changes in the molecule with an open ring B (Adam et al., 1991).

The importance of proper functionalities in ring B was demonstrated further (Kishi et al., 1986) with modified analogs with other heteroatoms in ring B (Table XXXV, no. 15, 16, and 33-36). The nitrogen analog (no. 33) showed 1% of the activity of Bl, the 6-deoxy analog (no. 35) had the same activity, and the activity of the thio analog (no. 34) was only 0.1% of that of HB1. On the other hand, the nitrogen (no. 15) and sulfur (no. 16) derivatives of (22S,23S)-HB1 showed the same level of activity, corresponding to 0.1% of the activity of Bl. The last compound in this series, a lactam (no. 36) with an unfunctionalized side chain, was practically inactive.

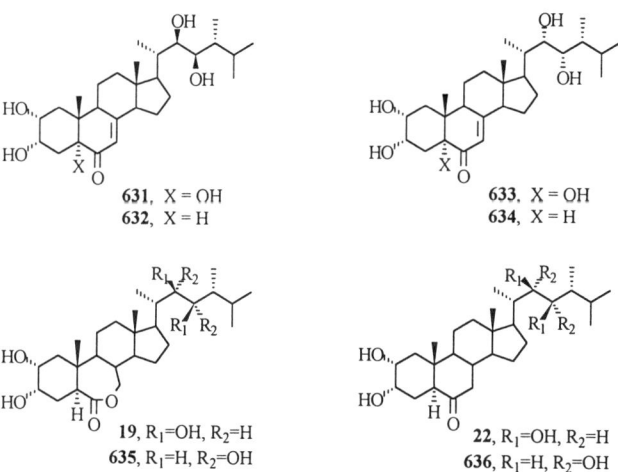

Fig. 65. Bl analogs with modified ring B.

A similar situation took place for the Bl analogs **626-630** modified in ring B (Fig. 65). The activities of lactam **626** and cyclic ether **628** were comparable to that of EBl in the rice lamina inclication test, whereas thiolactone **627** and ketone **629** were less active. When no oxygen functionality in ring B was present, the compound **630** had no activity at all (Back *et al.*, 1997b).

Compounds **631-634** (Fig. 66) with a double bond in ring B were investigated in the rice lamina inclination test, together with the corresponding lactones and ketones **19, 22, 635**, and **636** (Takatsuto *et al.*, 1987a). EBk **22** showed the same activity as EBl **19**, which is 10 times less active than Bl. Introduction of a double bond in ring B led to decreasing activity, and compound **632** was 10 times less active than **19**. The decrease of the activity was even more substantial (1000 times) when the hydrogen atom at C-5 was replaced by a hydroxyl group (compound **631**). Similar results were obtained for the 22*S*,23*S* isomers **633** and **634**, but the magnitude of the activity was lower. Thus, the introduction of a double bond at C-7 and of a hydroxyl group at C-5 (elements of the ecdysone structure) led to a remarkable diminution of the growth-promoting activity. Ecdysterone itself showed little or no activity in the selected plant growth bioassays (Dreier and Towers, 1988).

Wang, Y.Q. *et al.* (1994b) studied the biological activity of B1 analogs modified in the side chain (phenyl, furyl, isopentene, or carboxylic acid substituents at C-23) and in the AB rings (2-deoxy, 6-keto) of the molecule (Fig. 67). Some of these compounds, for example, the ketone **545** and the lactone **546** with a phenyl group in the side chain, proved to be more active than EB1, which was used as a standard. The activity of the compounds **544** and **637**, without the hydroxyl group at C-2, was less and approximately equal to that of EB1.

From the data above it can be concluded that the relative activity of BS is different in different bioassays. This is in particular the case in bioassays on plant segments compared with those on intact plants. Changes in the order of activity are sometimes observed even for the same compounds in the same

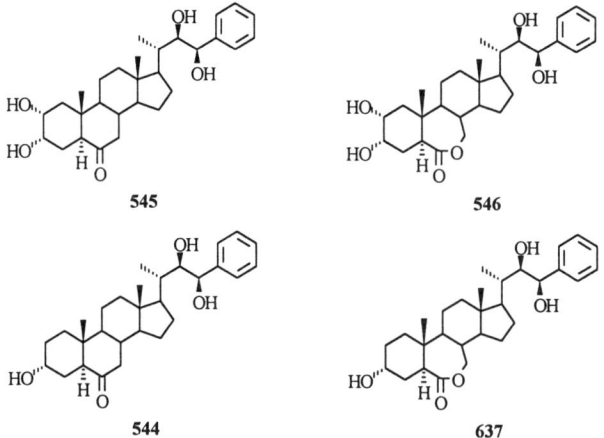

Fig. 67. BS analogs with a phenyl substituent in the side chain.

bioassays but obtained by different authors, due to the stringent dependence of the responses on the test conditions such as the environmental conditions of growth and treatment of plants, the mode and time of BS treatment, the plant cultivar, and the mode of estimation. For this reason the evaluation of the activity of new analogs is usually performed parallel with some known BS (Bl, EBl, HBl), which are used as reference compounds.

The data obtained in bioassays, especially with plant segments, should be considered with care when they have to serve as a basis for the selection of compounds for field trials. Metabolic transformations of exogenic steroids in plants can lead to the deactivation of the compound and to lack of activity in field conditions. On the other hand, some compounds can be transformed *in vivo* into active metabolites and show higher growth-promoting activity in field conditions than in test systems. An example of such a behavior is the compound **638** (Fig. 68) (Kakiuchi *et al.*, 1988). In the rice lamina inclination test it showed no activity, while Bl showed a high activity. The results of testing under field conditions were different. In 10 days after treatment of *Raphanus* plants, Bl showed higher activity and the increase of fresh weight was 128% compared to the control. The increase of fresh weight caused by epoxide **638** at that time was only 104%. In 25 days, however, compound **638** showed higher activity in the enhancement of fresh weight, up to 112%, while Bl induced an enhancement of 109%. These results could mean that epoxide **638** is transformed in a plant into a more active compound after some days, whereas Bl is deactivated in the

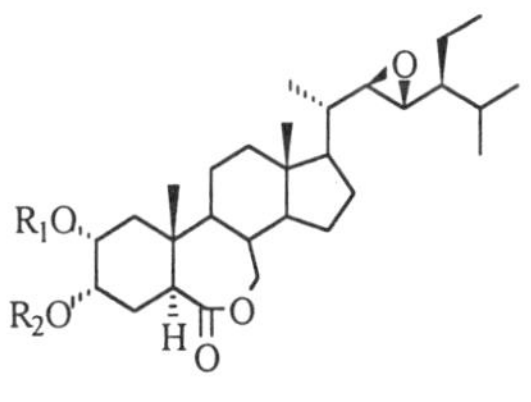

638, $R_1, R_2 = C(CH_3)_2$
639, $R_1 = R_2 = Ac$

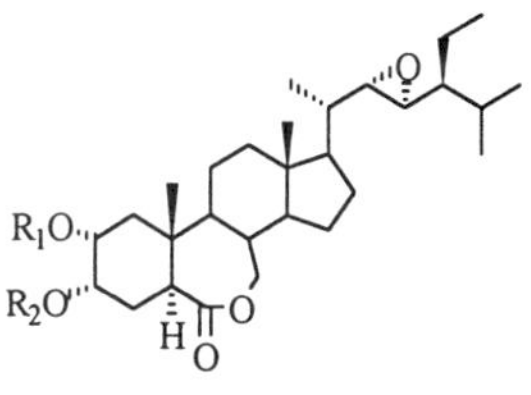

640, $R_1, R_2 = C(CH_3)_2$
641, $R_1 = R_2 = Ac$

Fig. 68. BS analogs with a 22,23-epoxy ring.

meantime. Later, a larger group of compounds, including Bl, HBl, the tetraacetate of HBl, the acetonides **638** and **640**, and the diacetate **639** were tested in a similar manner and comparable results were obtained (Kamuro and Takatsuto, 1991).

A comparative study of the influence of two 22,23-epoxy-BS **639** and **640** and Bl showed that all three BS effectively promoted the elongation of the first trifoliate leaf of soybean. Both epoxy-BS, but not Bl, also elongated the second internode and were more effective in plant height enhancement (Takatsuto *et al.*, 1989a).

The biological testing of a large number of BS and their synthetic analogs elucidated structural peculiarities for these compounds that are important for their activity as plant growth promoters (Mandava, 1988). Generally they are as follows: (1) In the side chain a *cis* configuration of two oxygen-containing substituents, usually hydroxyl groups, gives the highest activity, although minor deviations can be tolerated. Compounds with the $22R,23R$ configuration usually show higher activity. (2) The presence of an alkyl group at C-24 is important; the generally observed order of activity is $CH_3 > C_2H_5 > H$. (3) In the cyclic part of the molecule a 7-oxalactone or a 6-keto function in ring B is necessary. The lactones are usually more active. (4) The highest activity is observed when two α-orientated hydroxyl groups are present at C-2 and C-3 in ring A. (5) *trans*-AB ring junction is more common in active compounds.

These requirements are valid for the majority of the investigated BS and analogs, but as more natural BS are isolated and more new analogs synthesized, more refinements in this list will appear. Bl is the most active BS and traditionally it is considered as a standard for the estimation of the activity of other compounds. This molecule was considered to have the optimal structural parameters for growth-stimulating activity. Rather unexpected and difficult to explain in this connection is the high biological activity of a number of compounds whose structures differ substantially from that of Bl. Among these are the compounds with a shorter or without a side chain or with a modified cyclic part.

The importance of certain structural features for BS activity is in accordance with the hypothesis about the mediation of BS action in plants by the protein receptor. Zurek *et al.* (1994) investigated the ability of a number of BS to enhance the soybean epicotyl elongation and to regulate the expression of BRU1, a gene which is regulated specifically by active BS. It was shown that BS caused an increase in BRU1 expression in soybean epicotyls concomitant with an increase in the epicotyl length. Bl, EBl, (22S,23S)-EBl, and HBl evoked similar effects in both tests. The 22S,23S isomer of HBl was 10 times less active and other synthetic analogs with a 6-keto or a 5,6-seco ring B and a modified side chain were less active. An analysis of the interatomic distances in the calculated energy-minimized conformations of the investigated BS showed the similarity of the corresponding distances between C-16 on the ring and O-22 and O-23 in the side chain and C-16 and the carbon atoms of the side chain for all active BS, but not for (22S,23S)-HBl, which differed significantly. This means that the overall dimensions of the side chain are more important than the configuration of the individual chiral carbon atoms.

Recently a new approach based on molecular modeling was reported for the evaluation of structure-activity relationships (Brosa *et al.*, 1996a). It was assumed that the activity of BS depends mainly on the spatial position of the oxygen atoms. The mutual orientation of the oxygen atoms at C-2, C-3, C-22, C-23, and C-6, which might be involved in the receptor-ligand interaction (angles and distances), was considered as an enthalpic factor and the number of possible conformers of the molecule as an entropic factor, with both factors determining the activity. Since Bl is most active, its conformation was taken as a reference for other compounds. The higher activity should be observed for the compounds whose spatial arrangement is closer to that of Bl. This method can explain the high activity of some analogs that do not meet the structural requirements formulated above. An example is the 2β,3β-diols with an AB-*cis*-ring junction (Table XXXV, no. 27 and 29).

Nowadays the level of knowledge on structure-activity relationships allows one to explain the observed facts and mostly to predict qualitatively the activity of a particular compound, if data on its close analogs are available. The whole picture of the structure-activity relationships of BS is not yet complete and needs further improvement. The investigations on the biosynthesis of BS in plants and on the mechanism of their action contribute to further understanding the structural peculiarities that are responsible for the specific bioactivity of BS.

CHAPTER **XI**

PRACTICAL APPLICATIONS AND TOXICOLOGY

A. APPLICATIONS OF BS

In previous chapters it was shown that BS have a strong influence on plant growth and possess an exclusive combination of properties which are very promising for practical use. This is the reason why BS have been considered for application in agriculture since their discovery (Maugh, 1981) and their economic value as crop-increasing agents was predicted (Cutler, 1991). Moreover, by the time BS were discovered, the necessity of new types of growth regulators with a favorable action on plants and no interference with the environment became evident. The application of BS in agriculture and horticulture is based on their ability to increase crop yield and to stimulate physiological processes in plants. As a result it may become feasible to grow crops under unfavorable conditions, such as a high salinity, drought, or insufficient nutrients. The results of practical application of BS in field conditions were treated and summarized in a number of reviews especially dedicated to this subject (Fujita, 1985; Abe, 1989; Mandava, 1991; Ikekawa and Zhao, 1991; Takeuchi *et al.*, 1992; Khripach *et al.*, 1995c, 1997; Prusakova and Chizhova, 1996; Takeuchi and Kamuro, 1997).

BS possess some specific features which are of a great potential value for their practical application:

- BS are natural products and widespread in the plant kingdom. They are not strangers on the earth but are included in the food chains of men and mammals, with whom their biosynthetic and metabolic pathways were interconnected during a long common evolution.

- Plants respond to very small doses of BS (5-50 mg/ha), comparable with their natural content.

- As plant growth-promoting substances BS have a broad spectrum of stimulative and protective activity, which has a positive effect on the quantity and quality of crops.

- BS increase plant resistance against phytopathogens and can be used as substitutes (total or partial) for some traditional pesticides. In this way the unfavorable interaction of pesticides with the environment can be diminished.

-BS can be applied for the treatment of plants or seeds using existing equipment and technologies.

The influence of BS on the development and crop yield of different agricultural plants such as cereals, legumes, vegetables, fruits, and others was tested in laboratory and field conditions. For large-scale field application two modes of BS treatment are possible: soaking seeds and foliar spray. Both methods were investigated extensively, but the results with the latter method were found to be very dependent on the phase of plant development. Generally, better results were obtained when young plants were treated. The formulation of the spraying solution is very important. Additives are necessary to facilitate the spreading of the active substance and to prevent early drying to ensure the penetration of BS *via* cell walls.

Bl, EBl, HBl, and some BS analogs were tested in field trials to estimate their influence on plant growth and development and on crop yield in natural conditions. As pointed out elsewhere, the results obtained in field trials do not always coincide with those estimated from the bioassays. First of all, some bioassays deal with parts of a plant, which can react differently than the whole plant. Even when intact seedlings are used in the test, the period of time between the treatment and the recording of the responses is not long (from a few hours to a few days) whereas in field conditions the effect of the treatment may last the whole life of the plant. Moreover, in bioassays plants in a strictly defined phase of development are used, whereas in field conditions there are always plants that are more or less developed than the average.

The stability of BS is considered to be important for field trials. For this reason intensive investigations have been carried out to create compounds with long-lasting activity (Kamuro and Takatsuto, 1991). One such compound, termed TS303 [($22R,23R$)-epoxy-HBl 2,3-diacetate], revealed beneficial effects on germination, growth, fruit-setting, crop yield, cold stress resistance, and root formation for a number of agricultural plants (Takatsuto *et al.*, 1996c; Kamuro *et al.*, 1997) and is now being officially tested for registration in many countries. It also showed promising effects in combination with other plant growth regulators (Kamuro *et al.*, 1996).

Experiments to investigate the potential use of BS in agriculture were started in the seventies in the U.S. and in the early eighties in other countries where BS were available, mainly in Japan and in the USSR. Since that time numerous reports from all over the world have appeared and many findings for potential practical use are patented. In Belarus and in Russia production of the plant growth promoter Epin®, with EB1 as an active substance, has been organized. Epin® has been officially registered in Russia and in Belarus since 1992 (Chemicals for Plant Protection, 1995; List of Pesticides, 1997a,b) and recommended for treatment of different agricultural plants such as tomato, potato, cucumber, pepper, barley, and some others. Farmers in these countries have acknowledged benefits of its application as an effective environmentally friendly helper (Kositsina-Pinegina, 1995; Fedoseeva, 1996; Victorova, 1996; Ugarova, 1997; Pilneva and Shcherbina, 1997). Selected examples of BS trials on different agricultural plants are presented in Table XXXVI.

1. Cereals

Among the cereals rice was investigated mainly in Japan and China because of its significance for food production in these countries. An application of B1, HB1, or EB1 as a foliar spray or by feeding *via* roots gave an enhancement of crop yield by 10-15% (Nippon Kayaku Co., 1988). An acceleration of the ripening of rice plants treated with B1 was observed with an especially pronounced effect for plants kept in cold conditions (Fujii *et al.*, 1991; Hirai *et al.*, 1991). EB1 applied as a foliar spray at flowering time diminished cold and salt depression of plant growth (Takematsu and Takeuchi, 1989). An acceleration of rice growth by BS is described in patents (Hirai and Fujii, 1985; Igai *et al.*, 1985; Hirai *et al.*, 1986a; Takematsu *et al.*, 1986). B1 treatment reduced herbicidal injury (Takematsu, 1986a,b; Choi, C.D. *et al.*, 1990a) and improved the salt tolerance of rice plants (Hamada, 1986; Dong *et al.*, 1989). The combined application of BS with auxin (Genma, 1987a) or quinoline salts (Kajita *et al.*, 1985) was shown to enhance the influence of BS.

The BS influence on wheat has been extensively studied. Takematsu and Takeuchi (1989) investigated the action of EB1 and (22S,23S)-HB1, applied as a

TABLE XXXVI
Selected Applications of BS

Plant	Region	BS, dose, mode of treatment	Crop enhancement; other effects	References
rice	Japan	BI, 10^{-2}-10^{-4} ppm, spraying on leaves	10%	Nippon Kayaku Co., 1988
		EBI, 10^{-2} ppm, feeding *via* roots	12%	
wheat	Japan	BI, 10^{-4}-1 ppm, spraying at flowering	5-10%	Nippon Kayaku Co., 1988
		HBI, 10^{-3} ppm, spraying at flowering	24-37%	
	Kazachstan	EBI, 10 mg/ha, spraying at flowering	15%	CIASA, 1989
	Russia, Moscow region	EBI, 0.4 mg/ha, spraying at flowering	17%	CIASA, 1989
		EBI, 4 mg/ha, spraying at flowering	16%	
		EBI, 10 mg/ha + choline chloride	63%	
	The Ukraine	EBI, 2.5 mg/ha, spraying at heading	6%	Grinchenko and Belokon, 1991
	China	EBI	6-15%; disease suppression	Ikekawa and Zhao, 1991

(continues)

TABLE XXXVI (continued)

Plant	Region	BS, dose, mode of treatment	Crop enhancement; other effects	References
wheat	Russia, Kurgan region	EBl, 100 mg/t, soaking seeds	18%; diminishing chill injury	Nemchenko, 1993
rye	Belarus	EBl, 5 mg/ha, spraying at flowering	20%	CIASA, 1989
barley	Belarus	Bl, 10 mg/ha, spraying at flowering	13%	Streltsova, 1988
		EBl, 50 mg/ha, spraying at flowering	25%	
	Russia, Moscow region	EBl, 10 mg/ha, spraying seedlings (a)	17%	CIASA, 1989
		EBl, 10 mg/ha, spraying at flowering (b)	10%	
		EBl, 10 mg/ha, spraying (a) and (b)	15%	
	The Ukraine	EBl, 50 mg/ha, spraying at tillering	12%	Grinchenko and Belokon, 1991
		HBl, 50 mg/ha, spraying at tillering	9%	
	Belarus	EBl, 1 mg/ha, spraying at tillering and booting	21% (1990); 5% (1992); disease suppression	Volynets et al., 1997a
		EBl, 10 mg/ha, spraying as above	21% (1990); 15% (1992)	

(continues)

TABLE XXXVI (continued)

Plant	Region	BS, dose, mode of treatment	Crop enhancement; other effects	References
corn	The Ukraine	EBl, 50 mg/ha, spraying at 4-5 true leaves and at emergence of tassel	11%; growth stimulation, increase of lysine and tryptophan content	Tsibulko and Buriak, 1991
		HBl, 50 mg/ha, spraying as above	17%; effects as above	
	China	EBl, spraying before emergence of tassel	18%; reduction of kernel abortion at the tips of the ear	Ikekawa and Zhao, 1991
	The Ukraine	EBl, 10^{-3}-10^{-4}%, treatment of seeds	10%	Grinchenko and Belokon, 1993
		EBl, 50 mg/ha, spraying at emergence of tassel	10%	
		HBl, 50 mg/ha, spraying as above	15%	
buck-wheat	The Ukraine	EBl, 50 mg/ha, spraying at flowering	11% (1988); 11% (1990)	Tsibulko and Buriak, 1991
		HBl, 50 mg/ha, spraying at flowering	10% (1988)	
	Belarus	EBl, 10^{-8}%, treatment of seeds	29% (cv. Chernoplodnaya); 6% (cv. Smuglianka)	Pavlova and Deeva, 1995
		EBl, 50 mg/ha, spraying at 3 true leaves and budding	35% (cv. Chernoplodnaya); 22% (cv. Smuglianka)	

(continues)

TABLE XXXVI (*continued*)

Plant	Region	BS, dose, mode of treatment	Crop enhancement; other effects	References
buck-wheat	Russia, Moscow region	EBl, 10^{-8}%, soaking seeds and spraying at flowering	29%; ripening 10 days earlier	Runov et al., 1997
oats	The Ukraine	EBl, 20 mg/ha, spraying at tillering	7%	Grinchenko and Belokon, 1991
		HBl, 20 mg/ha, spraying at tillering	4%	
soybean	Japan	EBl, 10^{-1}-10^{-2} ppm, spraying on leaves	9-21%	Nakaseko and Yoshida, 1989
	The Ukraine	EBl, 50 mg/ha, spraying at flowering	3% (cv. Harkovskaya; 1988); 12% (cv. Belosnezhka; 1988)	CIASA, 1989
	The Ukraine	EBl, 50 mg/ha, spraying at flowering	5% (cv. Belosnezhka; 1989)	Tsibulko and Buriak, 1991
		HBl, 50 mg/ha, spraying at flowering	3% (cv. Belosnezhka; 1989)	
pea	The Ukraine	EBl, 50 mg/ha, spraying at bushing and budding	12% (1988); 6% (1989); 9% (1990)	Tsibulko and Buriak, 1991
		HBl, 50 mg/ha, spraying as above	5% (1989); 7% (1990)	
lupine	Belarus	Bl, 10^{-9} M, spraying at flowering	20%; increase of protein content	Mironenko et al, 1996

(*continues*)

TABLE XXXVI (continued)

Plant	Region	BS, dose, mode of treatment	Crop enhancement; other effects	References
lucerne	Russia, Voronezh region	EBl, 10 mg/ha, spraying at bushing and flowering	6%, increase of no. of clusters/sq. m and no. of seeds/pod	Kadyrov et al., 1995
		EBl, 100 mg/ha, spraying as above	15%	
potato	Russia, Krasnodar region	EBl, 10 mg/ha, spraying at budding	11%	Gonik and Mikhailova, 1992
		EBl, 20 mg/ha, spraying at budding	18%	
	Belarus	EBl, 20 mg/ha, spraying at budding	14% (1991); 13% (1992-1994); disease suppression, quality increase	Savelieva and Karas, 1993; Savelieva et al., 1995; Khripach et al., 1995g, 1996d
		HBl, 20 mg/ha, spraying at budding	16% (1989); 19% (1991); 8% (1992-1994); other effects as above	
tomato	Moldova	EBl, soaking seeds and spraying at flowering	30% (cv. Fakel); 15% (cv. Utro)	Kirillova et al., 1993

(*continues*)

TABLE XXXVI (continued)

Plant	Region	BS, dose, mode of treatment	Crop enhancement; other effects	References
tomato	Russia, Voronezh region	EBl, 25 mg/t, soaking of seeds (a)	26%; disease suppression	Churikova and Derevshchukov, 1997
		EBl, 25 mg/ha, spraying at flowering (b)	33%; disease suppression	
		EBl, twice: (a) and (b)	45%; disease suppression	
	Belarus	EBl, 20 mg/ha, spraying at budding	10%	Savelieva et al., 1997
		EBl, 20 mg/ha, spraying at setting	18%	
cucumber	Russia, Voronezh region	EBl, 25 mg/t, soaking of seeds (a)	43%; earlier flowering and as (b)	Churikova and Derevshchukov, 1997
		EBl, 25 mg/ha, spraying at flowering (b)	47%; disease suppression	
		EBl, 25 mg/ha, spraying twice: (a) and (b)	58%; same as (a) and (b)	
sugar beet	Germany	(22S,23S)-HBl, 10^3 mg/ha, spraying at 12 leaves	5% (1989); 7% (1990); sugar content increase 8% and 6%, resp.	Schilling and Schiller, 1991
	Belarus	EBl, 25 mg/ha, spraying at 10-13 leaves	8% (1991); 5% (1992); sugar content increase 10-13%	Kurganskii et al., 1993
		EBl, 50 mg/ha, spraying at 10-13 leaves	6% (1991); 8% (1992); sugar content increase 10-13%	

(continues)

TABLE XXXVI (continued)

Plant	Region	BS, dose, mode of treatment	Crop enhancement; other effects	References
sugar beet	Russia, Krasnodar region	EBl, 50 mg/ha, spraying at 2-3 leaves	7%; sugar content increase 10%	Gonik and Al-Zhadavi, 1993
	Belarus	EBl, 10^{-2} ppm, treatment of seeds	26%	Vedeneev et al., 1995
cotton	Uzbekistan	EBl, 10 mg/ha, spraying at budding	26%	CIASA, 1989
		EBl, 20 mg/ha, spraying at budding and flowering	32%	
rape	China	Bl, 0.01-0.5 ppm	25%; decreasing frost injury	Xia et al., 1992
tobacco	China	EBl, 0.01 ppm, spraying leaves	20%; increase of nicotine content	Ikekawa and Zhao, 1991
	Russia, Krasnodar region	EBl, 30 mg/ha, spraying at budding	19%; quality increase	Gonik and Mikhailova, 1992
grape	Moldova	EBl, 25 mg/ha, spraying at budding	27% (cv. Pino Black); 24% (cv. Traminer)	Chirilov et al., 1996

foliar spray in the flowering stage at a concentration of 0.01-0.1 ppm. The ear weight of the treated plants was 20-30% higher for EB1 and 10-20% for (22S,23S)-HB1. The number of seeds per ear was increased by 30% and 20%, respectively. In another paper the inhibition of main stem growth of wheat (up to 20%) is reported when sprayed at a dose of 10 mg/ha at flowering time (Prusakova and Chizhova, 1991). At the same time, the grain yield was 800 kg/ha higher due to the lengthening of the main shoot ear and increase in the number of ears and grains per ear. All this may be the consequence of the translocation of assimilates from the vegetative organs to the ears. The influence of the mode and time of B1 treatment on the growth, productivity, and physiological processes in wheat plants was studied by Luo *et al.* (1986, 1992). Treatment before anthesis decreased the yield mainly due to a decrease of productive tillers, but treatment from anthesis to maturity increased the grain yield by improvement of the grain set and the grain filling. Experiments conducted at research stations and in farmers' fields in India during 1989-1995 revealed that HB1 treatment of wheat and rice at the tillering and spike-panicle initiation phases significantly increased the grain yield (Ramraj *et al.*, 1997).

Some of the results on the influence of BS on wheat crop yield obtained in field trials at various places are presented in Table XXXVI. In several regions of China with different climate and with different wheat species, 5-15% increases in crop yield were obtained after application of HB1 (Ikekawa and Zhao, 1991). Yang *et al.* (1992), however, reported an insignificant influence of EB1 on the crop yield of winter wheat in field trials, while common application of EB1 together with triadimefon increased the yield by 18-20%. In 1986-1988 the Central Institute of Agrochemical Service for Agriculture (CIASA) organized field trials of EB1 in different regions of the former Soviet Union. Good results were obtained on wheat in places with drought conditions (North Kazachstan, cv. Vera; Moscow region, cv. Mironovskaya 808). In the Ukraine, where the soil is more rich, the crop increase was not so dramatic. A very high crop yield enhancement (up to 62%) was obtained when EBl was applied together with the growth retardant chlorocholine chloride. The latter alone increased the yield by 33%. The crop of winter rye was increased by 20% after treatment of the plants with EBl in the phase from heading till the beginning of flowering (CIASA, 1989).

The field trials with EBl on barley plants (cv. Zazersky 85) revealed the dependence of the crop enhancement on the time of the treatment. The best results were obtained when young seedlings were sprayed. The same cultivar of barley when treated with Bl or EBl was more sensitive to Bl at a lower concentration, but at the optimal dosage EBl gave a better crop increase (Streltsova, 1988; Khripach et al., 1991d). Spraying barley plants with EBl or HBl in the booting phase enhanced the stem strength and its resistance to lodgeability (Prusakova et al., 1991, 1993, 1995). The crop yield enhancement of barley was achieved together with diminishing leaf diseases (Volynets et al., 1997a). The advantage of the application of BS in areas that are polluted with heavy metals or radioactive debris is connected with their ability to diminish the accumulation of these contaminants in the crop (Khripach et al., 1995g, 1996a). Treatment of wheat and barley with BS enhances their resistance to stress conditions, such as salinity, heat shock (Khokhlova et al., 1990; Bokebayeva et al., 1990; Kulaeva et al., 1991), and moisture stress (Sairam, 1994).

Since corn is an important agricultural plant in many countries, the effects of BS on its cultivation were investigated. The results of trials with EBl in China from 1986 to 1988, summarized in a review by Ikekawa and Zhao (1991), showed 10-20% yield increase in a majority of the tests. The time of treatment was important, although spraying the ear and silk as well as spraying before emergence of the tassel of corn plants gave good results. Application of EBl as a foliar spray was shown to increase the productivity, and soaking the seeds improved germination. The results of field trials in the Ukraine have shown that treatment of plants with EBl or HBl solution resulted in a 5-21% yield enhancement with an increase of the content of lysine and tryptophan (Tsibulko et al., 1991; Grinchenko and Belokon, 1993).

An application of EBl and HBl to buckwheat gave approximately 10% enhancement of the seed yield (Tsibulko and Buriak, 1991; Tsibulko et al., 1993). Pavlova and Deeva (1995) reported the dependence of the crop increase on the cultivar of buckwheat and on the mode of EBl application. Spraying in the budding phase was preferable and gave a better effect than treatment of the seeds. Treatment with EBl resulted in up to 10% enhancement of the protein content in seeds. An application of BS to oat plants caused a modest crop increase (Grinchenko and Belokon, 1991).

2 Legumes

An application of EBl to soybean by spraying the leaves, onto the stems, or by injection into the soil resulted in increasing the leaf, pod, and total dry weight by 18, 40, and 10%, respectively. The seed yield of plants treated *via* the leaves or the soil was 9-21% higher, mainly due to an increase of the pod number and the weight of 100 seeds (Nakaseko and Yoshida, 1989). In contrast, Luo (1986) reported for soybean (cv. Enrei), treated with Bl a decrease in yield by 8% due to a decrease of the pod number on the main stem by 16-18%, while the pod number on the branches increased by 11-12%. In the application of EBl a dependence of the result on the cultivar of soybean was observed (Tsibulko and Popov, 1991; CIASA, 1989).

The influence of EBl and HBl on pea was studied also (Tsibulko and Popov, 1991). For the plants treated once in the budding phase or twice in the phase of 6-9 true leaves and in the budding phase a stimulation of growth was observed. The yield enhancement of seeds was achieved mainly due to an increase of the weight of 1000 seeds. The results were practically the same for both BS and independent of the number of treatments. Lucerne plants treated with HBl or EBl grew about 6-8% higher than untreated plants, the number of flowers was 4-8%

more, and the number of setting pods increased also, which resulted in a 7-26% seed yield enhancement. Employment of higher doses of EBl retarded the growth of the plants but increased the seed yield (Kadyrov *et al.*, 1995). The influence of EBl on greengram (*Vigna radiata* (L.) Wilczek) was investigated in two ways. When the seeds were soaked, the shoot growth was enhanced, but the root development retarded. Foliar spray at concentrations ranging from 0.0001 to 0.01 ppm increased the yield per pod, the seed test weight, and the yield per plant, whereas a 1 ppm foliar spray reduced the yield (Takahashi *et al.*, 1994). In experiments with lupine (cv. Narochanskii) spraying the plants with a 10^{-9} M Bl solution resulted in plant growth enhancement and gave a 20% increase of crop yield (Mironenko *et al.*, 1996). Field trials of BS on broad bean in Egypt over two growing seasons (1993-1995) revealed a significant increase in growth and yield. The content of growth promoters such as auxins and gibberellins in the treated plants was higher than in the control, whereas the amount of growth inhibitors diminished (Helmy *et al.*, 1997).

3. Potato and Vegetables

BS were shown to increase the size of potato tubers (Adam *et al.*, 1987). The treatment with (22*S*,23*S*)-HBl at a dose of 0.3 g/ha resulted in a 24% increase of the tuber fraction with a diameter of more than 6 cm. The growth-promoting effect of BS on potato is also claimed in a few patents (Genma, 1987b; Nat. Fed. Agric. Co-Op. Assoc., 1985). Spraying potato plants with a Bl solution (10^{-2}-10^{-4} ppm) 3 times at 1-week intervals led to an increase of the mean tuber weight from 100 to 145 g. It was shown that EBl treatment of potato plants by spraying in the budding phase or on the tubers before storage decreased the sprout length and diminished injury by phytopathogenic microorganisms. This effect may be caused by changes in abscisic acid and ethylene level in the treated tubers (Kazakova *et al.*, 1991; Korableva *et al.*, 1992, 1998; Platonova and Korableva, 1994). Field trials of EBl and HBl on potato plants from 1989 to 1994 revealed a steady crop enhancement due to the increasing number of big tubers (Savelieva et al., 1995). Crop was of better quality in regard to diminishing nitrate content and the enhancement of starch and vitamin C content; diminishing phytophtora infection was also observed (Savelieva and Karas, 1993, 1995; Khripach et al., 1996c,d; Vasyukova et al., 1994). BS were found to have a beneficial effect also on the reproduction of potato plants. When cuttings were grown in a nutrient medium that contained Bl, EBl or HBl, the yield of cuttings suitable for planting was increased by 25-50% depending on the cultivar. The stimulative effect lasted during the first generation of plants, resulting in up to 50% enhancement of the tuber yield (Bobrick, 1995).

An employment of Bl and its isomers for the acceleration of growth of tomato plants in greenhouses by soaking seeds for 4 h in a 1 ppm solution was described by Takematsu and Izumi (1985). Spraying plants with EBl resulted in a 10-18% yield enhancement (Savelieva et al., 1997). Treatment in the flowering phase led to crop increase due to enhancement of the number and weight of the tomatoes (Kirillova et al., 1993; Balmush et al., 1995). The highest crop enhancement in field conditions was obtained when tomato and cucumber were treated with EBl twice, first the seeds and then by spraying in the flowering phase. Treated plants were less damaged by phytophtora and had better consumer properties (Churikova and Derevshchukov, 1997).

Soaking spinach seeds in a 10^{-2} ppm aqueous solution of (22S,23S)-EBl for 8 h enhanced germination from 54% to 72% (Ikekawa and Akutsu, 1987). An acceleration of seed germination and a better development of roots and aerial organs in cabbage resulted in crop yield enhancement after seeds were soaked in EBl solution (Asatova, 1991).

4. Miscellaneous

The crop yield and quality of industrial plants was shown to increase under the action of BS. Treatment of cotton plants with EBl by spraying in the budding phase resulted in 10-15% growth enhancement and in an increase of crop yield by 26%. The second spraying in the flowering phase enhanced the yield by 32% (CIASA, 1989). When cotton plants were grown in the presence of BS, their strength, yield, and length were improved (Umarov and Kariev, 1991; Tajobo Co., 1995). An increase of seed and straw yield was observed in flax when plants were treated with EBl (Voskresenskaya, 1993).

An application of 0.01-0.5 ppm Bl to rape seedlings after transplanting them in autumn increased plant height and number of green leaves on the main stem and diminished frost injury to the plants (Xia et al., 1992). EBl promoted growth of roots and leaves of tobacco plants and increased the content of nicotine (Ikekawa and Zhao, 1991; Gonik and Mikhailova, 1992). Foliar application of BS to Indian mustard plants in the preflowering phase and at pod development diminished water-stress effects on the seed and oil yield (Kumavat et al., 1997). BS also increased the production of seeds in Cruciferae plants (Ikekawa et al., 1987).

BS were found to stimulate the growth of sugar beet; the aerial part of the plants increased by 18%, and the root crop increased by 12% (Genma, 1987e). Treatment of sugar beet plants with (22S,23S)-HBl under drought-stress conditions resulted in complete compensation of mild stress conditions and increased the sugar yield by 8% (Schilling and Schiller, 1990; Schilling et al., 1991). Field trials in Belarus and Russia in 1991-1992 also revealed an enhancement of crop yield and an increase in sugar content by 10-13% after the plants were sprayed with EBl (Kurganskii, 1993). Treatment of the seeds with 0.01 ppm EBl resulted in a better germination, development, and growth, and the crop yield increased by 26-33% (Vedeneev et al., 1995).

EBl promoted the fruit setting of melon (Ikekawa and Nagai, 1987) when applied as a foliar spray in the flowering phase, thereby increasing the yield by 10-20%. Spraying watermelon seedlings with 0.01 mg/l EBl markedly promoted growth, and increased plant height and stem thickness. Spraying in the flowering phase increased the percentage of fruit setting and the number of flowers, delayed leaf senescence, and increased yield by 20% (Wang, Y.Q. et al., 1994a).

Bl, EBl, and HBl were found to promote germination and growth of seedlings of groundnut (*Arachis hypogaea* L.) (Vidyarardhini and Rao, 1996, 1997). Spraying groundnut plants with a 5 ppm Bl solution promoted the transport of assimilates to the sink, and increased pod setting, well-filled seed percentage, and the pod yield by 11, 20, and 14%, respectively (Li et al., 1993). No influence on the yield or other parameters related to the size of seeds has been found when EBl or EBk were applied to two-year-old plants of *Coffea arabica* (Mazzafera and Zullo, 1990). Bl applied to one-year-old *Pinus elliottii* seedlings improved the resistance and tolerance to drought and temperature stress and increased height, diameter, branch, and root growth (Wang, A.L. et al., 1995). EBl increased germination of seeds of *Eucalyptus camaldulensis* in saline conditions (Sasse et al., 1995).

BS can be used to increase the efficiency of weed control *via* adding them to seed germination stimulants that are applied to the soil to cause suicidal germination of witchweed seeds (Takeuchi et al., 1991). BS were recommended for the enhancement and synchronization of the germination of other weeds (Takeuchi et al., 1995).

BS application in horticulture promotes fruit setting, reduces abscission, and enhances fruit yield. Spraying buds of the Japanese persimmon with 0.1 ppm Bl 3 days before flowering reduced abscission from 45% to 14% up to 30 days after flowering (Maotani et al., 1989). Retardation of the abscission of calamondin *(Citrus madurensis)* leaf and fruitlet explants was observed when a Bl solution was fed through the petiole (Iwahori et al., 1990). An enhancement of the fruit yield due to fruit set promotion and fruit abscission prevention was reported for apple (Yokota and Yamanaka, 1988; Handschack et al., 1990), peach (Yamanaka, 1988a), persimmon (Yamanaka, 1988b), citrus (Kuraishi et al.,

1989, 1991; Sugiyama and Kurashi, 1989; Takahashi *et al.*, 1985), and cherry (Shchekotova *et al.*, 1996).

Field trials of EB1 on grape in the years 1994-1995 revealed a substantial yield improvement together with a diminishing of frost damage of the buds (Chirilov *et al.*, 1996). Application of EB1 reduced the abscission of young flowers and berries in grape and caused earlier maturation (Xu *et al.*, 1994b). B1 promoted germination of grape when applied as a mixture with dehydrated lanolin, *N*-methylpyrrolidone, and EtOH at the cut surface of the branches (Sumi *et al.*, 1990). Investigations of the action of B1 on mulberry revealed enhancement of the dry weight of leaves and roots, alteration of nutrient translocation (Kuno, 1997), and promotion of mulberry callus proliferation (Kuno and Ji, 1996). BS increased the number of flowers, inflorescences, marketable berries, and total yield per plant of strawberries cv. "Miyoshi", but it had no influence on cv. "Enrai" (Pipattanawong *et al.*, 1996). An increase of production of cranberry plants by 9-15% was observed when they were treated with EB1 at the beginning of the development of the generative organs (Volodko and Zelenkevich, 1998). An enhancement of crop yield and resistance to chill and disease injury after treatment with EB1 were noticed for black and red currant and gooseberry (Malevannaya and Bednarskaya, 1995).

BS have good potentials for the application in flower growing (Runkova, 1991). Soaking gladiolus bulbs in EB1 solution resulted in an earlier emergence of floriferous shoots and flowers, in an increase of the number of flowers, and in a very high enhancement of the number (68%) and mass (85%) of bulbs and bulbiferous buds. Similar effects were observed for tulips. Spraying phlox with 0.5 mg/l EB1 resulted in the emergence of additional floriferous shoots and in an increase of inflorescence growth (Runkova, 1995). In experiments with lily EB1 stimulated the development of bulbs and callus (Ohkawa *et al.*, 1996; Kilchevskii and Frantsuzionok, 1997). An addition of 0.01 mg/l EB1 to the nutrient medium stimulated callus formation in *Rosa hybrida* L. when cultivated *in vitro* and shortened the breeding process (Markova and Getko, 1997). Spraying rose plants with EB1 solution resulted in growth promotion of the young shoots, earlier flowering, and enhancement of chill resistance of the plants (Malevannaya and Kositsina-Penegina, 1996).

Other examples of application of BS and their analogs as growth promoters are claimed in patents (Meudt *et al.*, 1980; Kumura *et al.*, 1984; Uesono *et al.*, 1985; Ikekawa and Nagai, 1987; Akhrem *et al.*, 1987c, 1988b; Khripach *et al.*, 1990f; Wang, 1992; Wu, 1993). In field trials in Cuba some BS analogs were found to increase the crop yield of tomato, potato, and tobacco (Nunez *et al.*, 1995a,b; Diz *et al.*, 1995). An employment of BS in combination with other plant growth promoters revealed a beneficial influence on the growth and development of agricultural plants (Genma, 1987a-f; Kobayashi and Nitani, 1989; Kuraishi *et al.*, 1989; Oritani, 1989; Hu, 1995). A number of studies show the efficiency of BS as growth-promoting components of nutrient medium, not only for growing plants but also for plant-cell cultures (Sala and Sala, 1985; Chen and Chi, 1986; Hirai *et al.*, 1986b; Bellincampi and Morpurgo, 1988, 1991; Chang and Cai, 1988). Based on the results of field trials in 1993-1997 with different agricultural plants (barley, oat, potato, winter rye, wheat), EB1 was recommended as an addition (10^{-4}-10^{-6}%) to the complex of nitrogen-phosphorus-potassium fertilizers to increase the yield and improve the quality of crops (Pirogovskaya *et al.*, 1996).

Growth stimulation effects of BS were found to be promising for practical application not only for higher plants but also for fungi and algae. Cultivation of the mycelium of *Psilocybe cubensis* in the presence of (22*S*,23*S*)-HB1 resulted in an increasing number of fruiting bodies in the first flush from 1-2 in the control to 4-7, and growth was accelerated 2-3 times (Gartz *et al.*, 1990; Adam *et al.*, 1991). EB1 stimulated also the growth of different algae (*Spirulina platensis, Euglena gracilis, Dunaliella salina, Chlorella vulgaris*) at concentrations of

10^{-8}-10^{-10} M, but it evoked inhibition at a concentration of 10^{-6} M (Melnikov *et al.*, 1996, 1998).

A chapter on application of BS would be not complete without mentioning the results obtained during studies of the action of EBl on the Russian sturgeon. The observed toxicoprotective and immunostimulative properties, especially in the earliest stages of fish development, were found to be very promising for application in fish breeding (Vitvitskaya *et al.*, 1997a-c). This is probably the first example of the stimulative action of BS outside the plant kingdom.

B. TOXICOLOGY OF BS

EBl is the most widely studied compound in this class with respect to toxicity, because it is probably the best candidate for practical application. Along with investigations of pure EBl, studies on the toxicity of formulations that can be applied in agriculture were carried out also. The data obtained for EBl showed its low toxicity. The acute toxicity (LD_{50}) in mice (female, *per os*) was higher than 1000 mg/kg, and the LD_{50} (*per os* and *cutaneously*) in rats (male/female) was more than 2000 mg/kg. In concentrations of 0.01%, EBl did not irritate mucous membranes of rabbit eyes. The fish toxicity, TML48, for carp was higher than 10 ppm, and the Ames test on mutagenicity was negative (Ikekawa and Zhao, 1991).

Similar results were obtained in other laboratories. The LD_{50} of EBl found in studies in mice and rats (*per os* and *cutaneously*) was 1000-2000 mg/kg; the LD_{50} of a formulation that contained 0.25% EBl was higher than 5000 mg/kg (Kuzmitskii and Mizulo, 1991). The LD_{50} for this formulation was higher than 15000 mg/kg (white rats, *per os* or *intranasally*) and it gave no irritation of the skin and mucous membranes. The inactive concentration of the formulation in chronic experiments (white rats, intranasal inhalation for 4 months) was equal to 29.49 mg/m^3, the threshold concentration was 294.87 mg/m^3, and the active concentration was 589.74 mg/m^3. The maximum permissible concentration was more than 70 mg/m^3. Cumulative properties could hardly be seen (K > 5) (Budnikov and Alexashina, 1995).

A study of the mutagenic effects of EB1 and its formulation carried out at the Scientific Research Center of Toxicologic and Hygienic Regulation of Biopreparations of Russia showed that the Ames test, with and without metabolic activation, was negative (*Salmonella typhimurium* TA1534, TA1537, TA1950, TA98, and TA100) (Onatskiy *et al.*, 1997). In micronuclear or chromosome aberration tests in mice both EB1 and its formulation did not show spontaneous mutations.

The study of effects of EB1 on the genetic structure and the processes in cell nuclei of several cultivars of barley revealed that it caused no breaches in meoisis, reduced the frequency of chromosome aberrations, and increased plant survival and fertility. Both in the anaphase and tetrade phase it decreased the number of cell breaches almost twofold when it was applied simultaneously with nitrosomethylurea (Khrustaleva *et al.*, 1991, 1995).

Thus all the results obtained to date on the toxicity of BS together with the other properties discussed above suggest that BS do not have negative influences in mammals, water organisms, soil microbiological processes, and plants. Because of their natural origin and application in doses comparable with the endogenous contents of phytohormones, they will not affect the environment in a negative way.

BS are mainly concentrated in plant pollen and they probably play a role in the beneficial healthy properties that are ascribed to pollen. Pollen are applied for biostimulation in folk medicine and also form the basis for the production of some antiinflammatory and metabolism stimulative medicines, which are especially recommended for children and elderly people with chronic infections (Mashkovskii, 1987). If so, a new direction concerning possible application in medicine will bring new important stimulus for research on BS in the near future.

APPENDIX

Structures, Occurrence and Spectral Characteristics of Natural BS

Brassinolide (Bl) *[(22R,23R,24S)-2α,3α,22,23-tetrahydroxy-24-methyl-B-homo-7-oxa-5α-cholestan-6-one, (2R,3S,22R,23R,24S)-2,3,22,23-tetrahydroxy-24-methyl-B-homo-7-oxa-5α-cholestan-6-one].*
$C_{28}H_{48}O_6$, MW 480.69.

Mp 274-278 °C MeOH (Aburatani *et al.*, 1985b); 273-278 °C MeOH-H_2O (Ishiguro *et al.*, 1980; Takatsuto *et al.*, 1984b); 279-281 °C EtOAc (Thompson *et al.*, 1981). α_D^{24} +41.9 c=0.34, $CHCl_3$-MeOH=9:1 (Sakakibara *et al.*, 1982; Mori *et al.*, 1982); α_D^{20} +41.9 c=0.253, MeOH (Zhou, W.-S. *et al.*, 1990a); α_D^{27} +47.0 c=1.02, $CHCl_3$-MeOH=9:1 (Aburatani *et al.*, 1985b); α_D^{27} +16 (Ishiguro *et al.*, 1980); α_D^{16} +16 c=0.985, CH_2Cl_2-MeOH=1:1 (Takatsuto *et al.*, 1984b).

OCCURRENCE: European alder *Alnus glutinosa* L. Gaertn. pollen (Plattner *et al.*, 1986); beeswax (Cao *et al.*, 1987); *Brassica campestris* var *pekinensis* L. immature seeds and sheaths < $0.3 \cdot 10^{-9}$% (Abe *et al.*, 1982), 10^{-9}% (Ikekawa *et al.*, 1983, 1984; Ikekawa and Takatsuto, 1984); rape *Brassica napus* L. pollen 10^{-5}% (Grove *et al.*, 1979); *Cassia tora* L. immature seeds $1.8 \cdot 10^{-9}$% (Park, K.-H. *et al.*, 1993b, 1994a); *Castanea crenata* insect galls (Arima *et al.*, 1984), 10^{-10}% (Ikekawa *et al.*, 1984), $6 \cdot 10^{-8}$% (Ikeda *et al.*, 1983); *Castanea* spp. insect galls $1.1 \cdot 10^{-7}$% (Ikekawa *et al.*, 1983); *Catharanthus roseus* Don (*Vinca rosea*

L.) cultured crown gall cells obtained by transformation with *Agrobacterium tumefaciens* Conn strains (Park, K.-H. *et al.*, 1989); orange *Citrus sinensis* Osbeck pollen (Motegi *et al.*, 1994); *Daucus carota ssp. sativus* seeds (Adam *et al.*, 1996b); buckwheat *Fagopyrum esculentum* Moench. (Takatsuto *et al.*, 1990b); *Helianthus annuuc* L. pollen (Takatsuto *et al.*, 1989b); *Lilium elegans* pollen (Suzuki *et al.*, 1994b); rice *Oryza sativa* whole plant $0.8 \cdot 10^{-6}$%, panicles $0.2 \cdot 10^{-7}$% (Shim *et al.*, 1996); *Phaseolus vulgaris* immature seeds (Yokota *et al.*, 1987b), young steam (Yokota *et al.*, 1991); Scots pine *Pinus silverstris* cambial region (Kim, S.-K. *et al.*, 1990); garden pea *Pisum sativum* shoot (Yokota *et al.*, 1991), fully grown seed (Yokota *et al.*, 1996a); tropical bean plant *Psophocarpus tetragonolobus* seeds (Yokota *et al.*, 1991); radish *Raphanus sativus* L. var. 'Remo' seeds $3 \cdot 10^{-8}$% (Schmidt *et al.*, 1991); *Rheum rhabarbarum* panicles (Schmidt *et al.*, 1995a); *Solidago altissima* L. stems $6.6 \cdot 10^{-9}$% (Tada *et al.*, 1987); *Thea sinensis* leaves (Ikekawa *et al.*, 1983), $4.6 \cdot 10^{-10}$% (Ikekawa *et al.*, 1984), $<1.5 \cdot 10^{-9}$% (Morishita *et al.*, 1983); *Vicia faba* L. pollen (Gamoh *et al.*, 1988), $1.9 \cdot 10^{-5}$% (Ikekawa *et al.*, 1988); immature seeds (Park, K.-H. *et al.*, 1987; Park, K.-H., 1988).

^{1}H NMR (500 MHz, CDCl$_3$): δ 0.71 (*s*, 3H, 18-H), 0.85 (*d*, 3H, *J* 7, 28-H), 0.90 (*d*, 3H, *J* 6.5, 21-H), 0.93 (*s*, 3H, 19-H), 0.95 (*d*, 3H, *J* 7, 26-H), 0.97 (*d*, 3H, *J* 7, 27-H), 1.22 (*m*, 1H, 24-H), 1.49 (1H, 20-H), 1.55 (1H, 1α-H), 1.65 (*m*, 1H, 25-H), 1.73 (*m*, 8-H), 1.87 (1H, *J* 4.8, 12.8, 1β-H), 1.95 (4α-H), 2.15 (*dd*, 1H, *J* 3, 12.2, 15.6, 4β-H), 3.11 (*dd*, 1H, *J* 5.0, 12), 3.53 (*d*, 1H, *J* 9.0, 22-H), 3.71 (*dd*, 1H, *J* 2.0, 9.0, 23-H), 3.72 (*br*, 1H, 2-H), 4.02 (*br*, 1H, 3-H), 4.09 (*m*, 1H, 7α-H) (Suzuki *et al.*, 1993b). ^{13}C NMR (CDCl$_3$, 75.5 MHz, C-1⇒C-28): δ 41.5, 68.0, 68.1, 31.3, 40.7, 176.6, 70.5, 39.3, 58.2, 38.1, 22.4, 39.7, 42.8, 51.1, 24.7, 27.5, 52.5, 11.6, 15.3, 37.0, 11.7, 74.6, 73.5, 40.1, 30.8, 21.0, 20.7, 9.9 (Porzel *et al.*, 1992). EI-MS (as bismethaneboronate) m/z (rel. int., %) 528 [M]$^+$ (1.8), 457 (4.0), 374 (20.9), 345 (9.6), 332 (12.4), 177 (53.0), 155 (100) (Ikekawa *et* *al.*, 1984). IR* (KBr): v_{max} 3450 (s), 2975 (s), 2945 (s), 2870 (m), 2850 (m), 1730 (m), 1700 (m), 1690 (sh), 1640 (w), 1463 (m), 1443 (m), 1410 (m), 1388 (m), 1335 (m), 1320 (m), 1300 (m), 1285 (m), 1260 (m), 1239 (m), 1190 (m), 1170 (w), 1148 (m), 1130 (m), 1120 (m), 1097 (m), 1070 (s), 1040 (m), 1030 (m), 990 (m), 970 (w) cm^{-1} (Takatsuto *et al.*, 1984b).

2-Deoxybrassinolide (Bl2d) *[(22R,23R,24S)-3α,22,23-trihydroxy-24-methyl-B-homo-7-oxa-5α-cholestan-6-one]*. $C_{28}H_{48}O_5$, MW 464.69.

OCCURRENCE: *Apium graveolens* L. (Schmidt *et al.*, 1995c, 1996b,c); garden pea *Pisum sativum* fully grown seed (Yokota *et al.*, 1996a).

EI-MS (as trimethylsilyl methaneboronate): m/z (rel. int., %) 560 [M]$^+$ (10), 545 (31), 531 (17), 490 (25), 470 (7), 404 (5), 376 (3), 375 (4), 332 (8), 287 (4), 211 (8), 195 (63), 177 (14), 156 (100), 121 (16), 85 (20) (Schmidt *et al.*, 1995c).

Epibrassinolide (EBl) *[24-epibrassinolide, (24R)-brassinolide, (22R,23R,24R)-2α,3α,22,23-tetrahydroxy-24-methyl-B-homo-7-oxa-5α-cholestan-6-one]*. $C_{28}H_{48}O_6$, MW 480.69.

Mp 256-258 °C (EtOAc) (Thompson *et al.*, 1979; Anastasia *et al.*, 1983c; Takatsuto and Ikekawa, 1984a; Akhrem *et al.*, 1989c). α_D^{21} +32.0 (Anastasia *et al.*, 1983c); α_D^{25} +30.0 (Thompson *et al.*, 1979).

OCCURRENCE: *Arabidopsis thaliana* seeds 2.2•10^{-8}% (Schmidt *et al.*, 1997; Kauschmann *et al.*, 1997); *Gypsophila perfoliata* L. seeds (Schmidt *et al.*, 1996a); *Vicia faba* L. pollen 5•10^{-7}% (Ikekawa *et al.*, 1988).

^{1}H NMR (300 MHz, CDCl$_3$-CD$_3$OD=95:5): δ 0.71 (*s*, 3H, 18-H), 0.85 (*d*, 3H, *J* 6.9, 28-H), 0.87 (*d*, 3H, *J* 6.6, 27-H), 0.92 (*d*, 3H, *J* 6.9, 26-H), 0.92 (*s*, 3H, 19-

---

* Here and below for IR spectra: s - strong, m - medium, w - weak, sh - shoulder.

H), 0.97 (*d*, 3H, *J* 6.6, 21-H), 1.18 (14-H), 1.30 (9-H), 1.34/1.25 (16-H), 1.46 (20-H), 1.48 (24-H), 1.56 (17-H), 1.68/1.22 (15-H), 1.72 (8-H), 1.80/1.40 (11-H), 1.86/1.55 (1-H), 1.90 (25-H), 1.99/1.22 (12-H), 2.10/1.93 (4-H), 3.12 (*dd*, 1H, *J* 12.0, 4.4, 5-H), 3.37 (*dd*, 1H, *J* 4.5, 4.5, 23-H), 3.65 (*ddd*, 1H, 2-H), 3.66 (*br dd*, 1H, 22-H), 3.98 (*br s*, 1H, 3-H), 4.10 (7-H) (Porzel *et al.*, 1992). ^{13}C NMR (75.5 MHz, CDCl$_3$-CD$_3$OD=95:5, C-1⇒C-28): δ 41.2, 67.9, 68.0, 31.1, 40.9, 176.8, 70.5, 39.1, 58.0, 38.2, 22.2, 39.6, 42.4, 51.2, 24.7, 27.6, 52.5, 11.5, 15.3, 40.1, 12.3, 72.4, 76.0, 41.4, 26.9, 22.1, 17.2, 10.8 (Porzel *et al.*, 1992).

Dolicholide **(BDl)** *[(22R,23R)-2α,3α,22,23-tetrahydroxy-B-homo-7-oxa-5α-ergost-24(28)-en-6-one]*. C$_{28}$H$_{46}$O$_6$, MW 478.67.

Mp 238-242 °C MeCN-H$_2$O (Takatsuto and Ikekawa, 1983b,c); 235-237 °C MeOH (Okada and Mori, 1983b; Mori *et al.*, 1984). α$_D^{22}$ +56.3 c=0.41, MeOH (Okada and Mori, 1983b; Mori *et al.*, 1984).

OCCURRENCE: *Dolichos lablab* immature seeds 4.7•10^{-6}% (Yokota *et al.*, 1982b); *Phaseolus vulgaris* immature seeds (Yokota *et al.*, 1983c,d, 1987b).

^{1}H NMR (400 MHz, CDCl$_3$): δ 0.65 (*s*, 3H, 18-H), 0.92 (*s*, 3H, 19-H), 0.95 (*d*, 3H, *J* 7, 21-H), 1.08 (*d*, 3H, *J* 8, 26-H), 1.11 (*d*, 3H, *J* 8, 27-H), 2.26 (*septet*, 1H, 25-H), 3.11 (*dd*, 1H, *J* 12.5, 5, 5-H), 3.62 (*d*, 1H, *J* 8, 22-H), 3.72 (*m*, 1H, W2 22, 2-H), 4.02 (*br s*, 1H, 3-H), 4.03 (*d*, 1H, *J* 8, 23-H), 4.09 (*m*, 2H, 7-H), 5.05 (*s*, 1H, 28-H), 5.08 (*s*, 1H, 28-H) (Takatsuto and Ikekawa, 1983b). IR (KBr): ν$_{max}$ 3420 (s), 1730-1695 (s), 1180 (m), 1060 (s), 1025 (m) cm^{-1} (Mori *et al.*, 1984). EI-MS (as bismethaneboranate): m/z (rel. int., %) 526 [M]$^+$ (17.9), 373 (21.1), 345 (21.1), 343 (100), 330 (2.1), 177 (8.4), 153 (70.5) (Ikekawa *et al.*, 1984).

1β-Hydroxycastasterone **(Bk1β-hydroxy)**
[(22R,23R,24S)-1β,2α,3α,22,23-pentahydroxy-24-methyl-5α-cholestan-6-one]. $C_{28}H_{48}O_6$, MW 480.69.

OCCURRENCE: *Phaseolus vulgaris* large-scale experiments with immature seeds (Kim, S.-K., 1991). Its occurrence in natural sources is doubtful (Yokota, personal communication). Sakurai and Fujioka (1993) and Takatsuto (1994a) considered it to be a natural compound.

3-Epi-1α-hydroxycastasterone (Bk3β1α-hydroxy)
[(22R,23R,24S)-1α,2α,3β,22,23-pentahydroxy-24-methyl-5α-cholestan-6-one]. $C_{28}H_{48}O_6$, MW 480.69.

OCCURRENCE: *Phaseolus vulgaris* large-scale experiments with immature seeds (Kim, S.-K., 1991). The situation with respect to the occurrence of this compound as a natural product is similar to that for 1β-hydroxycastasterone (see above).

Castasterone (Bk) *[(22R,23R,24S)-2α,3α,22,23-tetrahydroxy-24-methyl-5α-cholestan-6-one]*.
$C_{28}H_{48}O_5$, MW 464.69.

Mp 259-261 °C MeCN-H$_2$O (Yokota *et al.*, 1982a); 256-259 °C EtOAc-MeOH (Aburatani *et al.*, 1985b). α_D^{20} -4.0 (Anastasia *et al.*, 1983d); α_D^{27} -0.7 c=1.04, CHCl$_3$-MeOH=9:1 (Aburatani *et al.*, 1985b), α_D^{23} +0.08 c=0.85, CHCl$_3$-MeOH (Kametani *et al.*, 1988a); α_D^{25} +0.03 c=1.17, CHCl$_3$-MeOH=9:1 (Mori *et al.*, 1984; Sakakibara and Mori, 1983a); α_D^{25} -2 c=0.542, MeOH (Zhou, W.-S. *et al.*, 1990a); α_D^{25} -7.0 (Thompson *et al.*, 1981); α_D^{25} +0.92 c=1.46, CHCl$_3$-MeOH=9:1 (Honda *et al.*, 1990; Tsubuki *et al.*, 1992a).

OCCURRENCE: Castasterone was found in most plants studied including European alder *Alnus glutinosa* L. Gaertn. pollen (Plattner *et al.*, 1986);

Arabidopsis thaliana shoots (Fujioka *et al.*, 1996), seeds $3.6 \cdot 10^{-8}$% (Schmidt *et al.*, 1997; Kauschmann *et al.*, 1997); sugar beet *Beta vulgaris* L. seeds (Schmidt *et al.*, 1994); Chinese cabbage *Brassica campestris* var. *pekinensis* L. seeds and sheaths $<0.3 \cdot 10^{-9}$% (Abe *et al.*, 1982), 10^{-7}% (Ikekawa *et al.*, 1983), $1.6 \cdot 10^{-10}$% (Ikekawa *et al.*, 1984); *Cannabis sativa* L. seeds (Takatsuto *et al.*, 1996b); *Cassia tora* L. immature seeds $1.6 \cdot 10^{-8}$% (Park, K.-H. *et al.*, 1993b, 1994a); chestnut *Castanea crenata* Sieb. et Zucc. insect galls $1.1 \cdot 10^{-9}$% (Ikekawa *et al.*, 1984), $1.2 \cdot 10^{-6}$% (Ikeda *et al.*, 1983), shoots $2 \cdot 10^{-6}$%, leaves $6 \cdot 10^{-7}$%, buds $3 \cdot 10^{-7}$% (Arima *et al.*, 1984); chestnut *Castanea* spp. insect galls $2.4 \cdot 10^{-7}$% (Yokota *et al.*, 1982a), $1.1 \cdot 10^{-6}$% (Ikekawa *et al.*, 1983); $1.1 \cdot 10^{-10}$% (Abe *et al.*, 1983); *Catharanthus roseus* Don (*Vinca rosea* L.) cultured crown gall cells obtained by transformation with *Agrobacterium tumefaciens* Conn strains $3 \cdot 10^{-6}$% (Park, K.-H. *et al.*, 1989); orange *Citrus sinensis* Osbeck pollen (Motegi *et al.*, 1994); *Daucus carota* ssp. *sativus* seeds (Adam *et al.*, 1996b); Japanese evergreen tree *Distilium racemosum* Sieb. et Zucc. insect galls $2.5 \cdot 10^{-7}$%; leaves $1.3 \cdot 10^{-8}$% (Ikekawa *et al.*, 1984); *Dolichos lablab* immature seeds (Yokota *et al.*, 1983b,c); buckwheat *Fagopyrum esculentum* Moench. (Takatsuto *et al.*, 1990b); sunflower *Helianthus annuus* L. pollen (Takatsuto *et al.*, 1989b); *Lilium elegans* pollen (Suzuki *et al.*, 1994b); *Ornithopus sativus* seeds $5 \cdot 10^{-7}$% (Schmidt *et al.*, 1993a, 1996b,c), shoots (Spengler *et al.*, 1995); rice *Oryza sativa* shoots, whole plant $1.4 \cdot 10^{-9}$% (Abe *et al.*, 1984b), immature seeds (Park, K.-H. *et al.*, 1993c); *Phalaris canariensis* seeds (Shimada *et al.*, 1996); *Phaseolus vulgaris* young stem (Yokota *et al.*, 1991); *Perilla frutescens* immature seeds (Park, K.-H., 1993a); garden pea *Pisum sativum* shoot (Yokota *et al.*, 1991), fully grown seed (Yokota *et al.*, 1996a); tropical bean plant *Psophocarpus tetragonolobus* seeds (Yokota *et al.*, 1991); Scots pine *Pinus silverstris* cambial region (Kim, S.-K. *et al.*, 1990); radish *Raphanus sativus* L. var. 'Remo' seeds $8 \cdot 10^{-8}$% (Schmidt *et al.*, 1991); *Rheum rhabarbarum* panicles (Schmidt *et al.*, 1995a); rye *Secale cereale* L. var. 'Petka' seeds (Schmidt *et al.*,

1995b); green tea *Thea sinensis* leaves (Ikekawa *et al.*, 1983), <1.5•10⁻⁹% (Morishita *et al.*, 1983), 1.1•10⁻⁸% (Ikekawa *et al.*, 1984); wheat *Triticum aestivum* L. grain (Yokota *et al.*, 1994); broad bean *Vicia faba* L. pollen (Ikekawa *et al.*, 1988), 1.3•10⁻⁵% (Gamoh *et al.*, 1989a), immature seeds (Park, K.-H. *et al.*, 1987; Park, K.-H., 1988); corn *Zea mays* pollen 1.2•10⁻⁵% (Suzuki *et al.*, 1986), immature seeds (Park, K.-H. *et al.*, 1995).

¹H NMR (360 MHz, CDCl₃): δ 0.68 (*s*, 3H, 18-H), 0.76 (*s*, 3H, 19-H), 0.84 (*d*, 3H, *J* 6.6, 26-H), 0.91 (*d*, 3H, *J* 6.6, 27-H), 0.94 (*d*, 3H, *J* 6.3, 28-H), 0.97 (*d*, 3H, *J* 6.6, 21-H), 2.30 (*dd*, 1H, *J* 12.9, 4.2, 7β-H), 2.67 (*dd*, 1H, *J* 15.6, 3.0, 5-H), 3.55 (*d*, 1H, *J* 22-H), 3.72 (*d*, 1H, *J* 9, 23-H), 3.78 (*m*, 1H, 2-H), 4.05 (*br s*, 1H, 3-H) (McMorris *et al.*, 1996). ¹³C NMR (67.8 MHz, CDCl₃-CD₃OD=4:1, C-1⇒C-28): δ 40.0, 68.2, 68.4, 26.5, 51.1, 213.9, 46.9, 38.1, 53.9, 42.9, 21.4, 39.7, 43.0, 56.8, 24.0, 27.7, 52.5, 12.0, 13.6, 37.1, 12.1, 74.6, 73.4, 40.5, 30.8, 20.8, 21.0, 10.3 (Ando *et al.*, 1993). EI-MS (as bismethaneboronate): m/z (rel. int., %) 512 (38.3), 441 (6.4), 358 (24.5), 329 (7.4), 155 (100) (Ikekawa *et al.*, 1984). IR (KBr): ν$_{max}$ 3400, 2944, 2870, 1708, 1450, 1382, 1260, 1084, 1042, 992, 978 cm⁻¹ (McMorris *et al.*, 1996).

2-Epicastasterone **(Bk2β)** *[(22R,23R,24S)-2β,3α,22,23-tetrahydroxy-24-methyl-5α-cholestan-6-one]*. C$_{28}$H$_{48}$O$_5$, MW 464.69.

OCCURRENCE: *Phaseolus vulgaris* large-scale experiments with immature seeds (Takahashi *et al.*, 1987; Kim, S.-K., 1991).

3-Epicastasterone **(Bk3β)** *[(22R,23R,24S)-2α,3β,22,23-tetrahydroxy-24-methyl-5α-cholestan-6-one]*. C$_{28}$H$_{48}$O$_5$, MW 464.69.

OCCURRENCE: *Phaseolus vulgaris* large-scale experiments with immature seeds (Takahashi *et al.*, 1987; Kim, S.-K., 1991).

2,3-Diepicastasterone (Bk2β3β) *[(22R,23R,24S)-2β,3β,22,23-tetrahydroxy-24-methyl-5α-cholestan-6-one]*. $C_{28}H_{48}O_5$, MW 464.69.

OCCURRENCE: *Phaseolus vulgaris* large-scale experiments with immature seeds (Takahashi et al., 1987; Kim, S.-K., 1991).

Typhasterol (Bk2d) *[(22R,23R,24S)-2α,22,23-trihydroxy-24-methyl-5α-cholestan-6-one]*. $C_{28}H_{48}O_4$, MW 448.69.

Mp 231-233 °C MeCN-H₂O (Yokota et al., 1983a).

OCCURRENCE: *Arabidopsis thaliana* shoots (Fujioka et al., 1996); *Cassia tora* L. immature seeds $7 \cdot 10^{-10}$% (Park, K.-H. et al., 1994a); *Cupressus arizonica* pollen $4.58 \cdot 10^{-5}$% (Griffiths et al., 1995a); lily *Lilium elegans* Thumb. pollen $1-5 \cdot 10^{-9}$% (Suzuki et al., 1994b); sitka spruce *Picea sitchensis* Bong Carr shoots $7 \cdot 10^{-6}$% (Yokota et al., 1985); Japanese black pine *Pinus thunbergii* Parl. pollen $8.9 \cdot 10^{-6}$% (Yokota et al., 1983a); garden pea *Pisum sativum* fully grown seed (Yokota et al., 1996a); *Robinia pseudo-acacia* L. pollen (Abe et al., 1995b); rye *Secale cereale* L. var. 'Petka' seeds (Schmidt et al., 1995b); green tea *Thea sinensis* leaves $1.5 \cdot 10^{-9}$% (Abe et al., 1984b); wheat *Triticum aestivum* L. cv. Chihoku grains (Yokota et al., 1994); cat tail *Typha latifolia* L. pollen $6.8 \cdot 10^{-6}$% (Schneider et al., 1983; Suntry, Ltd., 1983); corn *Zea mays* L. pollen $6.6 \cdot 10^{-7}$% (Suzuki et al., 1986). It is a main BS (10^{-5}% or more) in pollen of orange *Citrus unshiu* Marcov., tulip *Tulipa gesneriana* L., lily *Lilium longiflorum* cv. Georgia and cat tail *Typha latifolia* L. (Abe, 1991).

¹H NMR (360 MHz, CDCl₃): δ 0.68 (*s*, 3H, 18-H), 0.73 (*s*, 3H, 19-H), 0.85 (*d*, 3H, *J* 6.7, 28-H), 0.91 (*d*, 3H, *J* 6.5, 21-H), 0.95 (*d*, 3H, *J* 6.7, 26-H), 0.98 (*d*, 3H, *J* 6.7, 27-H), 1.72 (*dd*, 1H, *J* 8.4, 2.4), 2.31 (*dd*, 1H, *J* 12.7, 4.8, 7β-H), 2.73 (*apparent t*, 1H, *J* 7.7, 5-H), 3.58 (*d*, 1H, *J* 8.4, 22-H), 3.72 (*d*, 1H, *J* 8.4, 23-H), 4.17 (*m*, 1H, W2 6.7, 3-H) (Takatsuto et al., 1984b). EI-MS (as trimethylsilyl

methaneboronate): m/z (rel. int., %) 544 [M]$^+$ (95), 529 (56), 526 (26), 515 (100), 454 (82), 439 (33), 155 (54) (Suzuki *et al.*, 1994b).

Teasterone (Bk2d3β) *[(22R,23R,24S)-2β,22,23-trihydroxy-24-methyl-5α-cholestan-6-one]*.

$C_{28}H_{48}O_4$, MW 448.69.

Mp 200-201 °C (EtOAc) (Aburatani *et al.*, 1987b; Kametani *et al.*, 1988a).

OCCURRENCE: *Cannabis sativa* L. seeds (Takatsuto *et al.*, 1996b); *Cassia tora* L. immature seeds 4•10^{-9}% (Park, K.-H. *et al.*, 1993b, 1994a); *Cupressus arizonica* pollen 5•10^{-7}% (Griffiths *et al.*, 1995a); *Ginkgo biloba* L. seeds (Takatsuto *et al.*, 1996a); lily *Lilium elegans* Thumb. pollen 1-5•10^{-10}% (Suzuki *et al.*, 1994b); rice *Oryza sativa* L. cv. Tongjinbyeo immature seeds (Park, K.-H. *et al.*, 1993c, 1994b); *Phalaris canariensis* seeds (Shimada *et al.*, 1996); radish *Raphanus sativus* L. var. 'Remo' seeds (Schmidt *et al.*, 1993b); rye *Secale cereale* L. var. 'Petka' seeds (Schmidt *et al.*, 1995b); green tea *Thea sinensis* leaves (Abe, 1991), 6•10^{-9}% (Abe *et al.*, 1984a); wheat *Triticum aestivum* L. grain (Yokota *et al.*, 1994); corn *Zea mays* L. pollen 4.1•10^{-7}% (Suzuki *et al.*, 1986), immature seeds (Park, K.-H. *et al.*, 1995). EI-MS (as trimethylsilyl methaneboronate): m/z (rel. int., %) 544 [M]$^+$ (39), 529 (83), 515 (100), 454 (13), 319 (4), 300 (3), 155 (10) (Park, K.-H. *et al.*, 1994a).

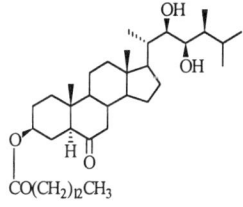

Teasterone 3-myristate (Bk2d3β3-myristate) *[(22R,23R,24S)-22,23-dihydroxy-2β-myristyloxy-24-methyl-5α-cholestan-6-one]*. $C_{42}H_{74}O_5$, MW 659.05.

OCCURRENCE: lily *Lilium longiflorum* cv. Georgia anthers (Asakawa *et al.*, 1994).

EI-MS (as methaneboronate): m/z (rel. int., %) 682 [M]$^+$ (6), 454 (100), 439 (30.9), 426 (9), 155 (21) (Asakawa *et al.*, 1994).

3-Oxoteasterone (Bk2d3k) *[3-dehydroteasterone, (22R,23R,24S)-22,23-dihydroxy-24-methyl-5α-cholestan-3,6-dione]*. $C_{28}H_{46}O_4$, MW 446.67.

Mp 178-181 °C (Abe *et al.*, 1994).

OCCURRENCE: lily *Lilium longiflorum* cv. Georgia anthers, *Distylium racemosum* leaves (Abe *et al.*, 1994); wheat *Triticum aestivum* L. grains (Yokota *et al.*, 1994).

^{1}H NMR (270 MHz, CDCl$_3$): 0.72 (*s*, 3H, 18-H), 0.85 (*d*, 3H, *J* 7.0, 28-H), 0.92 (*d*, 3H, *J* 6.5, 21-H), 0.95 (*d*, 3H, *J* 6.5, 26-H), 0.97 (*s*, 3H, 19-H), 0.98 (*d*, 3H, *J* 5.6, 27-H), 3.56 (*d*, 1H, *J* 7.8, 22-H), 3.73 (*dd*, 1H, *J* 8.6, 1.0, 23-H) (Abe *et al.*, 1994). ^{13}C NMR: 36.8 (C-1), 209.1 (C-3), 57.5 (C-5), 211.2 (C-6), 46.6 (C-7), 38.1 (C-8), 53.4 (C-9), 41.3 (C-10), 21.7 (C-11), 39.4 (C-12), 42.8 (C-13), 56.5 (C-14), 23.9 (C-15), 27.7 (C-16), 52.3 (C-17), 11.9 (C-18), 12.6 (C-19), 37.0 (C-20), 11.9 (C-21), 74.7 (C-22), 73.6 (C-23), 39.4 (C-24), 30.8 (C-25), 20.8 (C-26), 20.9 (C-27), 10.1 (C-28) (Abe *et al.*, 1994). EI-MS (as methaneboronate): m/z (rel. int., %) 470 [M]$^+$ (49), 399 (6), 339 (4), 316 (20), 287 (11), 260 (8), 245 (15), 155 (100) (Yokota *et al.*, 1994).

Secasterone (Bk2,3epoxy) *[(22R,23R,24S)-2β,3β-epoxy-22,23-dihydroxy-24-methyl-5α-cholestan-6-one]*. $C_{28}H_{46}O_4$, MW 446.67.

Mp 179-180 °C EtOAc-hexane (Voigt *et al.*, 1995).

OCCURRENCE: rye *Secale cereale* L. var. 'Petka' seeds (Schmidt *et al.*, 1995b).

^{1}H NMR (500 MHz, CDCl$_3$): 0.675 (*s*, 3H), 0.806 (*s*, 3H), 0.846 (*d*, 3H, *J* 6.8), 0.912 (*d*, 3H, *J* 6.4), 0.951 (*d*, 3H, *J* 7.3), 0.969 (*d*, 3H, *J* 7.3), 2.32 (*dd*, 1H, *J* 13.2, 3.9), 3.16 (*m*, 1H), 3.23 (*dd*, 1H, *J* 5.9, 3.8), 3.56 (*d*, 1H, *J* 7.7), 3.72 (*d*, 1H, *J* 7.7) (Voigt *et al.*, 1995). EI-MS (as methaneboronate): m/z (rel. int., %) 470 [M]$^+$ (66), 454 (70), 439 (76), 426 (25), 316 (23), 286 (10), 260 (12), 245 (19), 155 (100) (Voigt *et al.*, 1995).

Cathasterone (B23dk2d3β) *[22α-hydroxy-6-oxa-campestanol, (22S,24R)-3β,22-dihydroxy-24-methyl-5α-cholestan-6-one]*. $C_{28}H_{48}O_3$, MW 432.69.

Mp 176-177.5 °C MeOH (Fujioka *et al.*, 1995a).

OCCURRENCE: *Catharanthus roseus* cultured cells (1.96-3.91)•10^{-7}% (Fujioka *et al.*, 1995a). It is the first BS with one hydroxy group in the side chain.

^{1}H NMR (400 MHz, CDCl$_3$): δ 0.685 (*s*, 3H), 0.762 (*s*, 3H), 0.811 (*d*, 3H, *J* 6.8), 0.832 (*d*, 3H, *J* 6.3), 0.877 (*d*, 3H, *J* 6.8), 0.894 (*d*, 3H, *J* 6.8), 2.22 (*dd*, 1H, *J* 12.2, 2.9), 2.33 (*dd*, 1H, *J* 13.2, 4.4), 3.58 (*m*, 1H), 3.77 (*m*, 1H) (Fujioka *et al.*, 1995a). EI-MS: m/z (rel. int., %) 432 [M]$^+$ (3.6), 414 (4.1), 385 (2.5), 359 (5.9), 347 (3.9), 345 (4.9), 330 (10.8), 318 (100), 316 (45.3), 300 (11.3), 287 (18.9), 247 (10.8), 139 (16.0) (Fujioka *et al.*, 1995a).

6-Deoxocastasterone (Bd) *[(22R,23R,24S)-2α,3α,22,23-tetrahydroxy-24-methyl-5α-cholestane]*. $C_{28}H_{50}O_4$, MW 450.70.

Mp 225-226 °C, sinter at 218 °C (Mori *et al.*, 1984).

OCCURRENCE: *Arabidopsis thaliana* shoots (Fujioka *et al.*, 1996); chestnut *Castanea crenata* Sieb. et Zucc. gall, shoot (current and 2-year-old), leaf and flower bud 9-30•10^{-7}% (Arima *et al.*, 1984); *Cupressus arizonica* pollen 1.2•10^{-4}% (Griffiths *et al.*, 1995a); *Dolichos lablab* immature seeds (Yokota *et al.*, 1984); *Ornithopus sativus* Brot. (Spengler *et al.*, 1995); rice *Oryza sativa* L. cv. Koshihikari bran (Abe *et al.*, 1995a); rice *Oryza sativa* cv. Tongjinbyeo immature seeds (Park, K.-H. *et al.*, 1993c, 1994b); *Phaseolus vulgaris* cv. Kentucky Wonder immature seeds (Yokota *et al.*, 1983d, 1987b); garden pea *Pisum sativum* fully grown seed (Yokota *et al.*, 1996a); *Robinia pseudo-acacia* L. pollen (Abe *et al.*, 1995b); rye *Secale cereale* L. var. 'Petka' seeds (Schmidt *et al.*, 1995b); wheat *Triticum aestivum* L. grain (Yokota *et al.*, 1994); corn *Zea mays* immature seeds (Park, K.-H. *et al.*, 1995).

¹H NMR (400 MHz, CDCl₃): δ 0.68 (*s*, 3H), 0.81 (*s*, 3H), 0.85 (*d*, 3H, *J* 7), 0.90 (*d*, 3H, *J* 7), 0.95 (*d*, 3H, *J* 7), 0.97 (*d*, 3H, *J* 7), 3.56 (*d*, 1H, *J* 8), 3.72 (*dd*, 1H, *J* 8, 2), 3.76 (*ddd*, 1H, *J* 9, 4, 3), 3.96 (*br s*, 1H) (Mori *et al.*, 1984). EI-MS (as bismethaneboronate): m/z (rel. int., %) 498 [M]⁺ (51), 483 (16), 455 (3), 343 (7), 313 (11), 288 (18), 273 (100), 205 (22), 155 (42) (Yokota *et al.*, 1983d).

3-Epi-6-deoxocastasterone (Bd3β) *[(22R,23R,24S)-2α,3β,22,23-tetrahydroxy-24-methyl-5α-cholestane]*. $C_{28}H_{50}O_4$, MW 450.70.

OCCURRENCE: *Phaseolus vulgaris* large-scale experiments with immature seeds (Kim, S.-K., 1991).

6-Deoxotyphasterol (Bd2d) *[(22R,23R,24S)-2α,22,23-trihydroxy-24-methyl-5α-cholestane]*. $C_{28}H_{50}O_3$, MW 434.70.

OCCURRENCE: *Arabidopsis thaliana* shoots (Fujioka *et al.*, 1996); *Cupressus arizonica* mature pollen 6.4•10⁻⁴% (Griffiths *et al.*, 1995a).

¹H NMR (400 MHz, CDCl₃): 0.67 (*s*, 3H), 0.78 (*s*, 3H), 0.85 (*d*, 3H, *J* 7), 0.89 (*d*, 3H, *J* 6.5), 0.94 (*d*, 3H, *J* 7), 0.97 (*d*, 3H, *J* 7), 3.57 (*d*, 1H, *J* 8), 3.72 (*dd*, 1H, *J* 8, 2) (Griffiths *et al.*, 1995a). EI-MS (as trimethylsilyl methaneboronate): m/z (rel. int., %) 530 [M]⁺ (23), 440 (31), 425 (29), 369 (7), 305 (17), 285 (17), 230 (34), 215 (100), 155 (29) (Griffiths *et al.*, 1995a).

3-Dehydro-6-deoxoteasterone (Bd2d3k) *[(22R,23R,24S)-22,23-dihydroxy-24-methyl-5α-cholestan-3-one]*. $C_{28}H_{48}O_3$, MW 432.69.

OCCURRENCE: *Cupressus arizonica* mature pollen 2.3•10⁻⁴% (Griffiths *et al.*, 1995a). EI-MS (as methaneboronate): m/z (rel. int., %) 456 [M]⁺ (64), 385 (8), 301 (24), 246 (23), 231 (100), 155 (64) (Griffiths *et al.*, 1995a).

Epicastasterone (EBk) *[24-epicastasterone, (22R,23R,24R)-2α,3α,22,23-tetrahydroxy-24-methyl-5α-cholestan-6-one]*. $C_{28}H_{48}O_5$, MW 464.69.

Mp 241-245 °C EtOAc (Akhrem et al., 1989c). α_D^{23} +1.0 (Anastasia et al., 1983c); α_D^{25} +0.0 (Thompson et al., 1979).

OCCURRENCE: sugar beet *Beta vulgaris* L. seeds (Schmidt et al., 1994); green alga *Hydrodictyon reticulatum* (L.) Lagerheim whole plant $3 \cdot 10^{-8}$% (Yokota et al., 1987a); *Daucus carota* ssp. *sativus* seeds (Adam et al., 1996b); *Ornithopus sativus* Brot. seeds $2.5 \cdot 10^{-6}$% (Schmidt et al., 1993a, 1996b,c), shoots (Spengler et al., 1995); *Phoenix dactylifera* pollen (Adam et al., 1996b); rhubarb *Rheum rhabarbarum* L. panicles (Schmidt et al., 1995a).

^{1}H NMR (500 MHz, CDCl$_3$): δ 0.68 (*s*, 3H, 18-H), 0.76 (*s*, 3H, 19-H), 0.85 (*d*, 3H, *J* 7.0, 28-H), 0.87 (*d*, 3H, *J* 7.0, 27-H), 0.92 (*d*, 3H, *J* 7.0, 26-H), 0.98 (*d*, 3H, *J* 6.7, 21-H), 1.11 (15β-H), 1.28 (12β-H), 1.30 (16β-H), 1.31 (14-H), 1.34 (11β-H), 1.40 (9-H), 1.47 (20-H), 1.50 (24-H), 1.55 (1β-H), 1.56 (17-H), 1.58 (15α-H), 1.65 (11α-H), 1.72 (4β-H), 1.74 (1α-H); 1.76 (8-H), 1.90 (25-H), 1.92 (4α-H), 1.98 (16α-H), 2.00 (7β-H), 2.02 (12α-H), 2.30 (7α-H), 2.69 (*dd*, 1H, 11.9, 4.9, 5-H), 3.41 (*ddd*, 1H, *J* 5.0, 4.5, 4.5, 23-H), 3.70 (*br dd*, 1H, *J* 4.5, 4.5, 22-H), 3.77 (*br.m*, 1H, W2 23, 2-H), 4.05 (*ddd*, 1H, *J* 14.4, 9.5, 4.9, 3-H) (Hai et al., 1996). ^{13}C NMR (CDCl$_3$, 75.5 MHz, 1-C⇒28-C): δ 39.8, 67.9, 68.0, 26.2, 50.7, 213.1, 46.6, 37.7, 52.5, 42.5, 21.1, 39.3, 42.7, 53.6, 23.8, 27.5, 56.4, 11.7, 13.4, 40.0, 12.2, 72.3, 75.8, 41.4, 26.8, 17.1, 22.0, 10.7 (Voigt et al., 1993a). EI-MS (as bismethaneboronate): m/z (rel. int., %) 512 [M]$^+$ (43), 497 (3), 441 (13), 358 (27), 287 (22), 155 (100) (Yokota et al., 1987a).

3,24-Diepicastasterone (EBk3β) *[(22R,23R,24R)-2α,3β,22,23-tetrahydroxy-24-methyl-5α-cholestan-6-one]*. $C_{28}H_{48}O_5$, MW 464.69.

Mp 213-215 °C CHCl$_3$ (Levinson and Traven, 1996).

OCCURRENCE: *Phaseolus vulgaris* large-scale experiments with immature seeds (Takahashi *et al.*, 1987; Kim, S.-K., 1991).

^{1}H NMR (500 MHz, CDCl$_3$): δ 0.68 (*s*, 3H, 18-H), 0.81 (*s*, 3H, 19-H), 0.85 (*d*, 3H, *J* 7.0, 28-H), 0.87 (*d*, 3H, *J* 7.0, 27-H), 0.92 (*d*, 3H, *J* 7.0, 26-H), 0.98 (*d*, 3H, *J* 6.7, 21-H), 1.24 (1β-H), 1.46 (20-H), 1.51 (24-H), 1.60 (4β-H), 1.90 (25-H), 1.96 (4α-H), 2.06 (1α-H), 2.32 (*dd*, 1H, *J* 11.9, 4.9, 5-H), 3.39 (*br m*, 1H, W2 21, 3-H), 3.41 (*ddd*, 1H, *J* 5.0, 4.5, 4.5, 23-H), 3.60 (*br m*, 1H, W2 24, 2-H), 3.69 (*br dd*, 1H, *J* 4.5, 4.5, 22-H) (Hai *et al.*, 1996).

6-Deoxo-24-epicastasterone (EBd) *[(22R,23R,24R)-2α,3α,22,23-tetrahydroxy-24-methyl-5α-cholestane]*. $C_{28}H_{50}O_4$, MW 450.70.

Mp 216-217 °C EtOAc-hexane (Spengler *et al.*, 1995). α_D^{25} -6.6 MeOH, c=3.47 (Spengler *et al.*, 1995).

OCCURRENCE: *Ornithopus sativus* shoots (Spengler *et al.*, 1995; Schmidt *et al.*, 1996b,c).

^{1}H NMR (500 MHz, CDCl$_3$): δ 0.66 (*s*, 3H, 18-H), 0.80 (*s*, 3H, 19-H), 0.84 (*d*, 3H, *J* 7.02, 28-H), 0.87 (*d*, 3H, *J* 6.72, 26-H), 0.92 (*d*, 3H, *J* 6.72, 27-H), 0.92 (*d*, 3H, *J* 6.41, 21-H), 3.40 (*m*, 1H, 23-H), 3.70 (*m*, 1H, 22-H), 3.76 (*m*, 1H, 2-H), 3.96 (*br.s*, 1H, 3-H) (Spengler *et al.*, 1995). EI-MS (as bismethaneboronate): m/z (rel. int., %) 498 [M]$^+$ (18), 483 (9), 427 (2), 343 (4), 288 (12), 273 (100), 213 (20), 205 (24), 155 (45) (Spengler *et al.*, 1995).

Dolichosterone (BDk) *[(22R,23R)-2α,3α,22,23-tetrahydroxy-5α-ergost-24(28)-en-6-one]*. $C_{28}H_{46}O_5$, MW 462.67.

Mp 233-237 °C MeCN-H$_2$O (Baba et al., 1983). α_D^{22} +4.0 c=0.20, MeOH (Mori et al., 1984).

OCCURRENCE: *Dolichos lablab* immature seeds $1.5 \cdot 10^{-7}$% (Baba et al., 1983); strobilus *Equisetum arvense* L. $3.5 \cdot 10^{-8}$% (Takatsuto et al., 1990a); *Eucalyptus marinata* pollen (Takatsuto, 1994a); *Phaseolus vulgaris* immature seeds (Yokota et al., 1984, 1987b); rice *Oryza sativa* cv. Arborio J1 fresh shoots $8.4 \cdot 10^{-10}$% (Abe et al., 1984a; Ikekawa et al., 1984).

^{1}H NMR (400 MHz, CDCl$_3$): 0.62 (*s*, 3H, 18-H), 0.75 (*s*, 3H, 19-H), 0.96 (*d*, 3H, *J* 7, 21-H), 1.09 (*d*, 3H, *J* 7, 26-H), 1.11 (*d*, 3H, *J* 27-H), 1.92 (*dt*, 1H, *J* 15.7, 4.3), 2.26 (*septet*, 1H, 25-H), 2.29 (*dd*, 1H, *J* 12.9, 4.3), 2.69 (*dd*, 1H, *J* 13, 3, 5-H), 3.62 (*d*, 1H, *J* 8, 22-H), 3.77 (*m*, 1H, W2 23, 2-H), 4.03 (*d*, 1H, *J* 23-H), 4.05 (*br s*, 1H, W2 8.5, 3-H), 5.04 (*s*, 1H, 28-H), 5.07 (*s*, 1H, 28-H) (Takatsuto and Ikekawa, 1983b). EI-MS (as bismethaneboronate): m/z (rel. int., %) 526 (17.9), 373 (21.1), 345 (21.1), 343 (100), 330 (2.1), 177 (8.4), 153 (70.5) (Ikekawa et al., 1984). IR (KBr): ν_{max} 3350 (s), 1710 (s), 1640 (w), 1080 (m), 1010 (m), 985 (m) cm^{-1} (Mori et al., 1984).

6-Deoxodolichosterone (Bdd) *[(22R,23R)-2α,3α,22,23-tetrahydroxy-5α-ergost-24(28)-ene)]*. $C_{28}H_{48}O_4$, MW 448.69.

Mp 219-221 °C EtOAc-MeOH (Mori et al., 1984). α_D^{24} +33.2 c=0.51, MeOH (Mori et al., 1984).

OCCURRENCE: *Phaseolus vulgaris* immature seeds (Yokota et al., 1983d, 1984, 1987b).

^{1}H NMR (400 MHz, CDCl$_3$ + D$_2$O): δ 0.63 (*s*, 3H), 0.80 (*s*, 3H), 0.94 (*d*, 3H, *J* 6.7), 1.08 (*d*, 3H, *J* 8), 1.11 (*d*, 3H, *J* 8), 2.27 (*m*, 1H), 3.63 (*m*, 1H), 3.75 (*m*, 1H), 3.95 (*br s*, 1H), 4.03 (*dd*, 1H, *J* 8, 2), 5.03 (*s*, 1H), 5.06 (*s*, 1H) (Mori et al.,

1984). EI-MS (as bismethaneboronate): m/z (rel. int., %) 496 [M]$^+$ (23), 313 (84), 273 (18), 153 (51), 124 (94), 82 (100) (Yokota et al., 1984). IR (KBr): v_{max} 3600-3100 (s), 1670 (w), 1065 (m), 1040 (m), 1020 (m), 990 (m), 910 (m) cm^{-1} (Mori et al., 1984).

Homobrassinolide **(HBl)** *[(22R,23R,24S)-2α,3α,22,23-tetrahydroxy-24-ethyl-B-homo-7-oxa-5α-cholestan-6-one]*. $C_{29}H_{50}O_6$, MW 494.71.

Mp 268-271 °C dec., EtOAc (Takatsuto and Ikekawa, 1982). α_D^{30} +41.7 c=0.501, CHCl$_3$-MeOH=9:1 (Zhu, J.-L., and Zhou, W.-S., 1991), α_D^{24} +42.9 c=0.68, CHCl$_3$-MeOH=9:1 (Sakakibara and Mori, 1982).

OCCURRENCE: Chinese cabbage *Brassica campestris* var. *pekinensis* L. (Ikekawa et al., 1984); *Perilla frutescens* immature seeds (Park, K.-H., 1993a).

^{1}H NMR (400 MHz, C$_5$D$_5$N): δ 0.73 (*s*, 3H), 1.05 (*s*, 3H), 1.07 (*d*, 3H, *J* 6.8), 1.10 (*t*, 3H, *J* 7.6), 1.15 (*d*, 3H, *J* 6.8), 1.21 (*d*, 3H, *J* 6.8), 2.31 (*dt*, 1H, *J* 14.6, 4.4), 2.51 (*ddd*, 1H, *J* 14.6, 12.2, 2.0), 3.61 (*dd*, 1H, *J* 12.2, 4.4), 3.97 (*d*, 1H, *J* 8.8), 4.00-4.14 (*m*, 4H), 4.43 (*br s*, 1H) (Sakakibara and Mori, 1982). ^{13}C NMR (25 MHz, CDCl$_3$-CD$_3$OD): δ 11.9, 12.1, 13.9, 15.6, 19.3, 19.5, 19.7, 21.4, 22.6, 25.1, 27.8, 29.5, 31.7, 37.6, 38.6, 39.6, 40.1, 41.4, 42.9, 52.9, 58.5, 68.2, 68.2, 71.1, 72.8, 74.6, 178.2 (Sakakibara and Mori, 1982). EI-MS (as bismethaneboronate): m/z (rel. int., %) 542 [M]$^+$ (0.90), 457 (6.3), 374 (22.5), 346 (5.9), 345 (16.2), 177 (47.3), 169 (100) (Ikekawa et al., 1984). IR (KBr): v_{max} 3450 (s), 2972 (s), 2948 (s), 2880 (s), 2850 (m), 1732 (m), 1715 (sh), 1701 (s), 1650 (w), 1470 (m), 1460 (m), 1445 (w), 1409 (m), 1388 (m), 1333 (m), 1320 (sh), 1300 (w), 1282 (m), 1260 (w), 1230 (m), 1190 (m), 1147 (m), 1130 (m), 1080 (sh), 1067 (s), 1040 (sh), 1030 (m), 1018 (sh), 990 (w) cm^{-1} (Takatsuto and Ikekawa, 1984b).

Homodolicholide (HBDl) *[(22R,23R,24(28)E)-24(28)-ethylidene-2α,3α,22,23-tetrahydroxy-B-homo-7-oxa-5α-cholestan-6-one]*. $C_{29}H_{48}O_6$, MW 492.70.

Mp 227-228 °C MeCN-H$_2$O (Yokota *et al.*, 1983b, 1984). α_D^{21} +35.4 c=0.48, CHCl$_3$-MeOH=4:1 (Sakakibara and Mori, 1983b; Mori *et al.*, 1984).

OCCURRENCE: *Dolichos lablab* immature seeds $3.5 \cdot 10^{-8}$% (Yokota *et al.*, 1983b, 1984).

^{1}H NMR (400 MHz, CDCl$_3$): δ 0.65 (*s*, 3H, 18-H), 0.92 (*s*, 3H, 19-H), 0.93 (*d*, 3H, *J* 7, 21-H), 1.06 (*d*, 3H, *J* 7, 26-H), 1.14 (*d*, 3H, *J* 7, 27-H), 1.72 (*dd*, 3H, *J* 7, 1, 29-H), 2.77 (*septet*, 1H, *J* 7, 25-H), 3.11 (*dd*, 1H, *J* 12, 4, 5-H), 3.68 (*dd*, 1H, *J* 8, 4, 22-H), 3.72 (*br*, 1H, 2-H), 3.95 (*d*, 1H, *J* 8, 23-H), 4.02 (*br*, 1H, W2 10, 3-H), 4.09 (*m*, 2H, 7-H), 5.51 (*q*, 1H, *J* 7, 28-H) (Yokota *et al.*, 1983b). IR (KBr): ν$_{max}$ 3500 (s), 3425 (s), 2980 (s), 2900 (m), 1740 (s), 1705 (s), 1650 (w), 1470 (m), 1445 (m), 1410 (m), 1383 (m), 1370 (sh), 1335 (s), 1283 (m), 1235 (w), 1225 (w), 1182 (s), 1170 (sh), 1145 (w), 1120 (w), 1090 (w), 1070 (w), 1040 (w), 1025 (m), 995 (m), 960 (m), 940 (w), 878 (w), 840 (w), 825 (w), 775 (w), 705 (w), 670 (w), 603 (m) cm^{-1} (Sakakibara and Mori, 1983b).

Homocastasterone (HBk) *[(24S)-ethylbrassinone, 28-homocastasterone, (22R,23R,24S)-2α,3α,22,23-tetrahydroxy-24-ethyl-5α cholestan-6-one]*. $C_{29}H_{50}O_5$, MW 478.71.

Mp 253-256 °C (EtOAc or EtOH) (Takatsuto and Ikekawa, 1984b; Lakhvich *et al.*, 1985a,b). α_D^{25} +12.5 c=0.82, CHCl$_3$ (Huang *et al.*, 1993), α_D^{25} +15 c=0.51, MeOH (Zhou, W.-S. *et al.*, 1990a).

OCCURRENCE: Chinese cabbage *Brassica campestris* var. *pekinensis* immature seeds and sheaths $1.3 \cdot 10^{-11}$% (Abe *et al.*, 1983; Ikekawa *et al.*, 1984; Abe,

1991); *Cupressus arizonica* pollen $4 \cdot 10^{-7}$%) (Griffiths *et al.*, 1995a); green alga *Hydrodictyon reticulatum* L. Lagerheim whole plant $4 \cdot 10^{-7}$% (Yokota *et al.*, 1987a); rice *Oryza sativa* cv. Arborio J1 $1.36 \cdot 10^{-9}$% (Abe *et al.*, 1984b); rye *Secale cereale* L. var. 'Petka' seeds (Schmidt *et al.*, 1995b); green tea *Thea sinensis* leaves trace amounts (Abe *et al.*, 1983; Ikekawa *et al.*, 1984; Abe, 1991).

EI-MS (as bismethaneboronate): m/z (rel. int., %) 526 [M]$^+$ (18.1), 441 (9.6), 358 (17.0), 329 (5.3), 169 (100) (Ikekawa *et al.*, 1984). IR (KBr): ν_{max} 3440 (s), 2948 (s), 2910 (sh), 2875 (s), 1710 (s), 1690 (sh), 1642 (w), 1467 (m), 1390 (m), 1370 (sh), 1340 (w), 1330 (w), 1312 (w), 1280 (m), 1262 (w), 1237 (w), 1130 (w), 1010 (w), 1085 (m), 1060 (sh), 1045 (m), 1019 (m), 993 (w) cm^{-1} (Takatsuto and Ikekawa, 1984b).

Homotyphasterol (HBk2d) *[28-homotyphasterol, (22R,23R,24S)-3α,22,23-trihydroxy-24-ethyl-5α-cholestan-6-one]*. $C_{29}H_{50}O_4$, MW 462.71.

Mp 239-240 °C EtOAc (Takatsuto and Ikekawa, 1984b).

OCCURRENCE: rice *Oryza sativa* L. bran (Abe *et al.*, 1995a).

^{1}H NMR (100 MHz, CDCl$_3$): δ 0.69 (*s*, 3H, 18-H), 0.76 (*s*, 3H, 19-H), 2.76 (*dd*, 1H, *J* 13, 5, 5-H), 3.58 (*d*, 1H, *J* 9, 22-H), 3.72 (*d*, 1H, *J* 9, 23-H), 4.17 (*m*, 1H, W2 8, 3-H) (Takatsuto and Ikekawa, 1984b). EI-MS (as trimethylsilyl methaneboronate): m/z (rel. int., %) 558 [M]$^+$ (100), 543 (42), 540 (28), 529 (74), 468 (78), 169 (64) (Takatsuto and Ikekawa, 1986a).

Homoteasterone (HBk2d3β) *[28-homoteasterone, (22R,23R,24S)-3β,22,23-trihydroxy-24-ethyl-5α-cholestan-6-one]*. $C_{29}H_{50}O_4$, MW 462.71.

Mp 206-209 °C EtOAc-hexane (Takatsuto and Ikekawa, 1984b).

OCCURRENCE: rice *Oryza sativa* L. bran (Abe *et al.*, 1995a); radish *Raphanus sativus* L. var. 'Remo' seeds (Schmidt *et al.*, 1993b).

^{1}H NMR (100 MHz, CDCl$_3$): δ 0.69 (*s*, 3H, 18-H), 0.75 (*s*, 3H, 19-H), 3.52 (*m*, 1H, 3-H), 3.58 (*d*, 1H, *J* 9, 22-H), 3.72 (*d*, 1H, *J* 9, 23-H) (Takatsuto and Ikekawa, 1984b). EI-MS (as trimethylsilyl methaneboronate): m/z (rel. int., %) 558 [M]$^+$ (47), 543 (93), 529 (60), 468 (14), 169 (100) (Schmidt *et al.*, 1993b).

Homodolichosterone (HBDk) *[28-homodolichosterone, (22R,23R,24(28)E)-24(28)-ethylidene-2α,3α,22,23-tetrahydroxy-5α-cholestan-6-one].*

C$_{29}$H$_{48}$O$_5$, MW 476.70.

Mp 225-226 °C CHCl$_3$-MeOH (Mori *et al.*, 1984). α$_D^{22}$ -9.3 c=0.41, CHCl$_3$-MeOH=9:1 (Sakakibara and Mori, 1983c), α$_D^{22}$ -9.8 c=0.60, CHCl$_3$-MeOH=9:1 (Mori *et al.*, 1984).

OCCURRENCE: *Dolichos lablab* immature seeds 6•10^{-8}% (Baba *et al.*, 1983; Yokota *et al.*, 1984).

^{1}H NMR (400 MHz, CDCl$_3$): δ 0.62 (*s*, 3H), 0.75 (*s*, 3H), 0.93 (*d*, 3H, *J* 6.4), 1.06 (*d*, 3H, *J* 7.1), 1.14 (*d*, 3H, *J* 7.1), 1.71 (*d*, 3H), 2.69 (*dd*, 1H, *J* 12.7, 3.0), 2.77 (*septet*, 1H, *J* 7.1), 3.69 (*ddd*, 1H, *J* 8.5, 3.5, 0.8), 3.76 (*m*, 1H), 3.96 (*dd*, 1H, *J* 8.8, 4.2), 4.05 (*br s*, 1H), 5.52 (*q*, 1H, *J* 7.0) (Sakakibara and Mori, 1983c). IR (KBr): ν$_{max}$ 3525 (s), 2960 (s), 2900 (m), 1710 (s), 1650 (w), 1480-1410 (m), 1385 (m), 1370 (sh), 1360-1310 (w), 1280 (w), 1270-1200 (w), 1135 (w), 1110 (w), 1090 (m), 1060 (m), 1050 (m), 1045 (m), 1015 (m), 990 (m), 980 (w), 955 (w), 940 (w), 870 (w), 840 (w), 830 (w), 670 (w), 590 (w) cm^{-1} (Sakakibara and Mori, 1983c).

6-Deoxohomodolichosterone (HBDd) *[6-deoxo-28-homodolichosterone, (22R,23R,24(28)E)-24(28)-ethylidene-2α,3α,22,23-tetrahydroxy-5α-cholestane].* C$_{29}$H$_{52}$O$_4$, MW 464.73.

Mp 225-227 °C MeOH (Takatsuto and Ikekawa, 1986d). α_D^{25} + 13.0 (Takatsuto and Ikekawa, 1986d).

OCCURRENCE: *Phaseolus vulgaris* immature seeds (Yokota *et al.*, 1987b).

^{1}H NMR (400 MHz, CDCl$_3$): δ 0.61 (*s*, 3H, 18-H), 0.79 (*s*, 3H, 19-H), 0.92 (*d*, 3H, *J* 6.5, 21-H), 1.06 (*d*, 3H, *J* 7.1, 26-H), 1.14 (*d*, 3H, *J* 7.1, 27-H), 1.71 (*d*, 3H, *J* 7.1, 29-H), 2.76 (*sept*, 1H, *J* 7.1, 25-H), 3.69 (*ddd*, 1H, *J* 8.5, 3.5, 0.8, 22-H), 3.76 (*m*, 1H, 2-H), 3.95 (*dd*, 1H, *J* 8.5, 4.0, 23-H), 3.96 (*br s*, 1H, 3-H), 5.51 (*q*, 1H, *J* 7.1, 28-H) (Takatsuto and Ikekawa, 1986d). EI-MS (as bismethaneboronate): m/z (rel. int., %) 510 [M]$^+$ (28), 467 (76), 413 (6), 365 (2), 356 (18), 343 (20), 313 (74), 283 (12), 273 (14), 255 (4), 235 (6), 222 (28), 167 (78), 138 (100), 96 (38), 82 (16) (Takatsuto and Ikekawa, 1986d).

25-Methylcastasterone (25HBk) *[(22R,23R,24R)-2α,3α,22,23-tetrahydroxy-24,25-dimethyl-5α-cholestan-6-one]*. C$_{29}$H$_{50}$O$_5$, MW 478.71.

Mp 251-253 °C MeOH (Mori and Takeuchi, 1988). α_D^{22} +14.3 c=0.11, MeOH (Mori and Takeuchi, 1988).

OCCURRENCE: ryegrass *Lolium perenne* L. pollen (Taylor *et al.*, 1993).

^{1}H NMR (400 MHz, CDCl$_3$): δ 0.69 (*s*, 3H), 0.76 (*s*, 3H), 0.85 (*d*, 3H, *J* 7.0, 21-H), 0.92 (*d*, 3H, *J* 6.7, 28-H), 0.96 (*s*, 9H), 2.30 (*dd*, 1H, *J* 13.2, 4.8), 2.69 (*dd*, 1H, *J* 12.7, 3.0), 3.48 (*dd*, 1H, *J* 9.0, 1.2, 23-H), 3.77 (*ddd*, 1H, *J* 12.0, 5.2, 3.2, 3-H), 3.77 (*dd*, 1H, *J* 9.0, 0.9, 22-H), 4.05 (*dd*, 1H, *J* 5.2, 2.7, 2-H) (Mori and Takeuchi, 1988). IR (KBr): v_{max} 3400 (s), 2940 (s), 2920 (sh), 2870 (s), 1700 (s), 1460 (m), 1440 (m), 1380 (s), 1365 (m), 1325 (m), 1230 (m), 1210 (sh), 1200 (sh), 1120 (m), 1080 (s), 1055 (sh), 1040 (s), 1010 (s), 975 (s), 945 (m), 930 (m), 905 (w), 870 (m), 835 (w) (Mori and Takeuchi, 1988).

25-Methyldolichosterone (25HBDk) [(22R,23R)-2α,3α,22,23-tetrahydroxy-25-methyl-5α-ergost-24(28)-en-6-one)]. $C_{29}H_{48}O_5$, MW 476.70.

Mp 254 °C (Kim, S.-K. et al., 1987). α_D^{22} +4.3 c=0.13, MeOH (Mori and Takeuchi, 1988).

OCCURRENCE: *Phaseolus vulgaris* immature seeds (Takahashi et al., 1987), < $8.8 \cdot 10^{-8}$% (Kim, S.-K. et al., 1987).

^{1}H NMR (400 MHz, CDCl$_3$): 0.61 (s, 3H, 18-H), 0.75 (s, 3H, 19-H), 0.96 (d, 3H, J 6.9, 21-H), 1.06 (ddd, 1H, J 5.8, 12.5, 12.5), 1.11 (s, 9H), 1.92 (ddd, 1H, J 15.0, 3.5, 3.5), 1.95-2.05 (m, 3H), 2.29 (dd, 1H, J 12.5, 4.5), 2.68 (dd, 1H, J 12.5, 2.8), 3.71-3.82 (m, 2H, 22-H and 23-H), 4.02-4.11 (m, 2H, 2-H and 3-H), 5.09 (s, 1H, 28-H), 5.15 (d, J 1.2, 28-H) (Mori and Takeuchi, 1988). IR (KBr): v_{max} 3430 (s), 2970 (s), 2930 (sh), 2890 (m), 1715 (s), 1385 (m), 1370 (w), 1335 (w), 1330 (w), 1320 (w), 1280 (m), 1260 (w), 1235 (w), 1210 (w), 1155 (w), 1130 (w), 1105 (w), 1090 (m), 1060 (sh), 1050 (m), 1020 (m), 990 (m), 945 (w), 940 (w), 910 (w), 875 (w), 865 (w) cm^{-1} (Mori and Takeuchi, 1988).

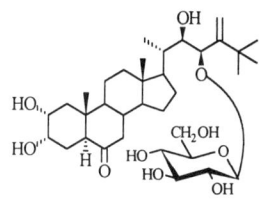

23-*O*-β-D-Glycopyranosyl-25-methyldolichosterone (25HBDk23-gly) [(22R,23R)-2α,3α,22-trihydroxy-23-O-β-D-glycopyranosyl-25-methyl-5α-ergost-24(28)-en-6-one)]. $C_{35}H_{58}O_{10}$, MW 638.84.

OCCURRENCE: *Dolichos lablab* immature seeds (Yokota et al., 1986).

^{1}H NMR (400 MHz, CDCl$_3$-CD$_3$OD=20:1): 0.64 (s, 3H, 18-H), 0.75 (s, 3H, 19-H), 1.01 (d, 3H, J 7, 21-H), 1.08 (s, 9H, 26-, 27- and 29-H), 2.68 (dd, 1H, J 12.1, 3.1, 5-H), 3.26 (ddd, 1H, J 9.0, 4.3, 3.4, 5'-H), 3.32 (dd, 1H, J 9.0, 8.0, 2'-H), 3.47 (dd, 1H, J 9.0, 9.0, 4'-H), 3.52 (dd, 1H, J 9.0, 9.0, 3'-H), 3.69 (d, 1H, J 6, 22-H), 3.71 (s, 1H, 2-H), 3.78 (dd, 1H, J 12.4, 4.3, 6'-H), 3.86 (dd, 1H, J 12.4, 3.4, 6'-H), 4.01 (br s, 1H, 3-H), 4.30 (d, 1H, J 6.0, 23-H), 4.32 (d, 1H, J 8.0, 1'-H), 5.15 (s, 1H, 28-H), 5.26 (s, 1H, 28-H) (Yokota et al., 1986).

2-Epi-25-methyldolichosterone (25HBDk2β)

[(22R,23R)-2β,3α,22,23-tetrahydroxy-25-methyl-5α-ergost-24(28)-en-6-one)]. $C_{29}H_{48}O_5$, MW 476.70.

OCCURRENCE: *Phaseolus vulgaris* large-scale experiments with immature seeds (Takahashi *et al.*, 1987; Kim, S.-K., 1991).

23-*O*-β-D-Glycopyranosyl-2-epi-25-methyldolichosterone (25HBDk2β23-gly)

[(22R,23R)-2β,3α,22-trihydroxy-23-O-β-D-glycopyranosyl-25-methyl-5α-ergost-24(28)-en-6-one)]. $C_{35}H_{58}O_{10}$, MW 638.84.

OCCURRENCE: *Dolichos lablab* immature seeds (Kim, S.-K., 1991).

2,3-Diepi-25-methyldolichosterone (25HBDk2β3β)

[(22R,23R)-2β,3β,22,23-tetrahydroxy-25-methyl-5α-ergost-24(28)-en-6-one)]. $C_{29}H_{48}O_5$, MW 476.70.

Mp 248 °C MeOH (Mori and Takeuchi, 1988). α_D^{22} +4.6 c=0.18, CHCl$_3$ (Mori and Takeuchi, 1988).

OCCURRENCE: *Phaseolus vulgaris* large-scale experiments with immature seeds (Takahashi *et al.*, 1987; Kim, S.-K., 1991).

^{1}H NMR (400 MHz, CDCl$_3$): 0.62 (*s*, 3H, 18-H), 0.96 (*d*, 3H, *J* 6.3, 21-H), 0.98 (*s*, 3H, 19-H), 1.11 (*s*, 9H), 1.38 (*dd*, 1H, *J* 13.8, 3.3), 1.48 (*dd*, 1H, *J* 12.0, 6.3, 20-H), 1.54 (*ddd*, 1H, *J* 10.0, 2.8, 2.8), 2.12 (*dd*, 1H, *J* 15.0, 2.8), 2.22 (*dd*, 1H, *J* 10.0, 2.8), 2.31 (*dd*, 1H, *J* 12.8, 5.0), 3.63 (*ddd*, 1H, *J* 12.0, 5.0, 3.5, 3-H), 3.77 (*d*, 1H, *J* 8.3, 22-H), 4.03 (*dd*, 1H, *J* 5.0, 2.2, 2-H), 4.06 (*d*, 1H, *J* 8.3, 23-H), 5.09 (*s*, 1H, 28-H), 5.15 (*d*, 1H, *J* 1.0, 28-H) (Mori and Takeuchi, 1988). IR (KBr): ν$_{max}$ 3450 (s), 2960 (s), 2880 (m), 1710 (s), 1385 (s), 1365 (m), 1280 (s), 1250 (s), 1160 (w), 1135 (w), 1060 (s), 1010 (m), 990 (s), 940 (w), 930 (w), 910 (w), 865 (w) cm^{-1} (Mori and Takeuchi, 1988).

2-Deoxy-25-methyldolichosterone (25HBDk2d) *[(22R,23R)-3α,22,23-trihydroxy-25-methyl-5α-ergost-24(28)-en-6-one)]*. $C_{29}H_{48}O_4$, MW 460.70.

OCCURRENCE: *Phaseolus vulgaris* large-scale experiments with immature seeds $4.7 \cdot 10^{-8}$% (Yokota and Takahashi, 1987; Kim, S.-K., 1991).

3-Epi-2-deoxy-25-methyldolichosterone (25HBDk2d3β) *[(22R,23R)-3β,22,23-trihydroxy-25-methyl-5α-ergost-24(28)-en-6-one)]*. $C_{29}H_{48}O_4$, MW 460.70.

OCCURRENCE: *Phaseolus vulgaris* large-scale experiments with immature seeds $6.3 \cdot 10^{-8}$% (Yokota and Takahashi, 1987; Kim, S.-K., 1991).

6-Deoxo-25-methyldolichosterone (25HBDd) *[(22R,23R)-2α,3α,22,23-tetrahydroxy-25-methyl-5α-ergost-24(28)-ene)]*. $C_{29}H_{50}O_4$, MW 462.71.

OCCURRENCE: *Phaseolus vulgaris* large-scale experiments with immature seeds (Kim, S.-K., 1991).

Norbrassinolide (NBl) *[28-norbrassinolide, (22R,23R)-2α,3α,22,23-tetrahydroxy-B-homo-7-oxa-5α-cholestan-6-one]*. $C_{27}H_{46}O_6$, MW 466.66.

Mp 257-260 °C (Takatsuto and Ikekawa, 1983c). α_D^{14} +32.0 c=1.15, MeOH (Takatsuto *et al.*, 1981, 1984b); α_D^{25} +24.0 (Thompson *et al.*, 1982).

OCCURRENCE: Chinese cabbage *Brassica campestris* var. *pekinensis* immature seeds and sheaths (Abe *et al.*, 1983; Ikekawa *et al.*, 1984; Abe, 1991); Japanese evergreen tree Isunoki *Distylium racemosum* Sieb. et Zucc. leaves $1.56 \cdot 10^{-8}$% (Abe, 1991).

^{1}H NMR (400 MHz, C$_5$D$_5$N): δ 0.69 (s, 3H, 18-H), 1.03 (d, 3H, J 7.1, 26-H), 1.05 (s, 3H, 19-H), 1.06 (d, 3H, J 7.1, 27-H), 1.21 (d, 3H, J 7.1, 21-H), 2.31 (dt, 1H, J 14.3, 4.3), 2.51 (t, 1H, J 14.3), 3.60 (dd, 1H, J 12.9, 4.3, 5-H), 3.74 (d, 1H, J 8.6, 22-H), 4.00-4.26 (m, 4H, 2-H, 7-H$_2$, and 23-H), 4.43 (m, 1H, W2 8.6, 3-H) (Takatsuto et al., 1984b). EI-MS (as bismethaneboronate): m/z (rel. int., %) 514 [M]$^+$ (2.5), 457 (0.83), 374 (10.4), 345 (5.8), 318 (27.9), 177 (57.5), 141 (100) (Ikekawa et al., 1984).

Brassinone (NBk) *[28-norcastasterone, (22R,23R)-2α,3α,22,23-tetrahydroxy-5α-cholestan-6-one].*
C$_{27}$H$_{46}$O$_5$, MW 450.66.
Mp 254-255 °C EtOAc (Takatsuto et al., 1984b).
α$_D^{25}$ -14.0 (Thompson et al., 1982).

OCCURRENCE: Chinese cabbage *Brassica campestris* var. *pekinensis* immature seeds and sheaths (Abe et al., 1983; Ikekawa et al., 1984; Abe, 1991); *Cassia tora* L. immature seeds 8•10^{-10}% (Park, K.-H. et al., 1994a); chestnut *Castanea* spp. insect galls 1.1•10^{-9}% (Abe et al., 1983) [however, Arima et al. (1984) could not detect brassinone in any of the tissues of *Castanea crenata* Sieb. et Zucc. and explained the corresponding SIM peak due to 6-deoxocastasterone]; Isunoki *Distylium racemosum* Sieb. et Zucc. 1.6•10^{-9}% (Abe, 1991); strobilus *Equisetum arvense* L. 3.49•10^{-8}% (Takatsuto et al., 1990a); sunflower *Helianthus annuus* L. pollen 2.1•10^{-6}% (Takatsuto et al., 1989b); rye *Secale cereale* L. var. 'Petka' seeds (Schmidt et al., 1995b); green tea *Thea sinensis* leaves 2•10^{-10}% (Abe et al., 1983); broad bean *Vicia faba* L. pollen 6.28•10^{-5}% (Gamoh et al., 1989a); corn *Zea mays* immature seeds (Park, K.-H. et al., 1995).

^{1}H NMR (400 MHz, C$_5$D$_5$N): δ 0.70 (s, 3H, 18-H), 0.85 (s, 3H, 19-H), 1.03 (d, 3H, J 7.1, 26-H), 1.06 (d, 3H, J 7.1, 27-H), 1.24 (d, 3H, J 7.1, 21-H), 2.32 (dt, 1H, J 15.7, 4.3), 2.36 (dd, 1H, J 12.9, 4.3), 3.13 (dd, 1H, J 12.9, 2.9, 5-H), 3.77 (dd, 1H, J 8.0, 1.4, 22-H), 4.01-4.10 (m, 2H, 2-H and 23-H), 4.43 (m, 1H, W2 8.6, 3-H) (Takatsuto et al., 1984b). EI-MS (as bismethaneboronate): m/z (rel.

int., %) 498 [M]$^+$ (100), 483 (13), 456 (9), 441 (7), 399 (5), 358 (23), 328 (10), 302 (15), 287 (55), 141 (59) (Park, K.-H. *et al.*, 1994a).

6-Deoxo-28-norcastasterone **(NBd)** *[(22R,23R)-2α,3α,22,23-tetrahydroxy-5α-cholestane]*. $C_{27}H_{48}O_4$, MW 436.68.

OCCURRENCE: *Ornithopus sativus* shoots (Spengler *et al.*, 1995; Schmidt *et al.*, 1996b,c).

EI-MS (as bismethaneboronate): m/z (rel. int., %) 484 [M]$^+$ (34), 469 (14), 343 (3), 313 (2), 288 (13), 273 (100), 213 (15), 205 (18), 141 (13) (Spengler *et al.*, 1995).

References

Abe, H. (1989). Advances in brassinosteroid research and prospects for its agricultural application. *Jpn. Pest. Inform.* 10-14.

Abe, H. (1991). Rice-lamina inclination, endogenous levels in plant tissues and accumulation during pollen development of brassinosteroids. *In* "Brassinosteroids. Chemistry, Bioactivity, and Applications. ACS Symposium Series" (H.G. Cutler, T. Yokota, and G. Adam, Eds.), Vol. 474, pp. 200-207. American Chemical Society, Washington.

Abe, H., and Marumo, S. (1991). Brassinosteroids in leaves of *Distylium racemosum* Sieb. et Zucc.: The beginning of brassinosteroids research in Japan. *In* "Brassinosteroids. Chemistry, Bioactivity, and Applications. ACS Symposium Series" (H.G. Cutler, T. Yokota, and G. Adam, Eds.), Vol. 474, pp. 18-24. American Chemical Society, Washington.

Abe, H., and Yuya, M. (1991). Preparation of 2-deoxybrassinosteroid derivatives as plant growth regulators. Jpn. Kokai Tokkyo Koho **JP 05,222,090 [93,222,090]** [*C. A.* **120**, 107478].

Abe, H., and Yuya, M. (1992). 2-Deoxy-3-acylbrassinosteroids. Jpn. Kokai Tokkyo Koho **JP 06 25,281 [94 25,281]** [*C. A.* **120**, 317801].

Abe, H., Morishita, T., Uchiyama, M., Marumo, S., Munakata, K., Takatsuto, S., and Ikekawa, N. (1982). Identification of brassinolide-like substances in chinese cabbage. *Agric. Biol. Chem.* **46**, 2609-2611.

Abe, H., Morishita, T., Uchiyama, M., Takatsuto, S., Ikekawa, N., Ikeda, M., Sassa, T., Kitsuwa, T., and Marumo, S. (1983). Occurrence of three new brassinosteroids: brassinone, 24(*S*)-24-ethylbrassinone and 28-norbrassinolide, in higher plants. *Experientia* **39**, 351-353.

Abe, H., Morishita, T., Uchiyama, M., Takatsuto, S., and Ikekawa, N. (1984a). A new brassinolide-related steroid in the leaves of *Thea sinensis*. *Agric. Biol. Chem.* **48**, 2171-2173.

Abe, H., Nakamura, K., Morishita, T., Uchiyama, M., Takatsuto, S., and Ikekawa, N. (1984b). Endogenous brassinosteroids of the rice plant: castasterone and dolichosterone. *Agric. Biol. Chem.* **48**, 1103-1104.

Abe, H., Honjo, C., Kyokawa, Y., Asakawa, S., Natsume, M., and Narushima, M. (1994). 3-Oxoteasterone and the epimerization of teasterone: identification in lily anthers and *Distylium racemosum* leaves and its biotransformation into typhasterol. *Biosci. Biotech. Biochem.* **58**, 986-989.

Abe, H., Takatsuto, S., Nakayama, M., and Yokota, T. (1995a). 28-Homotyphasterol, a new natural brassinosteroid from rice (*Oryza sativa* L.) bran. *Biosci. Biotech.*

Biochem. **59**, 176-178.

Abe, H., Takatsuto, S., Okuda, R., and Yokota, T. (1995b). Identification of castasterone, 6-deoxycastasterone, and typhasterol in the pollen of *Robinia pseudo-acacia* L. *Biosci. Biotech. Biochem.* **59**, 309-310.

Abe, H., Asakawa, S., and Natsume, M. (1996). Interconvertible metabolism between teasterone and its conjugate with fatty acid in cultured cells of lily. *Proc. Plant Growth Regul. Soc. Am.* **23**, 9.

Aburatani, M., Takeuchi, T., and Mori, K. (1984a). 6,6-Ethylenedioxy-22R-hydroxy-2R,3S-isopropylidenedioxy-5α-cholest-23-yne. UK Pat. Appl. **GB 2,156,353**.

Aburatani, M., Takeuchi, T., and Mori, K. (1984b). 6,6-Ethylenedioxy-22R-hydroxy-2R,3S-isopropylidenedioxy-5α-cholest-23-yne. Ger. Offen. **DE 3,511,716** [*C. A.* **105**, 43158].

Aburatani, M., Takeuchi, T., and Mori, K. (1985a). 3α,5-Cyclo-22,23-dihydroxy-5α-steroid compounds. Eur. Pat. Appl. **EP 201,042** [*C. A.* **106**, 102614].

Aburatani, M., Takeuchi, T., and Mori, K. (1985b). Structural revision of the acetal intermediates in brassinolide synthesis. *Agric. Biol. Chem.* **49**, 3557-3562.

Aburatani, M., Takeuchi, T., and Mori, K. (1986). Efficient synthesis of (22R,23R,24S)-22,23-isopropylidenedioxy-5α-ergost-2-en-6-one, a key intermediate in the preparation of brassinolide. *Agric. Biol. Chem.* **50**, 3043-3047.

Aburatani, M., Takeuchi, T., and Mori, K. (1987a). A simple synthesis of steroidal 3α,5-cyclo-6-ones and their efficient transformation to steroidal 2-en-6-ones. *Synthesis* 181-183.

Aburatani, M., Takeuchi, T., and Mori, K. (1987b). Facile syntheses of brassinosteroids: brassinolide, castasterone, teasterone and typhasterol. *Agric. Biol. Chem.* **51**, 1909-1913.

Adam, G. (1987). Synthese von Brassinosteroiden. *Z. Chem.* **27**, 41-49.

Adam, G., and Marquardt, V. (1986). Brassinosteroids. *Phytochemistry* **25**, 1787-1799.

Adam, G., Marquardt, V., Schönecker, B., and Hauschild, U. (1983). Preparation of 3β-(protected)hydroxypregna-5,7-diene-20-carboxaldehyde iron tricarbonyl complexes as intermediates for vitamin D derivatives, ecdysone analogs, and brassinolides. Ger. (East) **DD 247,904** [*C. A.* **109**, 73757].

Adam, G., Richter, K., and Vorbrodt, H.-M. (1986). Mittel zur Bekampfung von Insekten. Ger. (East) **DD 252,751** [*C. A.* **109**, 50273].

Adam, G., Bergmann, H., Lang, S., Meisgeier, G., and Vorbrodt, H.-M. (1987). Increase of potato tuber size by brassinosteroids. Ger. (East) **DD 265,316** [*C. A.* **111**, 189587].

Adam, G., Nguyen, T.K., and Vorbrodt, H.-M. (1988). Preparation of 5,6-secobrassinosteroid analogs as agrochemicals. Ger. (East) **DD 270,299** [*C. A.* **112**, 139654].

Adam, G., Vorbrodt, H.-M., Hörhold, C., Böhme, K.H., Dänhardt, S., Porzel, A., and Zeigan, D. (1989). Microbial 12β-hydroxylation of brassinosteroids. Ger. (East) **DD 287,957** [*C. A.* **115**, 112822].

Adam, G., Marquardt, V., Vorbrodt, H.-M., Hörhold, C., Andreas, W., and Gartz, J. (1991). Aspects of synthesis and bioactivity of brassinosteroids. *In* "Brassinosteroids. Chemistry, Bioactivity, and Applications. ACS Symposium Series" (H.G. Cutler, T. Yokota, and G. Adam, Eds.), Vol. 474, pp. 74-85. American Chemical Society, Washington.

Adam, G., Schneider, B., Kolbe, A., Hai, T., Porzel, A., and Voigt, B. (1996a). Progress in brassinosteroid metabolism. *Proc. Plant Growth Regul. Soc. Am.* **23**, 7.

Adam, G., Porzel, A., Schmidt, J., Schneider, B., and Voigt, B. (1996b). New developments in brassinosteroid research. *In* "Studies in Natural Product Chemistry" (Atta-ur-Rahman, Ed.), Vol. 18, pp. 495-549.

Ahmad, M.S., Shafiullah, M.M., and Asif, M. (1970). Baeyer-Villiger oxidation of $3\alpha,5$-cyclocholestan-6-one and its related 3β-halo derivatives. *Indian J. Chem.* **8**, 1062-1064.

Ahmad, M.S., Ansari, I.A., and Moinuddin, G. (1981). Beckmann rearrangement of some ketoximes of the stigmastane series: comparison of mass spectrum of 3β-acetoxy-7a-aza-B-homostigmast-5-en-7-one with its cholestane analogue. *Indian J. Chem., Sect. B* **20**, 602-604.

Akhrem, A.A., and Kovganko, N.V. (1989). "Ecdysteroids: chemistry and biological activity". Science and Technique, Minsk.

Akhrem, A.A., and Titov, Yu.A. (1965). "Microbiological transformations of steroids". Science, Moscow.

Akhrem, A.A., and Titov, Yu.A. (1967). "Total synthesis of steroids", Science, Moscow [Engl. transl. Akhrem, A.A., and Titov, Yu.A. (1970)."Total steroid synthesis". Plenum Press, New York].

Akhrem, A.A., Levina, I.S., and Titov, Yu.A. (1973). "Ecdysones - steroidal hormones of insects". Science and Technique, Minsk.

Akhrem, A.A., Lakhvich, F.A., Khripach, V.A., and Kovganko, N.V. (1981). Woodward's hydroxylation of steroidal 2,4-dienes. *Dokl. AN USSR* **257**, 1133-1135.

Akhrem, A.A., Lakhvich, F.A., Khripach, V.A., and Kovganko, N.V. (1982). Synthesis of $2\alpha,3\alpha$-diacetoxy-5α-cholest-7-en-6-one. *Vesti AN BSSR, Ser. Khim. Navuk*, 80-83.

Akhrem, A.A., Lakhvich, F.A., Khripach, V.A., and Kovganko, N.V. (1983a). A new way of introduction of functional groups characteristic of brassinosteroids into A and B cycles of Δ^5-sterols. *Dokl. AN USSR* **269**, 366-368.

Akhrem, A.A., Lakhvich, F.A., Khripach, V.A., and Kovganko, N.V. (1983b). Hydroxylation of steroidal 2,4-diene-6-ones. *Zh. Org. Khim.* **19**, 1249-1256.

Akhrem, A.A., Lakhvich, F.A., Khripach, V.A. and Kovganko, N.V. (1983c). Preparation of steroidal 2,4-diene-6-ketones. *Vesti AN BSSR, Ser. Khim. Navuk*, 65-70.

Akhrem, A.A., Lakhvich, F.A., Khripach, V.A., Kovganko, N.V., and Zhabinskii, V.N. (1984a). A new synthesis of (22S,23S)-28-homocastasterone. *Dokl. AN USSR* **275**, 1089-1091.

Akhrem, A.A., Lakhvich, F.A., Khripach, V.A., Kovganko, N.V., and Zhabinskii, V.N. (1984b). An unusual reaction of 24S-ethyl-3α,5-cyclo-5α-cholest-22-en-6-one with trifluoroperacetic acid. *Dokl. AN USSR* **275**, 626-628.

Akhrem, A.A., Lakhvich, F.A., Khripach, V.A., Zhabinskii, V.N., and Kovganko, N.V. (1984c). Bromination of 24S-ethyl-3α,5-cyclo-5α-cholest-22-en-6-one. *Zh. Org. Khim.* **20**, 2140-2143.

Akhrem, A.A., Kovganko, N.V., Lakhvich, F.A., and Khripach, V.A. (1985a). Advances in synthesis of ecdysones and related compounds. *Vesti AN BSSR, Ser. Khim. Navuk*, 96-117.

Akhrem, A.A., Lakhvich, F.A., Khripach, V.A., Kovganko, N.V., and Zhabinskii, V.N. (1985b). A new synthesis of (22S,23S)-28-homobrassinolide. *Dokl. AN USSR* **283**, 130-133.

Akhrem, A.A., Lakhvich, F.A., Khripach, V.A., Zhabinskii, V.N., and Kovganko, N.V. (1985c). Preparation of 3α,5-cyclo-24S-ethyl-5α-cholest-22-en-6-one. U.S.S.R. **SU 1,162,816**.

Akhrem, A.A., Khripach, V.A., Lakhvich, F.A., Zavadskaya, M.I., Drachenova, O.A., and Zorina, I.A. (1987a). A new approach to formation of steroidal polyfunctional side chain. *Dokl. AN USSR* **297**, 364-367.

Akhrem, A.A., Lakhvich, F.A., Khripach, V.A., Kovganko, N.V., and Zhabinskii, V.N. (1987b). Synthesis of 22S,23S-brassinosteroids starting from stigmasterol. *Zh. Org. Khim.* **23**, 762-770.

Akhrem, A.A., Lakhvich, F.A., Khripach, V.A., Kovganko, N.V., Zhabinskii, V.N., Bychovets, A.I., Borisova, M.P., Kuzmitskii, B.B., Misulo, N.A., Streltsova, V.A., Kandelinskaya, O.L., Bushueva, S.A., Mironenko, A.V., D'yakova, R.V., and Chodortsov, I.R. (1987c). Plant growth stimulator. U.S.S.R. **SU 1,577,104**.

Akhrem, A.A., Khripach, V.A., Kashkan, Zh.N., and Kovganko, N.V. (1988a). Woodward's hydroxylation of steroidal 2,4-diene-6-ketoximes. *Dokl. AN USSR* **301**, 878-880.

Akhrem, A.A., Lakhvich, F.A., Khripach, V.A., Kovganko, N.V., Zhabinskii, V.N., Bychovets, A.I., and Borisova, M.P. (1988b). 2α,3α,22S,23S-Tetrahydroxy-24S-ethylcholest-4-en-6-one, possessing plant growth promoting activity. U.S.S.R. **SU 1,433,005**.

Akhrem, A.A., Khripach, V.A., Lakhvich, F.A., Zavadskaya, M.I., Drachenova, O.A., and Zorina, I.A. (1989a). The isoxazole route to formation of polyfunctional steroidal side chain. *Zh. Org. Khim.* **25**, 2120-2128.

Akhrem, A.A., Khripach, V.A., Litvinovskaya, R.P., Baranovskii, A.V., Zavadskaya, M.I., Kharitonovich, A.N., Borisov, E.V., and Lakhvich, F.A. (1989b). 1,3-Dipolar cycloaddition of nitrile oxides to Δ^{22}-steroids. *Zh. Org. Khim.* **25**, 1901-1907.

Akhrem, A.A., Khripach, V.A., Zhabinskii, V.N., and Olkhovick, V.K. (1989c). Synthesis of (24R)-brassinosteroids starting from ergosterol. *Vesti AN BSSR, Ser. Khim. Navuk*, 69-73.

Akhrem, A.A., Lakhvich, F.A., Khripach, V.A., Kovganko, N.V., and Zhabinskii, V.N. (1989d). Hydroxylation of steroidal 2,4-diene-6-ketones of stigmastane and spirostane series. *Zh. Org. Khim.* **25**, 1661-1665.

Akhrem, A.A., Galitskii, N.M., Yasnitskii, G.A., Gritsuk, V.I., Lakhvich, F.A., Khripach, V.A., and Kovganko, N.V. (1990a). Hydroxylation of steroidal 2,4-diene-6-ketones. X-ray investigation of 2α,3α-diacetoxy-4-en-6-one. *Zh. Org. Khim.* **26**, 771-775.

Akhrem, A.A., Kananovich, O.P., and Kovganko, N.V. (1990b). Study on the reaction of 3α,5-cyclopregna-6,20-dione with trifluoroperacetic acid. *Zh. Org. Khim.* **25**, 1050-1051.

Allevi, P., Anastasia, M., Cerana, R., Ciuffreda, P., and Lado, P. (1988). 24-Epibrassinolide uptake in growing maize root segments evaluated by multiple-selected ion monitoring. *Phytochemistry* **27**, 1309-1313.

Amann, A., Ourisson, G., and Luu, B (1987). Stereospecific synthesis of the four epimers of 7,22-dihydroxycholesterol. *Synthesis* 1002-1005.

Amrhein, N. (1981). Growth [of plants]. *Prog. Bot.* **43**, 100-118.

Anastasia, M., and Fiecchi, A. (1981). Stereospecific synthesis of the marine sterol stellasterol, (22E,24S)-5α-ergosta-7,22-dien-3β-ol. *J. Org. Chem.* **46**, 1726-1728.

Anastasia, M., Scala, A., and Galli, G. (1976). Synthesis of 14β-cholest-5-en-3β-ol. *J. Org. Chem.* **41**, 1064-1067.

Anastasia, M., Allevi, P., Ciuffreda, P., and Fiecchi, A. (1983a). Stereoselective synthesis of crinosterol [(22E,24S)-ergosta-5,22-dien-3β-ol]. *J. Chem. Soc., Perkin Trans. I* 2365-2367.

Anastasia, M., Allevi, P., Ciuffreda, P., and Fiecchi, A. (1983b). A convenient protection of the 3β-hydroxy or 3β-chlorosubstituent of 6-oxo-steroids. *Synthesis* 123-124.

Anastasia, M., Ciuffreda, P., and Fiecchi, A. (1983c). A new synthesis of brassinosteroids: plant growth promoting steroids. *J. Chem. Soc., Perkin Trans. I* 379-382.

Anastasia, M., Ciuffreda, P., Del Puppo, M., and Fiecchi, A. (1983d). Synthesis of castasterone and its 22S,23S-isomer: two plant growth promoting ketones. *J. Chem. Soc., Perkin Trans. I* 383-386.

Anastasia, M., Allevi, P., Brasca, M.G., Ciuffreda, P., and Fiecchi, A. (1984a). Synthesis of (2R,3S,22S,23S)-2,3,22,23-tetrahydroxy-B-homo-6-aza-5α-stigmastan-7-one, an aza analog of brassinolide. *Gazz. Chim. Ital.* **114**, 159-161.

Anastasia, M., Allevi, P., Ciuffreda, P., Fiecchi, A., and Scala, A. (1984b). Direct transformation of ergosterol to (22S,23E)-6β-methoxy-3α,5-cyclo-5α-ergost-23-en-22-ol, a key intermediate for the synthesis of brassinolide. *J. Org. Chem.* **49**, 4297-4300.

Anastasia, M., Allevi, P., Ciuffreda, P., and Oleotti, A. (1985a). A convenient isomerization of 6-oxo-3α,5-cyclo-5α-steroids to 6-oxo-Δ^2-5α-steroids. *Steroids* **45**, 561-564.

Anastasia, M., Allevi, P., Ciuffreda, P., Fiecchi, A., and Scala, A. (1985b). Synthesis of (2R,3S,22R,23R)- and (2R,3S,22S,23S)-2,3,22,23-tetrahydroxy-B-homo-7a-oxa-5α-ergostan-7-ones, two new brassinolide analogues. *J. Org. Chem.* **50**, 321-325.

Anastasia, M., Allevi, P., Ciuffreda, P., Fiecchi, A., and Scala, A. (1986a). Synthesis of (2R,3S,22S,23S)-2,3,22,23-tetrahydroxy-B-homo-7-aza-5α-stigmastan-6-one, an aza-analogue of homobrassinolide. *J. Chem. Soc., Perkin Trans. I* 2117-2121.

Anastasia, M., Allevi, P., Ciuffreda, P., Fiecchi, A., and Scala, A. (1986b). Beckmann rearrangement of Δ^4-6-hydroxyimino steroids. *J. Chem. Soc., Perkin Trans. I* 2123-2126.

Ando, T., Aburatani, M., Koseki, N., Asakawa, S., Mouri, T., and Abe, H. (1993). ^{13}C NMR assignments of brassinosteroids by two-dimensional techniques. *Magn. Res. Chem.* **31**, 94-99 [*C. A.* **118**, 147870].

Arima, M., Yokota, T., and Takahashi, N. (1984). Identification and quantification of brassinolide-related steroids in the insect gall and healthy tissues of the chestnut plant. *Phytochemistry* **23**, 1587-1591.

Arteca, R.N. (1984). Calcium(2$^+$) acts synergistically with brassinosteroid and indole-3-acetic acid in stimulating ethylene production in mung bean hypocotyl segments. *Physiol. Plant.* **62**, 102-104.

Arteca, R.N., and Bachman, J.M. (1987). Light inhibition of brassinosteroid-induced ethylene production. *J. Plant Physiol.* **129**, 13-18.

Arteca, R., and Schlagnhaufer, C. (1984). The effect of brassinosteroid 2,4-D-a-aminoacid conjugates on ethylene production by etiolated mung bean segments. *Physiol. Plant.* **62**, 445-447.

Arteca, R.N., Tsai, D.S., Schlagnhaufer, C., and Mandava, N.B. (1983). The effect of brassinosteroid on auxin-induced ethylene production by etiolated mung bean segments. *Physiol. Plant.* **59**, 539-544.

Arteca, R.N., Bachman, J.M., and Yopp, J.H. (1984). The relation of brassinosteroid and its ability to promote ethylene production in etiolated mung bean segments. *Plant Physiol. (Suppl.)* **75**, 1046.

Arteca, R.N., Bachman, J.M., Yopp, J.H., and Mandava, N.B. (1985). Relationship of steroidal structure to ethylene production by etiolated mung bean segments. *Physiol. Plant.* **64**, 13-16.

Arteca, R.N., Bachman, J.M., and Mandava, N.B. (1988a). Effects of indole-3-acetic acid and brassinosteroid on ethylene biosynthesis in etiolated mung bean hypocotyl segments. *J. Plant Physiol.* **133**, 430-435.

Arteca, R.N., Bachman, J.M., Tsai, D.-S., and Mandava, N.B. (1988b). Fusicoccin, an inhibitor of brassinosteroid-induced ethylene production. *Physiol. Plant.* **74**, 631-634.

Arteca, R.N., Tsai, D.-S., and Mandava, N.B. (1991). The inhibition of brassinosteroid-induced ethylene biosynthesis in ethiolated mung bean hypocotyl segments by 2,3,5-triiodobenzoic acid and 2-(p-chlorophenoxy)-2-methylpropionic acid. *J. Plant Physiol.* **139**, 52-56.

Asakawa, S., Abe, H., Kyokawa, Y., Nakamura, S., and Natsume, M. (1994). Teasterone 3-myristate: a new type of brassinosteroid derivative in *Lilium longiflorium* anthers. *Biosci. Biotech. Biochem.* **58**, 219-220.

Asakawa, S., Abe, H., Nishikawa, N., Natsume, M., and Koshioka, M. (1996). Purification and identification of new acyl-conjugated teasterones in lily pollen. *Biosci. Biotech. Biochem.* **60**, 1416-1420.

Asatova, S.S. (1991). Effect of epibrassinolide on growth and development of vegetables. *In* "Conference on brassinosteroids", 2nd, p. 45, Minsk.

Baba, J., Yokota, T., and Takahashi, N. (1983). Brassinolide-related new bioactive steroids from *Dolichos lablab* seed. *Agric. Biol. Chem.* **47**, 659-661.

Back, T.G. (1995). Stereoselective synthesis of brassinosteroids. *In* "Studies in Natural Product Chemistry" (Atta-ur-Rahman, Ed.), Vol. 16, pp. 321-364.

Back, T.G., and Baron, D.L. (1996). Stereoselectivity of directed epoxidation of 22-hydroxy-Δ^{23}-sterol side chain. *Can. J. Chem.* **74**, 1857-1867.

Back, T.G., and Krishna, M.V. (1991). Synthesis of castasterone and formal synthesis of brassinolide from stigmasterol *via* a selenosulfonation approach. *J. Org. Chem.* **56**, 454-457.

Back, T.G., Brunner, K., Krishna, M.V., and Lai, E.K.Y. (1989). A selenosulfonation-based approach to (22R,23E)-6β-methoxy-3α,5-cyclo-5α-ergost-23-en-22-ol, a key intermediate for brassinolide synthesis. *Can. J. Chem.* **67**, 1032-1037.

Back, T.G., Blazecka, P.G., and Krishna, M.V. (1991). A concise synthesis of the brassinolide side chain. *Tetrahedron Lett.* **32**, 4817-4818.

Back, T.G., Blazecka, P.G., and Krishna, M.V. (1993). A new synthesis of castasterone and brassinolide from stigmasterol. A concise and stereoselective elaboration of the side chain from a C-22 aldehyde. *Can. J. Chem.* **71**, 156-163.

Back, T.G., Baron, D.L., Luo, W., and Nakajima, S.K. (1997a). Concise, improved procedure for the synthesis of brassinolide and some novel side-chain analogs. *J. Org. Chem.* **62**, 1179-1182.

Back, T.G., Baron, D.L., Luo, W., Nakajima, S.K., Janzen, L., and Pharis, R.P. (1997b). A concise synthesis of brassinolide and the preparation and bioactivity of some novel analogues. *Proc. Plant Growth Regul. Soc. Am.* **24**, 107-110.

Baggiolini, E.G., Iacobelli, J.A., Hennessy, B.M., and Uskokovich, M.R. (1982). Stereoselective total synthesis of 1α,25-dihydroxycholecalciferol. *J. Am. Chem. Soc.* **104**, 2945-2948.

Baiguz, A., and Czerpak, R. (1995). The occurrence and biological activity of brassinosteroids - a new type of plant hormones. *Kosmos (Warsaw)* **44**, 129-144 [*C. A.* **123**, 193557].

Baiguz, A., and Czerpak, R. (1996). Effect of brassinosteroids on growth and proton extrusion in the alga *Chlorella Beijerinck* (Chlorophyceae). *J. Plant Growth Regul.* **15**, 153-156 [*C. A.* **126**, 44788].

Balmush, G.T., Russu, M.M., and Karabadzhak. (1995). Effect of epibrassinolide on tomato growth and development. *In* "Brassinosteroids - biorational, ecologically safe regulators of growth and productivity of plants", 4th, pp. 22-23, Minsk.

Baranovskii, A.V., Litvinovskaya, R.P., and Khripach, V.A. (1993). Steroids with a side chain containing a heterocyclic fragment: synthesis and transformations. *Russ. Chem. Rev.* **62**, 661-682.

Barker, G.C., Mercer, J.G., Svoboda, J.A., Thompson, M.J., Rees, H.H., and Howells, R.E. (1989). Effects of potential inhibitors on *Brugia pahangi* in vitro: macrofilaricidal action and inhibition of microfilarial production. *Parasitology* **99**, 409-416.

Barton, D.H.R., Feakins, P.G., Poyser, J.P., and Sammes, P.G. (1970). A synthesis of the insect moulting hormone, ecdysone and related compounds. *J. Chem. Soc.(C)* 1584-1591.

Barton, D.H.R., Poyser, J.P., and Sammes, P.G. (1972). Some stereoselective and regioselective olefin additions: iodoacetoxylation and related electrophilic additions across the 22(23)-bond of $3\alpha,5\alpha$-cycloergosta-7,22-dien-6-one. *J. Chem. Soc., Perkin Trans. I* 53-55.

Barton, D.H.R., Lusinchi, X., Magdzinski, L., and Ramirez, J.S. (1984). Selective reduction of 7(8)-double bond in ergosterol. *J. Chem. Soc., Chem. Commun.* 1236-1238.

Batcho, A.D., Berger, D.E., Davoust, S.G., Wovkulich, P.M., and Uskokovic, M.R. (1981a). Stereoselective introduction of steroid side chains at C(17) and C(20). *Helv. Chim. Acta* **64**, 1682-1687.

Batcho, A.D., Berger, D.E., and Uskokovic, M.R. (1981b). C-20 Stereospecific introduction of a steroid side chain. *J. Am. Chem. Soc.* **103**, 1293.

Beinhauer, K., Andreas, W., Creuzburg, D., Lang, S., Vorbrodt, H.-M., and Adam, G. (1990). Brassinosteroid-induced stimulation of growth and invertase activity in radish cotyledons. *In* "Int. Workshop. Brassinosteroids: Chemistry, Bioactivity, Application", p. 22, Halle.

Bellincampi, D., and Morpurgo, G. (1988). Stimulation of growth in *Daucus carota* L. cell cultures by brassinosteroid. *Plant Sci.* **54**, 153-156.

Bellincampi, D., and Morpurgo, G. (1991). Stimulation of growth induced by brassinosteroid and conditioning factors in plant-cell cultures. *In* "Brassinosteroids. Chemistry, Bioactivity, and Applications. ACS Symposium Series" (H.G. Cutler, T. Yokota, and G. Adam, Eds.), Vol. 474, pp. 189-199. American Chemical Society, Washington.

Berthon, L., Tahri, A., and Uguen, D. (1994). A C-B-A-D approach to brassinosteroids; obtention of a C-ring precursor from pyridine. *Tetrahedron Lett.* **35**, 3937-3940.

Bobrick, A.O. (1993). Brassinosteroids in the potato breeding. *In* "Brassinosteroids - biorational, ecologically safe regulators of growth and productivity of plants", 3rd, p. 27, Minsk.

Bobrick, A.O. (1995). Application of brassinosteroids in the potato breeding. *In* "Brassinosteroids - biorational, ecologically safe regulators of growth and productivity of plants", 4th, p. 23, Minsk.

Bogoslovskii, N.A., Litvinova, T.E., and Samokhvalov, G.I. (1978). Study on vitamin D. Synthesis of vitamin D_3 analogues containing one or two double bonds in a side chain. *Zh. Obsch. Khim.* **48**, 908-913.

Bokebayeva, G.A., and Khripach, V.A. (1993). Effect of 24-epibrassinolide on seed germination and growrh of barley plants under salt stress. *In* "Brassinosteroids - biorational, ecologically safe regulators of growth and productivity of plants", 3rd, p. 21, Minsk.

Bokebayeva, G.A., Khokhlova, V.A., Khripach, V.A., Adam, G., and Kulaeva, O.N. (1990). Brassinosteroids' protective effect on barley leaves under saline stress conditions. *In* "Int. Workshop. Brassinosteroids: Chemistry, Bioactivity, Application", p. 45, Halle.

Bokebayeva, G.A., Khokhlova, V.A., Khripach, V.A., and Kulaeva, O.N. (1990). Method for assessment of antistress activity of brassinosteroids. U.S.S.R. **SU 1,750,493** [*C. A.* **119**, 44717].

Braun, P., and Wild, A. (1984a). The influence of brassinosteroid on growth and parameters of photosynthesis of wheat and mustard plants. *J. Plant Physiol.* **116**, 189-196 [*C. A.* **101**, 227169].

Braun, P., and Wild, A. (1984b). The influence of brassinosteroid, a growth-promoting steroidal lactone, on development and CO_2-fixation capacity of intact wheat and mustard seedlings. *In* "Advances in photosynthesis research", Vol. III, pp. 462-464.

Brewster, J.H. (1986). On the distinction of diastereoisomers in the Cahn-Ingold-Prelog (*RS*) notation. *J. Org. Chem.* **51**, 4751-4753.

Brooks, C.J.W., and Ekhato, I.V. (1982). Highly regioselective reduction of ring B seco-5α-steroid anhydrides to afford the lactone grouping characteristic of brassinolide. *J. Chem. Soc., Chem. Commun.* 943-944.

Brosa, C. (1997). Biological effects of brassinosteroids. *In* "Biochem. Funct. Sterols", (E.J. Parish, and W.D. Nes, Eds.), Chapt. 15, pp. 201-220. CRC Press.

Brosa, C., and Miro, X. (1997). Synthesis of new brassinosteroids with epoxy functions: the effect on the regioselectivity of the Baeyer-Villiger reaction. *Tetrahedron* **53**, 11347-11354.

Brosa, C., Peracaula, R., Puig, R., and Ventura, M. (1992). Use of dihydroquinidine 9-O-(9'-phenanthryl)ether in osmium-catalyzed asymmetric dihydroxylation in the synthesis of brassinosteroids. *Tetrahedron Lett.* **33**, 7057-7060.

Brosa, C., Nusimovich, S., and Peracaula, R. (1994). Synthesis of new brassinosteroids with potential activity as antiecdysteroids. *Steroids* **59**, 463-467.

Brosa, C., Capdevila, J.M., and Zamora, I. (1996a). Brassinosteroids: a new way to define the structural requirements. *Tetrahedron* **52**, 2435-2448.

Brosa, C., Puig, R., Comas, X., and Fernandez, C. (1996b). New synthetic strategy for the synthesis of 24-epibrassinolide. *Steroids* **61**, 540-543.

Brosa, C., Soca, L., Terricabras, E., and Zamora, I. (1996c). Brassinosteroids: looking for a practical solution. *Proc. Plant Growth Regul. Soc. Am.* **23**, 21-26.

Brosa, C., Zamora, I., Terricabras, E., Soca, L., Peracaula, R., and Rodriguez-Santamarta, C. (1997). Synthesis and molecular modeling: related approaches to progress in brassinosteroid research. *Lipids* **32**, 1341-1347.

Budnikov, D.A., and Alexashina, Z.A. (1995). Technical report "Toxicologo-hygienic evaluatin of Epin". Sanitary-Hygienic Institute of Belarus, Minsk.

Burchanova, E.A., Fedina, A.B., and Danilova, N.V. (1991). Action of homobrassinolide, human interferone and (2-5)-oligoadenilates on synthesis of proteins in leaves of wheat. *Biokhimiya* **56**, 1228-1234.

Caballero, G., Centurion, O.T., Galagovsky, L., and Gros, E. (1996). FAB mass spectrometry of 5α-H and 5α-OH derivatives of 28-homocastasterone and 28-homoteasterone. *Proc. Plant Growth Regul. Soc. Am.* **23**, 50-55.

Caballero, G.M., Gros, E.G., Centurion, O.T., and Galagovsky, L.R. (1997). Fast-atom bombardment mass spectrometry of brassinosteroid analogs. *J. Am. Soc. Mass Spectrom.* **8**, 270-274.

Cahn, R.S., Ingold, C.K., and Prelog, V. (1966). Specification of molecular chirality. *Angew. Chem., Int. Ed. Engl.* **5**, 385-414.

Calverley, M.J. (1987). Synthesis of MC 903, a biologically active vitamin D metabolite analogue. *Tetrahedron* **43**, 4609-4619.

Cao, H., and Chen, S. (1995). Brassinosteroid-induced rice lamina joint inclination and its relation to indole-3-acetic acid and ethylene. *Plant Growth Regul.* **16**, 189-196.

Cao, H., Chen, S., and Jiang, J. (1987). Isolation and biological activity of brassinolide-like substances in bees wax. *Zizan Zazhi* **10**, 952-953 [*C. A.* **108**, 200138].

Castedo, L., Granja, J.R., and Mourino, A. (1985). [2,3]Wittig sigmatropic rearrangements in steroid synthesis. New stereocontrolled approach to steroidal side chains at C-20. *Tetrahedron Lett.* **26**, 4959-4960.

Castedo, L., Granja, J., Maestro, M.A., and Mourino, A. (1987). Stereoselective synthesis of 25-hydroxyvitamin D_2 side chain *via* the acetal template route. *Tetrahedron Lett.* **28**, 4589-4590.

Centurion, O.T., Makler, F.S., Ramirez, J.A., Galagovsky, L.R., and Gros, E.G. (1996). Alternative synthesis of (22R,23R,24S)-2α,3α,22,23-tetrahydroxy-24-ethyl-5-cholestan-6-one. *Proc. Plant Growth Regul. Soc. Am.* **23**, 44-49.

Cerana, R., Bonetti, A., Marre, M.T., Romani, G., Lado, P., and Marre, E. (1983a). Effects of a brassinosteroid on growth and electrogenic proton extrusion in *Azuki bean* epicotyls. *Physiol. Plant.* **59**, 23-27.

Cerana, R., Colombo, R., and Lado, P. (1983b). Changes in mulate content and dark carbon dioxide fixation associated with brassinosteroid- and indoleacetic acid-induced changes in proton pump activity of maize roots. *Physiol. Veg.* **21**, 875-881 [*C. A.* **100**, 171646].

Cerana, R., Lado, R., Anastasia, M., Ciuffreda, P., and Allevi, P. (1984). Regulating effects of brassinosteroids and sterols on growth and proton secretion in maize roots. *Pflanzenphysiol.* **114**, 221-225 [*C. A.* **101**, 36008].

Cerana, R., Spelta, M., Bonetti, A., and Lado, P. (1985). On the effect of cholesterol on hydrogen ion extrusion and on growth in maize root segments: comparison with brassinosteroid. *Plant Sci.* **38**, 99-105.

Cerny, V. (1989). An alternative route to 2α,3α-diols from 2,3-unsaturated 5α-steroids avoiding the use of osmium tetroxide. *Collect. Czech. Chem. Commun.* **54**, 2211-2217.

Cerny, V., and Budesinsky, M. (1990). A route to 24-epibrassinolide from ergosterol avoiding the use of osmium tetraoxide. *Collect. Czech. Chem. Commun.* **55**, 2738-2755.

Cerny, V., Kaminek, M., and Strnad, M. (1984). Brassinolide analogs and their production as plant growth regulators. Czech. **CS 242,122** [*C. A.* **110**, 187809].

Cerny, V., Strnad, M., and Kaminek, M. (1986). Preparation of 2α,3α-dihydroxy-7-oxa-6-oxo-23,24-dinor-B-homo-5α-cholanic acid, its esters and amides, as brassinolide analogs. *Collect. Czech. Chem. Commun.* **51**, 687-697.

Cerny, V., Zajicek, J., and Strnad, M. (1987). 2α,3α-Dihydroxy-7-oxa-6-oxo-B-homo-5α-pregnan-21-oic acid and its derivatives as brassinolide analogs. *Collect. Czech. Chem. Commun.* **52**, 215-222.

Chang, J.-Q., and Cai, D.-T. (1988). The effects of brassinolide on seed germination and cotyledon tissue culture in *Brassica napus* L. *Oil Crops China* 18-22.

Charrois, G.J.R., Mao, H., and Kaufman, W.R. (1996). Impact on salivary gland degeneration by putative ecdysteroid antagonists and agonists in the ixodid tick amblyomma hebraeum. *Pestic. Biochem. Physiol.* **55**, 140-149 [*C. A.* **126**, 27992].

"Chemicals for Plant Protection". (1995). Uradzai, Minsk, p. 176.

Chen, J.-C., and Chi, J.-F. (1986). Bud differentiation in *Astragalus adsurgens* hypocotyl explants induced by epibrassinolide. *Plant Physiol. Commun.* 51-52.

Chen, J.-C., Wang, L.-F., and Zhao, Y.-J. (1990). Effects of 24-epibrassinolide on growth of tobacco root explants. *Acta Agric. Shanghai* **6**, 89-92.

Chirilov, A., Khripach, V., Toma, S., Scurtul, A., Zhabinskii, V., Cozmic, R., Zavadskaya, M., and Erlinman, J. (1996). The procedure of cultivation of grape. Pat. Appl. **9,600,258**, Moldova.

Choe, S., Dilkes, B.P., Fujioka, S., Takatsuto, S., Sakurai, A., and Feldmann, K.A. (1997). An *Arabidopsis* cell elongation mutant *dwf4* is defective in 22-α-hydroxylation steps in brassinosteroid biosynthetic pathways. *Proc. Plant Growth Regul. Soc. Am.* **24**, 98.

Choi, C., Takeuchi, Y., and Takematsu, T. (1986). The effects of brassinolides on the elongation of the radish hypocotyl. *Chem. Regul. Plants (Shokubutsu no Kagaku Chosetsu)* **21**, 134-143 [*C. A.* **107**, 93432].

Choi, C.-D., Kim, S.-C., and Lee, S.-K. (1990a). Agricultural use of plant growth regulators. 3. Effects of brassinolide on reducing herbicidal phytotoxicity of rice seedlings. *Res. Rep. Rural Dev. Admin.* **32**, 65-71.

Choi, C.-D., Kim, S.-C., and Lee, S.-K. (1990b). Interaction between brassinolide and auxins on bioassays. *Korean J. Crop Sci.* **35**, 58-64.

Choi, Y.-H., Inoue, T., Fujioka, S., Saimoto, H., and Sakurai, A. (1993). Identification of brassinosteroid-like active substances in plant-cell cultures. *Biosci. Biotech. Biochem.* **57**, 860-861.

Choi, Y.-H., Fujioka, S., Harada, A., Yokota, T., Takatsuto, S., and Sakurai, A. (1996). A brassinolide biosynthetic pathway *via* 6-deoxocastasterone. *Phytochemistry* **43**, 593-596.

Choi, Y.-H., Fujioka, S., Nomura, T., Harada, A., Yokota, T., Takatsuto, S., and Sakurai, A. (1997). An alternative brassinolide biosynthetic pathway *via* late C-6 oxidation. *Phytochemistry* **44**, 609-613.

Chory, J., Catterjee, M., Cook, R.K., Elich, T., Fankhauser, C., Li, J., Nagpal, P., Neff, M., Pepper, A., Poole, D., Reed, J., and Vitart, V. (1996). From seed germination to flowering, light controls plant development *via* the pigment phytochrome. *Proc. Natl. Acad. Sci. U.S.A.* **93**, 12066-12071.

Choudhry, S.C., Belica, P.S., Coffen, D.L., Focella, A., Maehr, H., Manchand, P.S., Serico, L., and Yang, R.T. (1993). Synthesis of a biologically active vitamin-D_2 metabolite. *J. Org. Chem.* **58**, 1456-1500.

Churikova, V.V., and Derevshchukov, S.N. (1997). Registration trials of growth regulator "Epin" on tomato and cucumber. Technical Report of Voronezh State University.

Churikova, V.V., and Vladimirova I.N. (1997). Effect of Epin on activity of enzymes of oxidative metabolism of cucumber in peronosporous epiphytotia conditions. *In* "Plant growth and development regulators", 4th, p. 78, Moscow.

CIASA (1989). Report on the results of field trials of epibrassinolide on agricultural plants. Technical Report of the Central Institute of Agrochemical Service for Agriculture, Moscow.

Clouse, S.D. (1996a). Molecular genetic studies confirm the role of brassinosteroids in plant growth and development. *Plant J.* **10**, 1-8.

Clouse, S.D. (1996b). Plant hormones: brassinosteroids in the spotlight. *Curr. Biol.* **6**, 658-661 [*C. A.* **125**, 53516].

Clouse, S.D. (1997). Molecular genetic analysis of brassinosteroid action. *Physiol. Plant.* **100**, 702-709.

Clouse, S.D., and Langford, M. (1995). A brassinosteroid-insensitive mutant in *Arabidopsis thaliana*. *Plant Physiol.(Suppl.)* **108**, 81.

Clouse, S.D., and Zurek, D. (1991). Molecular analysis of brassinolide action in plant growth and development. *In* "Brassinosteroids. Chemistry, Bioactivity, and Applications. ACS Symposium Series" (H.G. Cutler, T. Yokota, and G. Adam, Eds.), Vol. 474, pp. 122-140. American Chemical Society, Washington.

Clouse, S.D., Zurek, D.M., McMorris, T.C., and Baker, M.E. (1992). Effect of brassinolide on gene expression in elongating soybean epicotyls. *Plant Physiol.* **100**, 1377-1383.

Clouse, S.D., Hall, A.F., Langford, M., McMorris, T.C., and Baker, M.E. (1993). Physiological and molecular effects of brassinosteroids on *Arabidopsis thaliana*. *J. Plant Growth Reg.* **12**, 61-66 [*C. A.* **120**, 158952].

Clouse, S.D., Langford, M., and McMorris, T.C. (1996a). A brassinosteroid-insensitive mutant in *Arabidopsis thaliana* exhibits multiple defects in growth and development. *Plant Physiol.* **111**, 671-678.

Clouse, S.D., Langford, M., and McMorris, T.C. (1996b). A brassinosteroid-insensitive mutant in *Arabidopsis thaliana. Proc. Plant Growth Regul. Soc. Am.* **23**, 14.

Clouse, S., Oh, M.-H., and Torisky, B. (1997). Molecular genetic approaches to understanding brassinosteroid action. *Proc. Plant Growth Regul. Soc. Am.* **24**, 97.

Cohen, J.D., and Meudt, W.J. (1983). Investigation of the mechanism of the brassinosteroid response. I. Indole-3-acetic acid metabolism and transport. *Plant. Physiol.* **72**, 691-694.

Coll, F., Alonso, E., Iglesias, M., Marquardt, V., and Adam, G. (1992). Synthesis of spirostanic analogs of brassinosteroids. II. (25R)-5α-spirostan-2α,3α-diol. *Rev. Cubana Quim.* **6**, 7-12 [*C. A.* **121**, 9811].

Coll, M.F., Jomarron, R.I.M., Robaina, R.C.M., Alonso, B.E.M., and Cabrera, P.M.T. (1995). Polyhydroxyspirostanones as plant growth regulators. PCT Int. Appl. **WO 97 13,780** [*C. A.* **126**, 343720].

Criegee, R., Marchand, B., and Wannowins, H.W. (1942). Zur Kentnis der organischen Osmium-Verbindungen. *Liebigs Ann. Chem.* **550**, 99-133.

Cutler, H.G. (1991). Brassinosteroids through the looking glass. *In* "Brassinosteroids. Chemistry, Bioactivity, and Applications. ACS Symposium Series" (H.G. Cutler, T. Yokota, and G. Adam, Eds.), Vol. 474, pp. 334-345. American Chemical Society, Washington.

Dahiya, B.N., Singh, V.P., and Tripatki, I.D. (1976). Genetic study of synchrony of ear emergence in barley. *Indian J. Agric. Res.* **10**, 165-170.

Dahse, I., Sack, H., Bernstein, M., Petzold, U., Müller, E., Vorbrodt, H.-M., and Adam, G. (1990). Effect of (22S,23S)-homobrassinolide and related compounds on membrane potential and transport of Egeria leaf cells. *Plant Physiol.* **93**, 1268-1271.

Dahse, I., Petzold, U., Willmer, C.M., and Grimm, E. (1991). Brassinosteroid-induced changes of plasmalemma energization and transport and of assimilate uptake by plant tissues. *In* "Brassinosteroids. Chemistry, Bioactivity, and Applications. ACS Symposium Series" (H.G. Cutler, T. Yokota, and G. Adam, Eds.), Vol. 474, pp. 167-175. American Chemical Society, Washington.

Dauben, W.G., and Brookhart, T. (1981). Stereocontrolled synthesis of steroidal side chains. *J. Am. Chem. Soc.* **103**, 237-238.

Dauben, W.G., and Brookhart, T. (1982). Stereocontrolled synthesis of 20S steroidal side chain. *J. Org. Chem.* **47**, 3921-3923.

Dauben, W.G., and Deviny, E.J. (1966). Reductive opening of conjugated cyclopropyl ketones with lithium in liquid ammonia. *J. Org. Chem.* **31**, 3794-3798.

Deeva, V.P., Pavlova, I.P., and Khripach, V.A. (1996a). The effect of 24-epibrassinolide on ATPase activity of plasmalemma and cytoplasmatic components in different buckwheat genotypes. *Proc. Plant Growth Regul. Soc. Am.* **23**, 32-35.

Deeva, V.P., Sanko, N.V., and Khripach, V.A. (1996b). The specific features of 24-epibrassinolide action on RNA- and protein biosynthesis in genetically different barley plants. *Proc. Plant Growth Regul. Soc. Am.* **23**, 56-60.

Deeva, V.P., Mazets, Z.E., and Khripach, V.A. (1997a). Genetic control of brassinosteroid effects in wheat. *Proc. Plant Growth Regul. Soc. Am.* **24**, 127-132.

Deeva, V.P., Sanko, N.V., Vedeneev, A.N., and Pavlova, I.V. (1997b). Physiological and biochemical bases of adaptivity reactions of different barley genotypes under the action of exogenous physiologically active substances. *In* "Problems of experimental botany", pp. 355-367. Byelorussian Science, Minsk.

De Marino, S., Palagiano, E., Zollo, F., Minale, L., and Iorizzi, M. (1997). A novel sulfated steroid with a 7-membered 5-oxalactone B-ring from an Antarctic starfish of the family Asteriidae. *Tetrahedron* **53**, 8625-8628.

Diz, C.S., Perez, N., Nunez, M., and Torres, W. (1995). Effects of the synthetic brassinosteroid DAA-6 on tobacco (*Nicotiana tabacum* L.). *Cultivos Tropicales*, **16**, 53-55.

Diziere, R., Tahri, A., and Uguen, D. (1994). A C-B-A-D approach to brassinosteroids; obtention of an A-B-C ring system precursor. *Tetrahedron Lett.* **35**, 3941-3944.

Djerassi, C., and Fishman, J. (1955). Constitution and stereochemistry of samogenin, markogenin and mexogenin. *J. Am. Chem. Soc.* **77**, 4291-4297.

Donaubauer, J.R., Greaves, A.W., and McMorris, T.C. (1984). A novel synthesis of brassinolide. *J. Org. Chem.* **49**, 2833-2834.

Dong, J.W., Lou, S.S., Han, B.W., He, Z.P., and Li, P.M. (1989). Effects of brassinolide on rice seed germination and seedling growth. *Acta Agric. Univ. Pekinensis* **15**, 153-156.

Dreier, S.I., and Towers, G.H.N. (1988). Activity of ecdysterone in selected plant growth bioassays. *J. Plant Physiol.* **132**, 509-512.

Ecker, J.R. (1997). BRI-ghtening the pathway to steroid hormone signaling events in plants. *Cell* **90**, 825-827.

Egbert, W.H., Lauren, J.D., and David, M.B. (1981). The effects of brassinolide on growth and enzyme activity in mung bean (*Phaseolus aureus* Roxb.). *Proc. Plant Growth Regul. Soc. Am.* **8**, 146.

Eguchi, T.E., Yoshida, M., and Ikekawa, N. (1989). Synthesis and biological activities of 22-hydroxy and 22-methoxy derivatives of $1\alpha,25$-dihydroxyvitamin D_3: importance of side chain conformation to biological activities. *Bioorg. Chem.* **17**, 294-307.

Ershova, A.N. (1995). Effect of epibrassinolide on lipid peroxidation in plants. *In* "Brassinosteroids - biorational, ecologically safe regulators of growth and productivity of plants", 4th, pp. 12-13, Minsk.

Ershova, A.N., and Khripach, V.A. (1996). Effect of epibrassinolide on lipid peroxidation in *Pisum sativum* at normal aeration and under oxygen deficiency. *Russ. J. Plant Physiol.* **43**, 750-752.

Ershova, A.N., and Novikova, I.V. (1993). Effect of hypoxia and epibrassinolide on lipid peroxidation in pea seedlings. *In* "Brassinosteroids - biorational, ecologically safe regulators of growth and productivity of plants", 3rd, p. 25, Minsk.

Eun, J.-S., Kuraishi, S., and Sakurai, N. (1989). Changes in levels of auxin and abscisic acid and the evolution of ethylene in squash hypocotyl after treatment with brassinolide. *Plant Cell Physiol.* **30**, 807-810.

Faris, M.A., and Klink, H.R. (1982). Comparison between different methods of measuring synchrony of ear emergence in barley, oat and spring wheat cultivars. *Pflanzenzuchtg.* **88**, 79-88.

Fattorusso, E., Lanzotti, V., Magno, S., and Novellino, E. (1985). Cholest-5-ene-2α,3α,7β,15β,18-pentol 2,7,15,18-tetraacetate, a novel highly hydroxylated sterol from the marine hydroid *Eudendrium glomeratum*. *J. Org. Chem.* **50**, 2868-2870.

Fedoseeva, T.G. (1996). Epin-news. *Priusadebnoye choziaystvo*, 15.

Ferrer, J.C., Lalueza, R., Saavedra, O., and Brosa, C. (1990). Short step synthesis of (22E,24R)-5α-ergosta-2,22-dien-6-one, a key intermediate for the preparation of 24-epibrassinolide. *Tetrahedron Lett.* **31**, 3941-3942.

Fieser, L., and Fieser, M. (1959). "Steroids", Reinhold.

Frelek, J., Ikekawa, N., Takatsuto, S., and Snatzke, G. (1997). Application of [$Mo_2(OAc)_4$] for determination of absolute configuration of brassinosteroid vic-diols by circular dichroism. *Chirality* **9**, 578-582 [*C. A.* **127**, 307545].

Fuendjiep, V., Charles, G., and Chriqui, D. (1989). Hemisynthese et activite physiologique des methyl-28 brassinolides 22R,23S et 22S,23R. *Bull. Soc. Chim. Fr.*, 711-715 [*C. A.* **113**, 24330].

Fujii, S., Hirai, K., and Saka, H. (1991). Growth-regulating action of brassinolide in rice plants. *In* "Brassinosteroids. Chemistry, Bioactivity, and Applications. ACS Symposium Series" (H.G. Cutler, T. Yokota, and G. Adam, Eds.), Vol. 474, pp. 306-311. American Chemical Society, Washington.

Fujii, S., and Saka, H. (1992). Effect of brassinolide on the translocation of assimilate in rice plants during the ripening stage. *Jpn. J. Crop Sci.* **61**, 193-199.

Fujimoto, Y., Kimura, M., Terasawa, T., Khalifa, F.A.M., and Ikekawa, N. (1984). Stereocontrolled synthesis and determination of the C-24 and -25 stereochemistry of glaucasterol. *Tetrahedron Lett.* **25**, 1805-1808.

Fujimoto, Y., Ohhana, M., Terasawa, T., and Ikekawa, N. (1985). Stereocontrolled synthesis of petrosterol. *Tetrahedron Lett.* **26**, 3239-3242.

Fujioka, S., and Sakurai, A. (1995). Biosynthesis of brassinosteroids and its regulation. *Chem. Regul. Plants (Shokubutsu no Kagaku Chosetsu)* **30**, 137-141.

Fujioka, S., and Sakurai, A. (1997a). Biosynthesis and metabolism of brassinosteroids. *Physiol. Plant.* **100**, 710-715.

Fujioka, S., and Sakurai, A. (1997b). Brassinosteroids. *Nat. Prod. Rep.* **14**, 1-10.

Fujioka, S., Inoue, T., Takatsuto, S., Yanagisawa, T., Yokota, T., and Sakurai, A. (1995a). Identification of a new brassinosteroid, cathasterone, in cultured cells of *Catharanthus roseus* as a biosynthetic precursor to teasterone. *Biosci. Biotech. Biochem.* **59**, 1543-1547.

Fujioka, S., Inoue, T., Takatsuto, S., Yanagisawa, T., Yokota, T., and Sakurai, A. (1995b). Biological activity of biosynthetically-related congeners of brassinolide. *Biosci. Biotech. Biochem.* **59**, 1973-1975.

Fujioka, S., Choi, Y.-H., Takatsuto, S., Yokota, T., Li, J., Chory, J., and Sakurai, A. (1996). Identification of castasterone, 6-deoxocastasterone, typhasterol and 6-deoxotyphasterol from the shoots of *Arabidopsis thaliana*. *Plant Cell Physiol.* **37**, 1201-1203.

Fujioka, S., Li, J., Choi, Y.-H., Seto, H., Takatsuto, S., Noguchi, T., Watanabe, T., Kuriyama, H., Yokota, T., Chory, J., and Sakurai, A. (1997). The *Arabidopsis* deetiolated2 mutant is blocked early in brassinosteroid biosynthesis. *Plant Cell* **9**, 1951-1962.

Fujioka, S., Noguchi, T., Yokota, T., Takatsuto, S. and Yoshida, S. (1998). Brassinosteroids in *Arabidopsis thaliana*. *Phytochemistry* **48**, 595-599.

Fujita, F. (1985). Prospects for brassinolide utilization in agriculture. *Chem. Biol. (Kagaku to Seibutsu)* **23**, 717-725 [*C. A.* **104**, 64041].

Fukuzawa, A., Kumagai, Y., Masamune, T., Furusaki, A., Katayama, C., and Matsumoto, T. (1981). Acetyl pinnasterol and pinnasterol, ecdysone-like metabolites, from the marine red alga *Laurencia pinnata* Yamada. *Tetrahedron Lett.* **22**, 4085-4086.

Fung, S., and Siddall, J.B. (1980). Stereoselective synthesis of brassinolide: a plant growth promoting steroidal lactone. *J. Am. Chem. Soc.* **102**, 6580-6581.

Furuta, T., and Yamamoto, Y. (1992). Stereocontrol of three contiguous centers of brassinosteroid side chain by using α-alkoxy organoleads. *J. Org. Chem.* **57**, 2981-2982.

Galston, A.W., Davies, P.J., and Satter, R.L. (1983). "The life of the green plant". Mir, Moscow.

Gamburg, K.Z. (1986). Brassins - steroidal hormones of plants. *Uspekhi sovremennoi biologii* **102**, 314-320.

Gamoh, K., and Takatsudo, H. (1990). Separation and analysis of brassinosteroid C-24 epimers. Jpn. Kokai Tokkyo Koho **JP 04,145,098 [92,145,098]** [*C. A.* **117**, 251623].

Gamoh, K., and Takatsuto, S. (1989). A boronic acid derivative as a highly sensitive fluorescence derivatization reagent for brassinosteroids in liquid chromatography. *Anal. Chim. Acta* **221**, 201-204 [*C. A.* **112**, 154628].

Gamoh, K., and Takatsuto, S. (1994). Liquid chromatography assay of brassinosteroids in plants. *J. Chromatogr.* **658**, 17-25.

Gamoh, K., Kitsuwa, T., Takatsuto, S., Fujimoto, Y., and Ikekawa, N. (1988). Determination of trace brassinosteroids by high-performance liquid chromatography. *Anal. Sci.* **4**, 533-535.

Gamoh, K., Omote, K., Okamoto, N., and Takatsuto, S. (1989a). High-performance liquid chromatography of brassinosteroids in plants with derivatization using 9-phenanthreneboronic acid. *J. Chromatogr.* **469**, 424-428.

Gamoh, K., Takatsudo, H., and Tejima, I. (1989b). Aminophenolic acid derivative and its use in determination of glycols by HPLC. Jpn. Kokai Tokkyo Koho **JP 03 63,283 [91 63,283]** [*C. A.* **115**, 173873].

Gamoh, K., Okamoto, N., Takatsuto, S., and Tejima, I. (1990a). Determination of traces of natural brassinosteroids as dansylaminophenylboronates by liquid chromatography with fluorimetric detection. *Anal. Chim. Acta* **228**, 101-105.

Gamoh, K., Sawamoto, H., Takatsuto, S., Watabe, Y., and Arimoto, H. (1990b). Ferroceneboronic acid as a derivatization reagent for the determination of brassinosteroids by high-performance liquid chromatography with electrochemical detection. *J. Chromatogr.* **515**, 227-233.

Gamoh, K., Takatsuto, S., and Ikekawa, N. (1992). Effective separation of C-24 epimeric brassinosteroids by liquid chromatography. *Anal. Chim. Acta* **256**, 319-322.

Gamoh, K., Yamaguchi, I., and Takatsuto, S. (1994). Rapid and selective sample preparation for the chromatographic determination of brassinosteroids from plant material using solid-phase extraction method. *Anal. Sci.* **10**, 913-917 [*C. A.* **122**, 50408].

Gamoh, K., Abe, H., Shimada, K., and Takatsuto, S. (1996a). Liquid chromatography/mass spectrometry with atmospheric pressure chemical ionization of free brassinosteroids. *Rapid Commun. Mass Spectrom.* **10**, 903-906.

Gamoh, K., Prescott, M.C., Goad, L.J., and Takatsuto, S. (1996b). Analysis of brassinosteroids by liquid chromatography/mass spectrometry. *Bunseki Kagaku* **45**, 523-527.

Gao, B., Wang, Z., Zheng, Q., and Xu, Z. (1993). Stereoselective synthesis of 2β,3β-dihydroxy-6-cholesterone. *Yingyong Hyaxue* **10**, 12-15 [*C. A.* **119**, 28424].

Garbuz, N.I., Yankovskaya, G.S., Baranovskii, A.V., Litvinovskaya, R.P., and Khripach, V.A. (1994). Circular dichroism and stereochemistry of asymmetrical centers C-20 and C-22 of 17-isoxazolinyl- and 20-hydroxy-20-isoxazolinyl steroids. *Khim. Pripodn. Soed.* 391-397

Gartz, J., Adam, G., and Vorbrodt, H.-M. (1990). Growth promoting effect of a brassinosteroid in mycelial cultures of the fungus *Psilocybe cubensis*. *Naturwissenschaften* **77**, 388-389.

Genma, T. (1987a). Crop yield-increasing method for gramineous crops using brassinolides and auxins. Jpn. Kokai Tokkyo Koho **JP 01,106,802 [89,106,802]** [*C. A.* **111**, 169393].

Genma, T. (1987b). Method for cultivating potatoes with brassinolide-containing yield enhacer. PCT Int. Appl. **WO 88 04,890** [*C. A.* **110**, 187813].

Genma, T. (1987c). Plant growth promoters containing brassinolides, for Chinese yam. Jpn. Kokai Tokkyo Koho **JP 01,146,804 [89,146,804]** [*C. A.* **111**, 148915].

Genma, T. (1987d). Plant growth promoters containing brassinolides, for beans. Jpn. Kokai Tokkyo Koho **JP 01,146,805 [89,146,805]** [*C. A.* **111**, 148916].

Genma, T. (1987e). Plant growth promoters containing brassinolides, for sugar beet. Jpn. Kokai Tokkyo Koho **JP 01,146,806 [89,146,806]** [*C. A.* **111**, 148917].

Genma, T. (1987f). Plant growth stimulation on pasture, using brassinosteroids. Jpn. Kokai Tokkyo Koho **JP 01,163,105** [89,163,105] [*C. A.* **112**, 32150].

Gilhooly, M.A., Morris, D.S., and Williams, D.H. (1982). Synthesis of 25,26-dihydroxyvitamin D_2. *J. Chem. Soc., Perkin Trans. I* 2111-2116.

Goetz, M.A., Meinwald, J., and Eisner, T. (1981). New defensive steroids and a pterin from the firefly *Photinus pyralis* (Coleoptera:Lampyridae). *Experientia* **37**, 679-680.

Gonik, G.E., and Mikhailova, T. (1992). Results of field trials of "Epin" on potato and tobacco, Technical Report, Krasnodar.

Gonik, G.E., and Al-Zhadavi, S. (1993). Effect of epibrassinolide on the growth and productivity of sugar beet in Krasnodar region. *In* "Brassinosteroids - biorational, ecologically safe regulators of growth and productivity of plants", 3rd, pp. 32-33, Minsk.

Gonzalez, S.M., Bustos, D.A., Zudenigo, M.E., and Ruveda, E.A. (1986). Configurational assignment of epimeric 22,23-epoxides of steroids by carbon-13 NMR spectroscopy. *Tetrahedron* **42**, 755-758.

Granja, J.R. (1991). 2,3-Wittig sigmatropic rearrangement of steroidal 16β-propargyl ethers for the synthesis of 25-hydroxyvitamin D side chain analogues. *Synth. Commun.* **21**, 2033-2038.

Granja, J.R., Castedo, L., and Mourino, A. (1993). Studies on the opening of dioxanone and acetal templates and application to the synthesis of 1α,25-dihydroxyvitamin D_2. *J. Org. Chem.* **58**, 124-131.

Gregory, L.E. (1981). Acceleration of plant growth through seed treatment with brassins. *Am. J. Bot.* **68**, 586-588.

Gregory, L.E., and Mandava, N.B. (1982). The activity and interaction of brassinolide and gibberellic acid in mung bean epicotyls. *Physiol. Plant.* **54**, 239-243.

Griffiths, P.G., Sasse, J.M., Yokota, T., and Cameron, D.W. (1995a). 6-Deoxotyphasterol and 3-dehydro-6-deoxoteasterone, possible precursors to brassinosteroids in the pollen of *Cupressus arizonica*. *Biosci. Biotech. Biochem.* **59**, 956-959.

Griffiths, P.G., Sasse, J.M., Yokota, T., and Cameron, D.W. (1995b). 6-Deoxotyphasterol and 3-dehydro-6-deoxoteasterone, possible precursors to brassinosteroids in pollen of *Cupressus arizonica*. *Plant Physiol. (Suppl.)* **108**, 81.

Grinchenko, A.L., and Belokon, L.M. (1991). Effectiveness of application of brassinosteroids on cereals in the North Ukrainia. *In* "Conference on brassinosteroids", 2nd, p. 34, Minsk.

Grinchenko, A.L., and Belokon, L.M. (1993). Application of brassinosteroids for the corn crop enhancement. *In* "Regulators of plant growth and development", 2nd, p. 23, Moscow.

Gross, D., and Parthier, B. (1994). Novel natural substances acting in plant growth regulation. *J. Plant Growth Regul.* **13**, 93-114.

Grove, M.D., Spencer, G.F., Rohwedder, W.K., Mandava, N., Worley, J.F., Warthen, J.D., Jr., Steffens, G.L., Flippen-Anderson, J.L., and Cook J.C., Jr. (1979).

Brassinolide, a plant growth-promoting steroid isolated from *Brassica napus* pollen. *Nature* **281**, 216-217.

Guan, M., and Roddick, J.G. (1988a). Comparison of the effects of epibrassinolide and steroidal estrogens on adventitious root growth and early shoot development in mung bean cuttings. *Physiol. Plant.* **73**, 426-431.

Guan, M., and Roddick, J.G. (1988b). Epibrassinolide-inhibition of development of excised, adventitious and intact roots of tomato (*Lycopersicon esculentum*): comparison with the effects of steroidal estrogens. *Physiol. Plant.* **74**, 720-726.

Hai, T., Schneider, B., and Adam, G. (1995). Metabolic conversion of 24-epibrassinolide into pentahydroxylated brassinosteroid glucosides in tomato cell cultures. *Phytochemistry* **40**, 443-448.

Hai, T., Schneider, B., Porzel, A., and Adam, G. (1996). Metabolism of 24-epicastasterone in cell suspension cultures of *Lycopersicon esculentum*. *Phytochemistry* **41**, 197-201.

Hamada, K. (1986). Brassinolide in crop cultivation. *Plant growth regulators in agriculture, FFTC Book Ser.* **34**, 188-196.

Handschack, M., Adam, G., Vorbrodt, H.-M., Marquardt, V., and Beinhauer, K. (1990). Stone fruit quality improvement by synergistic mixtures of auxins with brassinosteroids. Ger. Offen. **DE 4,039,017** [*C. A.* **117**, 145322].

Hata, S., Takagishi, Egawa, Y., and Ota, Y. (1986). Effect of compactin, a 3-hydroxy-3-methylglutaryl coenzyme and reductase inhibitor on the growth of alfalfa (*Medicago sativa*) seedlings and the rhizogenesis of pepper (*Capsicum annuum*) explants. *Plant Growth Regul.* **4**, 335-346.

Hathout, T.A. (1996). Salinity stress and its counteraction by the growth regulator "brassinolide" in wheat plants (*Triticum aestivum* L. cultivar *Giza 157*). *Egypt. J. Physiol. Sci.* **20**, 127-152 [*C. A.* **127**, 275612].

Haun, J.R. (1973). Visual quantification of wheat development. *Agron. J.* **65**, 116-119.

Hayami, H., Sato, M., Kanemoto, S., Morizawa, Y., Oshima, K., and Nozaki, H. (1983). Transition-metal-catalyzed silylmetalation of acetylenes and its application to the stereoselective synthesis of steroidal side chain. *J. Am. Chem. Soc.* **105**, 4491-4492.

Hayashi, S., Hohjoh, T., Furuse, T., Kuriyama, H., Watanabe, T., and Takatsuto, S. (1986). A process for producing steroid derivatives. Eur. Pat. Appl. **EP 261,656** [*C. A.* **109**, 129462].

Hayashi, S., Hohjoh, T., Shida, A., and Ikekawa, N. (1987). Preparation of 23-phenylbrassinosteroids as plant growth regulators. Eur. Pat. Appl. **EP 282,984** [*C. A.* **110**, 95639].

Hazra, B.G., Joshi, P.L., and Pore, V.S. (1990). Resin catalysed ene reaction on 3β-toluene-*p*-sulfonoxy-(Z)-pregna-5,17(20)-diene: synthesis of (20S)-6β-methoxy-3α,5-cyclo-5α-pregnane-20-carbaldehyde. *Tetrahedron Lett.* **31**, 6227-6230.

Hazra, B.G., Argade, N.P., and Joshi, P.L. (1992). A stereoselective approach to the brassinolide side chain *via* Wittig reaction. *Tetrahedron Lett.* **33**, 3375-3376.

Hazra, B.G., Pore, V.S., and Joshi, P.L. (1993a). Short-step synthesis of 2α,3α,22-triacetoxy-23,24-dinor-5α-cholan-6-one: key intermediate for the preparation of 24-norbrassinolide, dolicholide and dolichosterone. *J. Chem. Soc., Perkin Trans. 1* 1819-1822.

Hazra, B.G., Pore, V.S., Joshi, P.L., Padalkar, S.N., Deshpande, S.A., and Rajamohanan, P.R. (1993b). Synthesis and configurational assignment of epimeric 22-hydroxy-, 23,24-acetylenic, olefinic or epoxy steroids using carbon-13 NMR spectroscopy. *Magn. Res. Chem.* **31**, 605-608 [*C. A.* **119**, 160634].

Hazra, B.G., Joshi, P.L., Bahule, B.B., Argade, N.P., Pore, V.S., and Chordia, M.D. (1994). Stereoselective synthesis of (22*R*,23*R*,24*S*)-3-β-hydroxy-5-ene-22,23-dihydroxy-24-methyl-cholestane: A brassinolide intermediate from 16-dehydropregnenolone acetate. *Tetrahedron* **50**, 2523-2532.

Hazra, B.G., Kumar, T.P., and Pore, V.S. (1996). Synthesis of 3β-hydroxy-24-methylcholesta-5,23-dien-23-one: a brassinolide intermediate. *J. Chem. Res., Synop.* 536-537 [*C. A.* **126**, 131691].

Hazra, B.G., Basu, S., Bahule, B.B., Pore, V.S., Vyas, B.N., and Ramraj, V. (1997a). Stereoselective synthesis of new hexanor (C_{23}-C_{28})castasterone-20,22-ethyl diether from 16-dehydropregnenolone acetate and its plant growth promoting activity. *Tetrahedron* **53**, 4909-4920.

Hazra, B.G., Kumar, T.P., and Joshi, P.L. (1997b). New synthesis of 28-homobrassinolide from stigmasterol. *Liebigs Ann./Recl.* 1029-1034.

He, R., Wang, G., and Wang, X. (1991). Effect of brassinolide on growth and chilling resistance of maize seedlings. *In* "Brassinosteroids. Chemistry, Bioactivity, and Applications. ACS Symposium Series" (H.G. Cutler, T. Yokota, and G. Adam, Eds.), Vol. 474, pp. 220-230. American Chemical Society, Washington.

Hellrung, B., Voigt, B., Schmidt, J., and Adam, G. (1997). Synthesis of new Δ^5-7-oxygenated and $\Delta^{5,7}$-unsaturated brassinosteroid analogs. *Steroids* **62**, 415-421.

Helmy, Y.I., Sawan, O.M.M., and Abdel-Halim, S.M. (1997). Growth, yield and endogenous hormones of broad bean plants as affected by brassinosteroids. *Egypt. J. Hortic.* **24**, 109-115.

Hetru, C., Roussel, J.P., Mori, K., and Nakatani, Y. (1986). Antiecdysteroid activity of brassinosteroids. *C. R. Acad. Sci., Ser. 2* **302**, 417-420 [*C. A.* **105**, 39706].

Hewitt, F.R., Hough, T., O'Neill, P., Sasse, J.M., Williams, E.G., and Rowan, K.S. (1985). Effect of brassinolide and other growth regulators on the germination and growth of pollen tubes of *Prunus avium* using multiple hanging-drop assay. *Aust. J. Plant. Physiol.* **12**, 201-211.

Hirai, Y., and Fujii, S. (1985). Acceleration of rice growth by brassinolide. Jpn. Kokai Tokkyo Koho **JP 62 67,005 [87 67,005]** [*C. A.* **107**, 72886].

Hirai, Y., Fujii, S., and Kamuro, Y. (1985). Plant growth acceleration by brassinolide under red light. Jpn. Kokai Tokkyo Koho **JP 62 63,502 [87 63,502]** [*C. A.* **107**, 72884].

Hirai, Y., Fujii, S., and Honojo, K. (1986a). Stimulation of rice growth with brassinolide at low temperature. Jpn. Kokai Tokkyo Koho **JP 63,135,303 [88,135,303]** [*C. A.* **109**, 224719].

Hirai, Y., Sasaki, H., and Katsura, N. (1986b). Brassinolide as differentiation promoter for plant tissue culture. Jpn. Kokai Tokkyo Koho **JP 62,259,513 [87,259,513]** [*C. A.* **109**, 21620].

Hirai, K., Fujii, S., and Honjo, K. (1991). The effect of brassinolide on grain ripening in rice plants under the low temperature conditions. *Jpn. J. Crop Sci.* **60**, 29-35.

Hirano, Y., and Djerassi, C. (1982). Stereoselective synthesis *via* Claisen rearrangement of the marine sterols occelasterol, patinosterol and 22,23-dihydroccelasterol. *J. Org. Chem.* **47**, 2420-2426.

Hirano, Y., Takatsuto, S., and Ikekawa, N. (1984). Further investigations of the stereochemistry of electrophilic addition reactions of the steroidal C-22 double bonds. *J. Chem. Soc., Perkin Trans. I* 1775-1779.

Hoffmann, S. (1985). Brassinolide/z-DNA structural complementarities. *Z. Chem.* **25**, 331-331.

Honda, T. (1990). Preparation of steroid derivatives. Jpn. Kokai Tokkyo Koho **JP 03,294,290 [91,294,290]** [*C. A.* **116**, 214780].

Honda, T., and Tsubuki, M. (1990). Development of a highly stereoselective construction of steroid side chains and their application to the synthesis of natural products. *Yuki Gosei Kagaku Kyokaishi* **48**, 43-55 [*C. A.* **112**, 217331].

Honda, T., Keino, K., and Tsubuki, M. (1990). A concise stereoselective synthesis of castasterone. *J. Chem. Soc., Chem. Commun.* 650-652.

Honda, T., Takada, H., Miki, S., and Tsubuki, M. (1993). Synthesis and structure elucidation of a novel ecdysteroid, gerardiasterone. *Tetrahedron Lett.* **34**, 8275-8278.

Hooley, R. (1996). Plant steroid hormones emerge from the dark. *Trend. Genet.* **12**, 281-283.

Horgen, P.A., Nakagawa, C.H., and Irvin, R.T. (1984). Production of monoclonal antibodies to a steroid plant growth regulator. *Can J. Biochem. Cell Biol.* **62**, 715-721 [*C. A.* **101**, 208803].

Houston, T.A., Tanaka, Y., and Koreeda, M. (1993). Stereoselective construction of 22-oxygenated steroid side chains by dimethylaluminum chloride-mediated ene reaction of aldehydes. *J. Org. Chem.* **58**, 4287-4292.

Hu, G. (1995). Compound fertilizer for plant growing and desease resistance. Faming Zhuanli Shenqing Gongkai Shuomingshu **CN 1,110,673** [*C. A.* **124**, 201114].

Huang, L.-F., and Zhou, W.-S. (1992). Enhanced product diastereomeric excesses in asymmetric dihydroxylation of the (22*E*,24*R*)- and (22*E*,24*S*)-24-alkyl steroidal unsaturated side chain by using the Sharpless improved chiral ligand. *Chin. Chem. Lett.* **3**, 969-970 [*C. A.* **119**, 95913].

Huang, L.-F., and Zhou, W.-S. (1994). Novel method for construction of the side-chain of 23-arylbrassinosteroids *via* Heck arylation and asymmetric dihydroxylation as key steps. *J. Chem. Soc., Perkin Trans. 1* 3579-3585.

Huang, L.-F., Zhou, W.-S., Sun, L.-Q., and Pan, X.-F. (1993). Osmium tetroxide catalyzed asymmetric dihydroxylation of the (22*E*,24*R*)- and the (22*E*,24*S*)-24-alkyl steroidal unsaturated side chain. *J. Chem. Soc., Perkin Trans. 1* 1683-1686.

Huang, L.-F., Hu, Q.-Y., Zhou, W.-S., Xia, L.-J., and Bi, M.-H. (1995). Effect of methanesulfonamide on the asymmetric osmylation of (22*E*)-steroidal unsaturated side chains. *Acta Chim. Sin. (Huaxue Xuebao)* **53**, 501-508 [*C. A.* **123**, 112507].

Huang, Y.-Z., Shi, L.-L., and Li, S.-W. (1988). An efficient and stereoselective synthesis of (*E*)-α-enones *via* arsonium salts. Preparation of key intermediates for the synthesis of brassinosteroid and prostaglandin. *Synthesis* 975-977.

Igai, T., Hirai, Y., and Fujii, S. (1985). Sowing of rice seeds coated with calcium peroxide and brassinolide in water-filled paddies. Jpn. Kokai Tokkyo Koho **JP 62 69,907 [87 69,907]** [*C. A.* **107**, 72888].

Iglesias, A.M.A., Perez, G.R., and Coll, M.F. (1996). Natural brassinosteroids and their synthetic analogs. *Rev. CENIC, Cienc. Quim.* **27**, 3-12 [*C. A.* **127**, 248270].

Iglesias, A.M.A., Lara, V.L., Martinez, C.P., and Manchado, F.C. (1997). Synthesis of spirobrassinosteroid analogs of 6-deoxocastasterone. *Quim. Nova* **20**, 361-364 [*C. A.* **127**, 95449].

Ihara Chem. Ind. Co., Ltd. (1982a). (22*S*)-B-Homo-7-oxa-5α-cholestan-6-one-3β,22-diol. Jpn. Kokai Tokkyo Koho **JP 59 27,886 [84 27,886]** [*C. A.* **101**, 130976].

Ihara Chem. Ind. Co., Ltd. (1982b). Stereoselective Grignard reaction of steroid aldehydes. Jpn. Kokai Tokkyo Koho **JP 59,113,000 [84,113,000]** [*C. A.* **102**, 95907].

Ihara Chem. Ind. Co., Ltd. (1983). 26,27-Bisnorbrassinosteroid derivatives. Jpn. Kokai Tokkyo Koho **JP 59,227,900 [84,227,900]** [*C. A.* **103**, 105225].

Ikeda, M., Takatsuto, S., Sassa, T., Ikekawa, N., and Nukina, M. (1983). Identification of brassinolide and its analogues in chestnut gall tissue. *Agric. Biol. Chem.* **47**, 655-657.

Ikekawa, N., and Akutsu, T. (1987). Culturing method for spinach using brassinosteroids as growth promoters. Jpn. Kokai Tokkyo Koho **JP 63,239,201 [88,239,201]** [*C. A.* **111**, 52465].

Ikekawa, N., and Aoki, S. (1987). Brassinosteroids as a corn growth regulators. Jpn. Kokai Tokkyo Koho **JP 63,239,202 [88,239,202]** [*C. A.* **111**, 52464].

Ikekawa, N., and Nagai, T. (1987). Brassinosteroid fruiting hormones for melons. Jpn. Kokai Tokkyo Koho **JP 63,243,002 [88,243,002]** [*C. A.* **111**, 129007].

Ikekawa, N., and Takatsuto, S. (1984). Microanalysis of brassinosteroids in plants by gas chromatography/mass-spectrometry. *Mass Spectrosc. (Shitsuryo Bunseki)* **32**, 55-70 [*C. A.* **101**, 68577].

Ikekawa, N., and Zhao, Y.-J. (1991). Application of 24-epibrassinolide in agriculture. In "Brassinosteroids. Chemistry, Bioactivity, and Applications. ACS Symposium Series" (H.G. Cutler, T. Yokota, and G. Adam, Eds.), Vol. 474, pp. 280-291. American Chemical Society, Washington.

Ikekawa, N., Takatsuto, S., Marumo, S., Abe, H., Morishita, T., Uchiyama, M., Ikeda, M., Sassa, T., and Kitsuwa, T. (1983). Identification of brassinolide and its 6-oxo analog in the plant kingdom by selected ion monitoring. *Proc. Jpn. Acad., Ser. B* **59**, 9-12 [*C. A.* **98**, 140552].

Ikekawa, N., Takatsuto, S., Kitsuwa, T., Saito, H., Morishita, T., and Abe, H. (1984). Analysis of natural brassinosteroids by gas chromatography and gas chromatography-mass spectrometry. *J. Chromatogr.* **290**, 289-302.

Ikekawa, N., Akutsu, T., and Kano, T. (1987). Increased production of Cruciferae plant seeds by brassinosteroids. Jpn. Kokai Tokkyo Koho **JP 63,307,803** [88,307,803] [*C. A.* **111**, 111016].

Ikekawa, N., Nishiyama, F., and Fujimoto, Y. (1988). Identification of 24-epibrassinolide in bee pollen of the broad bean, *Vicia faba* L. *Chem. Pharm. Bull.* **36**, 405-407.

Ishiguro, M., and Ikekawa, N. (1975). Stereochemistry of electrophilic reactions at the steroidal C-22 double bond. *Chem. Pharm. Bull.* **23**, 2860-2866.

Ishiguro, M., Takatsuto, S., Morisaki, M., and Ikekawa, N. (1980). Synthesis of brassinolide, a steroidal lactone with plant growth promoting activity. *J. Chem. Soc., Chem. Commun.* 962-964.

Iwahori, S., Tominaga, S., and Higuchi, S. (1990). Retardation of abscission of citrus leaf and fruitlet explants by brassinolide. *Plant Growth Regul.* **9**, 119-125.

Iwasaki, T., and Shibaoka, H. (1991). Brassinosteroids act as regulators of tracheary-element differentiation in isolated *Zinnia mesophyll* cells. *Plant Cell Physiol.* **32**, 1007-1014.

Jacobsen, E.N., Marko, I., Mungall, W.S., Schroder, G., and Sharpless, K.B. (1988). Asymmetric dihydroxylation *via* ligand-accelerated catalysis. *J. Am. Chem. Soc.* **110**, 1968-1970.

Jiang, Y., Xu, Z., and Guo, Q. (1989). Synthesis of brassinolide analogues containing ether group in a side chain. *J. Xiamen Univ. Nat. Sci.* 519-522.

Jin, F., Xu, Y., and Huang, W. (1993). 2,2-Difluoro enol silyl ethers: convenient preparation and application to the synthesis of novel fluorinated brassinosteroids. *J. Chem. Soc., Perkin Trans. I* 795-799.

Johnson, W.S., Elliott, J.D., and Hanson, G.J. (1984). A stereoselective approach to a key intermediate for the preparation of vitamin D metabolites. *J. Am. Chem. Soc.* **106**, 1138-1139.

Kadyrov, S.V., Evdokimova, A.A., and Bikov, Y.N. (1995). Effect of epibrassinolide and chitosan on the quality of lucerne seeds. *In* "Brassinosteroids - biorational, ecologically safe regulators of growth and productivity of plants", 4th, pp. 27-28, Minsk.

Kajita, T., Furushima, M., and Takematsu, T. (1985). Brassinosteroids and choline salts as synergistic plant-growth hormone. Jpn. Kokai Tokkyo Koho **JP 62 77,305** [87 77,305] [*C. A.* **107**, 129134].

Kakiuchi, T., Kamuro, Y., Takatsuto, S., and Kobayashi, K. (1988). A new brassinolide analog and its practical efficacy under field-cultivation conditions. *Agric. Biol. Chem.* **52**, 2381-2382.

Kalinich, J.F., Mandava, N.B., and Todhunter, J.A. (1985). Relationship of nucleic acid metabolism to brassinolide-induced responses in beans. *J. Plant. Physiol.* **120**, 207-214.

Kalinich, J.F., Mandava, N.B., and Todhunter, J.A. (1986). Relationship of nucleic acid metabolism in brassinolide-induced responces in beans. *J. Plant. Physiol.* **125**, 345-353 [*C. A.* **106**, 15850].

Kalituho, L.N. (1997). "Relation of processes of photosynthetic apparatus formation and growth on early stages of plant development", Ph.D. Thesis, Institute of Photobiology, Academy of Sciences of Belarus, Minsk.

Kalituho, L.N., and Kabashnikova, L.F. (1998). In press.

Kalituho, L.N., Chaika, M.T., Mazhul, V.M., and Khripach, V.A. (1996). Effect of 24-epibrassinolide on pigment apparatus formation. *Proc. Plant Growth Regul. Soc. Am.* **23**, 36-40.

Kalituho, L.N., Chaika, M.T., Kabashnikova, L.F., Makarov, V.N., and Khripach, V.A. (1997a). On the phytochrome mediated action of brassinosteroids. *Proc. Plant Growth Regul. Soc. Am.* **24**, 140-145.

Kalituho, L.N., Kabashnikova, L.F., Khripach, V.A., and Chaika, M.T. (1997b). Epibrassinolide action on the plasmatic membranes of the barley roots. *In* Int. Congr. "Stress of life. Stress and adaptation from molecules to man", p. 26, P57, Budapest.

Kametani, T. (1988). Stereocontrolled construction of steroid side chains. *Actual. Chim. Ther.* **15**, 131-147.

Kametani, T., and Honda, T. (1986). Derives γ-lactone, et leur procede de preparation en vue de la synthese du brassinolide. **Pat. FR. 2,597,482**.

Kametani, T., and Honda, T. (1987a). Castasterone derivatives as plant growth stimulants. Jpn. Kokai Tokkyo Koho **JP 01 29,396 [88 29,396]** [*C. A.* **111**, 134648].

Kametani, T., and Honda, T. (1987b). Preparation of brassinosteroids as plant growth stimulants. Jpn. Kokai Tokkyo Koho **JP 01 29,372 [89 29,372]** [*C. A.* **111**, 154225].

Kametani, T., and Honda, T. (1987c). Preparation of pregnan-20-one derivatives as intermediates for brassinolides and epibrassinolides. Jpn. Kokai Tokkyo Koho **JP 63,246,395 [88,246,395]** [*C. A.* **111**, 58159].

Kametani, T., and Honda, T. (1987d). Preparation of pregnane derivatives as intermediates for brassinolides and epibrassinolides. Jpn. Kokai Tokkyo Koho **JP 63,246,394 [88,246,394]** [*C. A.* **111**, 39675].

Kametani, T., Tsubuki, M., Higurashi, K., and Honda, T. (1985). Stereocontrolled synthesis of 2-deoxycrustecdysone and related compounds. *J. Org. Chem.* **51**, 2932-2939.

Kametani, T., Katoh, T., Tsubuki, M., and Honda, T. (1986). One-step stereochemical determination of contiguous four acyclic chiral centers on the steroidal side chain: a novel synthesis of brassinolide. *J. Am. Chem. Soc.* **108**, 7055-7060.

Kametani, T., Katoh, T., Tsubuki, M., and Honda, T. (1987). Stereoselective synthesis of 26,27-bisnorbrassinolide. *Chem. Pharm. Bull.* **35**, 2334-2338.

Kametani, T., Katoh, T., Fujio, J., Nogiwa, I., Tsubuki, M., and Honda, T. (1988a). An improved synthesis of plant growth regulating steroid brassinolide and its congeners. *J. Org. Chem.* **53**, 1982-1991.

Kametani, T., Kigawa, M., Tsubuki, M., and Honda, T. (1988b). Stereocontrolled synthesis of the brassinolide side-chain: formal synthesis of brassinolide. *J. Chem. Soc., Perkin Trans. I* 1503-1507.

Kametani, T., Keino, K., Kigawa, M., Tsubuki, M., and Honda, T. (1989). Stereocontrolled synthesis of the brassinolide side chain *via* a pyranone derivative. *Tetrahedron Lett.* **30**, 3141-3142.

Kamuro, Y., and Inada, K. (1987). Effect of light conditions on brassinolide induced mung bean epicotyl elongation and radish growth. *Proc. Plant Growth Regul. Soc. Am.* **14**, 221-224.

Kamuro, Y., and Inada, K. (1991). The effect of brassinolide on the light-induced growth inhibition in mung bean epicotyl. *Plant Growth Regul.* **10**, 37-43.

Kamuro, Y., and Takatsuto, S. (1991). Capability for and problems of practical uses of brassinosteroids. *In* "Brassinosteroids. Chemistry, Bioactivity, and Applications. ACS Symposium Series" (H.G. Cutler, T. Yokota, and G. Adam, Eds.), Vol. 474, pp. 292-297. American Chemical Society, Washington.

Kamuro, Y., Kakiuchi, T., and Takatsuto, S. (1987). Preparation of (22R,23R,24S)-22,23-epoxy-2α,3α-(isopropylidenedioxy)-B-homo-7-oxa-5α-stigmastan-6-one as plant growth regulator. Eur. Pat. Appl. **EP 322,639** [*C. A.* **112**, 36265].

Kamuro, Y., Kakiuchi, T., and Takatsudo, H. (1988a). Preparation of brassinosteroids as plant growth stimulants. Jpn. Kokai Tokkyo Koho **JP 02 88,580 [90 88,580]** [*C. A.* **114**, 24324].

Kamuro, Y., Kakiuchi, T., and Takatsuto, S. (1988b). (22R,23R,24S)-22,23-Epoxy-2α,3α-isopropylidenedioxy-B-homo-7-oxa-5α-stigmastan-6-one and plant growth regulating method containing the same. Pat. **US 4,959,093**.

Kamuro, Y., Watanabe, T., and Kuriyama, H. (1993). Preparation of brassinosteroid derivatives and plant growth regulator containing the same. PCT Int. Appl. **WO 94 28,011** [*C. A.* **122**, 291316].

Kamuro, Y., Takatsuto, S., Noguti, T., Watanabe, T., and Fujisawa, H. (1996). Application of long-lasting BRs in combination with other PGRs. *Proc. Plant Growth Regul. Soc. Am.* **23**, 27-31 [*C. A.* **127**, 216318].

Kamuro, Y., Takatsuto, S., Watanabe, T., Noguchi, T., Kuriyama, H., and Suganuma, H. (1997). Practical aspects of brassinosteroid compound [TS303]. *Proc. Plant Growth Regul. Soc. Am.* **24**, 111-116.

Kandelinskaya, O.L., and Khripach, V.A. (1990). Brassinosteroid influence on protein metabolism in lupine seeds. *In* "Int. Workshop. Brassinosteroids: Chemistry, Bioactivity, Application", p. 46, Halle.

Katsumi, M. (1985). Interaction of a brassinosteroid with IAA and GA_3 in the elongation of cucumber hypocotyl sections. *Plant Cell Physiol.* **26**, 615-625.

Katsumi, M. (1991). Physiological mode of brassinolide action in cucumber hypocotyl growth. *In* "Brassinosteroids. Chemistry, Bioactivity, and Applications. ACS

Symposium Series" (H.G. Cutler, T. Yokota, and G. Adam, Eds.), Vol. 474, pp. 246-254. American Chemical Society, Washington.

Kauschmann, A., Adam, G., Jessop, A., Koncz, C., Szekeres, M., Voigt, B., Willmitzer, L., and Altmann, T. (1996a). Genetic evidence for an essential role of brassinosteroids in plant development. *Proc. Plant Growth Regul. Soc. Am.* **23**, 13.

Kauschmann, A., Jessop, A., Koncz, C., Czekeres, M., Willmitzer, L., and Altmann, T. (1996b). Genetic evidence for an essential role of brassinosteroids in plant development. *Plant J.* **9**, 701-713.

Kauschmann, A., Adam, G., Lichtblau, D., Mussig, C., Schmidt, J., Voigt, B., Willmitzer, L., and Altmann, T. (1997). Molecular/genetic analysis of brassinosteroid synthesis and action. *Proc. Plant Growth Regul. Soc. Am.* **24**, 95-96.

Kawamura, A., Berova, N., Nakanishi, K., Voigt, B., and Adam, G. (1997a). Configurational assignment of brassinosteroid sidechain by exciton coupled circular dichroic spectroscopy. *Tetrahedron* **53**, 11961-11970.

Kawamura, A., Berova, N., Nakanishi, K., Voigt, B., and Adam, G. (1997b). Microscale stereochemical analysis of brassinosteroid sidechain. *Proc. Plant Growth Regul. Soc. Am.* **24**, 117-122.

Kazakova, V.N., Karsunkina, N.P., and Sukhova, L.S. (1991). The effect of brassinolide and fusicoccin on potato yield and the resistance of stored tubers to fungal diseases. *Izv. Timiryazevskoi Sel'skokhozyaistvennoi Akad.* 82-88.

Kerb, U., Eder, U., and Krähmer, H. (1982a). 28-Methyl brassinosteroid derivatives useful as plant growth regulators. Ger. Offen. **DE 3,234,606** [*C. A.* **101**, 152187].

Kerb, U., Eder, U., and Krähmer, H. (1982b). 28-Methyl-Brassinosteroid-Derivative enthaltende Mittel mit Wachstumsregulatorischer Wirkung fur Pflanzen. Ger. (East) **DD 213,346**.

Kerb, U., Eder, U., and Krähmer, H. (1982c). Brassinosteroid derivatives and their use in plant growth regulating compositions. Ger. Offen. **DE 3,234,605** [*C. A.* **101**, 152188].

Kerb, U., Eder, U., and Krähmer, H. (1982d). Brassinosteroid-Derivative enthaltende Mittel mit Wachstumsregulatorischer Wirkung fur Pflanzen. Ger. (East) **DD 213,347**.

Kerb., U., Eder, U., and Krähmer, H. (1982e). Brassinosteroid derivatives having a plant growth-regulating action and their manufacture and use. UK Pat. Appl. **GB 2,127,021**.

Kerb, U., Eder, U., and Krähmer, H. (1983a). Hexanobrassinolide 22-ethers with plant growth regulating action. Ger. Offen. **DE 3,305,747** [*C. A.* **102**, 6952].

Kerb, U., Eder, U., and Krähmer, H. (1983b). Hexanor-Brassinolid-22-Äther enthaltende Mittel mit Wachstumsregulatorischer Wirkung für Pflanzen. Ger. (East) **DD 214,053**.

Kerb, U., Eder, U., and Krähmer, H. (1983c). Brassinosteroides, procede pour les prepareraimsi que produits contenant de tels composes et ayant une action regulatrice sur la croissance des plantes. Pat. **FR. 2,532,941**.

Kerb, U., Eder, U., and Krähmer, H. (1983d). Ethers en 22 d'hexanor-brassinolides, procede pour les preparer, produits qui en contiennent et qui ont une action regulatrice sur la croissance de plantes, et application de ces produits. Pat. **FR. 2,541,290**.

Kerb, U., Eder, U., and Krähmer, H. (1983e). Methyl-28 brassinosteroides, procede pour les preparer, ainsi que produits contenant de tels composes et ayant une action regulatrice sur la croissance des plantes. Pat. **FR. 2,532,942**.

Kerb, U., Eder, U., and Krähmer, H. (1985). Prostfedek k regulaci rustu rostlin a zposob vyroby jehe ucinne latky. Czech. **CS 241,148**.

Kerb, U., Eder, U., and Krähmer, H. (1986). Synthesis of hexanor-brassinolide-22-ethers with plant growth-promoting activity. *Agric. Biol. Chem.* **50**, 1359-1360.

Khokhlova, V.A., Bokebayeva, G.A., Adam, G., and Kulaeva, O.N. (1990). Comparative study of antistress effect of brassinosteroid and its inactive analogue on barley leave in salinity. *In* "Int. Workshop. Brassinosteroids: Chemistry, Bioactivity, Application", p. 44, Halle.

Khripach, V.A. (1990). Synthesis of brassinosteroids. *Pure Appl. Chem.* **62**, 1319-1324.

Khripach, V.A., Zhabinskii, V.N., and Zhernosek, E.V. (1988). Preparation of 3β-hydroxy-24R-methylcholest-5-ene. U.S.S.R. Pat. Appl. **SU 4612746/14**.

Khripach, V.A., Litvinovskaya, R., Baranovsky, A., and Drach, S. (1990a). Highly stereoselective synthesis of 20-isosteroidal side chain by dehydration of (20R)-20-hydroxy-20-isoxazolinylsteroids. *Tetrahedron Lett.* **31**, 7065-7068.

Khripach, V.A., Litvinovskaya, R.P., and Baranovskii, A.V. (1990b). Highly effective dehydration of steroidal isoxazolines. *Khim. Geterocycl. Soed.*, 852.

Khripach, V.A., Litvinovskaya, R.P., and Baranovskii, A.V. (1990c). A new method of formation of polyfunctional steroidal side chains. *Bioorg. Khim.* **16**, 1700-1701.

Khripach, V.A., Litvinovskaya, R.P., and Ermolenko, E.A. (1990d).The effective synthesis of 20ξ-acetoxy-5α-pregn-2-en-6-one - a key intermediate in synthesis of brassinosteroids and analogues starting from pregnenolone. *Vesti AN BSSR, Ser. Khim. Navuk* 70-74.

Khripach, V.A., Litvinovskaya, R.P., Baranovskii, A.V., and Ermolenko, E.A. (1990e). Cleavage of heterocyclic nucleus of 20-isoxazolinylsteroids under the action of bases. *Khim. Geterocycl. Soed.*, 1389-1393.

Khripach, V.A., Litvinovskaya, R.P., Drach, S.V., and Streltsova, V.A. (1990f). (22R)-3α,5-Cyclo-5α-cholestane-6,24-dion-22-ol showing phytogrowth-stimulating activity and method for its synthesis. U.S.S.R. **SU 1,761,761** [*C. A.* **119**, 250250].

Khripach, V.A., Litvinovskaya, R.P., Drach, S.V., and Streltsova, V.A. (1990g). (22R)-2α,3α-Dihydroxy-20-(3-isopropylisoxazolin-5-yl)-5α-pregnan-6-one, showing phytogrowth-stimulating activity, and method of its preparation. U.S.S.R. **SU 1,747,455** [*C. A.* **119**, 271494].

Khripach, V.A., Litvinovskaya, R.P., Drach, S.V., and Streltsova, V.A. (1990h). Method for production of (22*R*)-20-(3'-isopropylisoxazolin-5'-yl)-3α,5-cyclo-5α-pregnan-6-one, possessing phyto growth stimulating activity. U.S.S.R. **SU 1,786,807** [*C. A.* **126**, 60208].

Khripach, V.A., Zhabinskii, V.N., and Olkhovick, V.K. (1990i). Highly stereoselective synthesis of steroidal 22α-allilyc alcohols *via* 22-aldehydes and 1-silyl-1-iodo-1-alkenes: a new efficient route to the side chain construction of brassinolide. *Tetrahedron Lett.* **37**, 4937-4940.

Khripach, V.A., Zhabinskii, V.N., Olkhovick, V.K., and Akhrem, A.A. (1990j). Synthesis of 28-nor analogues of brassinolide. *Vesti AN BSSR, Ser. Khim. Navuk* 69-74.

Khripach, V.A., Zhabinskii, V.N., Olkhovick, V.K., and Lakhvich, F.A. (1990k). Synthesis of brassinolide and its analogues. *Zh. Org. Khim.* **26**, 2200-2206.

Khripach, V.A., Litvinovskaya, R.P., and Baranovskii, A.V. (1991a). A novel approach to synthesis of brassinolide. *Dokl. AN USSR* **318**, 597-600.

Khripach, V.A., Litvinovskaya, R.P., Baranovskii, A.V., and Akhrem, A.A. (1991b). (20*S*,5)-20-(3'-Isopropylisoxazolin-5'-yl)-6β-methoxy-3α,5-cyclo-5α-pregnan as an intermediate in synthesis of (22*R*,23*R*)-3β-acetoxy-22,23-isopropylidenedioxy-24-methylcholest-5-ene. Rus. **RU 2,004,548**.

Khripach, V.A., Litvinovskaya, R.P., Baranovskii, A.V., and Akhrem, A.A. (1991c). (22ξ)-6β-Methoxy-3α,5-cyclo-5α-cholestan-24-one as an intermediate in synthesis of (22*R*,23*R*)-3β-acetoxy-22,23-isopropylidenedioxy-24-methylcholest-5-ene. Rus. **RU 2,024,540**.

Khripach, V.A., Zhabinskii, V.N., and Litvinovskaya, R.P. (1991d). Synthesis and some practical aspects of brassinosteroids. *In* "Brassinosteroids. Chemistry, Bioactivity, and Applications. ACS Symposium Series" (H.G. Cutler, T. Yokota, and G. Adam, Eds.), Vol. 474, pp. 43-55. American Chemical Society, Washington.

Khripach, V.A., Zhabinskii, V.N., and Zhernosek, E.V. (1991e). Reduction of ergosterol diene system. *Vesti AN BSSR, Ser. Khim. Navuk* 187-190.

Khripach, V.A., Zhabinskii, V.N., Olkhovick, V.K., and Zhernosek, E.V. (1991f). Method of preparation of 3α,5-cyclo-24*R*-methyl-5α-cholest-7,22-dien-6β-ol. Rus. **RU 2,024,541**.

Khripach, V.A., Zhabinskii, V.N., Zhernosek, E.V., Lakhvich, F.A., Olkhovick, V.K., and Ivanova, G.V. (1991g). Synthesis of epibrassinolide *via* 22,23-epoxides. *Vesti AN BSSR, Ser. Khim. Navuk* 71-75.

Khripach, V.A., Litvinovskaya, R.P., and Baranovskii, A.V. (1992a). A new method for preparation of 20-isosteroidal side chain. *Bioorg. Khim.* **18**, 964-968.

Khripach, V.A., Zhabinskii, V.N., and Zhernosek, E.V. (1992b). A new synthesis of brassicasterol. *Khim. Pripodn. Soed.* 90-92.

Khripach, V.A., Zhabinskii, V.N., Ivanova, G.V., and Olkhovick, V.K. (1992c). A new synthesis of homobrassinolide. *Vesti AN Belarusi, Ser. Khim. Navuk* 70-72.

Khripach, V.A., Lakhvich F.A., and Zhabinskii, V.N. (1993a). "Brassinosteroids". Science and Technique, Minsk.

Khripach, V.A., Litvinovskaya, R.P., and Drach, S.V. (1993b). Synthesis of brassinosteroid analogues, containing isoxazoline fragment in a side chain. *Zh. Org. Khim.* **29**, 717-723 [*C. A.* **120**, 299068].

Khripach, V.A., Litvinovskaya, R.P., Baranovskii, A.V., and Drach, S.V. (1993c). Synthesis and hydroxylation of Δ^{22}-24-oxosteroids. *Zh. Org. Khim.* **29**, 724-730 [*C. A.* **121**, 9807].

Khripach, V.A., Zavadskaya, M.I., Kotyatkina, A.I., and Drachenova, O.A. (1993d). A new synthesis of (20S)-3β-hydroxycholesta-5,22(E)-dien-24-one. *Zh. Org. Khim.* **29**, 1368-1371.

Khripach, V.A., Zhabinskii, V.N., and Kotyatkina, A.I. (1993e). Construction of polyfunctionalysed steroidal side chains *via* C-22 nitrile oxide. *Zh. Org. Khim.* **29**, 1573-1577.

Khripach, V.A., Zhabinskii, V.N., and Kotyatkina, A.I. (1993f). Synthesis of the steroidal nitrile oxide and its reactions with olefins. *Zh. Org. Khim.* **29**, 1569-1572.

Khripach, V.A., Zhabinskii, V.N., and Olkhovick, V.K. (1993g). Stereoselective formation of C_{22}-C_{24} fragment of steroidal side chain. *Zh. Org. Khim.* **29**, 2214-2225.

Khripach, V.A., Zhabinskii, V.N., and Kotyatkina, A.I. (1994a). Synthesis of α,β-unsaturated ketones of ergastane series *via* 22,23-epoxides. *Zh. Org. Khim.* **30**, 966-969.

Khripach, V.A., Zhabinskii, V.N., and Olkhovick, V.K. (1994b). Synthesis of (24R)-homobrassinolide. *Khim. Pripodn. Soed.*, 385-391.

Khripach, V.A., Zhabinskii, V.N., Olkhovick, V.K., Ivanova, G.V., Zhernosek, E.V., and Kotyatkina, A.I. (1994c). Improved synthesis of epibrassinolide. *Zh. Org. Khim.* **30**, 1650-1655.

Khripach, V.A, Zhabinskii, V.N., and Zhernosek, E.V. (1995a). A new route to allylic alcohols: application to the construction of brassinolide side chain. *Tetrahedron Lett.* **36**, 607-608.

Khripach, V.A., Zhabinskii, V.N., and Kotyatkina, A.I. (1995b). An approach to synthesis of C^1-hydroxybrassinosteroids. *Zh. Org. Khim.* **31**, 1866-1867.

Khripach, V.A., Zhabinskii, V.N., and Lakhvich, F.A. (1995c). Perspectives of practical application of brassionosteroids - a new class of phytohormones. *Sel'skokhosyaistvennaya biologiya* 3-11.

Khripach, V.A., Zhabinskii, V.N., and Zhernosek, E.V. (1995d). A new approach to formation of brassinosteroids cyclic fragment *via* $\Delta^{4,6}$-dienones. *Zh. Org. Khim.* **31**, 1823-1825.

Khripach, V.A., Zhabinskii, V.N., and Zhernosek, E.V. (1995e). Synthesis of some epibrassinolide analogues. *Vesti AN Belarusi, Ser. Khim. Navuk* 64-67.

Khripach, V.A., Zhabinskii, V.N., Zhernosek, E.V., and Khripach, N.B. (1995f). Synthesis of [5,7,7-$^2H_3$]-epicastasterone. *Vesti AN Belarusi, Ser. Khim. Navuk* 75-78.

Khripach, V.A., Zhabinskii, V.N., Litvinovskaya, R.P., Zavadskaya, M.I., Deeva, V.P., and Vedeneev, A.N. (1995g). Preparation for the diminishing of radionuclides accumulation by plants and method of its application. Pat. Appl. **BY 950,941**.

Khripach, V.A., Voronina, L.V., and Malevannaya, N.N. (1996a). Preparation for the dimishing of heavy metals accumulation by agricultural plants. Pat. Appl. **RU 96,101,850**.

Khripach, V.A., Zhabinskii, V.N., Litvinovskaya, R.P., Zavadskaya, M.I., Savelieva, E.A., Karas, I.I., and Vakulenko, V.V. (1996b). Method of enhancement of food value of potato. Pat. Appl. **BY 963,445**.

Khripach, V.A., Zhabinskii, V.N., Litvinovskaya, R.P., Zavadskaya, M.I., Savelieva, E.A., Karas, I.I., Kilchevskii, A.V., and Titova, C.H. (1996c). Method of protection of potato from phytophtora infection. Pat. Appl. **BY 960,346**.

Khripach, V.A., Zhabinskii, V.N., and Zhernosek, E.V. (1996d). A new method for preparation of allylic alcohols and its use for synthesis of brassinolide side chain. *Zh. Org. Khim.* **32**, 693-697 [*C. A.* **126**, 31536].

Khripach, V.A., Zhabinskii, V.N., and Malevannaya, N.N. (1997). Recent advances in brassinosteroids study and application. *Proc. Plant Growth Regul. Soc. Am.* **24**, 101-106.

Khripach, V.A., Zhabinskii, V.N., Pavlovskii, N.D., Lyakhov, A.S., and Govorova, A.A. (1998). Synthesis and X-ray investigation of (20S)-22-benzamido-6β-methoxy-3α,5-cyclo-5α-25,26-bisnorcholest-22-en-24-one. *Bioorg. Khim.* In press.

Khrustaleva, L.I., Andreeva, G.N., Golovnina, Yu.M., Zlobin, A.I., and Pogorilaya, E.V. (1991). Cytogenetic investigation in somatic and generative cells of barley, treated by epibrassinolide in field conditions. *In* "Workshop on Brassinosteroids", 2nd, p. 31, Minsk.

Khrustaleva, L.I, Pogorilaya, E.V., Golovnina, Yu.M., and Andreeva, G.N (1995). Effect of epibrassinolide 55 on mitotic activity and the frequency of chromosome aberrations in barley root tip cells under salt stress. *Sel'skokhozyaistvennaya Biologiya* 69-73.

Kihira, K., and Hoshita, T. (1985). Synthesis of α,β-unsaturated C_{24} bile acid. *Steroids* **46**, 767-774.

Kilchevskii, A.V., and Frantsuzionok, V.V. (1997). Effect of epibrassinolide on proliferation of lily explants. *In* "Regulators of plant growth and development", 4th, pp. 297-298, Moscow.

Kim, D., Han, G.H., and Kim, K. (1989). Stereoselective ester enolate alkylation and hydroxylation at C-22 of a steroid side chain. *Tetrahedron Lett.* **30**, 1579-1580.

Kim, S.-K. (1991). Natural occurrences of brassinosteroids. *In* "Brassinosteroids. Chemistry, Bioactivity, and Applications. ACS Symposium Series" (H.G. Cutler, T. Yokota, and G. Adam, Eds.), Vol. 474, pp. 26-35. American Chemical Society, Washington.

Kim, S.-K., Yokota, T., and Takahashi, N. (1987). 25-Methyldolichosterone, a new brassinosteroid with a tertiary butyl group from immature seed of *Phaseolus vulgaris*. *Agric. Biol. Chem.* **51**, 2303-2305.

Kim, S.-K., Akihisa, T., Tamura, T., Matsumoto, T., Yokota, T., and Takahashi, N. (1988). 24-Methylene-25-methylcholesterol in *Phaseolus vulgaris* seed: structural relation to brassinosteroids. *Phytochemistry* **27**, 629-631.

Kim, S.-K., Abe, H., Little, C.H.A., and Pharis, R.P. (1990). Identification of two brassinosteroids from the cambial region of Scots pine (*Pinus silverstris*) by gas chromatography-mass spectrometry, after detection using a dwarf rice lamina inclination bioassay. *Plant Physiol.* **94**, 1709-1713.

Kim, S.-K., Mizuno, K., Hatori, M., and Marumo, S. (1994a). A brassinolide-inhibitor KM-01, its isolation and structure elucidation from a fungus *Drechslera avenae*. *Tetrahedron Lett.* **35**, 1731-1734.

Kim, S.-K., Yokota, T., and Takahashi, N. (1994b). Identification of new 2-deoxy type brassinosteroids in immature seed of *Phaseolus vulgaris* by gas chromatography-mass spectrometry. *J. Plant Biol.* **37**, 411-415 [*C. A.* **122**, 310197].

Kim, S.-K., Asano, T., and Marumo, S. (1995). Biological activity of brassinosteroid inhibitor KM-01 produced by a fungus *Drechslera avenae*. *Biosci. Biotech. Biochem.* **59**, 1394-1397.

Kim, S.-G. (1995). Brassinosteroids. *Hwahak Sekye* **35**, 36-38 [*C. A.* **123**, 286370].

Kirillova, E.N., Balmush, G.T., Russu, M.M., Dochmila, A.S., Khripach, V.A., and Zhabinskii. V.N. (1993). Some aspects of epibrassinolide action on tomato. *In* "Brassinosteroids - biorational, ecologically safe regulators of growth and productivity of plants", 3rd, pp. 26-27, Minsk.

Kishi, T., Wada, K., Marumo, S., and Mori, K. (1986). Synthesis of brassinolide analogs with a modified ring B and their plant growth promoting activity. *Agric. Biol. Chem.* **50**, 1821-1830.

Kislin, E.N., and Semicheva, T.V. (1991). Effect of brassinosteroids on endogenic level of cytokinins in barley leaves. *In* "Workshop on Brassinosteroids", 2nd, pp. 26-27, Minsk.

Kitani, Y. (1994). Induction of parthenogenetic haploid plants with brassinolide. *Jpn. J. Genet.* **69**, 35-39.

Klahre, U., and Chua, N.-H. (1996). Characterisation of the putative brassinosteroid mutant diminuto. *Proc. Plant Growth Regul. Soc. Am.* **23**, 12.

Klahre, U., Fujioko, S., Yokota, T., and Chua, N.-H. (1997). Characterisation of the diminuto mutant and genes regulated by brassinosteroids. *Proc. Plant Growth Regul. Soc. Am.* **24**, 99-100.

Kobayashi, K., and Nitani, F. (1989). Crop yield enhancers containing brassinolides. Jpn. Kokai Tokkyo Koho **JP 03,184,903** [91,184,903] [*C. A.* **116**, 78592].

Kobayashi, N., Higashi, T., and Shimada, K. (1994). Synthesis of (24R)-11α-(4-carboxybutyryloxy)-24,25-dihydroxyvitamin D_3: a novel haptenic derivative producing antibodies of high affinity for (24R)-24,25-dihydroxyvitamin D_3. *J. Chem. Soc., Perkin Trans. I* 269-275.

Kocienski, P.J., Lythgoe, B., and Ruston, S. (1978). Scope and stereochemistry of an olefin synthesis from β-hydroxysulfones. *J. Chem. Soc., Perkin Trans. 1* 829-834.

Kohout, L. (1984). Brassiny, brassinolid a brassinosteroidy. *Chem. Listy* **78**, 1129-1156.

Kohout, L. (1989a). 2α,3α-Bis(acyloxy)-17β-(2-methylbutyroyloxy)-7-oxa-B-homo-5α-androstan-6-ones and method for their preparation. Czech. **CS 275,040** [*C. A.* **120**, 107466].

Kohout, L. (1989b). Method of preparing the brassinolide hormone 2α,3α-dihydroxy-17β-(3-methylbutyryloxy)-7-oxa-B-homo-5α-androstan-6-one. Czech. **CS 274,530** [*C. A.* **119**, 250244].

Kohout, L. (1989c). Preparation of 2α,3α-diacyloxy-17β-(3-methylbutyryloxy)-7-oxa-B-homo-5α-androstan-6-ones as intermediates for a plant growth regulator. Czech. **CS 270,777** [*C. A.* **116**, 106587].

Kohout, L. (1989d). Preparation of 2α,3α-isoalkylidenedioxy-17β-(3'-methylbutyryloxy)-7-oxa-B-homo-5α-androstan-6-one as an intermediate for brassinolide plant hormone. Czech. **CS 270,783** [*C. A.* **117**, 90590].

Kohout, L. (1989e). Synthesis of brassinosteroids with a five carbon atom ester functionality in position 17. *Collect. Czech. Chem. Commun.* **54**, 3348-3359 [*C. A.* **113**, 24318].

Kohout, L. (1989f). The brassinolide plant hormone 2α,3α-dihydroxy-17β-(2-methylbutyryloxy)-7-oxa-B-homoandrostan-6-one and method of its preparation. Czech. **CS 275,008** [*C. A.* **118**, 147878].

Kohout, L. (1994a). 7-Oxo-7a-oxa-brassinosteroids with cholestane side chain. *Collect. Czech. Chem. Commun.* **59**, 1219-1225 [*C. A.* **121**, 205789].

Kohout, L. (1994b). New method of preparation of brassinosteroids. *Collect. Czech. Chem. Commun.* **59**, 457-460 [*C. A.* **121**, 57774].

Kohout, L. (1997). D-Homo-17a-oxa-brassinosteroid analogues. *In* "XVII Conference on Isoprenoids, Abstracts of Papers", p. 79. Kraków.

Kohout, L., and Kasal, L. (1992). Preparation of brassinosteroids. Czech Rep. **CZ 281,040** [*C. A.* **126**, 157697].

Kohout, L., and Strnad, M. (1986). Brassinosteroids with a cholestane side chain. *Collect. Czech. Chem. Commun.* **51**, 447-458 [*C. A.* **105**, 209273].

Kohout, L., and Strnad, M. (1989a). Brassinolide analogs without a side chain. *Collect. Czech. Chem. Commun.* **54**, 1019-1027 [*C. A.* **112**, 7785].

Kohout, L., and Strnad, M. (1989b). Preparation of a new brassinolide analog 2α,3α-dihydroxy-17β-(3'-methylbutyryloxy)-7-oxa-B-homo-5α-androstan-6-one. Czech. **CS 270,776** [*C. A.* **117**, 90589].

Kohout, L., and Strnad, M. (1992). Brassinosteroids with ester function with five carbon atoms at the 20 position. *Collect. Czech. Chem. Commun.* **57**, 1731-1738 [*C. A.* **117**, 251619].

Kohout, L., Velgova, H., and Strnad, M. (1986). The brassinolide phytohormone 2α,3α,17β-trihydroxy-5α-androstan-6-one and a method for its preparation. Czech. **CS 252,605** [*C. A.* **111**, 134642].

Kohout, L., Cerny, V., and Strnad, M. (1987a). Alternative synthesis of 2α,3α,17β-trihydroxy-7-oxa-B-homo-5α-androstan-6-one and some androstane brassinolide analogs. *Collect. Czech. Chem. Commun.* **52**, 1026-1042.

Kohout, L., Velgova, H., Strnad, M., and Kaminek, M. (1987b). Brassinosteroids with androstane and pregnane skeleton. *Collect. Czech. Chem. Commun.* **52**, 476-486.

Kohout, L., Strnad, M., and Kaminek, M. (1991). Types of brassinosteroids and their bioassays. *In* "Brassinosteroids. Chemistry, Bioactivity, and Applications. ACS Symposium Series" (H.G. Cutler, T. Yokota, and G. Adam, Eds.), Vol. 474, pp. 56-73. American Chemical Society, Washington.

Kohout, L., Kasal, A., and Strnad, M. (1996). Pregnane type brassinosteroids with a four carbon ester functionality in position 20. *Collect. Czech. Chem. Commun.* **61**, 930-940.

Kolbe, A., Marquardt, V., and Adam, G. (1992). Synthesis of tritium labelled 24-epibrassinolide. *J. Labelled Compd. Radiopharm.* **31**, 801-805.

Kolbe, A., Schneider, B., Porzel, A., Voigt, B., Krauss, G., and Adam, G. (1994). Pregnane-type metabolites of brassinosteroids in cell suspension cultures of *Ornithopus sativus*. *Phytochemistry* **36**, 671-673.

Kolbe, A., Schneider, B., Porzel, A., Schmidt, J., and Adam, G. (1995). Acyl-conjugated metabolites of brassinosteroids in cell suspension cultures of *Ornithopus sativus*. *Phytochemistry* **38**, 633-636.

Kolbe, A., Schneider, B., Porzel, A., and Adam, G. (1996). Metabolism of 24-epi-castasterone and 24-epi-brassinolide in cell suspension cultures of *Ornithopus sativus*. *Phytochemistry* **41**, 163-167.

Kolbe, A., Porzel, A., Schneider, B., and Adam, G. (1997). Diglycosidic metabolites of 24-epi-teasterone in cell suspension cultures of *Lycopersicon esculentum* L. *Phytochemistry* **46**, 1019-1022.

Konai, Y., Hayashi, S., Kubota, Y., and Kodama, K. (1984a). Process for producing *i*-brassicasterol. Pat. Appl. **GB 2,154,590**.

Konai, Y., Hayashi, S., Kubota, Y., and Kodama, K. (1984b). Procede de preparation de l'i-brassicasterol. Pat. **FR. 2,560,199**.

Konai, Y., Hayashi, S., Kubota, Y., and Kodama, K. (1985). Process for producing *i*-brassicasterol. Pat. **US 4,614,620**.

Koncz, C., Czekeres, M., Altmann, T., and Mathur, J. (1996). Cloning of cDNA and gene for cytochrome P450-type hydroxylase involved in the brassinosteroid synthesis in plants and use of P450 for plant growth regulation. PCT Int. Appl. **WO 97 35,986** [*C. A.* **127**, 304118].

Kondo, M., and Mori, K. (1983). Synthesis of brassinolide analogs with or without the steroidal side chain. *Agric. Biol. Chem.* **47**, 97-102.

Koolman, J. (1989). "Ecdysone", Georg Thieme-Verlag, Stuttgart and New York.

Korableva, N.P., and Sukhova, L.S. (1991). Regulation of rest period of potato bulbs and their resistance to deseases with epibrassinolide. *In* "Workshop on Brassinosteroids", 2nd, p. 46, Minsk.

Korableva, N., Sukhova, L., and Dogonadse, M. (1990). Effect of brassinosteroids on the sprouting of potato tubers and their resistance to diseases during storage. *In* "Int. Workshop. Brassinosteroids: Chemistry, Bioactivity, Application", p. 32, Halle.

Korableva, N.P., Suchova, L.S., Muromtsev, G.S., Koreneva, V.M., Kazakova, V.I., Karsunkina, N.P., and Dogonadze, M.Z. (1992). Method of potato treatment for long-term preservation. U.S.S.R. **SU 1,794,261**.

Korableva, N.P, Dogonadse, M.E, and Platonova, T.A. (1995). Mechanism of brassinosteroid action in regulation of rest period of potato tubers and of their resistance to deseases. *In* "Brassinosteroids - biorational, ecologically safe regulators of growth and productivity of plants", 4th, pp. 11-12, Minsk.

Korableva, N.P, Platonova, T.A., and Dogonadse, M.E. (1998). Changes in ethylene biosynthesis in meristems of potato tubers (*Solanum tuberosum* L.) under the action of brassinolide. *Dokl. Akad. Nauk* (Russia). In press.

Koreeda, M., and Ricca, D.J. (1986). Chirality transfer in stereoselective synthesis. A highly stereocontrolled synthesis of 22-hydroxylated steroid side chain *via* the [2,3]Wittig rearrangement. *J. Org. Chem.* **51**, 4090-4092.

Koreeda, M., and Tanaka, Y. (1987). Stereoselective acyclic synthesis *via* allylmetals: threo vicinal diols from both *E*- and *Z*-γ-alkoxyallyltins and aldehydes. *Tetrahedron Lett.* **28**, 143-146.

Koreeda, M., and Wu, J. (1995). Stereoselective synthesis of the brassinolide side chain by the use of a 5-exo-α-silyl radical cyclization-protodesilylation sequence. *Synlett* 850-852.

Koreeda, M., Tanaka, Y., and Schwartz, A. (1980). Stereochemically controlled synthesis of steroid side chains: synthesis of desmosterol. *J. Org. Chem.* **45**, 1172-1174.

Kositsina-Pinegina, E. (1995). "Epin" became my real friend. *Sad i ogorod* 18-19.

Kovganko, N.V., and Ananich, S.K. (1991). Synthesis of 5α-hydroxy analogues of brassinosteroids starting from Δ^5-sterols. *Zh. Org. Khim.* **27**, 103-108.

Kovganko, N.V., and Ananich, S.K. (1995). Synthesis of the 3,22,23-triacetates of 28-homotyphasterol and its (22*S*,23*S*)-isomer. *Khim. Prirodn. Soed.* **31**, 584-588.

Kovganko, N.V., and Kashkan, Zh.N. (1990). Synthesis of brassinosteroids analogues from β-sitosterol. *Zh. Org. Khim.* **26**, 2545-2552 [*C. A.* **115**, 159510].

Kovganko, N.V., and Netesova, T.N. (1991). A new synthesis of 2α,3α-dihydroxy-B-homo-7-oxa-5α-cholestan-6-one. *Zh. Org. Khim.* **27**, 100-102.

Kozick, T.A., and Kislin, E.N. (1991). Effect of brassinosteroids on cytokinin and ABA level in barley plants. *In* "Workshop on Brassinosteroids", 2nd, pp. 29-30, Minsk.

Krizek, D.T., and Mandava, N.B. (1983a). Influence of spectral quality on the growth response of intact bean plants to brassinosteroids, a growth-promoting steroidal lactone. I. Stem elongation and morphogenesis. *Physiol. Plant.* **57**, 317-323.

Krizek, D.T., and Mandava, N.B. (1983b). Influence of spectral quality on the growth response of intact bean plants to brassinosteroid, a growth-promoting steroidal lactone. II. Chlorophyll content and partitioning of assimilate. *Physiol. Plant.* **57**, 324-329.

Krizek, D.T., and Worley, J.F. (1973). The influence of light intensity on the internodal response of intact bean plants to brassins. *Bot. Gaz.* **13**, 147-150.

Krizek, P.T., and Worley, J.F. (1981). The influence of spectral quality on the internodal response of intact bean plants to brassins. *Physiol. Plant.* **51**, 259-264.

Kulaeva, O.N., Burchanova, E.A., Fedina, A.B., Danilova, R.V., Adam, G., Vorbrodt, H.-M., and Khripach, V.A. (1989). Brassinosteroids in regulation of protein synthesis in wheat leaves. *Dokl. AN USSR* **305**, 1277-1279 [*C. A.* **111**, 93957].

Kulaeva, O.N., Burkhanova, E.A., Fedina, A.B., Khokhlova, V.A., Bokebayeva, G.A., Vorbrodt, H.M., and Adam, G. (1991). Effect of brassinosteroids on protein synthesis and plant-cell ultrastructure under stress conditions. *In* "Brassinosteroids. Chemistry, Bioactivity, and Applications. ACS Symposium Series" (H.G. Cutler, T. Yokota, and G. Adam, Eds.), Vol. 474, pp. 141-155. American Chemical Society, Washington.

Kumavat, B.L., Sharma, D.D., and Jat, S.C. (1997). Effect of brassinosteroid on yield and yield attributing characters under water deficit stress condition in mustard (*Brassica juucea* (L.) Czern. and Coss.). *Ann. Biol. Ludhiana* **13**, 91-93.

Kumura, A., Ishii, R., Luo, B.S., Adachi, M., Hamada, K., and Fujita, F. (1984). Enhancement of crop yield. Ger. Offen. **DE 3,533,633** [*C. A.* **104**, 181755].

Kuno, K. (1997). Effects of plant growth steroid brassinolide, on dry-weight growth and nutrient translocation in mulberry shoots. *Nippon Sanshigaku Zasshi* **66**, 57-58 [*C. A.* **126**, 302595].

Kuno, K., and Ji, D. (1996). Characteristics of mulberry callus proliferated by the application of brassinolide and coumarin. *Nippon Sanshigaku Zasshi* **65**, 347-351 [*C. A.* **126**, 71527].

Kuraishi, S., Yamaki, Y., and Yamanaka, Y. (1987). Brassinolides-containing plant growth-promoting compositions containing amide solvents and/or water-soluble polymers. Jpn. Kokai Tokkyo Koho **JP 01,258,604** [89,258,604] [*C. A.* **112**, 93946].

Kuraishi, S., Sugiyama, K., and Yamanaka, Y. (1989). Regulation of flower setting of citrus with brassinosteroids. Jpn. Kokai Tokkyo Koho **JP 03,173,804** [91,173,804] [*C. A.* **116**, 36264].

Kuraishi, S., Sakurai, N., Eun, J.-S., and Sugiyama, K. (1991). Effect of brassinolide on levels of indoleacetic acid and abscisic acid in squash hypocotyls. *In* "Brassinosteroids. Chemistry, Bioactivity, and Applications. ACS Symposium Series" (H.G. Cutler, T. Yokota, and G. Adam, Eds.), Vol. 474, pp. 312-319. American Chemical Society, Washington.

Kurapov, P.B. (1996). "Hormonal balance in plants", Doctor of Sciences Degree Thesis, Tymyryazev Agricultural Academy, Moscow.

Kurapov, P.B., Skorobogatova, I.V., Kozik, T.A., Bumazhnyi, B.E., and Kislin, E.N. (1992). Effect of brassinosteroids on content of ABA, cytokinins and gibberellins in spring barley. *In* "Plant growth regulators", pp. 144-155. Ukrainian Academy of Sciences, Kiev.

Kurapov, P.B., Siusheva, A.G., and Skorobogatova, I.V. (1995). Effect of epibrassinolide on transport of assimilates in plants of potato, barley, and wheat. *In* "Brassinosteroids - biorational, ecologically safe regulators of growth and productivity of plants", 4th, pp. 7-8, Minsk.

Kurganskii N.P. (1993). Application of "Epin" on sugar-beet in 1991-1992. Technical report of experimental station on sugar beet, Belarus.

Kuriyama, H., Tanaka, T., and Furuse, T. (1989). Separation and purification of brassicasterol from phytosterols. Jpn. Kokai Tokkyo Koho **JP 02,215,794** [90,215,794] [*C. A.* **114**, 24325].

Kutner, A., Perlman, K.L., Sicinski, R.S., Phelps, M.E., Schnoes, H.K., and DeLuca, H.F. (1987). Vitamin D C-22 aldehydes. New key intermediates for the synthesis of side chain modified vitamin D analogues. *Tetrahedron Lett.* **28**, 6129-6132.

Kutner, A., Chodynski, M., Halkes, S.J., and Brugman, J. (1993). Novel concurrent synthesis of side-chain analogues of vitamins D_2 and D_3: 24,24-dihomo-25-hydroxycholecalciferol and (22E)-22-dehydro-24,24-dihomo-25hydroxycholecalciferol. *Bioorg. Chem.* **21**, 13-23.

Kutschabsky, L., Adam, G., and Vorbrodt, H.-M. (1990). Molecul- und Kristallstruktur von (22S,23S)-Homobrassinolid. *Z. Chem.* **30**, 136-137.

Kuzmitskii, B.B., and Mizulo, N.A. (1991). Technical report "Study of acute toxicity of epibrassinolide and its preparative forms". Institute of Bioorganic Chemistry, Academy of Sciences of Belarus, Minsk.

Lakhvich, F.A., Khripach, V.A., Kovganko, N.V., and Zhabinskii, V.N. (1985a). (22R,23R,24S)-3β-Bromo-22,23-epoxy-24-ethyl-5α-cholestan-6-one as intermediate in synthesis of (24S)-ethylbrassinone. U.S.S.R. **SU 1,270,154**.

Lakhvich, F.A., Khripach, V.A., Kovganko, N.V., and Zhabinskii, V.N. (1985b). (22R,23R,24S)-22,23-Epoxy-24-ethyl-5α-cholestan-6-one as intermediate in synthesis of (24S)-ethylbrassinone. U.S.S.R. **SU 1,270,155**.

Lakhvich, F.A., Khripach, V.A., Kovganko, N.V., and Zhabinskii, V.N. (1985c). (22R,23R,24S)-22,23-Dihydroxy-24-ethyl-5α-cholestan-6-one as intermediate in synthesis of (24S)-ethylbrassinone and method of its preparation. U.S.S.R. **SU 1,363,830** [*C. A.* **123**, 228644].

Lakhvich, F.A., Khripach, V.A., and Zhabinskii, V.N. (1990). Steroidal plant hormones, their isolation and structure. *Vesti AN BSSR, Ser. Khim. Navuk* 99-116.

Lakhvich, F.A., Khripach, V.A., and Zhabinskii, V.N. (1991). Synthesis of brassinosteroids - a new class of plant hormones. *Usp. Khim.* **60**, 1299-1333 [*Russ. Chem. Rev. (Engl. Transl.)* **60**, 658-675].

Laman, N.A., Vlasova, N.N., Khripach, V.A., Stratilatova, E.V., and Putyrskii, I.N. (1997). Effects of epibrassinolide, kinetin and gibberellic acid on growth and development of spring barley plants. In "Plant growth and development regulators", 4th, p.193, Moscow.

Lehmann, M., Vorbrodt, H.M., Adam, G., and Koolman, J. (1988). Antiecdysteroid activity of brassinosteroids. *Experientia* **44**, 355-356.

Leshem, Y.Y. (1984). Interaction of cytokinins with lipid-associated oxy free radicals. Senescence: a prospective mode of cytokinin action. *Can. J. Bot.* **62**, 2943-2949.

Levene, P.A., and Marker, R.E. (1935). The configurational relationship of acids of the isopropyl and isobutyl series to those of the normal series. *J. Biol. Chem.* **111**, 299-313.

Levinson, E.E., and Traven, V.F. (1996). Synthesis of 3,24-diepicastasterone: a natural brassinosteroid with 2,3-trans-diol function. *J. Chem. Res., Synop.* 196-197.

Levinson, E.E., Kuznetsova, N.A., Podkhalyuzina, N.Ya., and Traven, V.F. (1994). Synthesis of 2,24-diepicastasterone and its 22S,23S-isomer: novel brassinosteroids with trans-2,3-diol function. *Mendeleev Commun.*, 96-97.

Li, J., and Chory, J. (1997). A putative leucine-rich repeat receptor kinase involved in brassinosteroid signal transduction. *Cell* **90**, 929-938 [*C. A.* **127**, 305325].

Li, J., Nagpal, P., Vitart, V., Chao, A., Fujioka, S., Choi, Y.-H., Guha-Biswas, M., McMorris, T.C., Takatsuto, S., Yokota, T., Russell, D.W., Sakurai, A., and Chory, J. (1996a). A role for brassinosteroids in plant development. *Proc. Plant Growth Regul. Soc. Am.* **23**, 11.

Li, J., Nagpal, P., Vitart, V., McMorris, T.C., and Chory, J. (1996b). A role for brassinosteroids in light-dependent development of *Arabidopsis*. *Science* **272**, 398-401 [*C. A.* **124**, 309107].

Li, J., Biswas, M.G., Chao, A., Russell, D.W., and Chory, J. (1997). Conservation of function between mammalian and plant steroid 5α-reductases. *Proc. Natl. Acad. Sci. U.S.A.* **94**, 3554-3559 [*C. A.* **126**, 327248].

Li, N.-H., Chen, R.M., Huang, Q.S., and Pan, R.Z. (1993). Effects of brassinolide application during pod development on assimilate distribution and pod yield of *Arachis hypogaea*. *Oil Crops China* 43-46.

Liang, Y., Xu, Z., and Guo, Q. (1989). Synthesis of A and B rings of brassinolide analogue. *J. Xiamen Univ. Nat. Sci.* **28**, 284-287 [*C. A.* **114**, 24308].

List of pesticides permitted for application in Russian Federation (1997a). *Zashchita i karantin rastenii* 28-29.

List of pesticides permitted for application in Russian Federation (1997b). "Chemicals for Plant Protection", Agrorus, Moscow, p. 141.

Litvinovskaya, R.P., Drach, S.V., and Khripach, V.A. (1994). A new method of 22,23-diol group introduction into steroidal side chain. *Zh. Org. Khim.* **30**, 304-305.

Litvinovskaya, R.P., Drach, S.V., and Khripach, V.A. (1995). Stereochemically unusual cycloaddition of nitrile oxides to Δ^{23}-steroids. *Mendeleev Commun.*, 215-216.

Litvinovskaya, R.P., Drach, S.V., and Khripach, V.A. (1996a). A new route to the side chain of 28-norbrassinolide. *Zh. Org. Khim.* **32**, 1279-1280 [*C. A.* **126**, 264240].

Litvinovskaya, R.P., Tereshko, V.A., Drach, S.V., and Khripach, V.A. (1996b). X-ray investigation of (22R)-22-acetoxy-22-(3-methylisoxazolin-5-yl)-3α,5-cyclo-23-nor-5α-cholestan-6-one. *Zh. Obsch. Khim.* **66**, 859-862.

Litvinovskaya, R.P., Baranovskii, A,V., and Khripach V.A. (1997a). Functionalisation of A,B-cycles of 17-isoxazolinyl- and 20-hydroxy-20-isoxazolinyl steroids. *Zh. Org. Khim.* **33**, 1350-1356.

Litvinovskaya, R.P., Drach, S.V., Koval, N.V., and Khripach, V.A. (1997b). Synthesis of steroids, containing a vicinal diol function in a side chain, *via* isoxazoline intermediates. *Khim. Geterocycl. Soed.* 542-548

Litvinovskaya, R.P., Lyakhov, A.S., Govorova, A.A., Drach, S.V., and Khripach, V.A. (1997c). X-ray structure of (20R,22S,5'S)-22-(3'-methylisoxazolin-5'-yl)-6,6-ethylenedioxy-3α,5-cyclo-23-nor-5α-cholestan-22-ol. *Bioorg. Khim.* **23**, 147-151 [*C. A.* **127**, 262906].

Liu, G.-Q., Zhang, H.-G., Liu, J.-P., and Li, L. (1995). Determination of brassinolide by high performance liquid chromatography. *Sepu* **13**, 290-291 [*C. A.* **123**, 163196].

Liu, X., Li, Y., and Liang, X. (1989). Stereoselective synthesis of 3,5-cyclopregnane-6-methoxy-20S-22-aldehyde. *Gaodeng Xuexiao Huaxue Xuebao* **10**, 650-652 [*C. A.* **112**, 217341].

Luo, B. (1986a). Brassinosteroids from higher plant and their application. *Zhiwu Shenglixue Tongxun* 11-14 [*C. A.* **105**, 39326].

Luo, B.-S. (1986b). Effect of brassinolide on growth and fruiting in soybean. *Plant Physiol. Commun.* 14-17.

Luo, B.-S., Kumura, A., Ishii, R., and Wada, Y. (1986). Effects of brassinolide treatments on growth and developmental processes in wheat plants. *Jpn. J. Crop Sci.* **55**, 291-298 [*C. A.* **106**, 62934].

Luo, B., Yu, D., and Zhou, D. (1988). Effects of brassinolide on the changes in IAA, ABA levels in young cotton bolls and boll shedding. *Plant Physiol. Commun.* 31-34.

Luo, B.-S., Qu, Y.-L., and Liu, D.-H. (1992). Application and physiological effects of brassinolide on crops and their appraisal. *J. Huazhong Agric. Univ.* **11**, 41-47.

Luu, B., and Werner, F. (1996). Sterols that modify moulting in insects. *Pestic. Sci.* **46**, 49-53 [*C. A.* **124**, 109580].

Machackova, I., Vagner, M., and Slama, K. (1995). Comparison between the effects of 20-hydroxyecdysone and phytohormones on growth and development in plants. *Eur. J. Entomol.* **92**, 309-316.

Maeda, E. (1965). Rate of lamina inclination in excised rice leaves. *Physiol. Plant.* **18**, 813-827.

Mai, Y., Lin, S., Zeng, X., and Ran, R. (1989). Effect of brassinolide on nitrate reductase activity in rice seedlings. *Plant Physiol. Commun.* 50-52.

Malevannaya, N.N., and Bednarskaya, I. (1995). Epin. *Priusadebnoye choziaystvo* 8-9.

Malevannaya, N.N., and Kositsina-Pinegina, E. (1996). Epin-antistress agent. *Tsvetovodstvo* 7-8.

Mandai, T., Matsumoto, T., Kawada, M., and Tsuji, J. (1992a). A novel method for stereospecific generation of either C-20 epimer in steroid side chains by palladium-catalyzed hydrogenolysis C-20 allylic carbonates. *J. Org. Chem.* **57**, 6090-6092.

Mandai, T., Suzuki, S., Murakami, T., Fujita, M., Kawada, M., and Tsuji, J. (1992b). A simple preparative method for isopropenyl and vinyl groups from ketones. *Tetrahedron Lett.* **33**, 2987-2990.

Mandai, T., Matsumoto, T., Kawada, M., and Tsuji, J. (1994). A novel method for stereospecific generation of natural C-17 stereochemistry and either C-20 epimer in steroid side chains by palladium-catalyzed hydrogenolysis of C-17 and C-20 allylic carbonates. *Tetrahedron* **50**, 475-486.

Mandava, N.B. (1988). Plant growth-promoting brassinosteroids. *Annu. Rev. Plant Physiol. Plant Mol. Biol.* **39**, 23-52 [*C. A.* **109**, 146248].

Mandava, N.B. (1991). Brassinosteroids. U.S. Department of Agriculture contributions and Environmental Protection Agency registration requirements. *In* "Brassinosteroids. Chemistry, Bioactivity, and Applications. ACS Symposium Series" (H.G. Cutler, T. Yokota, and G. Adam, Eds.), Vol. 474, pp. 320-332. American Chemical Society, Washington.

Mandava, N., and Mitchell, J.W. (1971). New plant hormones: chemical and biological investigations. *Indian Agr.* **15**, 19-31.

Mandava, N.B., and Thompson, M.J. (1984). Chemistry and functions of brassinolide. *In* "Isoprenoids in Plants: Biochemistry and Function" (W.D. Nes, G. Fuller, and T.S. Tsai, Eds.), pp. 401-431. Marcel Dekker, New York.

Mandava, N.B., Kozempel, M., Worley, J.F., Matthees, D., Warthen, J.D., Jr., Jacobson, M., Steffens, G.L., Kenney, H., and Grove, M.D. (1978). Isolation of brassins by extraction of rape (*Brassica napus* L.) pollen. *Ind. Eng. Chem. Prod. Res. Dev.* **17**, 351-354.

Mandava, N.B., Sasse, J.M., and Yopp, J.H. (1981). Brassinolide, a growth-promoting steroidal lactone. II. Activity in selected gibberellin and cytokinin bioassays. *Physiol. Plant.* **53**, 453-461.

Mandava, N.B., Thompson, M.J., and Yopp, J.H. (1987). Effects of selected inhibitors of RNA and protein synthesis on brassinosteroid-induced responses in mung bean epicotyls. *J. Plant. Physiol.* **128**, 53-65.

Maotani, T., Suzuki, A., Nishimura, T., Kumomato, O., Oshima, K., and Yamanaka, Y. (1989). Control of physiological fruit drop of Japanese persimmon 'Hiratanenashi'. *J. Jpn. Soc. Hortic. Sci.* **58**, 557-562.

Marino, J.P., de Dios, A., Anna, L.J., and Fernandez de la Pradilla, R. (1996). Highly stereocontrolled formal synthesis of brassinolide *via* chiral sulfoxide-directed S_N2' reaction. *J. Org. Chem.* **61**, 109-117.

Markova, I.V., and Getko, N.B. (1997). Effect of epibrassinolide on orthogenesis of *Rosa hybrida* L. at the earlier stages of cultivation *in vitro*. *In* "Regulators of plant growth and development", 4th, pp. 306-307, Moscow.

Marquardt, V., and Adam, G. (1991). Recent advances in brassinosteroid research. *In* "Chemistry of Plant Protection, Vol. 7", pp. 104-139. Springer-Verlag, Berlin, Heidelberg.

Marquardt, V., Coll, M.F., and Alonso, B.E.M. (1988). Preparation of brassinosteroid analogs of spirostans as plant growth regulators. Ger. (East) **DD 273,638** [*C. A.* **112**, 217353].

Marumo, S., Hattori, H., Abe, H., Nonoyama, Y., and Munakata, K. (1968). The presence of novel plant growth regulators in leaves of *Distylium racemosum* Sieb et Zucc. *Agric. Biol. Chem.* **32**, 528-529.

Marumo, S., Mizuno, K., Kin, S., and Asano, T. (1993). KM-01 as an inhibitor for brassinosteroids and its manufacture with *Drechslera* species. Jpn. Kokai Tokkyo Koho **JP 07 69,987 [95 69,987]** [*C. A.* **123**, 54293].

Mashkovskii, M.D. (1997). "Medicinal substances". Topsing, Kharkov.

Mazets, Zh.E. (1997). "Peculiarities of kvartazin and brassinosteroid influence on physiological-biochemical processes of DT-lines of the *Chinese Spring* wheat", Ph.D. Thesis, Institute of Experimental Botany, Academy of Sciences of Belarus, Minsk.

Mazets, Zh.E., and Deeva, V.P. (1996). Influence of epibrassinolide on intensity of synthesis and polypeptide composition of freely soluble proteins in DT-lines of *Chinese spring* wheat. *Vesti Akademii Navuk Belarusi, Ser. Biol. Navuk* 63-66.

Matsumoto, T., Shimizu, N., Shigemoto, T., Itoh, T., Iida, T., and Nishioka, A. (1983). Isolation of 22-dehydrocampesterol from seeds of *Brassica juncea*. *Phytochemistry* **22**, 789-790.

Maugh, T.H. (1981). New chemicals promise larger crops. *Science* **212**, 33-34.

Mayumi, K., and Shibaoka, H. (1995). A possible double role for brassinolide in the reorientation of cortical microtubules in the epidermal cells of azuki bean epicotyls. *Plant Cell Physiol.* **36**, 173-181.

Mazzafera, P., and Zullo, M.A.T. (1990). Effect of brassinosteroids on coffee. *Bragantia* **49**, 37-42.

McMorris, T.C. (1997). Recent developments in the field of plant steroid hormones. *Lipids* **32**, 1303-1308.

McMorris, T.C., and Patil, P.A. (1993). Improved synthesis of 24-epibrassinolide from ergosterol. *J. Org. Chem.* **58**, 2338-2339.

McMorris, T.C., Donaubauer, J.R., Silveria, M.H., and Molinski, T.F. (1991). Synthesis of brassinolide. *In* "Brassinosteroids. Chemistry, Bioactivity, and Applications. ACS Symposium Series" (H.G. Cutler, T. Yokota, and G. Adam, Eds.), Vol. 474, pp. 36-42. American Chemical Society, Washington.

McMorris, T.C., Patil, P.A., Chavez, R.G., Baker, M.E., and Clouse, S.D. (1994). Synthesis and biological activity of 28-homobrassinolide and analogues. *Phytochemistry* **36**, 585-589.

McMorris, T.C., Chavez, R.G., and Patil, P.A. (1996). Improved synthesis of brassinolide. *J. Chem. Soc., Perkin Trans. 1* 295-302.

Meadows, D.J., and Williams, D.H. (1980). Substrates to study the mechanism of vitamin D hydroxylation: synthesis of $[24R-^2H]$-25-hydroxyvitamin D_3. *Tetrahedron Lett.* **21**, 4373-4376.

Melnikov, S.S., Manankina, E.E., and Budakova, E.A. (1996). Productivity of the algae *Spirulina platensis* under different conditions. *Vesti AN Belarusi, Ser. Biol. Navuk* 38-42.

Melnikov, S.S., Manankina, E.E., and Budakova, E.A. (1998). Effect of epibrassinolide on productivity of algae (*Chlorella vulgaris, Spirulina platensis, Dunaliella salina, Euglena gracilis*). *Vesti AN Belarusi, Ser. Biol. Navuk* In press.

Meudt, W.J. (1987). Effect of brassinolide on gravitropism of bean hypocotyls. *Plant. Physiol.* **83**, 195-198.

Meudt, W.J., and Bennett, N.W. (1978). Rapid bioassay for auxin. *Physiol. Plant.* **44**, 422-428.

Meudt, W.J., and Nes, W.D. (1987). Chemical and biological aspects of brassinolide. *In* "Ecology and Metabolism of Plant Lipids. ACS Symposium Series" (G. Fuller, Ed.), Vol. 325, pp. 53-75. American Chemical Society, Washington.

Meudt, W.J., and Thompson, M.J. (1983). Investigations on the mechanism of the brassinosteroid response. II. A modulation of auxin action. *Proc. Plant Growth Regul. Soc. Am.* **10**, 306-311.

Meudt, W.J., Thompson, M.J., Mandava, N., and Worley, J.F. (1980). Method for promoting plant growth. Can. **CA 1,173,659** [*C. A.* **102**, 19625].

Meudt, W.J., Thompson, M.J., and Bennet, H.W. (1983). Investigations on the mechanism of the brassinosteroid response. III. Techniques for potential enhancement of crop production. *Proc. Plant Growth Regul. Soc. Am.* **10**, 312-318.

Midland, M.M., and Kwon, Y.C. (1984). Stereocontrolled synthesis of 22-hydroxy-23-acetylenic steroids; key intermediates in steroid side chain construction. Observation of a directive effect by an α-chiral site during asymmetric reduction with *B*-3-pinanyl-9BBN (alpine-borane). *Tetrahedron Lett.* **25**, 5981-5984.

Mikami, K., and Sakuda, S. (1993). Towards the synthesis of the side chain of brassinolides: diastereodivergent alkoxyaldehyde-ene reaction and nickel-catalyzed transformation of vinyl sulfide. *J. Chem. Soc., Chem. Commun.* 710-712.

Mikami, K., Kawamoto, K., and Nakai, T. (1985a). A new Claisen approach to the stereospecific introduction of a steroid side chain at C-20. A simple synthesis of 20-epicholesterol *via* the β-face rearrangement. *Chem. Lett.* 115-118.

Mikami, K., Kawamoto, K., and Nakai, T. (1985b). Stereocontrolled synthesis of either (22*S*)- or (22*R*)-hydroxy-23-acetylenic steroid side chains *via* [2,3]-Wittig stigmatropic rearrangement. *Tetrahedron Lett.* **26**, 5799-5802.

Mikami, K., Kawamoto, K., and Nakai, T. (1986). Application of [2,3]-Wittig and [3,3]-Claisen rearrangements in steroid side chain synthesis. A highly stereocontrolled entry to either (22*S*)- or (22*R*)-hydroxy-23-carboxylic acid. *Tetrahedron Lett.* **27**, 4899-4902.

Mikami, K., Loh, T.-P., and Nakai, T. (1988a). Diastereocontrol *via* Lewis acid-promoted ene reaction with glyoxylates and its application to stereocontrolled synthesis of a 22*R*-hydroxy-23-carboxylate steroid side chain. *Tetrahedron Lett.* **29**, 6305-6308.

Mikami, K., Loh, T.-P., and Nakai, T. (1988b). Ene approach for concurrent control over the chiral centres at C-20 and C-22 of steroid side chains: a highly stereocontrolled synthesis of (20*S*,22*R*)-(*erythro*-)22-hydroxy-23,24-acetylenic steroid side chains. *J. Chem. Soc., Chem. Commun.* 1430-1431.

Mikami, K., Loh, T.-P., and Nakai, T. (1990). Acyclic stereocontrol based on chelation-controlled ene reaction with chiral α- and β-alkoxyaldehydes. *Tetrahedron: Asymmetry* **1**, 13-16.

Milborrow, B.V., and Pryce, R.J. (1973). The brassins. *Nature (London)* **243**, 46-46.

Mironenko, A.V., Kandelinskaya, O.L., Bushueva, S.A., and Uralskaya, H.R. (1996). Brassinolide: influence on physiological processes in lupine plants. *Dokl. Ross. Acad. S-kh. Nauk* 11-13 [*C. A.* **126**, 314857].

Mironenko, A.V., Kandelinskaya, O.L., Chekhova A.N., Uralskaya, E.R., and Dombrovskaya, N.D. (1997). Peculiarities of protein complex of different genotypes of lupine. *In* "Problems of experimental botany", pp. 329-339. Byelorussian Science, Minsk.

Mitchell, J.W., and Gregory, L.E. (1972). Enhancement of overall plant growth, a new response to brassins. *Nature (London)* **239**, 253-254.

Mitchell, J.W., Mandava, N., Worley, J.F., Plimmer, J.R., and Smith, M.V. (1970). Brassins - a new family of plant hormones from rape pollen. *Nature (London)* **225**, 1065-1066.

Mitchell, J.W., Mandava, N., Worley, J.F., and Drowne, M.E. (1971). Fatty hormones in pollen and immature seeds of bean. *J. Agric. Food Chem.* **19**, 391-393.

Mitra, R.B., and Kapoor, V.M. (1985). Formation of Δ^4-6-keto isomer in the synthesis of Δ^2-6-keto steroids used as intermediates in the synthesis of brassinolide and analogous plant growth stimulators. *Synth. Commun.* **15**, 1087-1094.

Mitra, R.B., Hazra, B.G., and Kapoor, V.M. (1984a). Synthesis of analogs of brassinolide and castasterone. *Indian J. Chem., Sect. B* **23**, 106-109 [*C. A.* **101**, 130968].

Mitra, R.B., Kapoor, V.M., and Hazra, B.G. (1984b). A process for the preparation and separation of stigmasterol-derived dihydroxycyclostigmastanones, useful as intermediates for brassinolides, from phytosterols of sugar-cane wax. Indian **IN 160,748** [*C. A.* **109**, 93447].

Mitra, R.B., Kapoor, V.M., and Hazra, B.G. (1984c). Preparation of 3,5-cyclocholestan-6-ones from sugar cane wax steroids. Indian **IN 163,810** [*C. A.* **116**, 235966].

Mori, K. (1980a). Homobrassinolide and its synthesis. Eur. Pat. Appl. **EP 40,517**.

Mori, K. (1980b). Synthesis of brassinolide analog with high plant growth promoting activity. *Agric. Biol. Chem.* **44**, 1211-1212.

Mori, K. (1982). Homobrassinolide compounds which are steroids having plant growth promoting activity. **Pat. US 4,453,967**.

Mori, K., and Kishi, M. (1985). Preparation of brassinolide analogs as plant growth hormones. Jpn. Kokai Tokkyo Koho **JP 62,135,459 [87,135,459]** [*C. A.* **108**, 94840].

Mori, K., and Takeuchi, T. (1988). Synthesis of 25-methyldolichosterone, 25-methyl-2,3-diepidolichosterone, 25-methylcastasterone and 25-methylbrassinolide. *Liebigs Ann. Chem.*, 815-818.

Mori, K., Takematsu, T., Sakakibara, M., and Oshio, H. (1981). Homobrassinolide, and its production and use. Eur. Pat. Appl. **EP 80,381**.

Mori, K., Sakakibara, M., Ichikawa, Y., Ueda, H., Okada, K., Umemura, T., Yabuta, G., Kuwahara, S., Kondo, M., Minobe, M., and Sogabe, A. (1982). Synthesis of (22S,23S)-homobrassinolide and brassinolide from stigmasterol. *Tetrahedron* **38**, 2099-2109.

Mori, K., Sakakibara, M., and Okada, K. (1984). Synthesis of naturally occurring brassinosteroids emploing cleavage of 23,24-epoxides as key reactions. Synthesis of brassinolide, castasterone, dolicholide, dolichosterone, homodolicholide, homodolichosterone,6-deoxocastasterone and 6-deoxodolichosterone. *Tetrahedron* **40**, 1767-1781.

Mori, K., Takeuchi, T., and Yuya, M. (1987a). Preparation of 25-methylbrassinosteroid intermediates as plant growth hormones. Jpn. Kokai Tokkyo Koho **JP 1135793**.

Mori, K., Takeuchi, T., Yuya, M., Nishi, S., and Fujita, F. (1987b). Preparation of brassinosteroids as plant growth stimulants. Jpn. Kokai Tokkyo Koho **JP 01,168,696 [89,168,696]** [*C. A.* **112**, 21200].

Morishita, T., Abe, H., Uchiyama, M., Marumo, S., Takatsuto, S., and Ikekawa, N. (1983). Evidence for plant growth promoting brassinosteroids in leaves of *Thea sinensis*. *Phytochemistry* **22**, 1051-1053.

Morris, D.S., Williams, D.H., and Norris, A.F. (1981). Structure and synthesis of 25-hydroxycholecalciferol-26,23-lactone, a metabolite of vitamin D. *J. Org. Chem.* **46**, 3422-3428.

Motegi, C., Takatsuto, S., and Gamoh, K. (1994). Identification of brassinolide and castasterone in the pollen of orange (*Citrus sinensis* Osbeck) by high-performance liquid chromatography. *J. Chromatogr.* **658**, 27-30.

Mukaiyama, T., Takeda, T., and Fujimoto, K. (1978). Asymmetric synthesis based on (2R,3S)-3,4-dimethyl-2-phenylperhydro-1,4-oxazepine-5,7-dione. Synthesis of highly optically active β-substituted alkanoic acids. *Bull. Chem. Soc. Jpn* **51**, 3368-3372.

Muromtsev, G.S., and Danilina, E.E. (1996). Endogenic chemical signals of plants and animals. Comparative analysis. *Usp. Sovr. Biol.* **116**, 533-550.

Muromtsev, G.S., Chkannikov, D.I., Kulaeva, O.N., and Gamburg, K.Z. (1987). "The Fundamentals of Chemical Regulation of Plant Growth and Productivity". Agropromizdat, Moscow.

Nakai, T., and Mikami, K. (1986a). A process for the preparation of 24-norcholanic acids as intermediates for vitamin D and brassinosteroids. Jpn. Kokai Tokkyo Koho **JP 62,205,099 [87,205,099]** [*C. A.* **109**, 6798].

Nakai, T., and Mikami, K. (1986b). Preparation of steroids as intermediates for vitamin D compounds and brassinosteroids. Jpn. Kokai Tokkyo Koho **JP 63 14,798 [88 14,798]** [*C. A.* **109**, 110747].

Nakai, T., and Mikami, K. (1989a). Preparation of steroids as intermediates for plant growth regulators. Jpn. Kokai Tokkyo Koho **JP 02,180,895 [90,180,895]** [*C. A.* **114**, 43313].

Nakai, T., and Mikami, K. (1989b). Preparation of steroids as intermediates for plant growth regulators. Jpn. Kokai Tokkyo Koho **JP 02,180,896 [90,180,896]** [*C. A.* **114**, 43314].

Nakajima, N., and Toyama, S. (1995). Study on brassinosteroid-enhanced sugar accumulation in cucumber epicotyls. *Nippon Sakumotsu Gakkai Kiji* **64**, 616-621 [*C. A.* **123**, 278620].

Nakajima, N., Shida, A., and Toyama, S. (1996). Effects of brassinosteroid on cell division and colony formation of Chinese cabbage mesophyll protoplasts. *Nippon Sakumotsu Gakkai Kiji (Jpn. J. Crop. Sci.)* **65**, 114-118 [*C. A.* **124**, 310155].

Nakamura, E., and Kuwajima, I. (1985). Stereocontrolled construction of oxygenated steroidal side chain. Synthesis and stereochemistry of depresosterol. *J. Am. Chem. Soc.* **107**, 2138-2141.

Nakamura, T., Tanino, K., and Kuwajima, I. (1993). Highly stereoselective chelation controlled ene-reaction of 2-(alkylthio)-allyl silyl ethers. *Tetrahedron Lett.* **34**, 477-480.

Nakane, M., Morisaki, M., and Ikekawa, N. (1975). Stereoselectivity in the electrophilic addition reactions of stigmast-22(23)-ene derivatives. *Tetrahedron* **31**, 2755-2760.

Nakaseko, K., and Yoshida, K. (1989). The effect of epi-brassinolide applied at the flowering stage on growth and yield of soyabean and azuki bean. *Mem. Fac. Agric., Hokkaido Univ.* **16**, 347-352.

Nakatani, M., Takao, H., Miura, I., and Hase, T. (1985). Azedarachol, a steroid ester antifeedant from *Melia azedarach* var. Japonica. *Phytochemistry* **24**, 1945-1948.

Naren, A., Prasad, T.G., Kumar, M.U., and Sashidhar, V.R. (1996). Determination of IAA in brassinolide treated coleoptiles of wheat by a modified indirect ELISA with polyclonal antibodies. *Indian J. Exp. Biol.* **34**, 257-261.

Nat. Fed. Agric. Co-Op. Assoc. (1985). Increase in the yield of potatoes using brassinolide. Jpn. Kokai Tokkyo Koho **JP 62 67,006 [87 67,006]** [*C. A.* **107**, 72887].

Neeland, E.G., and Towers, G.H.N. (1989). The synthesis of a 3,5-cyclosteroidal aldehyde from the distillates of canola oil. *Synth. Commun.* **19**, 2603-2611 [*C. A.* **112**, 198887].

Nemchenko, V.V. (1993). Effect of brassinosteroids of the enhancement of the resistance of the cereals to unfavorable environmental factors. *In* "Brassinosteroids - biorational, ecologically safe regulators of growth and productivity of plants", 3rd, pp. 7-8, Minsk.

Nippon Kayaku Co., Ltd. (1988). JRDC-694 (epi-brassinolide). Technical information.

Nishikawa, N., and Abe, H. (1996). Epimerisation and conjugation of brassinolide in cucumber seedlings. *Proc. Plant Growth Regul. Soc. Am.* **23**, 277.

Nishikawa, N., Toyama, S., Shida, A., and Futatsuya, F. (1994). The uptake and the transport of ^{14}C-labeled epibrassinolide in intact seedlings of cucumber and wheat. *J. Plant Res.* **107**, 125-130 [*C. A.* **121**, 297174].

Nishikawa, N., Abe, H., Hatsume, M., Shida, A., and Toyama, S. (1995a). Epimerization and conjugation of ^{14}C labelled epibrassinolide in cucumber seedlings. *J. Plant Physiol.* **147**, 294-300.

Nishikawa, N., Shida, A., and Toyama, S. (1995b). Metabolism of ^{14}C labelled epibrassinolide in intact seedlings of cucumber and wheat. *J. Plant Res.* **108**, 65-69.

Nomenculature of steroids (Recommendations 1989) (1989) *Pure Appl. Chem.* **61**, 1783-1822 [corrections: (1993) *Eur. J. Biochem.* **2**, 213]. Supersedes "Definitive rules for nomenclature of steroids". (1972). *Pure Appl. Chem.* **31**, 285-322.

Nomura, T., Nakayama, M., Reid, J.B., Takeuchi, Y., and Yokota, T. (1997). Blockade of brassinosteroid biosynthesis and sensitivity causes dwarfism in garden pea. *Plant Physiol.* **113**, 31-37.

Nowak, P., Blaszczyk, K., and Paryzek, Z. (1994). The preparation of Δ^2-steroids. An improved procedure. *Org. Prep. Proc. Int.* **26**, 374-376 [*C. A.* **121**, 179973].

Nunez, M., Domingos, J.P., Torres, W., Coll, F., Alonso, E., and Benitez, B. (1995a). Influence of the brassinosteroid analogue Biobras-6 on growth of tomato cultivar INCA-17 plants. *Cultivos Tropicales*, **16**, 49-52.

Nunez, M., Torres, W., and Coll, F. (1995b). Effectiveness of a synthetic brassinosteroid on potato and tomato yields. *Cultivos Tropicales*, **16**, 26-27.

Ohkawa, M., Ohshiro, T., and Ikekawa, T. (1996). Effects of 24-epibrassinolide and NAA on the formation of regenerated bulblet of *Lilium japonicum* Thumb. by scale culture *in vitro*. *Environm. Contr. Biol.* **34**, 15-19.

Okada, K., and Mori, K. (1983a). Synthesis of brassinolide analogs and their plant growth-promoting activity. *Agric. Biol. Chem.* **47**, 89-95.

Okada, K., and Mori, K. (1983b). Stereoselective synthesis of dolicholide, a plant growth promoting steroid. *Agric. Biol. Chem.* **47**, 925-926.

Onatskiy, N.M., Marchenko, A.I., and Mikhina, L.V. (1997). Techical report "Evaluation of mutagenic activity of epibrassinolide (active ingredient of Epin) in Ames test, chromosome aberrations and in micronuclear tests". Scientific Research Center of Toxicologic and Hygienic Regulation of Biopreparations of Russia, Serpukhov.

Oritani, T. (1989). Plant growth promoter containing brassinolide and adenine derivative. Jpn. Kokai Tokkyo Koho **JP 03,109,305 [91,109,305]** [*C. A.* **116**, 36237].

Palladina, T.A., and Simchuk, E.E. (1993). Effect of brassinosteroids on the processes of active transport in plasma membranes of plant cells. *In* "Brassinosteroids - biorational, ecologically safe regulators of growth and productivity of plants", 3rd, pp. 4-5, Minsk.

Palladina, T.A., Simchuk, E.E., and Nasyrova, G.F. (1995). Effect of brassinosteroids on H^+-ATPase of plasma membranes in cells of corn seedlings roots. *In* "Brassinosteroids - biorational, ecologically safe regulators of growth and productivity of plants", 4th, pp. 5-6, Minsk.

Palladina, T., Simchuk, E., Nasyrova, G., and Belyaeva, N. (1996). Effect of brassinosteroids on H^+-pump of plasma membranes in corn seedlings roots. *In* "Int. FESPP Congress. From molecular mechanisms to the plant: an integrated approach", 10th, pp. 12-18, Florence.

Park, J.-D., and Park, K.-H. (1997). Brassinosteroid-like ativity of monoglyceride analogs. *Han'guk Nonghwa Hakhoechi* **40**, 357-360 [*C. A.* **127**, 244262].

Park, K.-H. (1988). Occurrence of castasterone, brassinolide and methyl 4-chloroindole-3-acetate in immature *Vicia faba* seeds. *J. Korean Agric. Chem. Soc.* **31**, 106-113.

Park, K.-H., Yokota, T., Sakurai, A., and Takahashi, N. (1987). Occurrence of castasterone, brassinolide and methyl 4-chloroindole-3-acetate in immature *Vicia faba* seeds. *Agric. Biol. Chem.* **51**, 3081-3086.

Park, K.-H., Saimoto, H., Nakagawa, S., Sakurai, A., Yokota, T., Takahashi, N., and Syono, K. (1989). Occurrence of brassinolide and castasterone in crown gall cells of *Catharanthus roseus*. *Agric. Biol. Chem.* **53**, 805-811.

Park, K.-H., Kim, S.-J., and Hyun, K.-H. (1993a). Brassinosteroid substances in immature *Perilla frutescens* seeds. *J. Korean Agric. Chem. Soc.* **36**, 197-202.

Park, K.-H., Kim, S.-J., and Hyun, K.-H. (1993b). Brassinosteroid substances in immature *Cassia tora* seeds. *J. Korean Agric. Chem. Soc.* **36**, 99-104.

Park, K.-H., Kim, S.-J., Park, J.-D., Lee, L.-S., and Hyun, K.-H. (1993c). Brassinosteroid substances in immature *Oryza sativa* seeds. *J. Korean Agric. Chem. Soc.* **36**, 376-380.

Park, K.-H., Park, J.-D., Hyun, K.-H., Nakayama, M., and Yokota, T. (1994a). Brassinosteroids and monoglycerides in immature seeds of *Cassia tora* as the active principles in the rice lamina inclination bioassay. *Biosci. Biotech. Biochem.* **58**, 1343-1344.

Park, K.-H., Park, J.-D., Hyun, K.-H., Nakayama, M., and Yokota, T. (1994b). Brassinosteroids and monoglycerides with brassinosteroid-like activity in immature seeds of *Oryza sativa* and *Perilla frutescens* and in cultured cells of *Nicotiana tabacum*. *Biosci. Biotech. Biochem.* **58**, 2241-2243.

Park, K.-H., Kim, S.-K., and Yokota, T. (1995). Identification of brassinosteroids in immature seeds of *Zea mays* by GC-MS analysis. *Han'guk Nonghwa Hakhoechi* **38**, 179-183 [*C. A.* **123**, 222799].

Paroda, R.S. (1971). Importance of synchrony of ear emergence in plant breeding programmes. *Nature* **233**, 351-352.

Pavlova, I.V., and Deeva, V.P. (1995). Effect of epibrassinolide on buckwheat growth and productivity. *In* "Brassinosteroids - biorational, ecologically safe regulators of growth and productivity of plants", 4th, p. 25, Minsk.

Petzold, U., Peschel, S., Dahse, I., and Adam, G. (1992). Stimulation of source-applied ^{14}C-sucrose export in *Vicia faba* plants by brassinosteroids, GA$_3$ and IAA. *Acta Bot. Neerl.* **41**, 469-479.

Piatak, D.M., and Wicha, J. (1978). Various approaches to the construction of aliphatic side chains of steroids and related compounds. *Chem. Rev.* **78**, 199-241.

Pilnova, E., and Shcherbina, I. (1997). Epin. *Novii zemledelets*, 31.

Pipattanawong, N., Fujishige, N., Yamane, K., and Ogata, R. (1996). Effects of brassinosteroid on vegetative and reproductive growth in two day-neutral strawberries. *Engei Gakkai Zasshi (J. Jpn. Soc. Hort. Sci.)* **65**, 651-654 [*C. A.* **126**, 196362].

Pirogovskaya, G.V., Bogdevitch, I.M., Naumova, G.V., Khripach, V.A., Azizbekyan, S.G., and Krul, L.P. (1996). New forms of mineral fertilizers with additives of plant growth regulators. *Proc. Plant Growth Regul. Soc. Am.* **23**, 146-151.

Platonova, T.A., and Korableva, N.P. (1994). Effect of 24-epibrassinolide on growth of apical meristem of potato tubers. *Prikl. Biokhim. Mikrobiol.* **30**, 923-930.

Platonova, T.A., and Korableva, N.P. (1998). Ultrastructural and morphometric investigation of intra-cell changes in apexes of potato tubers under the action of epibrassinolide. *Physiol. Plants* (Russ.). In press.

Platonova, T.A., Korableva, N.P., and Koreneva, V.M. (1993). Study of epibrassinolide effect on apical meristems of potato tubers. *In* "Brassinosteroids - biorational, ecologically safe regulators of growth and productivity of plants", 3rd, p. 19, Minsk.

Plattner, R.D., Taylor, S.L., and Grove, M.D. (1986). Detection of brassinolide and castasterone in *Alnus glutinosa* (European alder) pollen by mass spectrometry/mass spectrometry. *J. Nat. Prod.* **49**, 540-545.

Porzel, A. (1996). NMR-studies of plants and plant constituents. *Nova Acta Leopold., Suppl.*, 95-106 [*C. A.* **126**, 169044].

Porzel, A., Marquardt, V., Adam, G., Massiot, G., and Zeigan, D. (1992). ^{1}H and ^{13}C NMR analysis of brassinosteroids. *Magn. Res. Chem.* **30**, 651-657.

Porzel, A., Stoldt, M., Drosihn, S., Brandt, W., and Adam, G. (1997). Side chain conformation of brassinosteroids. *Proc. Plant Growth Regul. Soc. Am.* **24**, 123-124.

Preus, M.W., and McMorris, T.C. (1979). The configuration at C-24 in oogoniol (24*R*-3β,11α,15β,29-tetrahydroxystigmast-5-en-7-one) and identification of 24(28)-dehydrooogoniols as hormones in *Achlya*. *J. Am. Chem. Soc.* **101**, 3066-3071.

Prusakova, L.D., and Chizhova, S.I. (1991). New type of regulators of plant growth and development. *In* "Regulators of plant growth and development", 1st, p. 49, Moscow.

Prusakova, L.D., Chizhova, S.I., and Kefeli, V.I. (1991). Effect of brassinosteroids on activity of α-amilase, growth and productivity of barley. *In* "14th Conf. on Plant Growth Substances", p. 85 Amsterdam.

Prusakova, L.D., Khripach, V.A., Kefeli, V.I., and Chizhova, S.I. (1993). Substances for the enhancement of stem strength of cereals. Pat. Appl. **RU 93,037,452**.

Prusakova, L.D., Chizhova, S.I., and Khripach, V.A. (1995). Stability and productivity of barley and wheat under the action of brassinosteroids. *Sel'skochosyaistvennaya biologiya* 93-97.

Prusakova, L.D., and Chizhova, S.I. (1996). The role of brassinosteroids in growth, tolerance and productivity of plants. *Agrokhimiya* 137-150 [*C. A.* **127**, 217771].

Pshenichnaya, L.A., Khripach, V.A., Volynetz, A.P., Prokhorchik, R.A., Manzhelesova, N.E., and Morozik, G.V.(1997). Brassinosteroids and resistance of barley plants to leave deseases. *In* "Problems of experimental botany", pp. 210-217. Byelorussian Science, Minsk.

Quyen, L., Adam, G., and Schreiber, K. (1994a). Partial synthesis of nitrogen brassinosteroid analogs with 22,26-epiminocholestane and spirosolane skeleton. *Liebigs Ann. Chem.*, 1143-1147.

Quyen, L., Adam, G., and Schreiber, K. (1994b). Partial synthesis of nitrogenous brassinosteroid analogs with solanidane skeleton. *Tetrahedron* **50**, 10923-10932.

Radi, S.H., and Maeda, E. (1988). Effect of brassinolide on the cultured rice root growth as modified by figaron and gibberellic acid. *Jpn. J. Crop Sci.* **57**, 191-198 [*C. A.* **109**, 2403].

Ramraj, V.M., Vyas, B.N., Godrey, N.B., Mistry, K.B., Swami, B.N., and Singh, N. (1997). Effects of 28-homobrassinolide on yields of wheat, rice, groundnut, mustard and cotton. *J. Agric. Science* **128**, 405-413.

Redpath, J., and Zeelen, F.J. (1983). Stereoselective synthesis of steroid side chains. *Chem. Soc. Rev.* **12**, 75-98.

Riccio, R., Finamore, E., Santaniello, M., and Zollo, F. (1990). Stereoselective synthesis of (24S)- and (24R)-24-(hydroxymethyl)cholesta-5,22(E)-dien-3β-ol: model compounds for stereochemical assigments of polyhydroxylated marine steroids. *J. Org. Chem.* **55**, 2548-2552.

Richter, K., and Adam, G. (1991). Neurodepressing effect of brassinosteroids in the cockroach *Periplaneta americana*. *Naturwissenschaften* **78**, 138-139.

Richter, K., and Koolman, J. (1991). Antiecdysteroid effects of brassinosteroids in insects. *In* "Brassinosteroids. Chemistry, Bioactivity, and Applications. ACS Symposium Series" (H.G. Cutler, T. Yokota, and G. Adam, Eds.), Vol. 474, pp. 265-278. American Chemical Society, Washington.

Richter, K., Adam, G., and Vorbrodt, H.-M. (1987). Inhibiting effect of 22S,23S-homobrassinolide on the molt of the cockroach *Periplaneta americana* (L.) (Orhopt., Blattidae). *J. Appl. Entomol.* **103**, 532-534 [*C. A.* **107**, 170534].

Riediker, M., and Schwartz, J. (1981). A new synthesis of 25-hydroxy-cholesterol. *Tetrahedron Lett.* **22**, 4655-4658.

Roddick, J.G. (1994). Comparative root growth inhibitory activity of four brassinosteroids. *Phytochemistry* **37**, 1277-1281.

Roddick, J.G., and Guan, M. (1991). Brassinosteroids and root development. *In* "Brassinosteroids. Chemistry, Bioactivity, and Applications. ACS Symposium

Series" (H.G. Cutler, T. Yokota, and G. Adam, Eds.), Vol. 474, pp. 231-245. American Chemical Society, Washington.

Roddick, J.G., and Ikekawa, N. (1992). Modification of root and shoot development in monocotyledon and dicotyledon seedlings by 24-epibrassinolide. *J. Plant Physiol.* **140**, 70-74.

Roddick, J.G., Rijnenberg, A.L., and Ikekawa, N. (1993). Developmental effects of 24-epibrassinolide in excised roots of tomato grown *in vitro*. *Physiol. Plant.* **87**, 453-458.

Rodkin, A.I., Konovalova, G.I., and Bobrick, A.O. (1997). Efficiency of application of biologically actiive substances in primary breeding of potato. *In* "Plant growth and development regulators", 4th, pp. 317-318, Moscow.

Romani, G., Marre, M.T., Bonetti, A., Cerana, R., Lado, P., and Marre, E. (1983). Effects of a brassinosteroid on growth and electrogenic proton extrusion in maize root segments. *Physiol. Plant.* **59**, 528-532.

Rönsch, H., Adam, G., Matschke, J., and Schachler, G. (1993). Influence of (22S,23S)-homobrassinolide on rooting capacity and survival of adult Norway spruce cuttings. *Tree Physiol.* **12**, 71-80.

Roy, A.K., and Klark J.N. (1980). "Gene Regulation by Steroid Hormones". Springer-Verlag, Berlin - Heidelberg - New York,.

Runkova, L.V. (1991). Perspectives of application of brassinosteroids in ornamental flower growing. *In* "Conference on brassinosteroids", 2nd, pp. 10-11, Minsk.

Runkova, L.V. (1995). Effect of epibrassinolide on flowering of some ornamental plants. *In* "Brassinosteroids - biorational, ecologically safe regulators of growth and productivity of plants", 4th, pp. 10-11, Minsk.

Runov, S.A., Salnikov, A.I., and Prusakova, L.D. (1997). Effect of 24-epibrassinolide on the seeds quality and growth of buckwheat. *In* "Regulators of plant growth and development", 4th, p. 234, Moscow.

Saimoto, H., Otsuka, M., Yamamoto, M., Kawashima, M., Fujioka, S., Sakurai, A., Yokota, T., and Shono, K. (1989). Brassinosteroid manufacture with crown gall cell cultures. Eur. Pat. Appl. **EP 434,375** [*C. A.* **115**, 69971].

Sairam, R.K. (1994). Effects of homobrassinolide application on plant metabolism and grain yield under irrigated and moisture-stress conditions of two wheat varieties. *Plant Growth Regul.* **14**, 173-181.

Saka, H., Kogen, A., Okumura, M., and Watanabe, S. (1992). Fluctuation of ethylene production in excised panicles and flag leaf blades during grain ripening in rice (*Oryza sativa* L.). *Jpn. J. Crop Science* **61**, 285-291.

Sakakibara, M., and Mori, K. (1982). Facile synthesis of (22R,23R)-homobrassinolide. *Agric. Biol. Chem.* **46**, 2769-2779.

Sakakibara, M., and Mori, K. (1983a). Improved synthesis of brassinolide. *Agric. Biol. Chem.* **47**, 663-664.

Sakakibara, M., and Mori, K. (1983b). Short-step synthesis of homodolicholide. *Agric. Biol. Chem.* **47**, 1407-1408.

Sakakibara, M., and Mori, K. (1983c). Short-step synthesis of homodolichosterone. *Agric. Biol. Chem.* **47**, 1405-1406.

Sakakibara, M., and Mori, K. (1984). Short-step syntheses of homodolicholide and homodolichosterone. *Agric. Biol. Chem.* **48**, 745-752.

Sakakibara, M., Okada, K., Ichikawa, Y., and Mori, K. (1982). Synthesis of brassinolide, a plant growth promoting steroidal lactone. *Heterocycles* **17**, 301-304.

Sakurai, A., and Fujioka, S. (1993). The current status of physiology and biochemistry of brassinosteroids. *Plant Growth Regul.* **13**, 147-159.

Sakurai, A., and Fujioka, S. (1997a). Biosynthesis of steroidal plant hormones, brassinosteroids. *Baiosaiensu to Indasutori* **55**, 38-40 [*C. A.* **126**, 183753].

Sakurai, A., and Fujioka, S. (1997b). Studies on biosyntheis of brassinosteroids. *Biosci. Biotech. Biochem.* **61**, 757-762.

Sakurai, A., Fujioka, S., and Saimota, H. (1991). Production of brassinosteroids in plant-cell cultures. *In* "Brassinosteroids. Chemistry, Bioactivity, and Applications. ACS Symposium Series" (H.G. Cutler, T. Yokota, and G. Adam, Eds.), Vol. 474, pp. 97-106. American Chemical Society, Washington.

Sakurai, A., Fujioka, S., Choi, Y.-H., Takatsuto, S., and Yokota, T. (1996). Biosynthetic pathways of brassinosteroids. *Proc. Plant Growth Regul. Soc. Am.* **23**, 8.

Sala, C., and Sala, F. (1985). Effect of brassinosteroid on cell division and enlargement in cultured carrot (*Dancus carota* L.) cells. *Plant Cell Rep.* **4**, 144-147.

Salmond, W., and Barta, M.A. (1978). A stereoselective Wittig reagent and its application to the synthesis of 25-hydroxylated vitamin D metabolites. *J. Org. Chem.* **43**, 790-792.

Sardina, F.J., Mourino, A., and Castedo, L. (1983). Studies on the synthesis of side chain hydroxylated metabolites of vitamin D. Stereospecific synthesis of 25-hydroxy-7,8-dihydroergosterol and its C-24 epimer. *Tetrahedron Lett.* **24**, 4477-4480.

Sardina, F.J., Mourino, A., and Castedo, L. (1986). Studies on the synthesis of side-chain hydroxylated metabolites of vitamin D. 2. Stereocontrolled synthesis of 25-hydroxyvitamin D_2. *J. Org. Chem.* **51**, 1264-1268.

Sasse, J.M. (1985). The place of brassinolide in the sequential response to plant growth regulators in elongating tissue. *Physiol. Plant.* **63**, 303-308.

Sasse, J.M. (1987). Effects of brassinolide and other natural plant growth regulators on the morphology of pea stem tissue. *Proc. Plant Growth Regul. Soc. Am.* **14**, 30-39.

Sasse, J.M. (1990). Brassinolide-induced elongation and auxin. *Physiol. Plant.* **80**, 401-408.

Sasse, J.M. (1991a). Brassinolide-induced elongation. *In* "Brassinosteroids. Chemistry, Bioactivity, and Applications. ACS Symposium Series" (H.G. Cutler, T. Yokota, and G. Adam, Eds.), Vol. 474, pp. 255-264. American Chemical Society, Washington.

Sasse, J.M. (1991b). The case for brassinosteroids as endogenous plant hormones. *In* "Brassinosteroids. Chemistry, Bioactivity, and Applications. ACS Symposium

Series" (H.G. Cutler, T. Yokota, and G. Adam, Eds.), Vol. 474, pp. 158-166. American Chemical Society, Washington.

Sasse, J.M. (1994). Brassinosteroids and roots. *Proc. Plant Growth Regul. Soc. Am.* **21**, 228-232.

Sasse, J. (1996). Introducing brassinosteroids. *Proc. Plant Growth Regul. Soc. Am.* **23**, 2-6.

Sasse, J.M. (1997). Recent progress in brassinosteroid research. *Physiol. Plant.* **100**, 696-701.

Sasse, J.M., Yokota, T., Taylor, P.E., Griffiths, P.G., Porter, Q.N., and Cameron, D.V. (1992). Brassinolide induced elongation. *In* "Progress in plant growth regulation" (Current Plant Science and Biochemistry in Agriculture) (C.M. Karssen, L.C. van Loon, and D. Vreugdenhil, Eds.), vol. 13, pp. 319-325. Kluwer Academic Publishers, Dordrecht - Boston - London.

Sasse, J.M., Smith, R., and Hudson, I. (1995). Effect of 24-epibrassinolide on germination of seeds of *Eucalyptus camaldulensis* in saline conditions. *Proc. Plant Growth Regul. Soc. Am.* **22**, 136-141.

Savelieva, E.A., and Karas, I.I. (1993). Effect of brassinosteroids on potato crop yield and quality. *In* "Brassinosteroids - biorational, ecologically safe regulators of growth and productivity of plants", 3rd, p. 28, Minsk.

Savelieva, E.A., and Karas, I.I. (1995). Application of brassinosteroids on potato. *In* "Brassinosteroids - biorational, ecologically safe regulators of growth and productivity of plants", 4th, pp. 21-22, Minsk.

Savelieva, E.A., Sorotchenko, G.K., and Bodylev, V.R. (1995). Effect of phytohormones on potato crop yield and quality. Technical report of Gomel agricultural station, Belarus.

Savelieva, E.A., Goncharov, V.M., and Tseiko, Z.E. (1997). Effect of "Epin" on the crop and disease resistance of tomato in greenhouses. Technical report of Gomel agricultural station, Belarus.

Sathiyamoorthy, P., and Nakamura, S. (1990). In vivo root induction by 24-epibrassinolide on hypocotyl segments of soybean [*Glucine max* (L.) Merr.]. *Plant Growth Regul.* **9**, 73-76.

Sato, H., Hojo, T., Inada, S., Sanada, H., and Takatsudo, H. (1987a). Preparation of brassinosteroids from $2\alpha,3\alpha,22,23$-tetrahydroxycholestan-6-one derivatives. Jpn. Kokai Tokkyo Koho **JP 01,175,992 [89,175,992]** [*C. A.* **112**, 21202].

Sato, H., Hojo, T., Inada, S., Sanada, H., and Takatsuto, H. (1987b). Preparation of brassinosteroids from $2\alpha,3\alpha,22,23$-tetrahydroxycholestan-6-one derivatives. Jpn. Kokai Tokkyo Koho **JP 1175992**.

Sauer, G., and Takeda, K. (1970). Structures of chiograsterone and isochiograsterone, two new sterols from *Chionographis japonica* Maxim. *J. Chem. Soc. (C).*, 911-914.

Schilling, G., and Schiller, C. (1990). Influence of brassinosteroids on organ relations and enzyme activities of sugar-beet plants. *In* "Int. Workshop. Brassinosteroids: Chemistry, Bioactivity, Application", pp. 20-21, Halle.

Schilling, G., Schiller, C., and Otto, S. (1991). Influence of brassinosteroids on organ relations and enzyme activities of sugar-beet plants. *In* "Brassinosteroids. Chemistry, Bioactivity, and Applications. ACS Symposium Series" (H.G. Cutler, T. Yokota, and G. Adam, Eds.), Vol. 474, pp. 208-219. American Chemical Society, Washington.

Schlagnhaufer, C.D., and Arteca, R.N. (1985a). Inhibition of brassinosteroid-induced epinasty in tomato plants by aminooxyacetic acid and cobalt (2^+). *Physiol. Plant.* **65**, 151-155.

Schlagnhaufer, C.D., and Arteca, R.N. (1985b). The inhibition of brassinosteroid-induced ethylene production by chlorpromazine, a calmoduline inhibitor. *Plant. Physiol.* **77**, 157.

Schlagnhaufer, C.D., and Arteca, R.N. (1985c). Brassinosteroid-induced epinasty in tomato plants. *Plant. Physiol.* **78**, 300-303.

Schlagnhaufer, C.D., and Arteca, R.N. (1991). The uptake and metabolism of brassinosteroid by tomato (*Lycopersicon esculentum*) plants. *J. Plant Physiol.* **138**, 191-194.

Schlagnhaufer, C., Arteca, R.N., and Yopp, J.H. (1984a). A brassinosteroid - cytokinin interaction on ethylene production by etiolated mung bean segments. *Physiol. Plant.* **60**, 347-350.

Schlagnhaufer, C., Arteca, R.N., and Yopp, J.H. (1984b). Evidence that brassinosteroid stimulates auxin-induced ethylene synthesis in mung bean hypocotyls between *S*-adenosylmethionine and 1-aminocyclopropane-1-carboxylic acid. *Physiol. Plant.* **61**, 555-558.

Schlagnhaufer, C.D, Arteca, R.N., and Philips, A.T. (1988). Monoclonal antibodies against brassinosteroid. A new plant growth regulating compound. *Plant Physiol. (Suppl.)* **86**, 113.

Schlagnhaufer, C.D., Arteca, R.N., and Phillips, A.T. (1991). Induction of anti-brassinosteroid antibodies. *J. Plant Physiol.* **138**, 404-410.

Schmidt, J., Vorbrodt, H.-M., and Adam, G. (1986a). Mass-spectroscopic investigation of positive and negative ions of brassinosteroids. *Zfl-Mitt.* **115**, 165-168 [*C. A.* **105**, 191470].

Schmidt, J., Vorbrodt, H.-M., and Adam, G. (1986b). The negative ion, mass spectra of brassinosteroids. *Biomed. Environ. Mass Spectrom.* **13**, 663-666.

Schmidt, J., Yokota, T., Adam, G., and Takahashi, N. (1991). Castasterone and brassinolide in *Raphanus sativus* seeds. *Phytochemistry* **30**, 364-365.

Schmidt, J., Spengler, B., Yokota, T., and Adam, G. (1993a). The co-occurrence of 24-epi-castasterone and castasterone in seeds of *Ornithopus sativus*. *Phytochemistry* **32**, 1614-1615.

Schmidt, J., Yokota, T., Spengler, B., and Adam, G. (1993b). 28-Homoteasterone, a naturally occurring brassinosteroid from seeds of *Raphanus sativus*. *Phytochemistry* **34**, 391-392.

Schmidt, J., Kuhnt, C., and Adam, G. (1994). Brassinosteroids and sterols from seeds of *Beta vulgaris*. *Phytochemistry* **36**, 175-177.

Schmidt, J., Himmelreich, U., and Adam, G. (1995a). Brassinosteroids, sterols and lup-20(29)-en-2α,3β,28-triol from *Rheum rhabarbarum*. *Phytochemistry* **40**, 527-531.

Schmidt, J., Spengler, B., Yokota, T., Nakayama, M., Takatsuto, S., Voigt, B., and Adam, G. (1995b). Secasterone, the first naturally occurring 2,3-epoxybrassinosteroid from *Secale cereale*. *Phytochemistry* **38**, 1095-1097.

Schmidt, J., Voigt, B., and Adam, G. (1995c). 2-Deoxybrassinolide - a naturally occurring brassinosteroid from *Apium graveolens*. *Phytochemistry* **40**, 1041-1043.

Schmidt, J., Böhme, F., and Adam, G. (1996a). 24-Epibrassinolide from *Gypsophila perfoliata*. *Z. Naturforsch., C: Biosci.* **51**, 897-899.

Schmidt, J., Spengler, B., Voigt, B., and Adam, G. (1996b). New brassinosteroids from *Ornithopus sativus* and *Apium graveolens*. *Proc. Phytochem. Soc. Eur.* **40**, 229-233 [*C. A.* **126**, 196955].

Schmidt, J., Spengler, B., Voigt, B., and Adam, G. (1996c). New brassinosteroids from european cultivated plants. *Proc. Plant Growth Regul. Soc. Am.* **23**, 61.

Schmidt, J., Altmann, T., and Adam, G. (1997). Brassinosteroids from seeds of *Arabidopsis thaliana*. *Phytochemystry* **45**, 1325-1327 [*C. A.* **127**, 188206].

Schmittberger, T., and Uguen, D. (1996). A convenient synthesis of the side-chain of sterols. *Tetrahedron Lett.* **37**, 29-32.

Schmittberger, T., and Uguen, D. (1997). A formal synthesis of brassinolide. *Tetrahedron Lett.* **38**, 2837-2840.

Schneider, J.A., Yoshihara, K., Nakanishi, K., and Kato, N. (1983). Typhasterol (2-deoxycastasterone): a new plant growth regulator from cat-tail pollen. *Tetrahedron Lett.* **24**, 3859-3860.

Schneider, B., Kolbe, A., Porzel, A., and Adam, G. (1994). A metabolite of 24-epibrassinolide in cell suspension cultures of *Lycopersicon esculentum*. *Phytochemistry* **36**, 319-321.

Schneider, B., Kolbe, A., Hai, T., Porzel, A., and Adam, G. (1996). Metabolism of brassinosteroids in plant cell cultures. *Proc. Plant Growth Regul. Soc. Am.* **23**, 43.

Schneider, B., Kolbe, A., Winter, J., Porzel, A., Schmidt, J., Strack, D., and Adam, G. (1997). Pathways and enzymology of brassinosteroid metabolism. *Proc. Plant Growth Regul. Soc. Am.* **24**, 91-93.

Schönecker, B., Dröscher, P., Müller, C., and Hauschild, U. (1984a). Preparation of 3β-hydroxy-23,24-bisnor-5-cholen-22-oic acid as an intermediate for vitamin D derivatives, ecdysone analogs, and brassinolides. Ger. (East) **DD 251,139** [*C. A.* **109**, 73758].

Schönecker, B., Dröscher, P., Kirsch, E., Siemann, H.J., and Rau, M. (1984b). Selective preparation of acetoxy steroid carboxylates as intermediates for vitamin D, ecdysone analogs, and brassinosteroids. Ger. (East) **DD 250,538** [*C. A.* **109**, 6797].

Schönecker, B., Hauschild, U., Marquardt, V., Adam, G., and Walther, D. (1989). Darstellung von funktionalisierten C-22-Steroiden aus dem Ergosterylacetat-Eisentricarbonyl-Komplex. *Z. Chem.* **29**, 218-219.

Schröder, M. (1980). Osmium tetraoxide cis hydroxylation of unsaturated substrates. *Chem. Rev.* **80**, 187-213.

Seo, S., Nagasaki, T., Katsuyama, Y., Matsubara, F., Sakata, T., Yoshioka, M., and Makisumi, Y. (1989). Synthesis of (22R,23R)- and (22S,23S)-[4-^{14}C]-24-epibrassinolide. *J. Labelled Compd. Radiopharm.* **27**, 1383-1393.

Sergeev, P.V. (1984). "Steroidal Hormones". Nauka, Moscow.

Shchekotova, L.A., Popov, M.A., Zidehina, T.V., Kashirskaya, H.Y., Shchekotov, A.D., and Korshun, N.N. (1996). Effect of application of "Epin" in horticulture. Technical Report of Michurin Institute of Horticulture, Moscow.

Shen, Z.-W., and Zhou, W.-S. (1990). Highly stereoselective construction of the side-chain of brassinosteroids utilizing the β-alkylative 1,3-carbonyl transposition of the steroidal 22-en-24-one. *J. Chem. Soc., Perkin Trans. 1* 1765-1767.

Shen, Z.-W., and Zhou, W.-S. (1992). Stereoselective synthesis of 26,27-bisnorbrassinolide and 26,27-bisnortyphasterol. *Sci. China, Ser. B* **35**, 1181-1186 [*C. A.* **118**, 147868].

Shim, J.-H., Kim, I.-S., Lee, K.-B., Suh, Y.-T., and Morgan, E.D. (1996). Determination of brassinolide by HPLC equipped with fluorescence detector in rice (*Oryza sativa* L.). *Han'guk Nonghwa Hakhoechi* **39**, 84-88 [*C. A.* **125**, 52652].

Shimada, K., Abe, H., Takatsuto, S., Nakayama, M., and Yokota, T. (1996). Identification of castasterone and teasterone from seeds of canary grass (*Phalaris canariensis*). *Rec. Res. Dev. Chem. Pharmacol. Sci.* **1**, 1-5.

Shoppee, C.W., Jones, D.N., and Summers, G.H.R. (1957). The epimeric cholestane-2,3-diols. *J. Chem. Soc.* 3100-3107.

Shu, A.Y.L., and Djerassi, C. (1981). Stereospecific synthesis of dinosterol. *Tetrahedron Lett.* **22**, 4627-4630.

Singh, J., Nakamura, S., and Ota, Y. (1993). Effect of epi-brassinolide on gram (*Cicer arietinum*) plants grown under water stress in juvenile stage. *Indian J. Agric. Sci.* **63**, 395-397 [*C. A.* **120**, 73694].

Skorobogatova, I.V., Kurapov, P.B., and Bumazhniy, B.E. (1991). Study of action of brassinosteroids on gibberellin content in barley plants. In "Workshop on Brassinosteroids", 2nd, p. 30, Minsk.

Smith, P.M., Taylor, P.E., Sasse, J.M., and Yokota, T. (1992). Towards a brassinosteroid receptor. *Proc. Plant Growth Regul. Soc. Am.* **19**, 93.

Spengler, B., Schmidt, J., Voigt, B., and Adam, G. (1995). 6-Deoxo-28-norcastasterone and 6-deoxo-24-epicastasterone - two new brassinosteroids from *Ornithopus sativus*. *Phytochemistry* **40**, 907-910.

Steele, J.A., and Mosettig, E. (1963). The solvolysis of stigmasteryl tosylate. *J. Org. Chem.* **28**, 571-572.

Steffens, G.L. (1991). U.S. Department of agriculture brassins project: 1970-1980. In "Brassinosteroids. Chemistry, Bioactivity, and Applications. ACS Symposium

Series" (H.G. Cutler, T. Yokota, and G. Adam, Eds.), Vol. 474, pp. 2-17. American Chemical Society, Washington.

Stoldt, M., Porzel, A., Adam, G., and Brandt, W. (1997). Side chain conformation of the growth-promoting phytohormones brassinolide and 24-epibrassinolide. *Magn. Res. Chem.* **35**, 629-636 [*C. A.* **127**, 234480].

Streltsova, V.A. (1988). Field trials of brassinosteroids. Technical report of Institute of Bioorganic Chemistry, Minsk, Belarus.

Strnad, M., and Kaminek, M. (1985). Sensitized bean first internode bioassay for auxins and brassinosteroids. *Biol. Plant.* **27**, 209-215.

Sucrow, W., and Littmann, W. (1976). Die Synthese von 24-Methylen- und 24(28)-24-Äthyliden-5α-cholestan-3β-ol. *Chem. Ber.* **109**, 2884-2889.

Suge, H. (1986). Reproductive development of higher plants as influenced by brassinolide. *Plant Cell Physiol.* **27**, 199-205.

Sugiyama, K., and Kurashi, S. (1989). Stimulation of fruit set of 'Morita' navel orange with brassinolide. *Acta Hortic.* **239**, 345-348.

Sumi, T., Kuraishi, S., Yamanaka, Y., and Akyama, S. (1990). Germination promotion of grapes with brassinolide. Jpn. Kokai Tokkyo Koho **JP 03,206,007** [91,206,007] [*C. A.* **116**, 17201].

Sumimoto Chem. Co., Ltd. (1983a). Dolicholide. Jpn. Kokai Tokkyo Koho **JP 59,161,374** [84,161,374] [*C. A.* **102**, 167024].

Sumimoto Chem. Co., Ltd. (1983b). Homodolicholide. Jpn. Kokai Tokkyo Koho **JP 59,170,088** [84,170,088] [*C. A.* **102**, 167027].

Sumitomo Chem. Co., Ltd. (1981). Epoxycholestanones. Jpn. Kokai Tokkyo Koho **JP 58 80,000** [83 80,000] [*C. A.* **99**, 122766].

Sumitomo Chem. Co., Ltd. (1982). (2R,3S,22R,23R,24R)-6,6-Ethylenedioxy-22,23-dihydroxy-2,3-isopropylidenedioxy-24-methyl-5α-cholestane. Jpn. Kokai Tokkyo Koho **JP** [*C. A.* **101**, 11269].

Sumitomo Chem. Co., Ltd. (1983). (2R,3S,22R,23E)-24-Ethyl-22-hydroxy-2,3-isopropylidenedioxy-B-homo-7-oxa-5α-cholest-23-en-6-one. Jpn. Kokai Tokkyo Koho **JP 59,231,088** [84,231,088] [*C. A.* **103**, 123804].

Sumitomo Chem. Co., Ltd.; Meiji Seika Kaisha, Ltd. (1981). New steroid plant hormone. Jpn. Kokai Tokkyo Koho **JP 82,118,503** [*C. A.* **97**, 177029].

Sun, L.-Q., Zhou, W.-S., and Pan, X.-F. (1991). Osmium tetroxide catalyzed asymmetric dihydroxylation of the (22E,24R)- and the (22E,24S)-24-alkyl steroidal unsaturated side chain. *Tetrahedron: Asymmetry* **2**, 973-976.

Suntory, Ltd. (1980). 5α-Ergostan-6-ones. Jpn. Kokai Tokkyo Koho **JP 82 70,900** [*C. A.* **97**, 163330].

Suntry, Ltd. (1981). [22R,23R]-B-Homo-7-oxa-5α-cholestan-6-one-2α,3α,22,23-tetraol. Jpn. Kokai Tokkyo Koho **JP 57,163,400** [82,163,400] [*C. A.* **98**, 198606].

Suntry, Ltd. (1983). A new steroid having plant growth control activity from *Typha latifolia* pollen. Jpn. Kokai Tokkyo Koho **JP 60 11,498** [85 11,498] [*C. A.* **103**, 174038].

Suzuki, Y., Yamaguchi, I., and Takahashi, N. (1985). Identification of castasterone and brassinone from immature seeds of *Pharbitis purpurea*. *Agric. Biol. Chem.* **49**, 49-54.

Suzuki, Y., Yamaguchi, I., Yokota, T., and Takahashi, N. (1986). Identification of castasterone, typhasterol and teasterone from the pollen of *Zea mays*. *Agric. Biol. Chem.* **50**, 3133-3138.

Suzuki, H., Fujioka, S., Takatsuto, S., Yokota, T., Murofushi, N., and Sakurai, A. (1993a). Biosynthesis of brassinolide from castasterone in cultured cells of *Catharanthus roseus*. *J. Plant Growth Reg.* **12**, 101-106.

Suzuki, H., Kim, S.-K., Takahashi, N., and Yokota, T. (1993b). Metabolism of castasterone and brassinolide in mung bean explant. *Phytochemistry* **33**, 1361-1368.

Suzuki, H., Fujioka, S., Takatsuto, S., Yokota, T., Murofushi, N., and Sakurai, A. (1994a). Biosynthesis of brassinolide from teasterone *via* typhasterol and castasterone in cultured cells of *Catharanthus roseus*. *J. Plant Growth Regul.* **13**, 21-26 [*C. A.* **121**, 130006].

Suzuki, H., Fujioka, S., Yokota, T., Murofushi, N., and Sakurai, A. (1994b). Identification of brassinolide, castasterone, typhasterol, and teasterone from the pollen of *Lilium elegans*. *Biosci. Biotech. Biochem.* **58**, 2075-2076.

Suzuki, H., Inoue, T., Fujioka, S., Takatsuto, S., Yanagisawa, T., Yokota, T., Murofushi, N., and Sakurai, A. (1994c). Possible involvement of 3-dehydroteasterone in the conversion of teasterone to typhasterol in cultured cells of *Catharanthus roseus*. *Biosci. Biotech. Biochem.* **58**, 1186-1188.

Suzuki, H., Fujioka, S., Takatsuto, S., Yokota, T., Murofushi, N., and Sakurai, A. (1995a). Biosynthesis of brassinosteroids in seedlings of *Catharanthus roseus*, *Nicotiana tabacum*, and *Oryza sativa*. *Biosci. Biotech. Biochem.* **59**, 168-172.

Suzuki, H., Inoue, T., Fujioka, S., Saito, T., Takatsuto, S., Yokota, T., Murofushi, N., Yanagisawa, T., and Sakurai, A. (1995b). Conversion of 24-methylcholesterol to 6-oxo-24-methylcholestanol, a putative intermediate of the biosynthesis of brassinosteroids, in cultured cells of *Catharanthus roseus*. *Phytochemistry* **40**, 1391-1397.

Szekeres, M., Nemeth, K., Koncz-Kalman, Z., Mathur, J., Kauschmann, A., Altmann, T., Redei, G.P., Nagy, F., Schell, J., and Koncz, C. (1996). Brassinosteroids rescue the deficiency of CYP90, a cytochrome P450, controlling cell elongation and de-etiolation in *Arabidopsis*. *Cell* **85**, 171-182 [*C. A.* **124**, 309141].

Tachibana, Y., Yokoyama, S., and Tejima, T. (1992). Preparation of phenyl sulfones as intermediates for brassinolide or vitamin D. Jpn. Kokai Tokkyo Koho **JP 05,221,955 [93,221,955]** [*C. A.* **120**, 106537].

Tachimori, M. (1986a). A process for the preparation of brassinolide by reaction of ergostanone derivatives with peracid. Jpn. Kokai Tokkyo Koho **JP 62,167,797 [87,167,797]** [*C. A.* **107**, 198756].

Tachimori, M. (1986b). A process for the preparation of ergostanone derivatives as intermediates for brassinolide. Jpn. Kokai Tokkyo Koho **JP 62,167,795 [87,167,795]** [*C. A.* **107**, 198757].

Tada, E., Uchida, M., and Funayama, T. (1987). Extraction of brassinolide-like substances. Jpn. Kokai Tokkyo Koho **JP 01,117,899 [89,117,899]** [*C. A.* **111**, 229277].

Tajobo Co., Ltd; Texas Tech. Univ. (1995). Brassinosteroids for culturing of cotton plant for improvement of cotton fiber. Jpn. Kokai Tokkyo Koho **JP 09 00,097 [97 00,097]** [*C. A.* **126**, 197504].

Takahara, T., Hirose, K., Ono, Y., Iwagaki, I., and Hirai, Y. (1985). Growth acceleration of citrus trees by brassinolide. Jpn. Kokai Tokkyo Koho **JP 62 63,501 [87 63,501]** [*C. A.* **107**, 72885].

Takahashi, H., Jai, S., Koshio, K., Ota, Y., and Singh, J. (1994). Effect of epi-brassinolide application on plant growth, yield components and yield of greengram (*Vigna radiata* (L.) Wilczek). *Japan. J. Tropical Agric.* **38**, 227-231.

Takahashi, K., and Kaufman, P.B. (1992). Regulation of internodal elongation in rice seedlings by plant growth regulators. *Jpn. J. Crop Science* **61**, 34-40.

Takahashi, N., Yokota, T., and Kin, S. (1987). Isolation of brassinosteroids from bean seeds, as plant growth regulators. Jpn. Kokai Tokkyo Koho **JP 63,255,297 [88,255,297]** [*C. A.* **111**, 36804].

Takahashi, T., Ootahe, A., Yamada, H., and Tsuji, J. (1985). Stereoselective reduction of the steroidal 23-en-22-one: a route to the side chain of the plant growth promoter brassinolide. *Tetrahedron Lett.* **26**, 69-72.

Takatsudo, H., and Hayashi, S. (1987). Preparation of brassinosteroids as plant growth regulators. Jpn. Kokai Tokkyo Koho **JP 63,303,978 [88,303,978]** [*C. A.* **111**, 78503].

Takatsudo, H., Hayashi, S., and Ikekawa, N. (1986). Preparation of brassinosteroids as plant growth regulators. Jpn. Kokai Tokkyo Koho **JP 63,166,900 [88,166,900]** [*C. A.* **109**, 231363].

Takatsudo, H., Kamuro, Y., Watanabe, T., and Kuriyama, H. (1993). Preparation of brassinosteroid analogs as plant growth regulators. Jpn. Kokai Tokkyo Koho **JP 06,340,689 [94,340,689]** [*C. A.* **122**, 291318].

Takatsuto, H., Nitani, F., Saito, M., and Sato, H. (1988). Plant freshness-preserving agents, containing brassinosteroids. Jpn. Kokai Tokkyo Koho **JP 01,313,401 [89,313,401]** [*C. A.* **113**, 19459].

Takatsuto, S. (1986). Synthesis of teasterone and typhasterol, brassinolide-related steroids with plant-growth-promoting activity. *J. Chem. Soc. Perkin Trans. I* **10**, 1833-1836.

Takatsuto, S. (1988). Synthesis of dolicholide from (22S,24R)-3α,5-cyclo-5α-ergost-22-en-6-one. *Agric. Biol. Chem.* **52**, 2361-2363.

Takatsuto, S. (1991). Microanalysis of naturally occurring brassinosteroids. *In* "Brassinosteroids. Chemistry, Bioactivity, and Applications. ACS Symposium Series" (H.G. Cutler, T. Yokota, and G. Adam, Eds.), Vol. 474, pp. 107-120. American Chemical Society, Washington.

Takatsuto, S. (1994a). Brassinosteroids: distribution in plants, bioassays and microanalysis by gas chromatography-mass spectrometry. *J. Chromatogr.* **658**, 3-15.

Takatsuto, S. (1994b). Brassinosteroids: synthesis, structure-activity relationship and microanalysis. *Chem. Regul. Plants (Shokubutsu no Kagaku Chosetsu)* **29**, 23-30 [*C. A.* **122**, 10348].

Takatsuto, S., and Gamoh, K. (1990). HPLC analysis of trace levels of brassinosteroids using prelabelling reagents. *Chem. Regul. Plants (Shokubutsu no Kagaku Chosetsu)* **25**, 114-123.

Takatsuto, S., and Hayashi, S. (1987). Brassinosteroid derivatives as plant growth regulators and their preparation. Jpn. Kokai Tokkyo Koho **JP 63,301,895 [88,301,895]** [*C. A.* **111**, 7676].

Takatsuto, S., and Ikekawa, N. (1982). Synthesis of (22R,23R)-28-homobrassinolide. *Chem. Pharm. Bull.* **30**, 4181-4185.

Takatsuto, S., and Ikekawa, N. (1983a). Remote substituent effect on the regioselectivity in the Baeyer-Villiger oxidation of 5α-cholestan-6-one derivatives. *Tetrahedron Lett.* **24**, 917-920.

Takatsuto, S., and Ikekawa, N. (1983b). Stereoselective synthesis of the plant-growth-promoting steroids dolicholide and dolichosterone. *J. Chem. Soc., Perkin Trans. I* 2133-2137.

Takatsuto, S., and Ikekawa, N. (1983c). Stereoselective synthesis of plant growth-promoting steroids, dolicholide and 28-norbrassinolide. *Tetrahedron Lett.* **24**, 773-776.

Takatsuto, S., and Ikekawa, N. (1984a). Short-step synthesis of plant growth-promoting brassinosteroids. *Chem. Pharm. Bull.* **32**, 2001-2004.

Takatsuto, S., and Ikekawa, N. (1984b). Synthesis and activity of plant growth-promoting steroids, (22R,23R,24S)-28-homobrassinosteroids, with modifications in rings A and B. *J. Chem. Soc., Perkin Trans. I* 439-447.

Takatsuto, S., and Ikekawa, N. (1986a). Analysis of 2-deoxybrassinosteroids by gas chromatography - mass-spectrometry. *Chem. Pharm. Bull.* **34**, 3435-3439.

Takatsuto, S., and Ikekawa, N. (1986b). Synthesis of [26,28-^{2}H$_6$]brassinolide, [26,28-^{2}H$_6$]castasterone, [26,28]typhasterol and [26,28-^{2}H$_6$]teasterone. *Chem. Pharm. Bull.* **34**, 1415-1418.

Takatsuto, S., and Ikekawa, N. (1986c). Synthesis of [26,28-^{2}H$_6$]crinosterol, a synthetic intermediate of [26,28-^{2}H$_6$]brassinolide and [26,28-^{2}H$_6$]castasterone. *J. Chem. Soc., Perkin Trans. I* 591-593.

Takatsuto, S., and Ikekawa, N. (1986d). Synthesis of 6-deoxohomodolichosterone, a new plant-growth-promoting steroid. *J. Chem. Soc., Perkin Trans. I* 2269-2272.

Takatsuto, S., and Ikekawa, N. (1986e). Synthesis of deuterio-labelled brassinosteroids, [26,28-^{2}H$_6$]brassinolide, [26,28-^{2}H$_6$]castasterone, [26,28-^{2}H$_6$]typhasterol, and [26,28-^{2}H$_6$]teasterone. *Chem. Pharm. Bull.* **34**, 4045-4049.

Takatsuto, S., and Ikekawa, N. (1987a). Synthesis of 28-norbrassinolides and related 2-deoxysteroids. *Chem. Pharm. Bull.* **35**, 3006-3010.

Takatsuto, S., and Ikekawa, N. (1987b). Synthesis of homodolichosterone and related 2-deoxysteroids. *Chem. Pharm. Bull.* **35**, 829-832.

Takatsuto, S., and Muramatsu, M. (1988). Synthesis of 2,3-diacetate derivatives of (22*S*,23*S*)- and (22*R*,23*R*)-28-homobrassinolides. *Agric. Biol. Chem.* **52**, 2943-2945.

Takatsuto, S., and Shimazaki, K. (1992). An improved preparation of steroidal 2-ene compounds from the corresponding 3β-tosylates. *Biosci. Biotech. Biochem.* **56**, 163-164.

Takatsuto, S., Ying, B., Morisaki, M., and Ikekawa, N. (1981). Synthesis of 28-norbrassinolide. *Chem. Pharm. Bull.* **29**, 903-905.

Takatsuto, S., Ishiguro, M., and Ikekawa, N. (1982a). Chirality transfer in the cholesterol side chain; synthesis of (24*R*)- and (24*S*)-24-hydroxycholesterols. *J. Chem. Soc., Chem. Commun.* 258-260.

Takatsuto, S., Ying, B., Morisaki, M., and Ikekawa, N. (1982b). Microanalysis of brassinolide and its analogs by gas chromatography and gas chromatography-mass spectrometry. *J. Chromatogr.* **239**, 233-241.

Takatsuto, S., Yazawa, N., Ikekawa, N., Morishita, T., and Abe, H. (1983a). Synthesis of (24*R*)-28-homobrassinolide analogues and structure-activity relationships of brassinosteroids in the rice-lamina inclination test. *Phytochemistry* **22**, 1393-1397.

Takatsuto, S., Yazawa, N., Ikekawa, N., Takematsu, T., Takeuchi, Y., and Koguchi, M. (1983b). Structure-activity relationship of brassinosteroids. *Phytochemistry* **22**, 2437-2441.

Takatsuto, S., Yazawa, N., and Ikekawa, N. (1984a). Synthesis and biological activity of brassinolide analogues, 26,27-bisnorbrassinolide and its 6-oxo analogue. *Phytochemistry* **23**, 525-528.

Takatsuto, S., Yazawa, N., Ishiguro, M., Morisaki, M., and Ikekawa, N. (1984b). Stereoselective synthesis of plant growth-promoting steroids, brassinolide, castasterone, typhasterol, and their 28-nor analogues. *J. Chem. Soc., Perkin Trans. 1* 139-146.

Takatsuto, S., Ikekawa, N., Morishita, T., and Abe, H. (1987a). Structure-activity relationship of brassinosteroids with respect to the A/B ring functional groups. *Chem. Pharm. Bull.* **35**, 211-216.

Takatsuto, S., Sato, H., and Futatsuya, F. (1987b). Preparation of brassinosteroid derivatives as plant growth regulators. Eur. Pat. Appl. **EP 315,921** [*C. A.* **111**, 214813].

Takatsuto, S., Kobayashi, K., Watanabe, T., Kuriyama, H., and Furuse, T. (1988a). (22*E*,24*S*)-5α-Stigmasta-3,22-dien-6-one: an intermediate of the isomerisation of (22*E*,24*S*)-3α,5-cyclo-5α-stigmast-22-en-6-one into (22*E*,24*S*)-5α-stigmasta-2,22-dien-6-one. *Agric. Biol. Chem.* **52**, 3217-3218.

Takatsuto, S., Muramatsu, M., Ohya, Y., Hayashi, S., and Shida, A. (1988b). Synthesis of 24-epibrassinolide-related compounds with plant-growth promoting activity. *Agric. Biol. Chem.* **52**, 2059-2064.

Takatsuto, S., Futatsuya, F., Kobayashi, K., and Satoh, H. (1989a). Synthesis and biological activity of 22,23-epoxybrassinosteroid-2,3-diacetates. *Agric. Biol. Chem.* **53**, 263-265.

Takatsuto, S., Yokota, T., Omote, K., Gamoh, K., and Takahashi, N. (1989b). Identification of brassinolide, castasterone and norcastasterone (brassinone) in sunflower (*Helianthus annuus* L.) pollen. *Agric. Biol. Chem.* **53**, 2177-2180.

Takatsuto, S., Abe, H., and Gamoh, K. (1990a). Evidence for brassinosteroids in strobilus of *Equisetum arvense* L. *Agric. Biol. Chem.* **54**, 1057-1059.

Takatsuto, S., Omote, K., Gamoh, K., and Ishibashi, M. (1990b). Identification of brassinolide and castasterone in buckwheat (*Fagopyrum esculentum* Moench) pollen. *Agric. Biol. Chem.* **54**, 757-762.

Takatsuto, S., Abe, H., Shimada, K., Nakayama, M., and Yokota, T. (1996a). Identification of teasterone and 4-desmethylsterols in the seeds of *Ginkgo biloba* L. *J. Jpn. Oil Chem. Soc.* **45**, 1349-1351.

Takatsuto, S., Abe, H., Yokota, T., Shimada, K., and Gamoh, K. (1996b). Identification of castasterone and teasterone in seeds of *Cannabis sativa* L. *J. Jpn. Oil Chem. Soc.* **45**, 871-873.

Takatsuto, S., Kamuro, Y., Watanabe, T., Noguchi, T., and Kuriyama, H. (1996c). Synthesis and plant growth promoting effects of brassinosteroid compound TS303. *Proc. Plant Growth Regul. Soc. Am.* **23**, 15-20.

Takatsuto, S., Kuriyama, H., Noguchi, T., Suganuma, H., Fujioka, S., Sakurai, A. (1997a). Synthesis of cathasterone and its related putative intermediates in brassinolide biosynthesis. *J. Chem. Res., Synop.* 418-41 [*C. A.* **127**, 346556].

Takatsuto, S., Watanabe, T., Fujioka, S., and Sakurai, A. (1997b). Synthesis of new naturally occurring 6-deoxo brassinosteroids. *J. Chem. Res., Synop.* 134-135 [*C. A.* **126**, 277655].

Takeda, K., Shimaoka, A., Iwasaki, M., and Minato, H. (1965). The steroidal components of domestic plants. XLVIII. Components of *Chionographys japonica*. *Chem. Pharm. Bull.* **13**, 691-694.

Takematsu, T. (1986a). Brassinolides for diminution of salt or herbicide damage to crops. Jpn. Kokai Tokkyo Koho **JP 63 66,104 [88 66,104]** [*C. A.* **109**, 224716].

Takematsu, T. (1986b). Insecticide antidotes containing brassinolide derivatives for crops. Jpn. Kokai Tokkyo Koho **JP 63,130,504 [88,130,504]** [*C. A.* **110**, 130539].

Takematsu, T., and Izumi, K. (1985). Acceleration of plant growth in culture soil. Jpn. Kokai Tokkyo Koho **JP 62 04,205 [87 04,205]** [*C. A.* **107**, 72876].

Takematsu, T., and Takeuchi, Y. (1989). Effects of brassinosteroids on growth and yields of crops. *Proc. Jpn. Acad., ser. B* **65**, 149-152.

Takematsu, T., Ikekawa, N., Shida, A. (1986). Increasing the yields of cereals by means of brassinolide derivatives. Eur. Pat. Appl. **EP 218,945**.

Takeno, K., and Pharis, R.P. (1982). Brassinosteroid-induced bending of the leaf lamina of dwarf rice seedlings: an auxin-mediated phenomenon. *Plant Cell Physiol.* **23**, 1275-1281.

Takeuchi, Y. (1992). Studies on the physiology and applications of brassinosteroids. *Chem. Regul. Plants* **27**, 1-10.

Takeuchi, Y., Worsham, A.D., and Awad, A.E. (1991). Effects of brassinolide on conditioning and germination of witchweed (*Striga asiatica*) seeds. In "Brassinosteroids. Chemistry, Bioactivity, and Applications. ACS Symposium Series" (H.G. Cutler, T. Yokota, and G. Adam, Eds.), Vol. 474, pp. 298-305. American Chemical Society, Washington.

Takeuchi, Y., Ogasawara, M., Konnai, M., and Takematsu, T. (1992). Application of brassinosteroids in agriculture in Japan. *Proc. Plant Growth Regul. Soc. Am.* **194**, 169-170.

Takeuchi, Y., Omigawa, Y., Ogasawara, M., Yoneyama, K., Konnai, M., and Worsham, A.D. (1995). Effects of brassinosteroids on conditioning and germination of clover broomrape (*Orobanche minor*) seeds. *Plant Growth Regul.* **16**, 153-160.

Takeuchi, Y., Ogasawara, M., Konnai, M., and Kamuro, Y. (1996). Promotive effectiveness of brassinosteroid (TS303) and jasmonoid (PDJ) on emergence and establishment of rice seedlings. *Proc. Jpn. Soc. Chem. Regul. Plants* **31**, 100-101.

Tanabe, M., and Hayashi, K. (1980). Stereocontrolled synthesis of sterol side chain. *J. Am. Chem. Soc.* **102**, 862.

Taylor, P.E., Spuck, K., Smith, P.M., Sasse, J.M., Yokota, T., Griffiths, P.G., and Cameron, D.W. (1993). Detection of brassinosteroids in pollen of *Lolium perenne* L. by immunocytochemistry. *Planta* **189**, 91-100.

Teijin, Ltd. (1983). Brassinolide-type steroids. Jpn. Kokai Tokkyo Koho **JP 59,181,300** [84,181,300] [*C. A.* **102**, 149631].

Thompson, M.J., Cohen, C.F., and Lancaster, S.M. (1965). Brassicasterol and 22,23-dihydrobrassicasterol from ergosterol *via* *i*-ergosterol. *Steroids* **5**, 745-752.

Thompson, M.J., Mandava, N., Flippen-Anderson, J.L., Worley, J.F., Dutky, S.R., Robbins, W.E., and Lusby, W. (1979). Synthesis of brassinosteroids: new plant-growth-promoting steroids. *J. Org. Chem.* **44**, 5002-5004.

Thompson, M.J., Mandava, N., Worley, J.F., Dutky, S.R., Robbins, W.E., and Flippen-Anderson, J.L. (1980a). Plant growth-promoting brassinosteroids. Pat. **US 4,346,226**.

Thompson, M.J., Mandava, N., Worley, J.F., Meudt, W.J., Dutky, S.R., and Robbins, W.E. (1980b). Plant growth-promoting brassinosteroids. US Pat. Appl. **182,210** [*C. A.* **95**, 127425].

Thompson, M.J., Mandava, N.B., Meudt, W.J., Lusby, W.R., and Spaulding, D.W. (1981). Synthesis and biological activity of brassinolide and its 22β,23β-isomer: novel plant growth-promoting steroids. *Steroids* **38**, 567-580.

Thompson, M.J., Meudt, W.J., Mandava, N.B., Dutky, S.R., Lusby, W.R., and Spaulding, D.W. (1982). Synthesis of brassinosteroids and relationship of structure to plant growth-promoting effects. *Steroids* **39**, 89-105.

Tian, W. (1984). Synthesis of new steroidal plant growth hormone brassinolide. *Youji Huaxue*, 259-264 [*C. A.* **101**, 130961].

Tian, W. (1992). Study on the rational use of steroidal sapogenin. I. Synthesis of sapogenin with A/B ring structure unit of brassinolide. *Acta Chim. Sin. (Huaxue Xuebao)* **50**, 72-77 [*C. A.* **116**, 214773].

Tian, W.-S., Zhou, W.-S., Jiang, B., and Pan, X.-F. (1989). Preparation of 22*R*- and 22*S*,24,25,26,27,28-pentanorbrassinolides. *Acta Chim. Sin.* **47**, 1017-1021 [*C. A.* **113**, 78812].

Tokuda, S. (1986). Extraction of brassinolides from soybean oil cake. Jpn. Kokai Tokkyo Koho **JP 63,115,804 [88,115,804]** [*C. A.* **112**, 50607].

Tominaga, R., Sakurai, N., and Kuraishi, S. (1994). Brassinolide-induced elongation of inner tissues of segments of squash (*Cucurbita maxima* Duch.) hypocotyls. *Plant Cell Physiol.* **35**, 1103-1106.

Torneiro, M., Fall, Y., Castedo, L., and Mourino, A. (1997). A short, efficient copper-mediated synthesis of 1α,25-dihydroxyvitamin D_2 (1α,25-dihydroxyergocalciferol) and C-24 analogs. *J. Org. Chem.* **62**, 6344-6352.

Torsell, K.B.G. (1988)."Nitrile oxides, nitrones and nitronates". VCH Verlagsgesellschaft, Weinheim.

Toyobo Co., Ltd. (1996). Brassinosteroids for culturing of cotton plant for improvement of cotton fiber. Jpn. Kokai Tokkyo Koho **JP 09,000,097** [*C. A.* **126**, 197504].

Traven, V.F., Kuznetsova, N.A., Levinson, E.E., and Podkhalyuzina, N.Ya. (1991a). The shortest way from ergosterol to 24-epibrassinolide and its 22*S*,23*S*-isomer. *Dokl. AN USSR* **317**, 901-904.

Traven, V.F., Levinson, E.E., and Kuznetsova, N.A. (1991b). Methods of selective reduction of (22*E*,24*R*)-3α,5-cyclo-5α-ergosta-7,22-dien-6-one. *Dokl. AN USSR* **320**, 1132-1136.

Trost, B.M., and Verhoeven, T.R. (1978). Stereocontolled approach to steroid side chain *via* organopalladium chemistry. Partial synthesis of 5α-cholestanone. *J. Am. Chem. Soc.* **100**, 3435-3443.

Tsai, D.S., and Arteca, R.N. (1985). Effects of auxin inhibitors on brassinosteroid and indole-3-acetic acid-induced ethylene production. *Plant. Physiol.* **77**, 157.

Tsibulko, V.S., and Buriak., Y.I. (1991). Effect of brassinosteroids on seeds of corn and lucerne. *In* "Conference on brassinosteroids", 2nd, p. 40, Minsk.

Tsibulko, V.S., and Popov, S.I. (1991). Influence of brassinosteroids on seeds of pea, soybean and buckwheat. *In* "Conference on brassinosteroids", 2nd, p. 42, Minsk.

Tsibulko, V.S., Buriak., Y.I., and Popov, S.I. (1993). Application of brassinosteroids for the crop enhancement of lucerne, pea, soybean, and buckwheat. *In* "Brassinosteroids - biorational, ecologically safe regulators of growth and productivity of plants", 3rd, pp. 30-31, Minsk.

Tsubuki, M., Keino, K., and Honda, T. (1992a). Stereoselective synthesis of plant-growth-regulating steroids: brassinolide, castasterone and their 24,25-substituted analogues. *J. Chem. Soc., Perkin Trans. 1* 2643-2649.

Tsubuki, M., Keino, K., Kakinuma, N., and Honda, T. (1992b). A facile construction of withanolide side chains: synthesis of minabeolide-3. *J. Org. Chem.* **57**, 2930-2934.

Uesono, T., Adachi, A., Hamada, K., Nishi, S., Fujita, F., and Fujiwara, S. (1985). Enhacement of plant tolerance to salts by brassinolide. Jpn. Kokai Tokkyo Koho **JP 62 48,602 [87 48,602]** [*C. A.* **107**, 91907].

Ugarova, T.Y. (1997). "Family vegetable growing on narrow ridges". Marketing, Moscow.

Umarov, A.A., and Kariev, A.I. (1991). Effect of epibrassinolide on cotton plants. *In* "Conference on brassinosteroids", 2nd, p. 51, Minsk.

Van Rheenen, V., Kelly, R.C., and Cha, D.Y. (1976). An improved catalytic OsO_4 oxidation of olefins to cis-1,2-glycols using tertiary amine oxides as the oxidant. *Tetrahedron Lett.*, 1973-1976.

Vanderah, O.J., and Djerassi, C. (1978). Marine natural products. Synthesis of four naturally occuring 20β-H cholanic acid derivatives. *J. Org. Chem.* **43**, 1442-1448.

Vasyukova, N.I., Khripach, V.A., Chalenko, G.I., and Kaneva, I.M. (1993). Effect of brassinosteroids on phytophthora infection of potato. *In* "Brassinosteroids - biorational, ecologically safe regulators of growth and productivity of plants", 3rd, p. 20, Minsk.

Vasyukova, N.I., Chalenko, G.I., Kaneva, I.M., Khripach, V.A., and Ozeretskovskaya, O.L. (1994). Brassinosteroids and potato late blight. *Prikl. Biokhim. Mikrobiol.* **30**, 464-470 [*C. A.* **121**, 78177].

Vedeneev, A.N., and Deeva, V.P. (1997). Effects of growth regulators on lipid peroxidation in different genotypes of barley. *In* "Plant growth and development regulators", 4th, pp.133-134, Moscow.

Vedeneev, A.N., Deeva, V.P., and Khripach, V.A. (1995). Effect of epibrassinolide on sugar beet. *In* "Brassinosteroids - biorational, ecologically safe regulators of growth and productivity of plants", 4th, p. 24, Minsk.

Vedeneev, A.N., Deeva, V.P., Apanovich, T.M., and Khripach, V.A. (1997a). Effect of epibrassinolide on growth and development of barley, and on accumulation of cesium and strontium. *In* "Plant growth and development regulators", 4th, p. 276, Moscow.

Vedeneev, A.N., Deeva, V.P., Volkov, S.M., and Khripach, V.A. (1997b). Epibrassinolide action on fatty acid composition in different genotypes of barley. *Plant Growth Regul. Soc. Am. Quat.* **25**, no. 2, p. 83.

Victorova, E. (1996). Wonderful preparation. *Kartofel i ovoschi*, 31.

Vidyarardhini, B., and Rao, S.S.R. (1996). Effect of brassinosteroids on germination of ground-nut (*Arachis hypogaea* L.) seeds. *Indian J. Plant Physiol.* **1**, 223-224 [*C. A.* **127**, 157921].

Vidyarardhini, B., and Rao, S.S.R. (1997). Effect of brassinosteroids on salinity induced growth inhibition of groundnut seedlings. *Indian J. Plant Physiol.*, **2**, 156-157.

Vitvitskaya, L.V., Nikonorov, S.I., Tikhomirov, A.M., Zagriichuk, V.P., Abtakhi, B., and Vorob'eva, E.I. (1997a). Effect of some toxicants on the behaviour of Russian sturgeon fingerlings and toxicoprotector effects of biologically active substances. *Dokl. Akad. Nauk* **352**, 842-844 [*C. A.* **126**, 313380].

Vitvitskaya, L.V., Tikhomirov, A.M., Nikonorov, S.I., Egorov, M.A., Malevannaya, N.N., Khripach, V.A., and Zhabinskii, V.N. (1997b). The method of survival increase of larvae and fingerlings of sturgeons in fish breeding. Pat. Appl. **RU 97,117,958**.

Vitvitskaya, L.V., Tikhomirov, A.M., Nikonorov, S.I., Egorov, M.A., Malevannaya, N.N., Khripach, V.A., and Zhabinskii, V.N. (1997c). The method for estimation and improvement of activity and viability of spermatozoons in animal breeding. Pat. Appl. **RU 97,117,959**.

Vlasova, N.N., Stratilatova, E.V., and Laman, N.A. (1998). The peculiarities of development of side shoots and productivity of spring barley in water deficiency and in optimal conditions in dependence of terms of plant treatment by epibrassinolide. In press.

Voigt, B., Porzel, A., Naumann, H., Hörhold-Schubert, C., and Adam, G. (1993a). Hydroxylation of the native brassinosteroids 24-epicastasterone and 24-epibrassinolide by the fungus *Cunninghamella echinulata*. *Steroids* **58**, 320-323.

Voigt, B., Porzel, A., Undisz, K., Hörhold-Shubert, C., and Adam, G. (1993b). Microbial hydroxylation of 24-epicastasterone by the fungus *Cochliobolus lunatus*. *Nat. Prod. Lett.* **3**, 123-129.

Voigt, B., Takatsuto, S., Yokota, T., and Adam, G. (1995). Synthesis of secasterone and further epimeric 2,3-epoxybrassinosteroids. *J. Chem. Soc., Perkin Trans. 1* 1495-1498.

Voigt, B., Porzel, A., Golsch, D., Adam, W., and Adam, G. (1996a). Regioselective oxyfunctionalization of brassinosteroids by methyl(trifluoromethyl)dioxirane: synthesis of 25-hydroxy-brassinolide and 25-hydroxy-24-epibrassinolide by direct C-H insertion. *Tetrahedron* **52**, 10653-10658.

Voigt, B., Porzel, A., Golsch, D., Adam, W., and Adam, G. (1996b). Synthesis of biologically highly active 25-hydroxy-brassinolide and 25-hydroxy-24-epibrassinolide. *Proc. Plant Growth Regul. Soc. Am.* **23**, 41-42.

Voigt, B., Schmidt, J., and Adam, G. (1996c). Synthesis of 24-epiteasterone, 24-epityphasterol and their B-homo-6a-oxalactones from ergosterol. *Tetrahedron* **52**, 1997-2004.

Voigt, B., Bruhn, C., Wagner, C., Merzweiler, K., and Adam, G. (1997). Synthesis of 24-epicathasterone and related new brassinosteroids. *Proc. Plant Growth Regul. Soc. Am.* **24**, 138-139.

Volodko, I.K., and Zelenkevich, A.V. (1998). In press.

Volynets, A.P., Pshenichnaya, L.A., Manzhelesova, N.E., Morozik, G.V., and Khripach, V.A. (1997a). Method of protection of barley plants from leave diseases. Pat. Appl. **BY 970,680**.

Volynets, A.P., Pschenichnaya, L.A., Manzhelesova, N.E., Morozik, G.V., and Khripach, V.A. (1997b). The nature of protective action of 24-epibrassinolide on barley plants. *Proc. Plant Growth Regul. Soc. Am.* **24**, 133-137.

von Arnim, A.G., Osterlund, M.T., Kwok, S.F., and Deng, X.-W. (1997). Genetic and developmental control of nuclear accumulation of COP1, a repressor of photomorphogenesis in *Arabidopsis*. *Plant Physiol.* **114**, 779-788.

Vorbrodt, H.-M., Adam, G., Porzel, A., Hörhold, C., Dänhardt, S., and Böhme, K.-H. (1991). Microbial degradation of $2\alpha,3\alpha$-dihydroxy-5α-cholestan-6-one by *Mycobacterium vaccae*. *Steroids* **56**, 586-588.

Voskresenskaya, L.G. (1993). Estimation of effect of "Epin" on growth and development of flax. Technical Report of the Institute of Flax, Belarus.

Wada, K., and Marumo, S. (1981). Synthesis and plant growth-promoting activity of brassinolide analogues. *Agric. Biol. Chem.* **45**, 2579-2585.

Wada, K., Marumo, S., Ikekawa, N., Morisaki, M., and Mori, K. (1981). Brassinolide and homobrassinolide promotion of lamina inclination of rice seedlings. *Plant Cell. Physiol.* **22**, 323-325.

Wada, K., Marumo, S., Mori, K., Takatsuto, S., Morisaki, M., and Ikekawa, N. (1983). The rice lamina inclination-promoting activity of synthetic brassinolide analogs with a modified side chain. *Agric. Biol. Chem.* **47**, 1139-1141.

Wada, K., Marumo, S., Abe, H., Morishita, T., Nakamura, K., Uchiyama, M., and Mori, K. (1984). A rice lamina inclination test - a microquantitative bioassay for brassinosteroids. *Agric. Biol. Chem.* **48**, 719-726.

Wada, K., Kondo, H., and Marumo, S. (1985). A simple bioassay for brassinosteroids: a wheat leaf-unrolling test. *Agric. Biol. Chem.* **45**, 2249-2251.

Wang, S.-G., and Deng, R.-F. (1992). Effects of brassinosteroid (BR) on root metabolism in rice. *J. Southwest Agric. Univ.* **14**, 177-181.

Wang, T.-W., Cosgrove, D.J., and Arteca, R.N. (1993). Brassinosteroid stimulation of hypocotyl elongation and wall relaxation in pakchoi (*Brassica chinensis* cv Lei-Choi). *Plant Physiol.* **101**, 965-968.

Wang, Y., Luo, W., Zhao, Y., and Ikekawa, N. (1988). Effect of epi-brassinolide on growth of celery. *Zhiwu Shenglixue Tongxun*, 29-31 [*C. A.* **108**, 219079].

Wang, Y.-Q., and Zhao, Y.-J. (1989). Relationship between structure and biological activity of brassinolide analogs. *Acta Phytophysiol. Sinica* **15**, 18-23 [*C. A.* **111**, 130875].

Wang, Y.-Q., Luo, W.-H., and Zhao, Y.-J. (1994a). Effect of epibrassinolide on growth and fruit quality of watermelon. *Zhiwu Shenglixue Tongxun* **30**, 423-425.

Wang, Y.-Q., Luo, W.-H., Hu, R.-J., Zhao, Y.-J., Zhou, W.-S., Huang, L.-F., and Shen, J.-M. (1994b). Biological activity of brassinosteroids and relationship of structure to plant growth promoting effects. *Chin. Sci. Bull.* **39**, 1573-1577 [*C. A.* **122**, 154040].

Wang, Z. (1992). Nutritive plant leaf fertilizer. Faming Zhuanli Shenqing Gongkai Shuomingshu CN **1,088,194** [*C. A.* **122**, 80264].

Watanabe, T., Kuriyama, H., Furuse, T., Kobayashi, K., and Takatsuto, S. (1988). An efficient preparation of $(22E,24S)$-5α-stigmasta-2,22-dien-6-one from $(22E,24S)$-$3\alpha,5$-cyclo-5α-stigmast-22-en-6-one. *Agric. Biol. Chem.* **52**, 2117-2118.

Watanabe, T., Takatsuto, S., Fukioka, S., and Sakurai, A. (1997). Improved synthesis of castasterone and brassinolide. *J. Chem. Res., Synop.* 360-361.

Werner, F., Parmentier, G., Luu, B., and Dinan, L. (1996). A convenient stereoselective synthesis of castasterone and its analogues using arsenic ylides. *Tetrahedron* **52**, 5525-5532.

Wiersig, J.R., Waespe-Sarcevic, N., and Djerassi, C. (1979). Stereospecific synthesis of the side chain of the steroidal plant sex hormone oogoniol. *J. Org. Chem.* **44**, 3374-3382.

Wilen, R.W., Sacco, M., Gusta, L.V., and Krishna, P. (1995). Effects of 24-epibrassinolide on freezing and thermotolerance of bromegrass (*Bromus inermis*) cell cultures. *Physiol. Plant.* **95**, 195-202.

Williams, D.E., Ayer, S.W., and Andersen, R.J. (1986). Diaulusterols A and B from the skin extracts of the dorid nudibranch *Diaulula sandiegensis*. *Can. J. Chem.* **64**, 1527-1529.

Wilson, S.R., Davey, A.E., and Guazzaroni, M.E. (1992). Two new approaches to the 25-hydroxy-vitamin D_2 side chain. *J. Org. Chem.* **57**, 2007-2012.

Winter, J., Schneider, B., Strack, D., and Adam, G. (1996). Involvement of cytochrome P450 in the metabolism of 24-epi-brassinolide in tomato cell cultures. *Proc. Plant Growth Regul. Soc. Am.* **23**, 290.

Winter, J., Schneider, B., Strack, D., and Adam, G. (1997). Role of a cytochrome P450-dependent monooxygenase in the hydroxylation of 24-epi-brassinolide. *Phytochemistry* **45**, 233-237.

Woodward, R.B., and Brutcher, F.V.Jr. (1958). cis-Hydroxylation of a synthetic steroid intermediate with iodine, silver acetate and wet acetic acid. *J. Am. Chem. Soc.* **80**, 209-211.

Worley, J.F., and Krizek, D.T. (1972). Influence of brassins on the growth of woody plants. *Hortic. Sci.* **7**, 480-481.

Worley, J.F., and Mitchell, J.W. (1971). Growth responses induced by brassins (fatty plant hormones) in bean plants. *J. Am. Hortic. Soc.* **96**, 270-273.

Wu, M. (1993). Low temperature protectant for plant. Faming Zhuanli Shenqing Gongkai Shuomingshu CN **1,090,970** [*C. A.* **122**, 154174].

Xia, W.-F., and Zhou, R.M. (1992). Techniques for applying brassinolide to rape. *Jiangsu Agric. Sciences*, 20-21.

Xu, R.-J., Guo, Y.S., and Zhao, Y.-J. (1990). Epibrassinolide induction of changes in enlargement, content of exogenic gibberellinic acid, ABA and starch in cucumber hypocotyls. *Acta Phytophysiol. Sinica* **16**, 125-130.

Xu, R.-J., He, Y.-J., Wang, Y.-Q., and Zhao, Y.-J. (1994a). Preliminary study of brassinosterone binding sites from mung bean epicotyls. *Acta Phytophysiol. Sinica* **20**, 298-302 [*C. A.* **122**, 101761].

Xu, R.-J., Li, X.-D., He, Y.-J, Wang, Y.-Q., and Zhao, Y.-J. (1994b). Effects of treatments with epibrassinolide and chololic lactone on the fruit-set and ripening in some grape cultivation. *J. Shanghai Agric. Coll.* **12**, 90-95.

Xu, R.-J., Wang, Y.-Q., Wu, D.-R., He, Y.-J., Zhao, Y.-J., and Fang, J. (1994c). Preparation of ^{125}I-brassinone and its biological activity. *Acta Phytophysiol. Sinica* **20**, 121-127 [*C. A.* **122**, 76780].

Xu, W., Purugganan, M.M., Polisensky, D.H., Antosiewicz, D.M., Fry, S.C., and Braam, J. (1995). *Arabidopsis* TCH4, regulated by hormones and the environment, encodes a xyloglucan endotransglucosylase. *Plant Cell* **7**, 1555-1567.

Xu, W., Campbell, P., Vargheese, A.K., and Braam, J. (1996). The *Arabidopsis* XET-related gene family: environmental and hormonal regulation of expression. *Plant J.* **9**, 879-889.

Yamada, S., Shiraishi, M., Ohmori, M., and Takayama, H. (1984). Facile and regioselective synthesis of 25-hydroxy vitamin D_2. *Tetrahedron Lett.* **25**, 3347-3350.

Yamaji, M., Saimoto, H., and Yamamoto, Y. (1992). Brassinosteroids manufacture with photosynthetic bacteria. Jpn. Kokai Tokkyo Koho **JP 06 86,691 [94 86,691]** [*C. A.* **120**, 321536].

Yamamoto, R., Demura, T., and Fukuda, H. (1997). Brassinosteroids induce entry into the final stage of tracheary element differentiation in cultured *Zinnia* cells. *Plant Cell Physiol.* **38**, 980-983.

Yamamoto, Y. (1985a). A process for the preparation of steroid alcohols *via* anti-Cram's rule. Jpn. Kokai Tokkyo Koho **JP 62 51,629 [87 51,629]** [*C. A.* **107**, 217934].

Yamamoto, Y. (1985b). A process for the stereoselective preparation of steroid alcohols as intermediates for plant growth hormones *via* Cram's rule. Jpn. Kokai Tokkyo Koho **JP 62 51,628 [87 51,628]** [*C. A.* **107**, 198754].

Yamamoto, Y. (1985c). Stereoselective preparation of steroids *via* Cram's rule. Jpn. Kokai Tokkyo Koho **JP 62 51,696 [87 51,696]** [*C. A.* **107**, 198755].

Yamamoto, Y., and Yamada, J. (1986). Asymmetric synthesis using organometallic compounds. Its application to the side chains of steroids. *Kagaku to Seibutsu* **24**, 515-522 [*C. A.* **106**, 176724].

Yamamoto, Y., Nishii, S., and Maruyama, K. (1986a). The titanium tetrachloride-mediated alkynylation and allylation of steroidal aldehyde *via* stannylacetylenes and allylstannanes with high Cram selectivity. *J. Chem. Soc., Chem. Commun.*, 102-103.

Yamamoto, Y., Nishii, S., and Yamada, J. (1986b). Importance of the timing of bond breaking and bond making in acetal templates. Enantiodivergent synthesis of steroidal side chains. *J. Am. Chem. Soc.* **108**, 7116-7117.

Yamamoto, Y., Abe, H., and Yamada, J. (1991). Enantiodivergent synthesis of steroidal side chains. Stereocontrol *via* S_N1 vs. S_N2 type cleavage of acetal templates. *J. Chem. Soc., Perkin Trans. I* 3253-3257.

Yamanaka, Y. (1988a). Brassinolide-based fruit set-promoting agents for peach. Jpn. Kokai Tokkyo Koho **JP 02 67,206 [90 67,206]** [*C. A.* **113**, 110921].

Yamanaka, Y. (1988b). Cultivation of persimmon with brassinolide. Jpn. Kokai Tokkyo Koho **JP 02,152,905 [90,152,905]** [*C. A.* **113**, 128086].

Yang, Z.S., Shi, G.A., and Jin, J.H. (1992). Effects of epibrassinolide and triadimefon on winter wheat yield and its physiological response. *Acta Univ. Agric. Boreali Occidentalis* **20**, 47-50.

Yasuta, E., Terahata, T., Nakayama, M., Abe, H., Takatsuto, S., and Yokota, T. (1995). Free and conjugated brassinosteroids in the pollen and anthers of *Erythronium japonicum* Decne. *Biosci. Biotech. Biochem.* **59**, 2156-2158.

Yokota, K., and Yamanaka, Y. (1988). Brassinolide for apple fruit abscission prevention. Jpn. Kokai Tokkyo Koho **JP 02 42,002 [90 42,002]** [*C. A.* **113**, 19473].

Yokota, T. (1984). Brassinosteroids from higher plants. *Chem. Regul. Plants (Shokubutsu no Kagaku Chosetsu)* **19**, 102-109 [*C. A.* **103**, 34821].

Yokota, T. (1985). Discovery and physiology of new plant hormone, brassinolide. *Gendai Kagaku* **167**, 14-20 [*C. A.* **103**, 3612].

Yokota, T. (1987a). Brassinosteroid and its utilization. *Nippon Nogei Kagaku Kaishi* **61**, 385-388.

Yokota, T. (1987b). Study on bio-organic chemistry of natural brassinosteroids. *Chem. Regul. Plants (Shokubutsu no Kagaku Chosetsu)* **22**, 10-17 [*C. A.* **108**, 34741].

Yokota, T. (1995). Chemical structures of plant hormones. *Chemical Regul. Plants* **30**, 185-196.

Yokota, T. (1997). The structure, biosyntheis and function of brassinosteroids. *Trends Plant Sci.* **2**, 137-143.

Yokota, T., and Takahashi, N. (1987). Isolation of brassinosteroids as plant growth regulators from kidney beans. Jpn. Kokai Tokkyo Koho **JP 63,216,896 [88,216,896]** [*C. A.* **111**, 4522].

Yokota, T., Arima, M., and Takahashi, N. (1982a). Castasterone, a new phytosterol with plant-hormone potency from chestnut insect gall. *Tetrahedron Lett.* **23**, 1275-1278.

Yokota, T., Baba, J., and Takahashi, N. (1982b). A new steroidal lactone with plant growth-regulatory activity from *Dolichos lablab* seed. *Tetrahedron Lett.* **23**, 4965-4966.

Yokota, T., Arima, M., Takahashi, N., Takatsuto, S., Ikekawa, N., and Takematsu, T. (1983a). 2-Deoxycastasterone, a new brassinolide-related bioactive steroid from *Pinus* pollen. *Agric. Biol. Chem.* **47**, 2419-2420.

Yokota, T., Baba, J., and Takahashi, N. (1983b). Brassinolide-related bioactive sterols in *Dolichos lablab*: brassinolide, castasterone and a new analog, homodolicholide. *Agric. Biol. Chem.* **47**, 1409-1411.

Yokota, T., Baba, J., Arima, M., Morita, M., and Takahashi, N. (1983c). Isolation and structures of new brassinolide-related compounds in higher plants. *Tennen Yuki Kagob.Toronkai Koen Yoshishu* **26**, 70-77 [*C. A.* **100**, 48616].

Yokota, T., Morita, M., and Takahashi, N. (1983d). 6-Deoxocastasterone and 6-deoxodolichosterone - putative precursors for brassinolide-related steroids from *Phaseolus vulgaris*. *Agric. Biol. Chem.* **47**, 2149-2151.

Yokota, T., Baba, J., Koba, S., and Takahashi, N. (1984). Purification and separation of eight steroidal plant growth regulators from *Dolichos lablab* seed. *Agric. Biol. Chem.* **48**, 2529-2534.

Yokota, T., Arima, M., Takahashi, N., and Crozier, A. (1985). Steroidal plant growth regulators, castasterone and typhasterol (2-deoxycastasterone) from the shoots of *Sitka spruce* (*Picea sitchensis*) (Pinaceae). *Phytochemistry* **24**, 1333-1335.

Yokota, T., Kim, S.-K., Kosaka, Y., Ogino, Y., and Takahashi, N. (1986). Conjugation of brassinosteroids. *In* "Conjugated Plant Hormones. Structure, Methabolism and Function. Proceedings of the International Symposium" (K. Schreiber. H.R.Schütte, and G. Sembdner, Eds.), pp. 288-296. VEB Deutscher Verlag der Wissenschaften, Berlin.

Yokota, T., Kim, S.-K., Fukui, Y., Takahashi, N., Takeuchi, Y., and Takematsu, T. (1987a). Brassinosteroids and sterols from a green alga, *Hydrodictyon reticulatum*: configuration at C-24. *Phytochemistry* **26**, 503-506.

Yokota, T., Koba, S., Kim, S.-K., Takatsuto, S., Ikekawa, N., Sakakibara, M., Okada, K., Mori, K., and Takahashi, N. (1987b). Diverse structural variations of the brassinosteroids in *Phaseolus vulgaris* seed. *Agric. Biol. Chem.* **51**, 1625-1631.

Yokota, T., Ogino, Y., Takahashi, N., Saimoto, H., Fujioka, S., and Sakurai, A. (1990a). Brassinolide is biosynthesized from castasterone in *Catharanthus roseus* gall cells. *Agric. Biol. Chem.* **54**, 1107-1108.

Yokota, T., Watanabe, S., Ogino, Y., Yamaguchi, I., and Takahashi, N. (1990b). Radioimmunoassay for brassinosteroids and its use for comparative analysis of brassinosteroids in stems and seeds of *Phaseolus vulgaris*. *J. Plant Growth Regul.* **9**, 151-159 [*C. A.* **113**, 148287].

Yokota, T., Ogino, Y., Suzuki, H., Takahashi, N., Saimoto, H., Fujioka, S., and Sakurai, A. (1991). Metabolism and biosynthesis of brassinosteroids. *In* "Brassinosteroids. Chemistry, Bioactivity, and Applications. ACS Symposium Series" (H.G. Cutler, T. Yokota, and G. Adam, Eds.), Vol. 474, pp. 86-96. American Chemical Society, Washington.

Yokota, T., Higuchi, K., Kosaka, Y., and Takahashi, N. (1992). Transport and metabolism of brassinosteroids in rice. *Curr. Plant Sci. Biotechnol. Agric.* **13**, 298-305.

Yokota, T., Nakayama, M., Wakisaka, T., Schmidt, J., and Adam, G. (1994). 3-Dehydroteasterone, a 3,6-diketobrassinosteroid as a possible biosynthetic intermediate of brassinolide from wheat grain. *Biosci. Biotech. Biochem.* **58**, 1183-1185.

Yokota, T., Matsuoka, T., Korarai, T., and Nakayama, M. (1996a). 2-Deoxybrassinolide, a brassinosteroid from *Pisum sativum* seed. *Phytochemistry* **42**, 509-511.

Yokota, T., Nomura, T., Nakayama, M., Takeuchi, Y., and Reid, J.B. (1996b). Dwarf pea *lka* and *lkb*, possible brassinosteroid biosynthesis and sensitivity mutants. *Proc. Plant Growth Regul. Soc. Am.* **23**, 10.

Yokota, T., Nomura, T., Kitasaka, Y., Takatsuto, S., and Reid, J.B. (1997). Biosynthetic lesions in brassinosteroid-deficient pea mutants. *Proc. Plant Growth Regul. Soc. Am.* **24**, 94.

Yopp, J.H., Colclasure, G.C., and Mandava, N. (1979). Effect of brassincomplex on auxin and gibberellin mediated events in the morphogenesis of the etiolated bean hypocotyl. *Physiol. Plant.* **46**, 247-254.

Yopp, J.H., Mandava, N.B., and Sasse, J.M. (1981). Brassinolide, a growth-promoting steroidal lactone. I. Activity in selected auxin bioassays. *Physiol. Plant.* **53**, 445-452.

Yuya, M., and Takeuchi, T. (1986). Preparation of plant-growth hormone brassinosteroid intermediates. Jpn. Kokai Tokkyo Koho **JP 62,201,897 [87,201,897]** [*C. A.* **109**, 38060].

Yuya, M., and Takeuchi, T. (1987). Preparation of brassinosteroids as plant growth stimulants. Jpn. Kokai Tokkyo Koho **JP 01 75,500 [89 75,500]** [*C. A.* **111**, 214814].

Yuya, M., Takeuchi, T., Adachi, A., Hamada, K., Fujita, F., Kamuro, Y., Hirai, Y., and Fujii, S. (1984a). Brassinolide derivatives as plant growth regulators. Jpn. Kokai Tokkyo Koho **JP 61,103,894 [86,103,894]** [*C. A.* **106**, 50552].

Yuya, M., Takeuchi, T., and Mori, K. (1984b). Steroids. Jpn. Kokai Tokkyo Koho **JP 61 69,790 [86 69,790]** [*C. A.* **105**, 153400].

Yuya, M., Takeuchi, T., and Mori, K. (1985a). 6,6-Ethylenedioxy-3-hydroxy-20S-formyl-5α-pregnane compounds. Jpn. Kokai Tokkyo Koho **JP 61,251,697 [86,251,697]** [*C. A.* **106**, 84947].

Yuya, M., Takeuchi, T., and Mori, K. (1985b). A process for the preparation of cholest-2-en-6-one, ergost-2,22-diene-6-one, and stigmast-2,22-diene-6-one as intermediates for plant growth hormone brassinolide. Jpn. Kokai Tokkyo Koho **JP 62 99,396 [87 99,396]** [*C. A.* **107**, 217936].

Zabolotny, A.I., Budkevich, T.A., and Yakushev, B.I. (1997). Technical Report "Exogenic regulation of radio-nuclides intake by wild and cultural plants". Institute of Experimental Botany, Minsk.

Zadoks, J.C., Chang, T.T., and Konzak, C.F. (1974). A decimal code for the growth stages of cereals. *Weed Res.* **14**, 415-421.

Zeelen, F.J. (1984). Stereochemistry of electrophilic addition reactions to 22E-unsaturated sterols; a Barton rule revised. *Rec. Trav. Chim. Pays-Bas* **103**, 245-251.

Zeelen, F.J. (1990). "Medicinal Chemistry of Steroids". Elsevier, Amsterdam - Oxford - New York - Tokyo.

Zhabinskii, V.N., Olkhovick, V.K., and Khripach, V.A. (1996). Methods of stereoselective construction of steroidal side chains. *Zh. Org. Khim.* **32**, 327-363.

Zhabinskii, V., Zhernosek, E., and Khripach, V. (1997). Synthesis of castasterone. In "XVII Conference on Isoprenoids, Abstracts of Papers", p. 77. Kraków.

Zhang, H., Zhang, H., and Pan, B. (1988). Preparation of brassinosteroids as plant growth stimulants. Faming Zhuanli Shenqing Gongkai Shuomingshu **CN 1,039,594** [*C. A.* **114**, 24323].

Zhang, H., Li, L., and Dai, X. (1989). Preparation of steroidal lactones from rapeseed oil as plant growth stimulant. Faming Zhuanli Shenqing Gongkai Shuomingshu **CN 1,037,710** [*C. A.* **113**, 132585].

Zhao, Y.-J., and Wu, D. (1990). Effects of 24-epibrassinolide on stem elongation and resistance to environmental stress in mung been seedlings. In "Int. Workshop. Brassinosteroids: Chemistry, Bioactivity, Application", p. 26, Halle.

Zhao, Y.-J., Luo, W.-H., Wang, Y.-Q., and Xu, R.-J. (1987). Retarding effects of epibrassinolide on maturation and senescence of hypocotyl segments of mung bean seedlings. *Acta Phytophysiol. Sinica* **13**, 129-135 [*C. A.* **107**, 233237].

Zhou, A.-Q. (1987). Effect of brassinolide on seed germination and growth of coleoptile in rice. *Plant Physiol. Communs.*, 19-22.

Zhou, W.-S. (1989). The synthesis of brassinosteroid. *Pure Appl. Chem.* **61**, 431-434.

Zhou, W.-S. (1994). Methods for construction of side chain of brassinosteroids and application to the syntheses of brassinosteroids. *Prog. Nat. Sci.* **4**, 129-136 [*C. A.* **121**, 83717].

Zhou, W.-S., and Ge, C.-S. (1989). A novel method for construction of side chain of brassinosteroids. *Sci. China, Ser. B* **32**, 1290-1299 [*C. A.* **113**, 59679].

Zhou, W.-S., and Huang, L.-F. (1992). Concise stereoselective construction of side chain of brassinosteroid from the intact side chain of hyodeoxycholic acid: formal syntheses of brassinolide, 25-methylbrassinolide, 26,27-bisnorbrassinolide and their related compounds. *Tetrahedron* **48**, 1837-1852.

Zhou, W.-S., and Shen, Z.-W. (1991). Formal synthesis of brassinolide via sulphenate-sulphoxide transformation. *J. Chem. Soc., Perkin Trans. 1* 2827-2830.

Zhou, W.-S., and Tian, W.-S. (1984). Synthesis of steroids from hyodeoxycholic acid containing structural units of A/B rings of brassinolide and β-ecdysone. *Acta Chim. Sin.* **42**, 1173-1177 [*C. A.* **102**, 221085].

Zhou, W.-S., and Tian, W.-S. (1985). Stereoselective synthesis of (22S,23S)-typhasterol from hyodeoxycholic acid. *Acta Chim. Sin. (Huaxue Xuebao)* **43**, 1060-1067.

Zhou, W.-S., and Tian, W.-S. (1987). Stereoselective synthesis of typhasterol from hyodeoxycholic acid. *Tetrahedron* **43**, 3705-3712.

Zhou, W.-S., and Wang, Z.-Q. (1991). Synthesis of bioactive steroids starting from hyodeoxycholic acid. *Chinese J. Pharm.* **22**, 30-36.

Zhou, W.-S., Jiang, B., and Pan, X.-F. (1988a). Studies on steroidal plant-growth regulators. A new route for the efficient synthesis of the 2α,3α-dihydroxy-7-oxa-6-oxo-B-homo structural unit of brassinolide. *J. Chem. Soc., Chem. Commun.* 791-793.

Zhou, W.-S., Jiang, L., Tian, W., Zhao, X., and Zheng, H. (1988b). Synthesis of 2α,3α-dihydroxy-7-oxa-6-oxo-23,24-bisnor-B-homo-5α-cholanic acid and 2α,3α-dihydroxy-7-oxa-6-oxo-24-nor-B-homo-5α-cholanic acid. *Acta Chim. Sin. (Huaxue Xuebao)* **46**, 1212-1218 [*C. A.* **111**, 78496].

Zhou, W.-S., Biao, J., and Pan, X.-F. (1989a). A novel synthesis of brassinolide and related compounds. *J. Chem. Soc., Chem. Commun.* 612-614.

Zhou, W.-S., Jiang, B., and Pan, X.-F. (1989b). Regioselective synthesis of methyl 3α-hydroxy-7-oxa-6-oxo-B-homo-5α-cholanate. *Acta Chim. Sin. (Huaxue Xuebao)* **47**, 182-185 [*C. A.* **111**, 78497].

Zhou, W.-S., Zhou, Y.-P., and Jiang, B. (1989c). Studies on steroidal plant-growth regulators: a new synthesis of brassinosteroids. *Synthesis* 426-427.

Zhou, W.-S., Jiang, B., and Pan, X.-F. (1990a). Stereoselective synthesis of the brassinolide side chain: Novel syntheses of brassinolide and related compounds. *Tetrahedron* **46**, 3173-3188.

Zhou, W.-S., Zhou, H.-Q., and Wang, Z.-Q. (1990b). Studies on synthesis of plant growth hormone steroids. Part 16. Stereoselective synthesis of 26,27-dinorbrassinolide. *J. Chem. Soc., Perkin Trans. 1* 2281-2286.

Zhou, W.-S., Huang, L.-F., Sun, L.-Q., and Pan, X.-F. (1991a). A stereoselective construction of brassinosteroid side chain: a new practical synthesis of brassinolide and its analogues. *Tetrahedron Lett.* **32**, 6745-6748.

Zhou, W.-S., Sun, L.-Q., and Pan, X.-F. (1991b). Stereocontrolled construction of the side chain of brassinolide and homobrassinolide *via* a tandem vicinal dialkylation of pyranone moiety. *Chin. Chem. Lett.* **2**, 929-932 [*C. A.* **116**, 235952].

Zhou, W.-S., Huang, L.-F., Sun, L.-Q., and Pan, X.-F. (1992a). A stereoselective construction of brassinosteroid side chain: a new practical synthesis of brassinolide analogues from hyodeoxycholic acid. *J. Chem. Soc., Perkin Trans. 1* 2039-2043.

Zhou, W.-S., Sun, L.-Q., and Pan, X.-F. (1992b). Stereoselective synthesis of crinosterol and brassicasterol from hyodeoxycholic acid. *Acta Chim. Sin.* **50**, 1192-1199.

Zhou, W.-S., Sun, L.-Q., and Pan, X.-F. (1993). Stereoselective construction of the side chain of brassinosteroid by employing a separated dialkylation and a tandem dialkylation of the pyranone moiety. *Chin. J. Chem.* **11**, 376-384.

Zhu, G.-D., and Okamura, W.H. (1995). Synthesis of vitamin D (calciferol). *Chem. Rev.* **95**, 1877-1952.

Zhu, J.-L., and Zhou, W.-S. (1991). A new synthesis of homobrassinolide and (22S,23S)-22,23-epi-homobrassinolide. *Sci. China, Ser. B* **34**, 706-711 [*C. A.* **116**, 59729].

Zoller, T., Uguen, D., De, Cian, A., Fischer, J., and Sable, S. (1997). A C-B-A-D approach to brassinosteroids; generation of the cis-anti-trans A-B-C ring system. *Tetrahedron Lett.* **38**, 3409-2412.

Zurek, D.M., and Clouse, S.D. (1994). Molecular cloning and characterization of a brassinosteroid-regulated gene from elongating soybean (*Glycine max* L.) epicotyls. *Plant Physiol.* **104**, 161-170.

Zurek, D.M., Rayle, D.L., McMorris, T.C., and Clouse, S.D. (1994). Investigation of gene expression, growth kinetics, and wall extensibility during brassinosteroid-regulated stem elongation. *Plant Physiol.* **104**, 505-513.

Index

A

Abbreviations
  BS names, 14
  protecting groups and ligands, xiv
  substances, reagents, and solvents, xiii
Abscisic acid, 1; 238; 239; 276
Acetonitrile oxide, 133
Acid secretion, 241; 242; 243; 245; 246
Adam's catalyst, 208
Adventitious roots, 222; 223; 306; 391
Alanates, 140; 172
Aldosterone, 313
Algae, 22
  source of BS, 2; 30, 359, 364
  growth responses, 241; 242; 344
Alkylation, 113; 121; 124; 140; 183
Alkylative cleavage, 145
R-Alpine-Borane, 110
S-Alpine-Borane, 110
Aluminum isopropoxide, 95; 172
Aluminum organic reagents, 113; 171; 178
Amberlyst-15, 163
Amino acid composition, 273
α-Aminoisobutyric acid, 255
m-Aminophenylboronic acids, 29
Amyloplast, 293
Ancymidol, 237
Androstenolone, 75
Antibodies, 294
  monoclonal, 29
  polyclonal, 293
Apical dominance, 231
Apple, 342
Arsenic ylides, 115

Arsonium organic compounds, 120
Arsonium ylide, 121; 153
Assimilates
  transport, 257
  uptake, 255
Asterasterol A, 22
Asymmetric hydroxylation, 92; 117; 164; 203
Auxin-oxidase, 282
Auxins
  accumulation, 269; 270; 282; 339
  application with BS, 328
  growth responses, 240; 256; 269; 270
  in biotests, 221; 302-308; 311
  in mutants, 63
  interaction with phytohormones, 236-240
  receptor, 292
  structure, 1
(22S,23S)-7-Azahomobrassinolide, 208
7-Azahomobrassinolide, 208
Azedarachol, 21
Azuki bean, 221; 236; 241; 306

B

Barley
  BS application in agriculture, 328; 330; 344
  disease resistance, 280-282
  fatty acids, 249
  glucose accumulation, 257
  growth responses, 221; 231-234; 243
  isoplasmatic lines, 249
  lipid peroxidation, 252
  lipids, 250-252
  metal absorption, 257; 258

445

osmoregulating metabolites, 274; 275
phytohormones, 238
radionuclide absorption, 258-260; 262; 263
RNA and proteins, 265-268
stress resistance, 276; 337
Bean first-internode bioassay, 302; 307; 310; 311; 314
Bean second-internode bioassay, 8; 25; 301; 307; 310; 311; 314; 315
Beckmann rearrangement, 206; 375; 378
Betaine, 274
26,27-Bisnorbrassinolide, 197; 198; 314; 396; 426; 431; 443
26,27-Bisnorcastasterone, 314
Brassicasterol
   in BS biosynthesis, 15; 57
   in BS synthesis, 75; 85; 88; 151; 154; 156; 157; 170; 173; 183
   labeled, 183
Brassinolide, 225; 227; 327
   application, 329-335
   biosynthesis, 55-61
   discovery, 2; 8
   early oxidation pathway, 60
   effect on nucleic acids and proteins, 263; 264; 272
   growth responses, 221-223; 241
   in bioassays, 306; 308; 310-312; 321
   in mutants, 62–64; 289; 290
   interaction with phytohormones, 236-238
   isolation, 19
   labeled, 183; 294
   late C6-oxidation pathway, 60–62
   mechanism of action, 291-294
   metabolism, 65
   2α-monoacetate, 217
   nomenculature, 11-16
   occurrence, 30; 347
   plant protective properties, 276; 278; 283; 286
   radioimmunoassay, 30
   relative activity, 60
   side chain synthesis, 88; 95; 97; 101; 111; 113; 114; 115; 116; 121; 123; 125; 126; 129; 130; 132; 135
   spectral properties, 39; 40; 42; 43; 45; 46; 347
   structure, 2; 347
   synthesis, 139–153
   x-ray analysis, 49; 51
Brassinone, 9
   (20R)-, 196
   (20R,22S,23S)-, 196
   biosynthesis, 15; 56
   isolation, 19
   occurrence, 370
   spectral properties, 370
   structure, 17; 370
   synthesis, 169
Brassinosteroids
   conjugates, 15; 59; 69-71
   content, 30; 31
   crystallographic data, 51; 52
   fluorinated, 51; 203
   nomenclature, 8
   shorthand description, 14
   trivial names, 10
Brassins, 8; 220; 221; 302
Broad bean, 353; 370
   source of epibrassinolide, 154
   BS application in agriculture, 339
2-Bromo-3-methylbut-1-ene, 171
Bromohydrins, 95; 97; 160; 163; 179; 193
B-seco derivatives, 83
Buckwheat
   BS application in agriculture, 331; 338
   H^+-ATPase activity, 247
   source of BS, 348; 352

C

Cabbage
   BS application in agriculture, 340
   cell reproduction, 224
   source of BS, 352; 362; 363; 369; 370
Calamondin, 342

Campestanol, 63; 64
Campesterol, 15; 56; 63; 64
Camphor-10-sulfonic acid, 149
Castasterone
  antibodies, 293
  antiecdysone effect, 295
  biosynthesis, 56; 57; 59; 60; 61; 63
  isolation, 2; 19; 25
  labeled, 183; 294
  metabolism, 65
  occurrence, 31; 60; 351
  radioimmunoassay, 30
  spectral properties, 38; 45; 53; 353
  structure, 17
  synthesis, 139–53
  Wolff-Kishner reduction, 174
Catalytic hydroxylation, 135; 157; 175
Cathasterone
  isolation, 20
  occurence, 357
  spectral properties, 357
  biosynthesis, 57; 59; 63; 64
CD spectroscopy, 53
Celery, 222; 271
Cell wall, 220; 327
  mechanical properties, 235, 240
  regulation of components, 291
Cherry, 305; 343
Chiograsterol A, 22
Chiograsterone, 22
Chiral acetals, 108
Chlorination, 121
Chlorocholine chloride, 337
m-Chloroperbenzoic acid, 140; 173; 192
(p-Chlorophenoxy)isobutyric acid, 236
Chlorophyll, 251; 273; 276; 286-289
N-Chlorosuccinimide, 130
Chlorotrimethylsilane, 200
Cholest-5-ene-2α,3α,7β,15β,18-pentol 2,7,15,18-tetraacetate, 21
Cholesterol, 75; 186; 248; 313
  labeled, 57
Choline, 274; 329
(R)-(+)-Citronellic acid, 122
Citrus, 342
Claisen rearrangement, 118; 150; 176; 183

Clotrimazole, 68
Cockroach, 295; 296
2,4,6-Collidine, 212
Collins reagent, 149
Conjugate addition, 124
Conjugated dienes, 254
Copper sulfate, 86
Corn
  BS application in agriculture, 331; 337
  source of BS, 353-355; 357; 370
Cotton, 335; 340; 341
Cranberry, 229; 343
Crinosterol, 75; 88; 139; 151; 180
  labeled, 183
  synthesis, 118
Cucumber
  bioassay, 308
  BS application in agriculture, 328; 334; 340
  disease and stress resistance, 276; 283; 286
  GA and ABA content, 238
  growth responses, 221
  metabolism of epibrassinolide, 65
  transport of BS, 294
Cumene, 148
Cumene hydroperoxide, 120
1-Cyanoisoindole-2-m-phenylboronic acid, 29
Cyclic part
  chemical shifts, 40
  synthesis, 76–88
Cyclization, 84; 128; 129; 208; 209
1,3-Cycloaddition, 201
Cytochrome P450, 68
Cytokinins, 240
  bioassays, 236, 306, 308
  changes in composition, 238
  structure, 1
Cytoplasmic metabolism, 241

D

(Dansylamino)phenylboronic acid, 29
Deacetylation, 81; 153; 206
Debromination, 132; 183
Decarbonylation, 118; 149; 150; 153

Dehydration, 43; 79; 102; 131; 152; 183; 195; 209
22-Dehydrocampesterol, *see* Crinosterol
3-Dehydro-6-deoxoteasterone
  isolation, 20
  occurrence, 358
  spectral properties, 358
  structure, 16
  synthesis, 177
Dehydrobromination, 97; 143; 158; 179; 205
22-Dehydrocholesterol, 15; 57; 165; 168
Dehydrogenation, 134
Dehydrohalogenation, 79
Dehydromesylation, 213
16-Dehydropregnenolone, 198
3-Dehydroteasterone, *see* 3-Oxoteasterone
6-Deoxo-24-epicastasterone
  isolation, 20
  occurrence, 360
  spectral properties, 360
  structure, 16; 360
  synthesis, 174
6-Deoxo-25-methyldolichosterone
  isolation, 20
  occurrence, 369
  spectral properties, 369
  structure, 16; 369
6-Deoxo-28-norcastasterone
  isolation, 20
  occurrence, 371
  spectral properties, 371
  structure, 16; 371
6-Deoxocastasterone
  biosynthesis, 60; 61
  isolation, 19
  occurrence, 357
  spectral properties, 357
  structure, 16; 357
  synthesis, 174; 177
6-Deoxodolichosterone
  isolation, 19
  occurrence, 361
  spectral properties, 361
  structure, 16; 361
  synthesis, 175
6-Deoxohomobrassinolide, 209
6-Deoxohomodolichosterone
  isolation, 20
  occurrence, 365
  spectral properties, 365
  structure, 16; 365
6-Deoxotyphasterol
  biosynthesis, 61
  isolation, 20
  occurrence, 358
  spectral properties, 358
  structure, 16; 358
22-Deoxy-24-epiteasterone, 193
2-Deoxy-25-methyldolichosterone, 20; 369
2-Deoxybrassinolide
  isolation, 20
  occurrence, 349
  spectral properties, 349
  structure, 16; 349
Deoxygenation, 86; 96
Desilylation, 115
Desulfonization, 114
Desulfurization, 120; 127; 174; 175
Diaulusterol B, 22
1,5-Diazabicyclo[4.3.0]non-5-ene, 100
Diazomethane, 207; 209
1,1-Dibromo-3-methylbutene, 145
$2\alpha,3\alpha$-Diacetoxycholest-4-en-6-one, 51
1,3-Dicyclohexylcarbodiimide, 209
Diels-Alder adduct, 85
2,3-Diepi-25-methyldolichosterone
  isolation, 19
  occurrence, 368
  spectral properties, 34; 38; 368
  structure, 18; 368
  synthesis, 178
2,24-Diepicastasterone, 216
2,3-Diepicastasterone
  isolation, 19
  occurrence, 354
  spectral properties, 354
  structure, 18; 354
3,24-Diepicastasterone, 18; 19
L-(+)-Diethyl tartrate, 120, 148

INDEX

Diethyl(3-methyl-2-oxobutyl)phosphonate, 121
Dihydroquinidine 1,4-phthalazinediyl diether, xiii; 91; 93
Dihydroquinidine 9-(1'-naphthyl) ether, 93
Dihydroquinidine 9-phenanthryl ether, 91; 93; 215
Dihydroquinidine p-chlorobenzoate, 91; 93; 94
Dihydroquinine p-chlorobenzoate, 91; 93; 94
Diisobutylaluminum hydride, 101; 110
2,3-Dimethoxybrassinolide, 313
22,23-Dimethoxybrassinolide, 313
1,2-Dimethoxyethane, 179
Dimethyl sulfoxide, 78
N,N-Dimethylacetamide, 156
N,N-Dimethylformamide, 79; 156; 158; 179; 205
Diosgenin, 75, 104
1,3-Dipolar cycloaddition, 101; 130
1,3-Dithiane, 127; 166; 198
DNA polymerase, 264
Dolicholide
  isolation, 19
  occurrence, 350
  spectral properties, 39; 45; 350
  structure, 16; 350
  synthesis, 171-174
Dolichosterone
  isolation, 19
  occurrence, 361
  reduction, 183
  spectral properties, 39; 45; 361
  structure, 17; 361
  synthesis, 171; 172
DT-lines, 226; 227; 239; 268; 269; 270; 271
Dwarf pea, 221; 240; 306; 308

E

Ecdysone, 22; 68; 320
Ecdysteroids, 22; 37; 74; 77; 79; 135; 154; 204; 295; 296; 297
Ecdysterone, 22; 68; 296; 320
EI-MS fragment ions, 45

Electrochemical detection, 29
Elimination, 79; 88; 100; 111; 165; 175; 205; 206; 211
Enamino ketones, 102; 132
Ene reaction, 98; 99; 127; 128
3-Epi-1α-hydroxycastasterone, 20; 351
2-Epi-25-methyldolicho-sterone, 18
2-Epi-25-methyldolichosterone, 20; 368
3-Epi-2-deoxy-25-methyldolichosterone, 20; 369
3-Epi-2-deoxy-25-methyldolichosterone, 17
3-Epi-6-deoxocastasterone, 20; 358
3-Epi-6-deoxo-castasterone, 17
Epibrassinolide
  antibodies, 30
  application, 328-344
  effect on cell membranes, 241; 243-247
  effect on chlorophyll content, 286-289
  effect on fish, 297-299
  effect on lipid content, 249-254
  effect on metal absorption, 257-260
  effect on nucleic acids and proteins, 266-268; 271-274
  effect on osmoregulating metabolites, 275
  effect on radionuclide absorption, 260; 262
  20-epi-, 196
  growth responses, 222-224; 227; 229-235
  in bioassays, 302; 311; 312
  interaction with phytohormones, 238-240; 270
  isolation, 19
  labeled, 65; 182; 183
  metabolism, 65; 66; 68; 70
  micribiological oxidation, 72
  occurrence, 349
  plant protective properties, 276-284
  spectral properties, 35; 35; 42-44; 46; 49; 51; 53; 349
  structure, 16; 349
  synthesis, 76; 154-160
  toxicology, 345; 346

2-Epicastasterone
  isolation, 19
  occurrence, 353
  spectral properties, 353
  structure, 18; 353
20-Epicastasterone, 196
24-Epicastasterone
  isolation, 20
  microbiological transformation, 72
  occurrence, 359
  spectral properties, 359
  structure, 17; 359
3-Epicastasterone
  biosynthesis, 59
  isolation, 19
  occurrence, 353
  spectral properties, 353
  structure, 18; 353
24-Epicathasterone, 193
Epin, 328
Ergosterol, 74; 75; 76; 85–87; 95; 154; 156; 157; 158; 178; 193; 205; 211
Ethylene, 1
  BS-influenced production, 237; 238; 278
  bioassays, 308; 313
  level in potato, 339

F

Fatty acid composition, 248; 249
FD-MS fragment ions, 46
Ferroceneboronic acid, 29
Fluorimetric detection, 29
Fungi, 2
  growth responses, 222; 344
  phytopathogens, 277-283
Fusicoccin, 237; 242; 243; 255

G

GC analysis, 28; 249
GC-MS, 4; 28; 29; 57
GC-MS-SIM, 28
Gene expression, ix; 2; 265; 275; 291
Gerardiasterone, 22
Gibberellic acid, *see* Gibberellins
Gibberellins, 1; 240
  growth responses, 232
  interaction wth BS, 236-238
  effect on assimilate transport, 256
  bioassays, 303; 304; 306
Gladiolus, 343
Glucocorticoids, 256
Glucose accumulation, 257
23-O-β-D-Glycopyranosyl-25-methyldolichosteron, 19; 27; 367
23-O-β-D-Glycopyranosyl-2-epi-25-methyldolichosterone, 20
25-O-β-D-Glycopyranosyloxy-24-epibrassinolide, 66
26-O-β-D-Glycopyranosyloxy-24-epibrassinolide, 66
23-O-β-D-Glycopyranosyloxy-25-methyldolichosterone, 65
23-O-β-D-Glycopyranosyloxy-brassinolide, 65
Gooseberry, 343
Grape, 335; 343
Greengram, 339
Grignard reaction, 166; 183; 198
Grignard reagent, 119; 127; 171; 176; 192
Groundnut, 342

H

H^+-ATPase activity, 246; 247; 256
Heavy metals, 257; 337
Homobrassinolide
  (24R)-, 196
  application, 329–335
  biosynthesis, 56
  growth responses, 222, 283
  in bioassays, 303; 310; 313; 318; 322
  interaction with phytohormones, 239
  isolation, 19
  occurrence, 362
  plant protective properties, 278-280
  reduction, 209
  spectral properties, 39; 362
  structure, 16; 362
  synthesis, 160–165
(22S,23S)-Homobrassinolide, 214; 217
  antyecdysone effect, 296

bioeffects, 223; 242; 243; 255; 256; 276; 295; 315; 316; 318; 323; 328; 334; 339; 341; 344
metabolism, 68
synthesis, 161; 189
x-ray analysis, 49; 52
Homocastasterone
biosynthesis, 56
isolation, 19
occurrence, 363
spectral properties, 363
structure, 17; 363
synthesis, 164
(22S,23S)-Homocastasterone
antyecdysone effect, 295
bis-acetonide, x-ray analysis, 50
irradiation, 210
synthesis, 190
Homodolicholide
isolation, 19
occurrence, 363
spectral properties, 39; 363
structure, 16; 363
synthesis, 172; 174
Homodolichosterone
isolation, 19
occurrence, 365
spectral properties, 41; 365
structure, 17; 365
synthesis, 172; 173
Homopropargylic alcohols, 104; 111
Homoteasterone
isolation, 20
occurrence, 365
spectral properties, 365
structure, 17; 365
Homotyphasterol
isolation, 20
occurrence, 364
spectral properties, 364
structure, 17; 364
Horner-Wadsworth-Emmons reaction, 121
HPLC, 26; 29
Hydride reduction, 82; 84; 109; 115; 149; 165; 183; 205
Hydroboration, 77
Hydrobromic acid, 79; 158; 179; 180

Hydrochloric acid, 79
Hydrogenation, 100; 102; 112; 125; 134; 135; 145; 152; 183; 204
Hydrogenolysis, 103; 132; 141; 411
Hydrolytic enzymes, 273
Hydroxy aldehydes, 105; 127; 129
Hydroxy ethers, 105; 129
Hydroxy furans, 104; 122; 148
Hydroxy lactones, 37; 43; 47; 69; 104; 125; 165; 193
25-Hydroxy-24-epibrassinolide, 67; 68; 193
25-Hydroxy-24-epicastasterone, 68
6α-Hydroxy-24-methylcholestanol, 57
(23S)-Hydroxy-6-oxocampestanol, 57
25-Hydroxybrassinolide, 68
1β-Hydroxycastasterone, 20
25-Hydroxyepicastasterone, 194
Hydroxylation
OsO_4, 80; 89; 90; 91; 93; 94; 135; 150; 157; 158; 161; 162
Woodward, 80; 89
Hyodeoxycholic acid, 75; 76; 87; 97; 104; 134; 135; 139; 153; 161

I

Imaginal disks, 295
Immunological methods, 29
Immunomodulators, 284
Indole-3-acetic acid, see Auxins
Insects, 2; 23; 256; 295; 296; 297
Iodine, 81; 89; 209; 215
Iodobenzene, 203
IR spectroscopy, 53
Isobutylcarbonylarsonium ylide, 121
Isobutylene, 127
Isobutylmagnesium bromide, 168
Isobutyraldoxime, 130
Isobutyronitrile oxide, 102; 202
Isofucosterol, 41; 56; 64
Isoplasmatic lines, 249; 265; 266
Isopropylmagnesium bromide, 148
3-Isopropyltetronic acid, 100; 152
Isosteroidal rearrangement, 77; 85; 151; 155; 156
Isoxazoles, 102; 105; 130; 132; 201

Isoxazolines, 53; 102; 105; 130; 131; 133; 201; 202

J

Julia olefination, 121; 122; 142

K

Ketoconazole, 68
Kinetine, see Cytokinins
Kvartazin, 267

L

Lactonization, 83; 141; 182
Lettuce, 221
Lewis acid, 127; 203; 413
Lily
　source of BS, 31; 59; 354-356
　growth responses, 343
Lindlar catalyst, 96; 112; 183
Lipid peroxidation, 250; 252; 253; 254
Lipoxygenase, 254
3-Lithio-1-(trimethylsilyl)propyne, 111
1-Lithio-2-methyl-1-propene, 129
2-Lithiofurans, 122
Lithium bromide, 79; 156
Lithium diisopropylamide, 113
Lithium dimethylcuprate, 124
Lithium in liquid ammonia, 85; 86; 114; 155; 175; 212
Lucerne, 333
Lucibufagenin, 21
Lupine
　BS application in agriculture, 332; 339
　^{137}Cs-accumulation, 260; 261
　photosynthetic capacity, 332; 339
　protein accumulation, 271-274

M

Maize
　growth responses, 222; 236; 306
　acid secretion, 241
　activity of H$^+$-ATPase, 246; 247
　chilling resistance, 276
　photosynthetic capacity, 286
Malonic dialdehyde, 252; 253; 254
Maturation, 224; 293; 343
Melon, 341
Membrane potential, 2; 242; 243
Mesylation, 109; 165; 177; 213
Methaneboronates, in BS analysis, 28; 46; 48
Methaneboronic acid, 28
Methoxymethyl chloride, 152
Methyl cyanocuprate, 120
Methyl(trifluoromethyl)dioxirane, 194
(3-Methyl-2-oxobutyl)triphenyl-arsonium bromide, 121
(2-Methylbut-1-ynyl)lithium, 141
(3-Methylbutynyl)magnesium bromide, 150
25-Methylcastasterone, 17; 20; 366
(2S,3R)-25-Methyldolichosterone, see 2,3-Diepi-25-methyldolichosterone
25-Methyldolichosterone
　isolation, 20; 26
　occurrence, 367
　spectral properties, 367
　structure, 17; 367
24-Methylcholestanol, 57
24-Methylenecholesterol, 64
(D)-(+)-N-Methylephedrine, 110
[^{14}C]-Methylmagnesium iodide, 183
N-Methylmorpholine N-oxide, 80
N-Methylpyrrolidone, 343
MH-inhibiting effect, 296
Microfibrils, 235
Microtubules, 235; 412
Mitsunobu reaction, 82; 109
Molecular oxygen, 96; 192
Mung bean
　bioassays, 304-306; 307; 313; 314
　BS metabolism, 65
　growth responses, 221; 222; 224; 236; 237
　proteins and nucleic acids, 263-265
　stress, 276
Mung bean epicotyl bioassay, 304
Mustard, 221; 271; 286; 341
Mutagenic effects, 346

N

Naphthaleneboronic acid, 29
Neurotropic effect, 296
Neutral metabolites, 245
Neutral protease, 273
Nitrile oxides, 51; 102; 130; 131; 133; 134; 201; 202
Norbrassinolide
    biosynthesis, 56
    isolation, 19
    occurrence, 369
    spectral properties, 45; 369
    structure, 16; 369
    synthesis, 127; 129; 133; 134; 165-169
Nucleophilic substitution, 82; 84; 95; 109; 160; 180

O

Oats, 332
Osmoregulating metabolites, 274; 275
Oxidation, 109; 111; 120
    asymmetric, 120; 147
    Baeyer-Villiger, 77; 82; 148; 150; 151; 153; 156; 157; 160; 163; 165; 170; 182; 187; 192; 194; 196; 210; 212
    epoxidation, 81; 115; 116; 142; 157; 173
    in biosynthesis, 57-60
    Jones oxidation, 78; 88; 97; 98; 151; 163; 179
    micribiological, 56; 71; 72
    Oppenauer, 85; 175; 183
    periodate, 84
Oxidative cleavage, 71; 103; 164; 207
6-Oxocampestanol, 65
Δ^{22}-6-Oxocampestanol, 57
6-Oxo-24-methylcholestanol, 57
3-Oxoteasterone, 17; 20; 181; 356
Ozonolysis, 84; 89; 104; 139; 140; 183; 196

P

Palladium acetate, 203
Pea
    bioassays, 306; 308
    BS application in agriculture, 332; 339
    growth responses, 221; 236
    interaction with phytohormones, 238; 240
    lipids, 253; 254
    source of BS, 348-352; 354; 357
Peach, 342
Periodic acid, 84; 207
Peronosporosis, 283
Peroxidase, 282; 283
Persimmon, 342
9-Phenanthreneboronic acid, 29
Phenyl selenide anion, 95; 173
4-Phenyl-1,2,4-triazoline-3,5-dione, 156; 211
Phenylboronic acid, 29
Phlox, 343
Phosphoenolpyruvate carboxylase, 241
Photomorphogenesis, 285; 290
Photosynthetic assimilates, 286
Phyllochrone, 232; 234
Phytosterol composition, 248
Pigment apparatus, 287
Pinnasterol, 22
Pinto beans, 263; 264
Polyphenoloxidase, 283
Polyphenols, 283
Polysaccharide hydrolases, 240
Ponasterone A, 295
Ponasterone B, 22; 295
Poriferasterol, 196
Potassium acetate, 78
Potassium bicarbonate, 78
Potassium maleate, 303; 305
Potato
    ABA level, 238
    BS application in agriculture, 333; 339; 340; 344
    growth responses, 223
    resistance to diseases, 277-279; 284

Pregnenolone, 75; 76; 101; 139; 152; 194
Propargylic alcohols, 104; 111; 112
Propenyltriphenylarsonium tetrafluoroborate, 115
Protein synthesis, 241; 263; 264; 268; 269; 270; 271; 272; 273; 275; 276; 284
Protiodesilylation, 128; 129
Protochlorophyllide, 288
Proton pump, 240; 241; 243; 245; 247; 256
Pyridinium bromide, 79; 156
Pyridinium chloride, 156
Pyridinium chlorochromate, 123
Pyridinium tosylate, 79

R

Radioimmunoassay, 30
Radionuclides, 257; 258; 261; 263
Radish
   bioassays, 305; 307; 312; 314
   growth responses, 221; 224
   metal accumulation, 257
   protein metabolism, 271
   source of BS, 348; 352; 355; 365
Ramberg-Bäcklund rearrangement, 121; 122
Raney nickel, 131; 133; 175; 208
Rape
   BS application in agriculture, 335; 341
   source of brassicasterol, 154
   source of brassinolide, 7; 8; 295; 347
Receptor-ligand interaction, 323
Red currant, 343
Reduction
   anti-Markovnikov, 116; 141; 146
   bio-, 63; 64; 70
   catalytic, 98; 100; 102; 125; 144; 183; 195; 198; 206; 208
   $\Delta^{5,7}$-dienes, 85
   dissolving metals, 85; 86
   $2\alpha,3\alpha$-epoxides, 82
   ergosterol, 76
   hydride, 152

   hydride, 82; 87; 96; 101; 109; 115; 118; 123; 132; 149; 152; 164; 178; 183; 205; 209
   Δ^7-6-ketones, 86
   sodium amalgam, 143
   Wolff-Kishner, 174
   zinc, 153; 209

Reductive amination, 208
Reductive cleavage, 102
Reductive opening, 101; 153; 163
Relactonization, 182; 212
Reproduction, 224; 340
Rice
   bioassays, 7; 25; 59; 65; 66; 217; 303; 304; 310; 312-318; 319; 321
   BS application in agriculture, 329; 336; 337
   growth responses, 221; 223; 224; 237; 238; 276
   photosynthetic capacity, 286
   proteins, 271
   source of BS, 348; 352; 355; 357; 361; 364; 365
   transport of BS, 294
Rice lamina inclination bioassay, 303; 307
RNA polymerase, 263; 264
Root system, 222; 223
Rose, 344
Russian sturgeon, 297; 299; 345
Rye, 330
Ryegrass, 293; 366

S

Samogenin, 21
Saponifiable fractions, 251; 252
Schmidt reaction, 207
Secasterone, 18; 20; 179
K-Selectride, 110
L-Selectride, 101; 110; 206
Selenosulfonation, 111
Senescence, 224; 342
Sharpless epoxidation, 148

Sharpless hydroxylation, 91; 157; 162; 176
Side chain
  chemical shifts, 42
  conformation, 41
  synthesis, 88–135
Sigmatropic rearrangements, 99
Silver acetate, 82; 89; 215
Silylation, 84; 111; 207
β-Sitosterol, 15; 56; 57; 64; 75; 186; 248
Sodium amalgam, 121; 143; 169
Sodium borohydride, 205; 208
Sodium bromide, 79
Solanidine, 188
Solasodine, 75; 188
Soybean
  BS application in agriculture, 332; 338
  growth responses, 222; 225
  in bioassays, 322; 323
  source of sterols, 75
Spinach, 340; 394
Stabilizing effect, 245; 246
Starred sturgeon, 297; 298
Stereochemistry
  epoxidation, 95
Stereochemistry of
  hydride reduction, 110
  hydroxylation, 94
  side chain, 10
Stereoselective
  addition, 96
  alkylation, 124; 129
  ene reaction, 98; 99
  epoxidation, 115; 163
  reduction, 124; 198
  substitution, 119
  synthesis of 22-alcohols, 105
Stigmasterol, 56; 74; 75; 88; 89; 139; 146; 160; 163; 166; 169; 170; 175; 178; 191; 213; 242; 248
Strawberry, 343
Stress, 224; 245; 246; 252; 253; 254; 273; 274; 275; 276; 277; 284; 286; 298; 327; 337; 341; 342
Structural requirements, 309
Sucrose, 243; 255; 256

Sugar beet
  BS application in agriculture, 334; 335; 341
  metal accumulation, 227
  proteins, 271
  source of BS, 352; 359
  stress resistance, 277
Sulfenate-sulfoxide rearrangement, 113
Sulfolane, 79; 157
Sulfones, 111; 114; 121; 122; 142; 143; 145; 169; 196; 197
Sulfonic acid, 79
Sunflower, 221; 352; 370
Synchrony, 231; 232; 233
Synchrony measure, 232
Synchrony range, 232

T

Teasterone
  biosynthesis, 61
  fatty acid conjugates, 69
  in mutants, 63; 64
  isolation, 19
  labeled, 57; 183
  occurrence, 59; 60; 355
  relative activity, 60
  spectral properties, 355
  structure, 17; 355
  synthesis, 180; 181
Teasterone 3-laurate, 15
Teasterone 3-myristate, 14; 15; 20; 355
Teasterone 3-palmitate, 217
Tertiary ammonium compounds, 274
Tetrabutyl ammonium fluoride, 111
Tetradecyltrimethylammonium permanganate, 162
Tetramethoxybrassinolide, 313
Thermostability of membranes, 276
7-Thiahomobrassinolide, 209
Thionyl chloride, 102
Titanium isopropoxide, 120
Tobacco, 223; 335; 341; 344
Tomato
  bioassays, 237; 305; 307; 312
  BS application in agriculture, 333; 340
  growth responses, 222

Tosylation, 151
Toxicity, 345
Transfer factor, 261
Transmembrane electric potential, 241; 245; 292
Triethyl orthoacetate, 150
Triethyl orthopropionate, 118; 183
Trimethyl orthoformate, 124
Triphenylphosphine, 203
Triticale
　chlorophyll accumulation, 286-289
　growth responses, 243; 244
TS303, 191; 327
Typhasterol
　biosynthesis, 57-61
　content, 31
　fatty acid conjugates, 69
　in mutants, 63; 64
　isolation, 19
　labeled, 57; 183
　occurrence, 354
　relative activity, 60
　spectral properties, 36; 37; 53; 354
　structure, 17; 354
　synthesis, 179; 180

U

Unsaponifiable fractions, 251

V

Vinyl anions, 115
Vinyl silanes, 115
Vinylmagnesium bromide, 202

Virus infection, 283; 284
Vitamin D, 67; 88; 121; 135; 154

W

Watermelon, 341; 437
Wheat
　BS application in agriculture, 328; 329; 336
　growth responses, 221; 222; 226-228; 257; 269
　metabolism of BS, 65
　phytohormones, 239; 270
　proteins, 271; 275
　source of BS, 353-357
Wheat leaf unrolling bioassay, 305; 307
Wheat leaf unrolling test, 305
Wilkinson's catalyst, 118; 150
Winter rye, 337; 344
Wittig
　olefination, 120; 121; 133; 153; 174; 203
　rearrangement, 99
Woody plants, 222

X

Xylem, 294
Xyloglucan endotransglycosylase, 291

Z

Zeatin-riboside, 238; 239